MELLIFERA BEEKEEPING AND POLLINATION

Four species of *Apis* ; **A.** *Apis florea* ; **B.** *Apis cerana* ; **C.** *Apis mellifera* ; **D.** *Apis dorsata*

MELLIFERA BEEKEEPING AND POLLINATION

PROF. AVTAR S. ATWAL
Ph.D. (Adel.)

KALYANI PUBLISHERS
LUDHIANA • NEW DELHI • NOIDA (U.P.) • HYDERABAD • CHENNAI
KOLKATA • CUTTACK • GUWAHATI • KOCHI • BENGALURU

KALYANI PUBLISHERS

Head Office

B-I/1292, Rajinder Nagar, **Ludhiana**-141 008 • Ph. : 0161-2760031, 2745756

E-mail : kalyanibooks@yahoo.co.in

Administration Office

4779/23, Ansari Road, Daryaganj, **New Delhi**-110 002 • Ph. : 011-23271469, 23274393

E-mail : kalyani_delhi@yahoo.co.in

Works

B-16, Sector-8, **NOIDA** (U.P.)

Branches

1, Mahalakshmi Street, T. Nagar, **Chennai**-600 017 • Ph. : 044-24344684

110/111, Bharatia Towers, Badambadi, **Cuttack**-753009 (Odisha) • Ph. : 0671-2311391

3-5-1108, Narayanaguda, **Hyderabad**-500 029 • Ph. : 040-24750368

10/2B, Ramanath Mazumdar Street, **Kolkata**-700 009 • Ph. : 033-22416024

Arunalaya, 1st Floor, Saraswati Road, Pan Bazar, **Guwahati**-781 001 • Ph. : 0361-2731274

Koratti Parambil House, Convent Road, **Kochi**-682035 • Ph. : 0484-2367189

No. 24 + 25, Ist Floor, Hameed Shah Complex, Cubbonpet Main Road, **Bengaluru**-560 002

KPPJ12568 05

First Edition, 2000
Second Revised Edition 2013

Typesetting at TIMES PRINTOGRAPHIC

ISBN 978-93-272-3077-2

PRINTED IN INDIA
At Process World India C-90 Sec-63 Noida
and published by Mrs Usha Raj Kumar for
Kalyani Publishers, New Delhi-110 002

Dr Avtar S. Atwal (1965-Photo)

The founder of *Apis mellifera* beekeeping in India

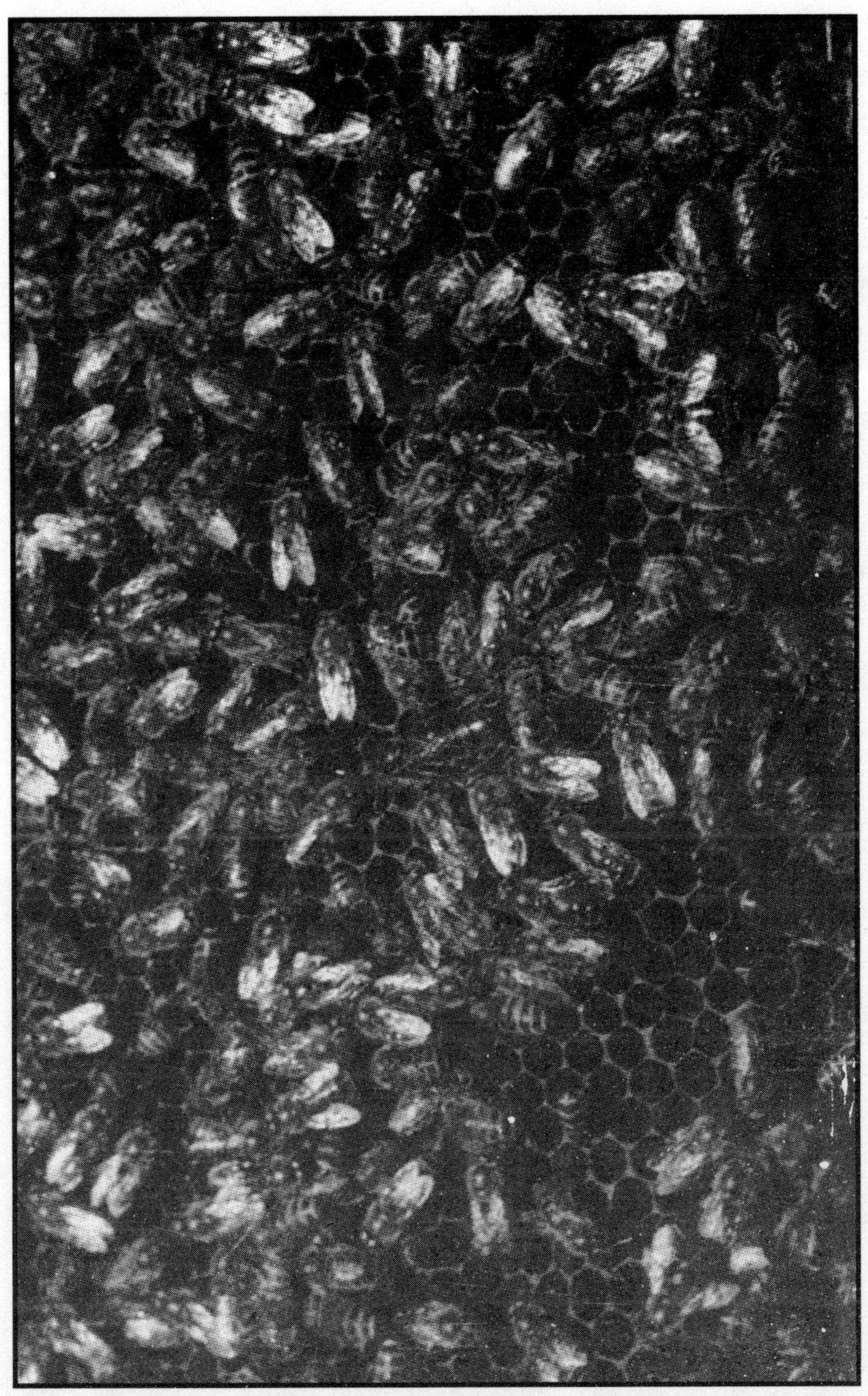

Apis mellifera queen introduced in the *Apis cerana* colony and is seen attended by the latter workers.

Acknowledgements

SECOND EDITION

Since the publication of first edition of this book ten years ago, the science of Apiculture has made great strides and so has the globalization of exchange of literature through internet. So much knowledge has been generated on every sub-topic of the honey bee behaviour that it can best be understood through specialized monographs which are now being published. Therefore, my effort in combining the 1st and 2nd editions of this book means to provide continuity in thought, as far as possible.

For the preparation of this volume I received the greatest help from Mr. David Collan who has green fingers for working on computers. He on his initiative gave me prints of papers on pheromones and bee-behaviour. The possible association between the mycoflora (fungi and mycotoxins) was searched and interpreted by my daughter Ms Chanda Siddoo Atwal. I benefitted a great deal from the news letters and papers of Prof. Mark L. Winston of the Simon Fraser University, City Campus, Vancouver, B.C.

Above all, I thank the farmers of Punjab who so naturally learnt the habits and techniques of handling honey bees even though not used to them in their usual life style. It is still the more creditable achievement that having demonstrated their skill and determination they adapted to learning the intelligence and habits of honey bees who are so small and organized to live in their house (hive). Thus the people have earned two tributes; firstly the "Harolds of Green Revolution" and now the "Pioneers of Honey Revolution" in India.

Dr G.S. Dhaliwal, Ex-Professor of Ecology, Punjab Agricultural University, Ludhiana and an eminent entomologist, thoroughly read the manuscript and the proofs for which work I express my heartfelt gratitude.

AVTAR S. ATWAL
51, Sector 4, Chandigarh, India

Preface

SECOND EDITION

The College of Agriculture, Ludhianá was established and brought upto the level of University with the competence and hard work of faculty for forty years. It was a life of devotion for learning, setting up trimester system of education under new rules, new curriculum, new discipline of internal assessment and teaching by short courses of 13 weeks of duration. It was to be supported by relevant research and extension. That is how teaching started both for under-graduate and post-graduate students. It is a remarkable story as the New Institution was raised from a refugee start to the use of latest scientific instruments for research. The manpower shortage was met by starting in-service training of old students and employees of the Department of Agriculture, Punjab. According to Dr Borma, Director-General, Food and Agriculture Organization of the United Nations, the institution was raised to the level of one of the best in Asia.

The start was made from the Faculty of Agriculture which gradually developed to include Home Science. Agricultural Engineering, Animal Science and Veterinary Science. With time many innovations were made to solve problems of plant and animal health. The major problems to be solved included self sufficiency in food with new varieties of wheat and other grains, oilseeds, poultry, milk, fruits and vegetables introduced first to feed the uprooted population of Punjab and then to meet 80 per cent shortages of food for the whole of India. Some limited quantities were even exported. Some of the elite articles of food like honey also became surplus and were exported to meet their shortages caused by diseases of honey bees in USA and Europe.

This volume deals with the remarkable story of introduction of European honey bee which became highly productive in Punjab and India. As chance would have it, a new ailment CCD (Colony Collapse Disorder) of the bees suddenly appeared in Americas and Europe in 2005-06 thus the abundance of honey there changed to shortages. Surprisingly countries like China and India, became exporters to those regions. Thus Punjab, followed by other states of India became the home of surpluses.

South Asia was always considered the home of wild and semi-wild honey bees which in nature played prominent role in the pollination of food and other crops. In India the introduction of European honey bee was banned for fear of introducing new disease like Acarine. So, the author devised a new method to overcome this difficulty. Only gravid queens of *Apis mellifera* were imported and released in de-queened colonies of *Apis cerana* (the local bee). There came a change in the fortunes of Indian beekeepers after years of research when the number of converted colonies became available since 1962-64 and were given to the anxiously waiting beekeepers here in 1976. For the first time India exported 10,000 tonnes of honey in 1994-95 by the hard work of Indian beekeepers who helped one another as a fraternity.

March 2013 **AVTAR S. ATWAL**

Acknowledgements

FIRST EDITION

It would be extremely difficult to thank each and everyone of the numerous beekeepers, eminent scientists and Institutions, who have helped so much in this Apis mellifera project over the past forty years, especially during the 60's and 70's when foreign exchange funds were so limited. One such aid was given by the family of my wife, Dr Sarjit K. Siddoo–Atwal.

While most of the photographs were taken by the author himself, particular mention must be made of the photographs of beekeeping field operations taken by Mr. Peter Satwant Singh and of the accurate line drawings, original and reproductions, made by Mr. V. Dalmotra.

Finally, I owe special thanks to Dr R.C. Mishra of ICAR at Haryana Agricultural University and Dr G.S. Dhaliwal at Punjab Agricultural University, Ludhiana, for providing reprints of old papers so promptly and willingly.

The author wishes to thank library personnel at the University of British Columbia, Simon Fraser University and the Vancouver Public Library who were most helpful and courteous.

Mr. S.P. Anand and his staff at the Kalyani Publishers did a satisfactory job in publishing this book.

Thank you all

AVTAR S. ATWAL

Preface

FIRST EDITION

"Essentials of Beekeeping and Pollination" is basically a summary and not an exhaustive review of literature. Eminent apiculturists who have written reviews know how difficult it is to attempt a summary for use in a country the size of India with a climate ranging from the tropical to that of the arctic.

The introduction of *Apis mellifera* to various continents is not a mere story but an epic which has been masterfully documented by Dr Eva Crane. Like the dedicated master beekeepers of the world, some in India have been practicing this art and science since the end of 19th century. India is a country of mysteries and it is amazing how many phenomena in nature appear and disappear for no apparent reason. Persistence is the password for insects (including honeybees) and we have to deal with them with equal patience. The lure and love for honeybees and honey has been around for thousands of years. The European beekeepers who moved honeybees across the oceans made remarkable efforts centuries ago.

The renewed efforts to introduce *Apis mellifera* into India in 1962-64 was a fresh effort and there was no report of the presence of *A. mellifera* in any part of India at that time and the events are well documented in literature. Towards the middle of the 20th century *Apis mellifera* was introduced to many countries of Asia, including Japan, Korea, China, etc. and it was proven that it produced more honey than the indigenous *Apis cerana*.

Strangely enough there was opposition to the introduction of *Apis mellifera* into India and it was believed by some that only *Apis cerana* should be propagated and encouraged in this country. However, the successful introduction of *A. mellifera* into India was inspite of the opposition. A great deal of credit for this success must be given to the farmers, particularly those in subtropical India, especially the cropped areas of the Indo-Gangetic plains of the Indian subcontinent.

The pollination of crops by bees is known to be essential for obtaining good crop yields (except grasses) and abundant fruit and vegetable production. The role of solitary bees, the wild honeybees, and the domesticated honeybees, has been studied off and on in India for almost a century. Apart from the monumental works of Howard, Ali Mohammad, and S.M. Sikka, the pollination studies have been inconsistent. The remarkable role of wild honeybees and of solitary bees has been studied to some extent but not as thoroughly as required.

The concise review of literature in chapter 15 is an effort to focus attention on the work done in India. However, a great deal of research on the pollination of crops and fruits still needs to be undertaken. It is an open field in agriculture that needs attention in the future. The author is hopeful that this book will draw attention to the unexplored avenues of crop sciences and encourage other scientists to devote their attention to the role of various species of bees, including *Apis mellifera,* in the pollination of crops.

Chandigarh
10th March, 2000

AVTAR S. ATWAL

Abbreviations Used

1. *Acre ft*	of water
2. *am*	apre meridian
3. *A.c.*	*Apis cerana*
4. *A.d.*	*Apis dorsata*
5. *A.f.*	*Apis florea*
6. *A.m.*	*Apis mellifera*
7. AD	Christian Year
8. BC	Christian Year
9. °C	degree centigrade
10. cm$^{cm^2, cm^3}$	centimeter
11. CC	cubic centimeter
12. *e.g.*	for example
13. Ram *et al.*,	Ram and others
14. etc.	etcetra
15. ft	feet
16. °F	degree f
17. g	gram
18. h	hour(s)
19. ha	hectare
20. *i.e.,*	that is
21. India, North, East, West, South etc.	
22. inc	inch
23. kg.	kilogram
24. km	kilometer
25. kmph	kilometer per hour
26. l	liter
27. lb	pound

28. m/m^2/m^3	meter, square, cube
29. min	minute
30. ml	milliliter
31. μ	micron
32. mm	millimeter
33. mg	milligram
34. nm	nanometer
35. *M.r.*	*Megachile rotundata*
36. *N.m.*	*Nomia melandari*
37. °	degree
38. %	per cent
39. P.s.	Pithitis smaragdula
40. pm	poste meridian
41. ppm	parts per million
42. q	quintal
43. var	variety
44. *viz.*	namely
45. vs	versus
46. r	rate
47. @	at the rate of
48. s	second
49. sp. spp.	species, one, more
50. sq	square
51. yd	yard
52. X. spp.	Xylocopa spp.
53. million	million
54. lac	lakh
55. te.	temperate
56. str.	subtropical
57. tr.	tropical

Contents

BOOK ONE
MELLIFERA BEEKEEPING

5. FUNCTIONAL BEHAVIOUR OF HONEYBEE 71–109

BOOK TWO
POLLINATION OF CROPS

List of Figures

List of Plates

Book One

MELLIFERA BEEKEEPING

Introduction

1.1 HONEYBEE

The honeybee is one of the most fascinating species in the animal kingdom. Honeybees collect nectar from flowers and use it to produce honey as a food for themselves and for their young brood. Sometimes in nature and, quite often with human aid, they collect large quantities of honey which can be extracted for our own consumption. What we extract in one year from a colony is only 1/6th of the total honey collected by them. In addition to honey, the bees also collect or produce other products such as pollen, propolis and beeswax which are put to good use in our daily life. The honeybees are also the most important among insects as pollinators of plants and hence they are a great help in producing fruits, vegetables, and seeds of crops like, clovers, *sarson,* sunflowers, etc.

Honeybees are social insects that live in large colonies in which there are many castes, namely, a single queen which is the mother of the entire colony, a few hundred drones (the helpless, lazy and awkward males), around 30,000 workers which are sterile females, and are the very life of the colony. It is the workers who feed the drones and the queen; the former doing practically nothing in the colony and the latter serving as a "slave" queen, laying eggs for the propagation of the colony. In a passive manner the queen controls the life activity of the colony, and if she dies, and there is no replacement for her, the colony will also die.

The wonder that is honeybee has been studied for hundreds of years and honey of the bees has been exploited by man for thousands of years. Honey was considered a food for gods and kings in ancient Egypt and it was found in original aroma when some of the pyramids were opened in this century after 5,000 years. It is mentioned in Rig Veda (2000-3000 BC) that Vishnu, the Indian god took the form of a blue bee on a lotus flower. It is also mentioned in the Rig Veda that Kama, the Indian God of love, carries a bow, that has a string made of nothing but a chain of bees. According to Joshi *et al*. (1980) the Vedic hymns describe that a colony of bees is ruled by a "bee-king" who cannot tolerate a rival in his hive. The use of smoke was prescribed for subduing the bees.

The honeybee is related to wasps and to solitary or subsocial bees commonly seen on flowers. Bumblebees are its very close relatives. The honeybee belongs to the family Apidae and the genus *Apis,* which has four species of truly social bees. The honeybees individually are cold blooded animals but being social creatures living in large colonies they maintain temperatures around 92° F and simulate the warmth of warm blooded animals; hence they can remain active throughout the year and raise their brood. They cover their brood and maintain more or less uniform conditions of temperature and humidity so that the young ones develop from eggs to the adult stage within a uniform duration.

Within the colony they have an elaborate system of communication, but once out in the field they work individually and are subject to environmental conditions of temperature, sunlight, cloudiness and rain. When temperatures are mild in nature they live longer and, in the hot summer their life is shortened. They always work hard for the maintenance of the colony. Scientific curiosity was created in honeybees by many interesting books (Ribbands, 1953; Materlink, 1960).

Of the four species of *Apis,* two are completely wild and two are subject to semidomestication. The wild species are *Apis dorsata* F., the giant honeybee, which makes its nests in the open on limbs of tall trees, on spurs of rocks, or under the cantilevers of tall buildings in cities; *Apis florea* F., the little honeybee, which makes its open nests on dwarf trees, in bushes and hedges, or in stacks of dry cotton or other sticks. The semidomesticated species are : *Apis cerana* F., the Asian honeybee, which makes its nests in hidden places such as hollows of trees, rocks, and walls of mud buildings; *Apis mellifera* L., the Afro-European honeybee, which also makes its nests in hollows of trees and rocks. The first three species are widely spread in Asia whereas the fourth one is widely spread in Russia, Middle East, Europe and Africa. It is considered to be more prolific, and a better gatherer of honey than *Apis cerana,* hence it has been introduced in all continents of the world and more recently in most of the Asian countries including India, Pakistan, Thailand, China, Japan. This book gives an account of *Apis mellifera,* unless mentioned otherwise.

1.2 BEEKEEPING

As a species the honeybee is totally wild. By confining it inside a hive it is subject to some management and manipulation but it is not domesticated in the sense domestic and farm animals are. The art and practice of keeping honeybees in hives for the production of honey (and other products) is called beekeeping. Beekeeping has been practised in Europe and Asia since time immemorial. Aristotle (384-322 BC) even observed the constancy of bees on flowers of certain crops. Before scientific beekeeping became popular in the last two hundred years or so, honey from the bee colonies in nature was exploited by man at least for the last 10,000 years in Eurasia.

In the nature, honeybees make nests in confined spaces such as hollows in rocks, or hollowed logs of wood with small entrances in the front. Man learnt to rob the bees of their honey, not caring what happened to the nest. Then he learnt to save the bees and protect them for future honey harvests. With this care and supervision started beekeeping which further developed into keeping colonies in an apiary. Colonies were kept in the primitive type of hives known as skeps which were made of hollowed out logs, woven wicker boxes, round baskets made of coiled straw or fired clay. In these skeps, there was a bee entrance in the front and a system of opening from the back. The combs had to be broken for harvesting honey. In India, villagers, while constructing mud houses, left a hollow in the outside wall leaving a small hole. On the inside a little window was made which could be opened for honey extraction and then shut at will. Crane (1998, 1999) gave an extensive survey of front- and back-opening wall hives found in Asia, Europe, and Africa. Their performance as compared with other hives were also discussed.

The honeybees made parallel combs which were broken at the time of honey harvest. The working bees could not be observed in the confined space and nothing was known of their castes and the role of their queen. Then hives made out of wooden planks nailed together were developed. Towards the middle of the last century, individual wooden frames were made to fit in a wooden box and the cut out combs made by the bees in nature were tied in the frames. Thus the movable frames replaced the fixed combs found inside the skeps.

1.3 HIVE AND EQUIPMENT

In 1851, Langstroth in USA observed that bees can move freely on moveable wooden frames if the space in between them was one quarter to three eight's of an inch wide (Naile, 1976). This discovery made it possible to devise movable frames which could be taken out of the hive (Fig. 1), observed, and replaced by shifting them suitably. By 1871, three other important inventions were made; these included comb foundation sheet, the honey extraction machine, and the smoker (Figs. 2 and 3). This was the advent of modern beekeeping and it became possible to manipulate the colonies of bees. The smoker, a type of bellows, was used for giving puffs of smoke after lifting the top cover. It was not known how smoke acted on bees but it was used regularly to subdue them before opening a hive. The comb foundation sheet made of beeswax having hexagonal impressions, one fifth of an inch in diameter when fixed on cross wire supports in a frame, were used by bees to raise combs of a uniform

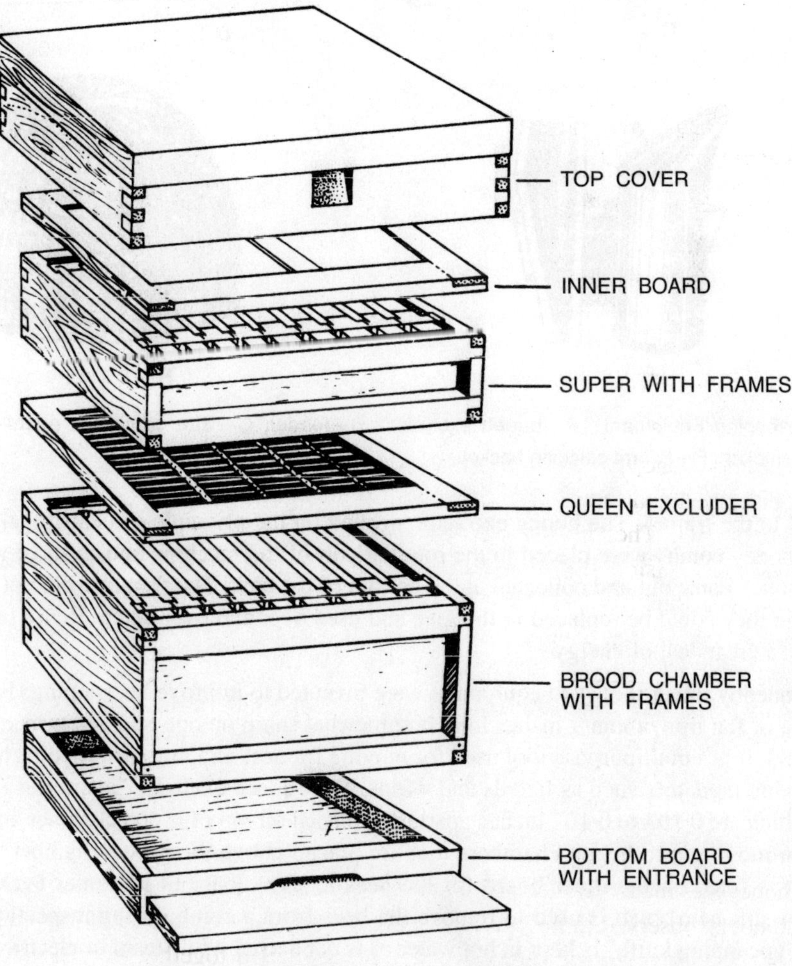

Fig. 1. Bee Hive, showing its parts separated.

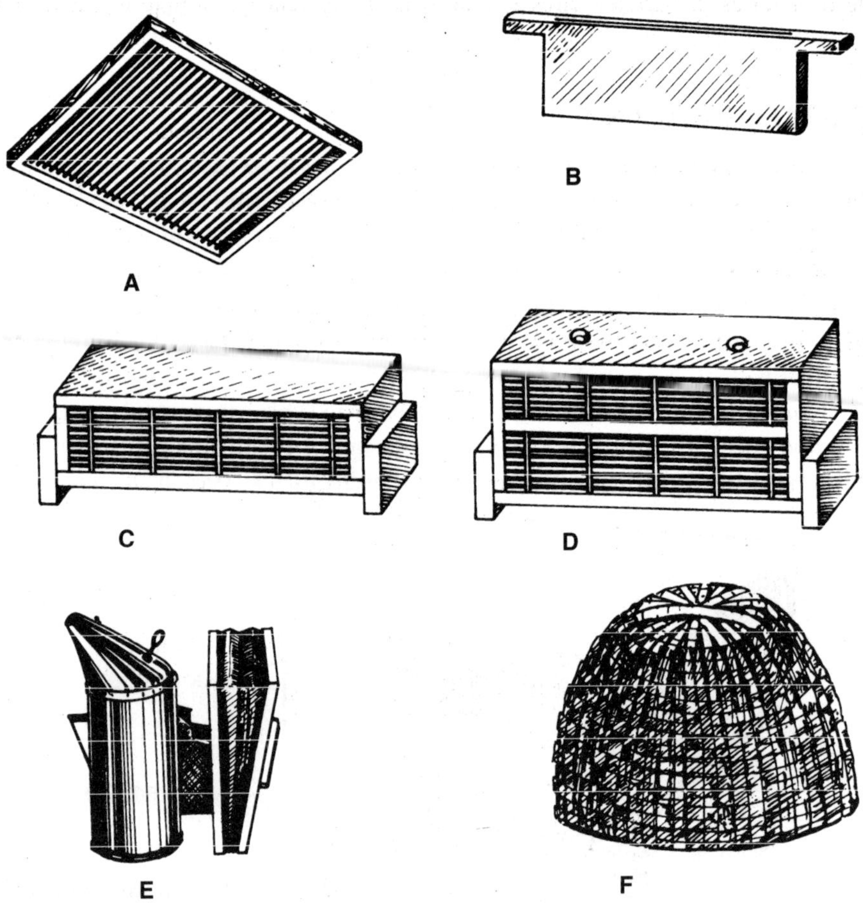

Fig. 2. Beekeeping Equipment : A—queen excluder ; B—feeder, C—wire enterance guard ; D—drone trap ; E—smoker ; F—swarm catching basket.

size (equal to the frame). The honey extractor worked on the principle of centrifugal force and when uncapped honey combs were placed in the rotating part of the machine and as force was applied pure and clear honey came out and collected at the bottom of the tank. The frames were not damaged in that process, and they could be replaced in the hive and used over and over again. Reuse of the combs thus saved bees, a great deal of energy.

Subsequently, other tools and equipment were invented to improve beekeeping (Fig. 3). The "hive tool", made of flat iron, about 9 inches long is somewhat sharp on one end and turned at 90 degrees at the other end. It is a multipurpose tool used for moving frames, scraping off propolis, hammering nails, and destroying predators such as lizards and wasps. The "queen excluder" is a series of wires fixed on a frame, which are 0·163 to 0·167 inches apart. It is placed on top of brood chamber and it prevents the queen from moving into honey chambers that are placed above the brood chamber. Bee escape is a funnel that is fixed on the inner board for the bees to move out but not enter back into the honey chamber. A soft hair brush is used to remove the bees from a comb during inspection. A long sharp knife, the "uncapping knife" is kept in hot water or is connected with steam or electric heating element and is used for cutting away the caps of sealed honey cells at the time of honey extraction. A metal or plastic lined cup-like feeder frame is used for feeding sugar syrup to bees when needed. The "drone trap" made of a series of parallel wires is placed at the entrance of a hive to catch extra drones, particularly during the shortage of food in the colony; the worker bees can pass through the wires. A wicker type "swarm basket" is used for catching a bee swarm that might escape from the hive and rest on a nearby bush or a tree limb. A "queen cell protector" is a wire cone fixed around the queen cells to be protected. A queen cage is used for imprisoning a queen before introducing it into a queenless colony; workers of the colony feed the queen through wire mesh and thus become accustomed to the odour of the new queen. A shoulder length veil of cloth netting, a pair of gloves made of leather or vinyl and an overall tucked in boots give protection from bee stings.

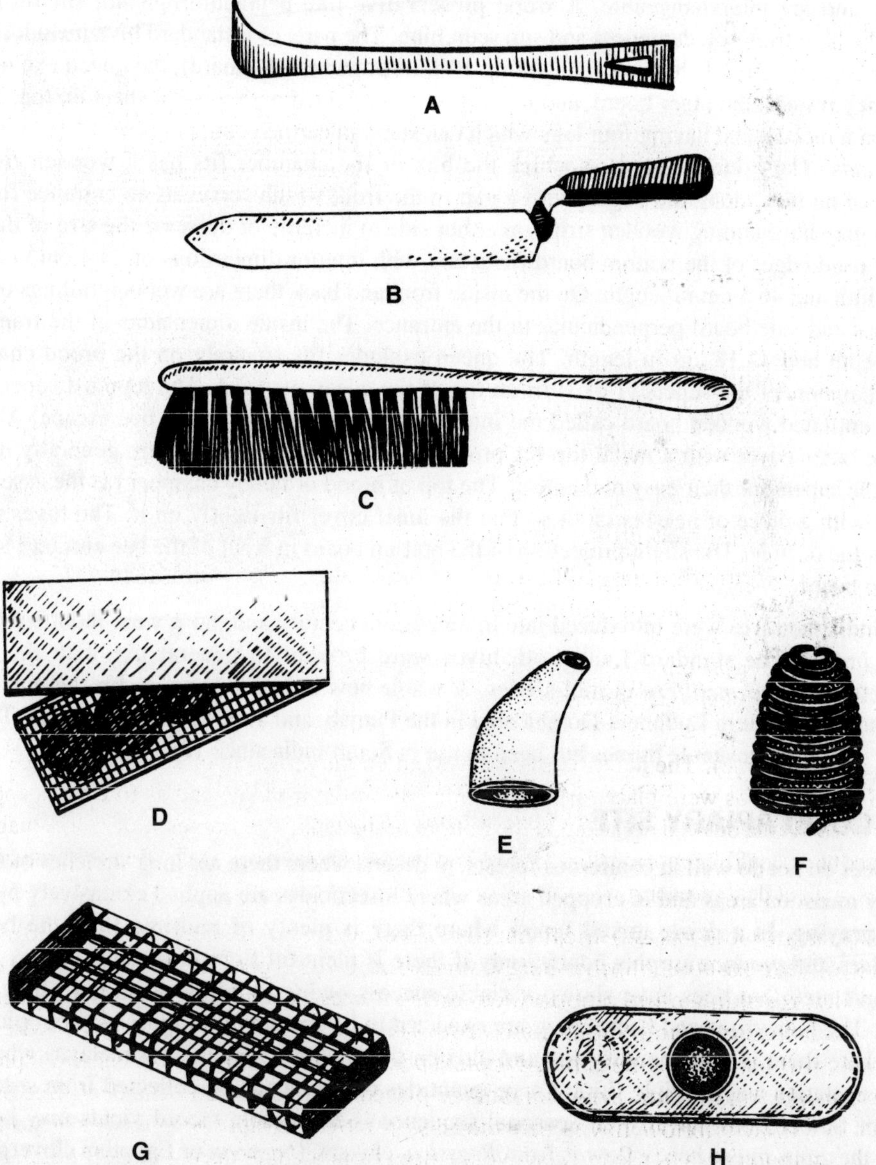

Fig. 3. Apiary Tools : A—hive tool ; B—uncapping knife ; C—bee brush ; D—queen cage ; E—conical bee-escape ; F—queen cell protector ; G—cage for queen introduction in colony ; H—flat bee escape.

The description, dimensions, and use of various types of A.m. hives namely Langstroth, Modified Dadant, British Standard etc. are given in detail by Crane (1990), Graham ed. (1992) and others. For *Apis cerana indica* bees the well known hives are Newton's, ISI, Jeolikote Villager, which have been described by various authors (ISI, 1970; Chandran and Shah, 1974; Bisht *et al.*, 1982; Mishra, 1995). The standard Langstroth hive generally made of light strong wood is now in common use in most of the countries where *Apis mellifera* L. has been introduced (Fig. 1). Such hives can be made locally by carpenters or they can be purchased from a factory. It is important to ensure that bee spaces are correct and that the frames and the box are of a standard size so that frames, covers, and queen excluder fit properly and are interchangeable. A wood preservative like pentachlorophenol should be used to protect the hive from rot, dampness and sun scorching. The parts of a standard hive include, the bottom board or base, the brood chamber (containing 10 frames and a side board), the queen excluder, supers with honey frames, the inner board, and the top cover protected with a metal sheet on top. The hive is placed on a metal stand having four legs which can stand in earthen containers of water as a protection against ants. The bottom board on which the box or the chamber fits has a wooden rise of three quarters of an inch along the edges, with a gap in the front which serves as an entrance for the bees. The gap may have sliding wooden strips on either side to increase or decrease the size of the entrance. On the raised edges of the bottom board fits a box with interior dimensions of 24.4 cm in depth, 37.3 cm in width and 46.5 cm in length. On the inside front and back there are wooden railings on which fit 10 frames and side board perpendicular to the entrance. The inside dimensions of the frame are 20.3 cm in depth and 43.18 cm in length. The queen excluder fits squarely on the brood chamber. The honey chambers or the supers (1 to 3) fit on top of the queen excluder. The topmost super is covered with a ventilated wooden board called the inner cover in which can fit the bee escape. Above it is a roof-like outer cover with a metal top for protection from weather. There are generally nine frames kept in the supers for their easy movement. The top of brood or honey chamber (as the case may be) is covered with a piece of hessian cloth so that the inner cover fits tightly on it. The hives are painted white on the outside. The slight projection of the bottom board in front of the bee entrance serves as an alighting board.

In India the hives were introduced late in nineteenth century and those were modified to suit *Apis cerana indica*. The standard Langstroth hives were brought to general use with the successful introduction of *Apis mellifera* in mid-sixties. A whole new industry came up for the manufacture of hives and equipment in Ludhiana-Doraha area in the Punjab, and at other places in India. The Newton hive for *A.c.* with moveable frames has been in use in South India since 1880 A.D.

1.4 A GOOD APIARY SITE

Honeybees do not do well in coniferous forests, in deserts where there are long stretches of sand dunes, in heavy monsoon areas and in cropped areas where insecticides are applied extensively by ground or aerial spraying. In a dense mixed forest where there is plenty of multiple flora the bees sustain themselves and produce surplus honey, only if there is plentiful flora in certain seasons. In tropical areas *Apis mellifera* bees have shown variable success owing to humid heat and heavy prolonged rainfall. The honeybees and beekeeping are expected to be successful in most of subtropical cropped areas where there is open countryside and also on the southern slopes of Himalayas where there is mixed vegetation and multiple flora. Large quantities of honey can be collected from specific crops, weeds or timber trees and if their seasonal sequence is favourable record yields may be obtained. Among the crops major honey flow is from *Brassica,* clovers, (*barseem* or Egyptian clover), chick pea, oil seeds particularly sunflower, niger, and cotton in non-spraying areas. *Plectranthus* is a rich weed in valleys around Kulu in Himachal Pradesh, Gharwal and Kashmir. Among the forest trees the most outstanding are Eucalyptus, *tun (Cydrus tuna), sheesham, Acacia, Khair, (Acacia catuja), mahwa,* rubber, coconut and other palms. The rich fruits, though of short floral duration, include pome fruits, citrus, guava, litchie, kaju, ber, and date palm.

For selecting an actual site of an apiary the following points should be kept in mind :

1. The land should remain dry all the year and be easily accessible by road.
2. The site should not be on a main road so that flying bees do not crash against vehicular traffic, telephone wires and other obstacles.
3. Fresh water should be available nearby but not in the form of a fast running stream in which bees are likely to be washed away.
4. The site should be near open fields where there are isolated trees for shade.
5. Hives should be kept in an enclosure protected from grazing animals which might topple the boxes over.
6. The hives should be kept behind a natural or artificial wind-break and their entrances should be away from the prevailing winds.
7. Hives should not be placed in a grove of trees, or in a forest where the atmosphere becomes close and muggy. Moreover, the swarms that issue from hives in such locations may rest high up in the trees and thus would be difficult to catch.
8. The hives should be located where there is shade during the midday sun and open during morning and evening hours.
9. Preferably, the hives should face the East and they should be 6 to 10 feet apart, with row-to-row distance not less than 12-15 feet.
10. As a protection against ants the legs of the hive stands should remain dipped in earthen pots full of water.

Up to ten colonies can be kept practically anywhere but a good site should have the carrying capacity of 40 to 50 colonies. In a mountainous area, slopes towards the East or the South, should be preferred for taking advantage of the sun. If there is a high hedge around the apiary bees will fly high and are not likely to strike against people moving in the area, thus minimizing the fear of bee stings and creating goodwill for beekeeping.

1.5 POLLINATION OF CROPS AND HONEY GATHERING

There are a large numbers of wild bees which are solitary, sub-social and social, and play an important role in the pollination of crops. Without their activity some crops and fruit trees may not set seeds or the yield would be poor. *Nomia, Pithitis, Megachile* are efficient pollinator of lucern (alfalfa), other clovers and oilseed crops. *Nomioides* and *Xylocopa* are abundant pollinators of cucurbits. On *Brassica* crops, *Apis florea* is the most abundant pollinator, followed by *Apis dorsata, Adndrena, Halictine* bees and *Colletes nursei*. The little honey bee *Apis florea* like the giant or the rock honeybee, *Apis dorsata,* is wild and plays an important role in the production of cruciferous oilseeds. These bees are by far the most abundant pollinators.

The hived honeybees, *Apis cerana* and *Apis mellifera* are important pollinators of a number of crops including clovers, Brassica, sunflower, citrus fruits and pome fruits etc. In contrast to the wild bees they can be moved to the crop to be pollinated, thus their role is crucial in the production of high yields of sunflower and pome fruits. In most western countries orchardists rent hives from beekeepers to assist in the pollination of their crops. It has been estimated that income from pollination activity of honeybees is fourteen times that of the honey value. The world production of honey is close to one million tonnes a year collected by 50 to 60 million colonies.

The highest density of bee hives, an average of seven per square mile, is in Europe (excluding Russia and East European countries) and there is an estimated total of 13 million colonies. In Russia and East Europe there are 10 million hives. There are 10 million colonies, both of *Apis cerana* and *Apis mellifera,* in Asia, including 2 million of *Apis mellifera* in China and Japan, and nearabouts 2-3 lacs in India. Africa has some 12 million *Apis mellifera* hives in all. In North America there are 5 million and in central and South America there are 4 million colonies. In New Zealand, the estimated number of colonies is four lacs and in Australia about 7 lacs.

TABLE 1. *Beekeeping in the World*

Order No.	*Continent/Country*	*No. of Colonies*
1.	Europe (excluding former Russia)	13 million
2.	East Europe (former Russia)	10 million
3.	Asia (including 2 m. *Apis mellifera* colonies)	10 million
4.	Africa	12 million
5.	North America	5 million
6.	South and Central America	4 million
7.	New Zealand	4 lacs
8.	Australia	7 lacs

1.6 STINGING BY THE BEES

Honeybees sometimes become excited and give a painful sting when they are disturbed in their hive. Sting causes inflammation around the wound site which lasts for a few days. One person in 10,000 gets a violent reaction which can be dangerous and if the victim is not rushed to a hospital for proper injections, he may die. For this reason some people are afraid of bees and beekeeping is not as popular as it could be otherwise. An understanding of the bee venom and of bee behaviour as a defense mechanism can minimize the fear of bees. As a bee stings a complex of pheromones are released which alarm other workers and they also attack the victim.

If a person opens the hive at wrong time of the day without first applying smoke the bees will be disturbed. The wrong timings are early in the morning or late in the evening, when it is windy and cold or when the operator smells of hot perspiration or perfume. Bees do not sting out of anger or revenge but it is their instinctive behavior to defend their house. When a hive is approached wrongly stinging starts from the guard bees in the vicinity of the hive entrance. These bees are sensitive to vibrating objects, dark or red color, abrupt movements and exhaled breath which contains chemicals disliked by them. The odour of hair and of leather also excites the bees.

Stinging by bees away from the nest is very unnusal and is an accident caused by sudden physical contact, a collision, an obstruction in the passage of the bees while riding a bicycle or by stepping on a bee in a lawn. The bees that are actively foraging on flowers normally do not sting.

Once a bee has stung and the sting has been deposited, alarm pheromone is liberated which tags the victim and more bees are attracted to the site. The body part having a sting should be washed with water and have ice applied to it. Hydocortisone ointment helps in suppressing the inflammation. The sting should be immediately removed with the sharp edge of a knife, but not pulled out with thumb and finger because in that case it goes deeper in the flesh. Serious multiple stinging is observed in the African races of *Apis mellifera* and the Giant honeybee *Apis dorsata*. The highest recorded number of stings on a man in Central Africa was 2243 and with timely treatment the victim was saved (Murray, 1964). A person who got 8000 stings in a cave in South America died and his body was removed next morning.

1.6.1 Bee Venom

When a worker bee stings a victim venom is injected into the muscles that causes pain followed by inflamation. Venom is a mixture of proteins and peptides, melittin being a major part of the proteins. It also, contains other compounds namely, hyalusonidase, phosphodipase A, acid phosphotase, histamine (Habermann 1972; Owen, 1978; Schmidt, 1982). At the base of sting inside the body of a honeybee there are a number of groups of glandular cells which produce alarm substance, a group of chemicals, that alarm, the bees and look for victims. It was thought at first isoamyl acetate was the main chemical but since then many more compounds inducing alarm behaviour have been identified; these are 2-nonanol, n-butyle acetate, n-hexyl acetate, benzyle acetate, isopentyl alcohol, non-octyl acetate, (z)-II-eicosen-I-ol etc. Their individuals role is being studied (Boch *et al.*, 1962, Blum, 1992). Earlier Banks and Shiplini (1986) studied the basic chemistry and pharmacology of bee venom.

Bee venom is considered to be a cure for arthritis (Broadman, 1962) and techniques for its collection have been devised (Benton, 1963). Moreover the nature and action of venoms from *A.c.* and *A.m.* have been studied separately (Inoue, 1985). An improved device, a new 'closed box' venom collector has been developed for handling a large number of bees from many colonies (Fakhim-Zadeh 1998).

1.7 FOOD OF THE BEES

The diet of the bees consists of proteins, carbohydrates, lipids, vitamins, minerals and water. Nectar and pollen of flowers form the primary food of the bees. In addition the young worker bees secrete brood food in their mouth which is fed to larvae, young workers, drones and queen. Nectar contains sucrose, glucose and fructose in varying quantities and is the source of carbohydrates. Pollen provides protein, lipids, vitamins and minerals. Brood food is rich in protein and certain other chemicals, including enzymes. The nutritional requirements in the larval stage of the three castes (worker, drone, queen) are basically the same but the quantities of food given vary a great deal, so that worker larvae given less food of certain type remain small whereas the drone and queen larvae given copious quantities of that food become large in size.

Adults of workers, drones and queens when newly emerged have many internal organs and glands that are not fully developed. They are given rich diet for the development of these organs. The wax glands and brood food glands of the workers develop in the first ten days of their life. In the beginning they are given some brood food but their primary food consists of nectar and pollen for the supply of energy and protein. The male reproductive organs in drones and the female organs in queens develop to the full size after emergence. During the first week drones are fed a mixture of brood food, pollen and honey by the nurse bees. Then they feed themselves out of the honey cells and large quantities are eaten by them before every flight. When energy is spent up they come back to the hive for more. The queen may feed herself on honey at first but the nurse bees soon start her with brood food and some honey. For her entire life she is looked after by the worker bees particularly of the age at which they secrete brood food. The size of the ovaries in the queen grow very large. A queen is also fed at intervals during egg-laying to provide energy for that activity. She is generally fed by one worker at a time, but up to five bees have been seen feeding her.

To tide over pollen shortage or to speed up brood rearing in the spring, pollen supplements and pollen substitutes are often provided to colonies. Pollen manually or mechanically collected from the field can be dried and stored at room temperature. While preparing pollen supplement it is mixed with soybean flour and is given to the bees. The pollen substitute is nutritionally adequate for brood rearing and for the development of brood food glands of the adult worker bees. The most accepted and widely used pollen substitute is a mixture of soybean flour, dried brewer's yeast, and dry skim milk. Addition of dried egg yolk, honey essence, and dark rum increase acceptability of the substitute.

During dearth of nectar in the field, bees feed on honey reserves in the colony. It may result in the shortage of food which can be critical for the survival of the colony. Before the onset of winter, such colonies should be fed sugar syrup in the ratio of 1 : 1, given in a feeder frame, a feed-can, or a jar.

Honeybees and their Relatives

2.0 HYMENOPTERA

The honeybee is an insect belonging to the class Insecta order Hymenoptera in the animal kingdom. Among the 2.5 million animal species on earth, three fourths are insects which are placed in thirty orders. The order Hymenoptera (membranous wings) with a strength of one quarter million includes the honey bee. Hymenoptera are very variable in body structure and behaviour and for the convenience of study they are divided into many super families including Formicoidea (ants), Apoidea (bees), Vespoidea (wasps and hornets), and Sphecoidea (solitary or digger wasps).

The super family Apoidea (Fig. 4a) has nine families namely; Andrenidae, Colletidae, Fidelidae, Xylocopidae, Megachilidae, Halictidae, Melittidae, Anthrophoridae, and Apidae. The family Apidae (Fig. 4b) has three subfamilies and many tribes including the Apini. The genus *Apis* apart from other species includes the four honeybees *Apis mellifera* L. (the common or the Euroafrican), *Apis cerana* L., (the Asian), *Apis dorsata,* (the giant), and *Apis florea* (the dwarf bee). Apart from the honeybees

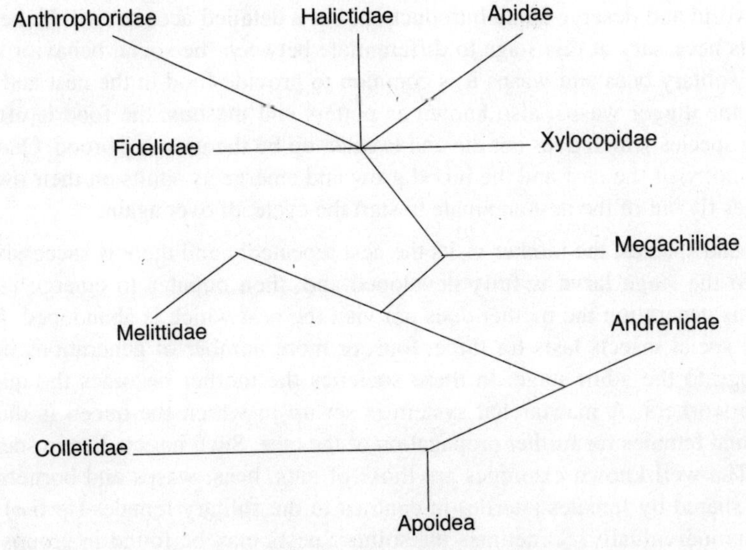

Fig. 4a. Families of Apoidea (Hymenoptera).

Apoidea includes many other solitary sub-social, and social bees that visit flowers and use pollen as food for rearing their brood.

Hymenoptera is a very diverse group of insects and is considered a super order by some authors. It includes such familiar insects as sawflies, gall forming wasps, predatory wasps and hornets, bumblebees, solitary bees, honeybees and ants. Some of them are pests as they feed on plants including valuable crops. Others are useful animals, and even friends, since they pollinate flowers, produce wax and honey. A large number of them are indirectly beneficial as they parasitize other animals injurious to forests and crops, notwithstanding their small size, innocuous appearance and silent activity. In size, the Hymenoptera vary from 0.2 mm to 5.0 cm, and in colour they vary from gray to brilliant shades or bands of all the colours of the rainbow and some are even shining metallic and spectacular in colour. They have two pairs of wings, transparent and clear like glass, which lie flat on the body when at rest. The females have a hollow spike-like structure at the tip of their abdomen which is used for depositing eggs in plant or animal tissues or it may be modified to form a sting with which to inject venom from their poison glands, causing pain or paralysis of the victim. When sting is present, the female deposits eggs with the help of accessory structures at the end of genital pore.

The common developmental stages in Hymenoptera are; egg, larva, pupa, adult. The larva has many stages of growth which are known as instars. Some larvae may be free living and eat plant tissue or they may mine inside and eat through the leaf parenchyma. Others may produce galls on plants or may live as parasites on many insect pests. They may develop in brood cells provisioned with pollen, honey or with paralyzed or killed spiders and insect larvae. They may grow on their own in the cell of a nest or they may be fed by the parents. Some larvae develop in the nests of other Hymenoptera as parasites. In the last larval stage it spins a cocoon and on pupation transforms into an adult. Most of the species are solitary but the instinct to form colonies has developed independently in many groups like ants, bees and hornets. The social species, among all insects, display the most advanced form of brood care.

2.1 SOCIAL BEHAVIOUR

Many of the relatives of the honeybee have plant and animal associations of importance to agriculture throughout the world and deserve some introduction, but a detailed account will be beyond the scope of this book. It is necessary at this stage to differentiate between the social behavior and the solitary life. Among the solitary bees and wasps it is common to provide food in the nest and deposit an egg nearby. Among the digger wasps, also known as potters and masons, the food is often a paralyzed larva of the prey species which does not die and is eaten up by the growing brood. Once food is given the mother does not visit the nest and the brood grow and emerge as adults on their own. The solitary males and females fly out of the nest and mate to start the cycle all over again.

In the subsocial species, the mother visits the nest repeatedly and there is successive provisioning of the nest up to the stage larva is fully developed and, then pupates to emerge as adult. On the completion of this generation the mother does not visit the nest which is abandoned. Parental care in the case of fully social insects lasts for three, four, or more number of generations in the same nest right from the egg to the adult stage. In these societies the mother becomes the queen and sterile females become workers. A matriarchal system is set up in which the queen is the head and she produces males and females for further propagation of the race. Such insects live in social groups or in large colonies. The well known examples are those of ants, bees, wasps and hornets. Work among social insects is shared by females (sterile) in contrast to the solitary females (fertile) who build and defend their nests individually. Sometimes the solitary nests may be found in groups in a particular locality or a site but this nesting habit does not make them social species.

2.2 WASPS AND HORNETS

They are sometimes called papermakers as they use wood cellulose for making their nests. The two groups are different in their body structure and habits. All hornets are wasps but only a few wasps are hornets. The hornets have thick-set and robust bodies and their abdomen is quite close to the thorax; the pedicel is not easily visible. They make multistorey nests generally in tree hollows but sometimes in roof thatchings and in the ground. The wasps have slender body, somewhat shiny, and the waist or the pedicel is visibly distinct. The yellow wasp, *Polistes haebreus* L., is quite common in North India. It makes round horizontal multicell nests on ceilings in barns and houses. The banded Oriental hornet, *Vespa orientalis* F., is distributed from the Middle East to India and is a common sight on sweets in bazaars. It makes a series of overlapping nests hidden in the hollows, rafters, and rocks.

2.3 BEES, BUMBLEBEES, HONEYBEES

The solitary bees, bumblebees, stingless and honeybees belong to Apoidea and they visit flowers to collect nectar and pollen as food for themselves and for their brood. Among the social insects they are the most important agents for plant and crop pollination. The social bees are present in the family Apidae and some of them in Halictidae but their role in pollination is not related to the social status. The structure of flowers and the adaptation of a bee species are the important characteristics that determine the effectiveness of pollination. The subject has been studied in much detail by Mitchener (1974, 1979), Bohart (1967, 1972), Free (1993, 1982), McGregor, (1976).

Biology of some of the important groups is described here. Except for the bees of family Apidae the other bees of various families of Apoidea are generally mentioned as solitary bees, although some species of Halictidae live a social life.

2.4 SOLITARY BEES

The solitary bees are placed in eight families of the super family Apoidea. They have varied habits of nesting and brood rearing performed by the females. They collect pollen and nectar from flowers and in this process they pollinate plants. Generally the fertilized females construct nests, collect provisions and lay eggs, working as solitary individuals. In some cases two or more females may have a common entrance to their nests or they may even work together in the same nest. Some of them are gregarious and make large clusters of nests close together maintaining the solitary habits. The common examples are alkali bee (Nomia), leaf cutting bees (Megachile), the sweat bees (Halictus), burrowing bees (Andrena), carpenter bees (Ceratina, Pithitis, Xylocopa), the cuckoo bees (Nomadia) and the digger bees (Anthrophorids).

Halictus is a ground nesting bee and completes two or three generations in a year. A fertilized female remains in hibernation during the winter and comes out of its burrow in the spring. She constructs a nest in the ground with side branches at the lower end. Pollen balls are deposited in these cells on which eggs are laid. From the fertilized eggs develop females which start brood rearing in the same nest and raise male brood. Mating takes place outside the nest. In autumn the fertilized females make new nests and hibernate there.

The Megachile are known as leaf-cutting bees. They cut out leaf pieces of regular patterns and use them to line their shallow nests made in earth mounds or hollows in trees. In USA and Canada the introduced species *Megachile rotundata* has been manipulated to accept artificial nesting sites in order to increase its population in alfalfa fields and thus to improve pollination and seed setting. The bees are attracted to make their nests in round holes of a specific diameter in wooden blocks. They line these holes with pieces of leaves and make nests to rear their brood. They raise a number of generations in a year and the population is increased many folds.

The *Nomadia* or cuckoo bees are wasp like scant haired Anthrophorids which are basically parasites. They do not raise their brood themselves, but lay eggs in the nests of other bees where the offspring grows as parasites and emerge as adults. The closely related digger bees are robust and hairy. They burrow in the ground or into earth mounds where they make their nests. The nests are lined with a thin layer of wax or a varnish like substance. Many of these bees collect pollen and raise their own brood but there are some species which are parasitic.

Among the carpenter bees in India the well known genera are *Pithitis, Ceratina* and *Xylocopa*. The former two are small bluish green bees 6 mm in body length. They make their nests in stems of pithy plants by cutting out pith and making linear cells one above the other. The latter are robust wasp-like bees having dark, bluish black shine on their body and attain a length of 25 mm. They make galleries in bamboo sticks used in roof thatchings over huts in the countryside or on the outskirts of villages. They can also excavate wood holes made by the larvae of wood-wasps. They make linear cells in which they rear their brood, raising more than one generation a year.

The Andrenids are small to medium size bees and they have nesting habits similar to that of *Halictus*. They burrow in the ground and make nests in sparsely wooded lands. Among Andrena the most common species in North-West India is *A. ilerda*. It is the size of a honeybee and has yellow and rust bands across its abdomen. They are persistent and noisy bees who hover around an intruder in their nesting areas which is a defensive activity. The typical life-cycle in Andrena has one generation in a year. The male and the female emerge in the spring, they mate, and the fertilized female makes an earthern tubular nest in the ground where she lays eggs besides the provisions collected by her for the brood. The larvae develop in their cells within a month, pupate and transform into adults. They remain imprisoned in the cells for many winter months and emerge the next spring.

Bees of the family Colletidae are known as Plaster Bees or the Yellow-face. They burrow in the ground to make their nests. The burrow has a thin translucent lining on the inside. The bees are moderate in size and are quite hairy and are not so common.

The Chimney Building Bees belong to the genus Anthrophora of the family Anthrophoridae. They are stocky furry bees having a long tongue. They are solitary bees and make nests on stream banks by burrowing downwards. Their entrance is marked by a chimney that is bent parallel to the ground with a groove running along its length. It is believed that the chimney helps in a controlled circulation of air inside, creating a system of natural airconditioning.

2.5 SOCIAL BEES

The social ways of life is evolved to varying degrees among the bumblebees, cuckoo bumblebees, stingless bees and honeybees. The social order is highly developed among the honeybees and they have quite a sophisticated system of communication among the members of a colony, and among them the caste system is well developed.

2.5.1 The Bumblebees

The bumblebees belong to the sub-family Bombinae of the family Apidae and are placed in the genus Bombus. They are social bees, have comparatively a long tongue, robust body, covered over with fuzzy hair which are often black, yellow or orange in color. Their wings are comparatively small, and their metatibia are flat and concave which are used as pollen baskets. They are seen hovering over flowers on warm sunny days. In a locality, not more than three or four species may be present.

The bumblebees are distributed throughout Eurasia, Africa, and the Americas and have also been introduced as pollinators in Australia and New Zealand. They prefer the cool climate of the temperate regions or the higher altitudes in the subtropical areas. If the temperatures in their surroundings becomes above 30°C it is too hot for them to visit flowers. In flight they appear to be more like a helicopter than an airplane and in order to achieve this they have to beat their wings very fast, around 200 times per second.

The bumblebees have various castes, namely, worker, queen, male or drone. A fertilized queen passes winter in hibernation all by herself in some hole or in one of the old nests. In the spring she emerges and initiates building a nest in a hollow, such as a rat burrow in the ground. It makes a cluster of pot-like largish cells made out of mixture of wax, pollen and resins. The cells are of different size and shape which are provisioned with pollen and honey by the queen herself. After laying fertilized eggs in the brood cells the queen keeps providing food to the larvae which develop till they spin a cocoon and pupate and ultimately emerge as workers. The first brood consists of half a dozen workers that take over the colony duties such as making more cells, foraging and warming the brood. The bumblebees have a unique system of generating energy and heat by beating the wings or by shivering. The heat from the thorax is passed on to the ventral side of the abdomen to warm the brood cells just as the hens brood on their eggs to warm them up. The size of the worker depends upon the size of the cell in which it is raised and foraging bees vary in size so much that it may be confusing to decide whether they belong to same species.

A number of generations are completed from the spring to the autumn and the colony strength may reach 400. In a season, the males or drones are also produced. A given colony may produce, apart from the workers, only males, only virgin queens or both. The males after emergence stay in the colony for three or four days and then fly out, never to return. They forage on flowers, fly around, and wait on foliage to mate with virgin queens from other colonies in the area; thus there is genetical intermixing of colonies. The mated queens return to the mother colony and feed on the stored pollen and nectar that are always available. Their fat bodies develop in size and thus they prepare themselves for the coming cold weather in which only the mated queens survive. The old nesting sites are generally abandoned except in the tropical areas where the nests may become perennial.

After the sexual forms are developed a colony starts to decline in strength because only a few workers are produced and the old ones are exhausted in raising the brood. There may be disturbances of various kinds in a colony. A queen emerging from hibernation late in the season may force herself into a colony and may come in direct contact with the active queen who has already made cells and gathered food. One of them is killed in the fight and the other starts the colony duties. Predators and parasites may invade the colony; make a direct attack or enter the colony in a sly manner and gradually replace it with their own race. It has however, been observed that every colony has its own distinctive odor. The workers recognize their own within a colony but they do not seem to have a communication system by which to direct the others to a foraging site. In the field the bumblebees work as individuals but they do find the nest on their return and can perceive any intruders in a colony.

2.5.2 The Cuckoo Bumblebees

These social bees are evolved to live as parasites on the bumblebees just as the koel lives as a parasite in the nest of a crow. The cuckoo bumblebee belongs to the genus *Psithyrus* of the subfamily Bombinae. They are social bees and compared to the bumblebees have a lighter body, are less hairy and are not so bright in colours; their flight sound is like murmuring rather than booming as in the latter. Their metatibia are convex and are not used as pollen baskets; in fact they have no use for the baskets as they do not collect pollen to feed their brood. The cuckoo bumblebees fool the real bumblebees and feed on their provisions and with treachery make them raise their brood like the cuckoo birds do. The cuckoo bumblebees do not build their own nests nor do they have a worker caste.

The queen of a cuckoo bumblebee on emergence out of hibernation in the spring enters a Bombus nest by force, but mostly on the sly since she is cryptic in colour. In case there is a fight, the cuckoo is often victorious since it has strong mandibles, has a harder and stronger body covering and a more powerful sting. When the queen enters in a sly manner she may hide for a while, acquire odor of the colony and start laying eggs on the pollen balls already collected by the host bee. She may eat the host-bee eggs and let the workers in the nest rear the cuckoo eggs, and even provide warmth to the brood. Thus more of the cuckoo brood is raised. The parent colony becomes weak and its females are eliminated. In the end the drones and virgin queens of the cuckoo bumblebee predominate in number. Naturally, there is a reversal sometimes and the host species pushes the parasite out but quite often the two queens work in the same nest, at least for some time in a season.

2.6 HONEYBEES AND THE STINGLESS BEES

These groups are highly evolved as social bees. The subfamily Apinae of the family Apidae includes three main groups of bees, the Golden bumblebees or the Orchid Bees (Euglossina), the Stingless Bees (Meliponini, including Melipona and Trigona) and the Honeybees (*Apis* spp.); the three tribes are named Euglossioni, Maliponini and Apini (Fig. 4b).

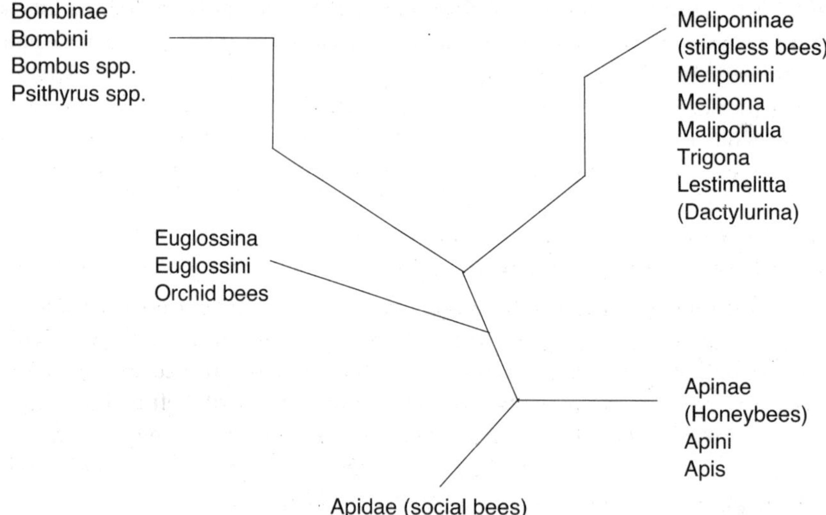

Fig. 4b. Family Apidae (Hymenoptera).

2.6.1 The Golden Bumblebees or the Orchid Bees

They are the most spectacular shiny, metallic bees, in body form somewhat like bumblebees but anatomically closer to honeybees, hence they are placed in genus Euglossina of the subfamily Apinae. They are social bees of the American tropics and are closely associated with the pollination of orchid flowers. They have very long tongue and brilliantly colored bodies ranging from blue, green and gold to red, orange and purple. There is most interesting color and odor mimicry associations between those of the orchid flowers, the female bees and the males. Some orchids mimic female bees in shape, size, color pattern and scent, and are so inviting that males are often fooled into mating with flowers as if they were females, and in the process, they pollinate the flowers. Some of these bees have huge tongues, twice the body length, and when not in use for collecting nectar it is carried under the body reaching beyond the tip of the abdomen. Associated with the perfume of flowers which the males acquire is the strange behaviour of congregation of other males of that species where the females are attracted for mating. The plant-bee association is quite unique.

These bees have a social life very similar to that of the bumblebees and being inhabitants of the tropics have perennial nests. Their behavior of male congregation is like that of the honeybee drones.

2.6.2 The Stingless Bees

The stingless bees belong to the subfamily Apinae and their tribes are distinct from the bumblebees and honeybees. In these bees the sting is present but it is not developed to inject poison, so the bees protect themselves by biting with their mandibles. Their social position is between that of bumblebees and honeybees and the organization of the colony is like that of the latter. Earlier these bees, misnamed as they are (stingless), were placed in a separate family Meliponidae but now they are considered tribes under Meliponinae. They are found throughout the tropical world and the two well known genera are *Melipona* and *Trigona,* the latter being smaller in size (2-8 mm).

The stingless bees build nests in tree trunks, crevices in rocks or in the ground, and quite often in abandoned termite nests. The bee nests may be protected from bad weather by making walls of wax or resin. They build a series of horizontal wax combs, the oldest being at the bottom and the new ones made above them separated by wax pillars. The cells resemble those of bumblebees but are not used for any purpose other than rearing brood. They make mass provision of brood cells which they seal with wax and the larvae develop inside, pupate there, and emerge as adults. Their life and social behaviour is similar to that of the honeybee. The colony strength varies from 20 to 80 thousand and they have the habit of swarming in the spring. When disturbed they crawl all over the body of the intruder and smear it with a caustic secretion that causes irritation to the human skin.

The swarm invariably carries a young queen as the old queen is heavy, full of eggs, has broken down wings. She is left behind in the old colony. The queens do not have pollen baskets and cannot produce wax. Therefore, a queen alone cannot start a colony as is the case with bumblebees. The division of labour and communication among workers does exist but it is not as fully understood as in the case of the honeybee. There are no parasitic stingless bees but some related species who's workers do not have pollen baskets attack the established colonies and the war of theft and robbing continues over long periods. Eventually the colony may be taken over and destroyed, but sometimes the odds may be reversed.

The stingless bees store honey in certain parts of their combs but the system is not as organized as in the Eastern and the Western honeybees, *Apis cerana* L., and *Apis mellifera* L., respectively. These stingless bees were kept for honey gathering by the Mayas long ago. The colonies were kept in hollow logs and honey was drained out twice a year. Out of interest some people even today try to keep the stingless bees but the honey crop is never as bumper or clean and pure as that of the honeybees (Weaver and Weaver, 1981).

2.6.3 The Honeybees

In honeybees caste system is perfectly developed. They belong to the family Apidae and the tribe Apini. There are four species of genus Apis. Two of them are wild and make their nests in the open. They are *Apis florea* F. and *Apis dorsata* F. The other two make their nests in hidden places and can be domesticated to live in hives. They are the Asian honeybee, *Apis cerana* F. and the Western honeybee, *Apis mellifera* L. (**Plate 1**).

The Eastern or the Asian honeybee, further has geographical races; those from Afghanistan, Himalayas, China are larger than those from the tropical areas of South India, Ceylon, and Indonesia. It is basically a frugal honeybee, maintains small colony strength and sub-divides easily. The bees do well in small hives, although with proper management they can sustain in Langstroth size hives also (smaller cell size). Owing to their poor colony strength and smaller body size their performance in gathering honey is not so impressive. The honey extraction is only a fraction of that collected by the superior *Apis mellifera* L. For that reason the Western honeybee has been introduced in most of the Asian countries, in the second half of the twentieth century.

An account of the three Asiatic species relatives of *Apis mellifera* L. is given first. Their geography and toxonomy has been studied by Ruttner (1986, 1988) and communication dances by Punchihewa *et al.* (1985) in Shri Lanka.

A. *Apis florea* Fabricius (1787)

The Dwarf honeybee, *Apis florea* F. (**Plate 1**) lives in nature in the subtropical region of India, Pakistan, as far west as the Persian Gulf, Oman, or may even be North Africa (Lord and Nagi, 1987), in Malaysia, Thailand, Indo-China, Indonesia. The ability to survive in a very hot and dry climate is its special trait and it can live in deserts without any harm at places where temperatures may touch 50°C. In areas of its distribution it lives along with *Apis cerana* and *Apis dorsata;* the ratio of their size is 1.0 : 1.27 : 2.09 respectively (Ruttner, 1988; Free, 1981). Compared to the other species of honeybees, its workers have a long life of 61.2 days which is 2.5 times that of an *Apis mellifera* worker (Michener, 1979).

The heat tolerance ability is further demonstrated by the fact that its daily activity starts at 18°C and continues up to 40°C. It is no wonder that she is a relentless visitor of flowers of crops, trees, shrubs and the annuals. Her small size restricts her to a shorter flight and hence she exploits the flowers intensively (Whitcombe, 1982, 1984). In one study, on Brassica crops this bee constituted 73-74% of the insect visitors observed (Atwal, 1970).

The Dwarf honeybees construct their nest around the stem of a bush, or a dried thick stick. At the top the nest encircles the strong stem so as to give it a good strength. The portion of the comb that encircles around the stem (or on the flat of ceiling in a building) is thick but as the comb is built further down it becomes thin in depth, although as broad as the upper portion. On the top the comb is flattened so that, it serves as a sort of platform on which the communication dances are performed (Whitcombe, 1982, 1984; Koeniger *et al.*, 1982; Dyer, 1985a). The long and deep cells in the upper portions are used for storing honey and the smaller standard size cells are used for rearing brood. Even when the comb is built on a flat surface like a ceiling in a verandah the upper portion is broad. Sometimes the combs are constructed in a protected place in the hollow of a hedge, a stack of sticks or even a hollow in a building structure (Sandhu & Singh, 1960; Thakar & Tonapi, 1962).

The worker brood cells are made on the flat surface in the middle of the comb. The larger drone cells are 4.2-4·8 mm in diameter, and are found on the lower part of the comb (**Plate 2**). The queen cells are at the lowest end of the comb and are quite large in size. The comb itself is just under one foot in length and contains about 14,000 cells. The size of all the three types of cells varies with geography and becomes smaller as we go from North to South. This honeybee also has a habit of nest-shifting within short distances. The bees may shift within two months in dry season and within five months in the rainy season. When they shift they carry away all the provisions, including wax (Free, 1981; Sakagami *et al.*, 1973; Ruttner, 1988).

In a colony there are about 6,000 bees and the queen lays 350 eggs per day in a brood area of 600 cm^2 (Kashirsagar *et al.*, 1980). Since this honeybee shifts a lot, it would appear its biology is adapted to multiplication, swarming and dissolution of the colony with the purpose of reassembling under a new situation. During a short span of 2-5 months the nest is built, brood is reared, honey is stored, and 12-16 queen cells are built for further propagation. When the first virgin queen emerges, the colony swarms and alongwith the mother queen half the number of workers leave. Subsequently, the virgin queens mate and smaller swarms leave every few days. In the end a handful of bees remain on the comb and eventually they also swarm away or simply disperse, leaving the comb bare and deserted (Sandhu & Singh, 1960; Kshirsagar *et al.*, 1980; Koeniger *et al.*, 1982).

The swarms forming the new colonies generally settle at a distance of less than 100 metres, after shifting their site a few times. Some swarms do go far but not more than a few hundred metres. It has been observed that if a queen is removed or it dies naturally, the workers build a queen cell by modifying a worker cell somewhere in the middle of the comb.

The communication among bees is through dances on the horizontal plane at the top of the comb. Making use of the position of the sun the dancing bees are able to convey to the others the direction and distance of the food source. They make the waggle dance by raising the tip of abdomen upwards and creating dorso-ventral motions (Lindauer, 1957, 1971). The Giant honeybee also takes the same posture but it dances on the vertical plane of the comb. The closed nest species, *Apis cerana* and *Apis mellifera,* also dance on the vertical plane but while dancing they bend their abdomen downwards. *Apis florea,* among the four species of honeybees is considered more primitive in dancing behaviour as it cannot transpose the direction and the distance of the food source on the vertical plane, which seems to be a more complicated manipulation (Lindauer, 1957). Pheromones also play important part in communication among these bees (Free & Williams, 1979).

The honeybee produces no more than half a kilo of honey which is dark in colour and has a strong flavour. It is not much of a commercial product but some people value it for its medicinal value. The natural enemies of the bee include wasps, ants, bee-eating birds, wax moth, and some mammals. A new species of Nosema has been isolated from this honeybee from Sri Lanka which can be transmitted to *Apis mellifera* experimentally in the laboratory (Ruttner, 1988).

Apis florea even though quite small in body size can be quite aggressive and resort to robbing the nests of *Apis mellifera,* much bigger in body size and greater in colony strength (Chahal *et al.*, 1986; Koeniger, 1976 b). Lately, some researches on this honeybee have also been conducted in Oman (Whitcombe, 1981, 1982, 1984) and in Iran showing interest in its preservation as a species (Tigori, 1971, 1987).

More recently, an allied species or may be a subspecies, *Apis andreniformis,* has been recorded in Borneo (Koeniger *et al.*, 1999) and it is believed natural hybridization does not occur.

B. *Apis dorsata* Fabricius (1793)

The Giant honeybee (**Plate 2**) is the most spectacular and ferocious of the four *Apis* species. The workers are the size of a wasp and the queen is slightly bigger. The bees are a common sight on sweet meat shops in Indian bazaars and are seen sucking sugar syrup and also juices of grapes and other over ripe fruits. In this process they remain quite unmindful of their occasional brushing away with a piece of cloth. They rarely sting the shopkeeper and the customers unless accidentally crushed or trampled upon. The geographical distribution of this honeybee ranges from the Philippines, through the entire Indo-Malayan region, as far west as the Indus river (Ruttner, 1988).

The nesting behaviour of this wild honeybee has been studied with much interest and efforts have been made to hive it also (Mahindre, 1968, 1977; Thakar, 1973). The honeybees build single combs on the projections of rocks or those of tall buildings in cities, and more commonly on tall trees on the roadside, in gardens or in forests. More than a hundred colonies can be counted in the Taj Mahal gardens at Agra. They can make nests on any tree species including the thorny *Acacia katechu,* but they seem to show preference for the tall *Ficus* trees which have thick limbs wide apart, allowing free flight to the bees. They ignore the people who may walk or sit under the trees during their travel or at the time of siesta. Someone out of mischief has to first throw a stone at the colony and its wrath is let loose. The bees attack the suspected enemy (quite often an innocent traveller) in scores, within a radius of hundreds of yards or even up to 1 km. After stinging men, cattle and animals for hours it takes them a long time, in fact an overnight, to calm down moderately. Even the next morning some excited bees are seen scouting around. It appears that colony defense in the open is an evolutionary behaviour. During this state of excitement of the colony many a passers by are stung by multiples of bees or by a cloud of bees and the victim writhes on the ground in agony and even looses the presence of mind to call for help. In many cases in the vicinity of the colony 80-100 stings per victim is a common occurrence. The very first step to be taken is to save the victim from shock and subsequent death, by administering avil or cortisone, through regulated injections (see Lindzon and Silver, 1988). It would appear they have a system of orientation at night (Dyer, 1985; Dyer and Gould, 1983).

The combs made by this honeybee are large and semicircular, quite thick on top, where honey is stored and honey harvesters go for it. Combs are attached to the strong limbs of a tree and the attachment is quite wide and strong to be able to carry the weight. According to the stage of growth of the comb the surface area varies from 0.1 m^2 to 1.0 m^2. On an average there are 23,000 cells, constructed bilaterally on either side of the comb. The cells are hexogonal in shape, 5.35-5.64 mm in diameter and they slope upwards. The depth of a brood cell is 16 mm (Diwan & Satvi, 1965; Nandi & Mahabel, 1974; Deodikar *et al.*, 1977).

In a comb, honey is stored on the top with thicker sides, and pollen is stored below, followed by the brood area. The queen cells are made at the lower end. In a colony, 80-90% of the bees arrange themselves in a layer of 3-6 bees thick and they are oriented in the same direction with their heads upwards and their wings half spread. This cover of bees is like a bag that touches the comb at the edges, leaving a hollow space of 1-2 cm. between the comb surface and the cover of the bees. This space is used by the nurse bees to move around and feed the brood. At the lower end of the bag there is the mouth, open for the worker bees to enter and to come out (Thakar & Tonapi, 1961; Reddy 1980; Qayyum & Ahmed, 1967). The field bees perform dances at the exposed portion of the comb on the vertical plane (Lindauer, 1957, 1971; Morse, 1969, 1970; Morse & Laigo, 1969).

In the normal working state of the colony it is not at all aggressive. Early in seventies I visited Pune and in the forest nearby Mahindre demonstrated to me a colony confined within a wooden frame brought down to the ground level with the help of a rope and a pulley. He made me put my hand on the curtain of bees. A strong wave of "shimmering" passed downwards and it happened everytime I repeated the act. The feeling I got was that I was patting them and they were reciprocating. It is reported that most of the small animals that might approach the colony as enemies are repelled by this shimmering (Mahindre, 1983a; Mahindre *et al.*, 1977; Koeniger *et al.*, 1975).

In case a single alarming bee makes a run as many as 5,000 bees may be alerted out of the curtain of bees. They will fall downwards and head straight at the intruders or the suspected enemies. Only one bee has to sting and the smell of isopentyl acetate released as a result of stinging will convey to the other bees where to locate the "enemy". The smell lasts for 10-15 minutes and within that time the victim is at a great peril. The bees chase him menacingly and the only escape for him is to decoy and hide behind a hedge, lie on the ground with face downwards and to cover himself with a sheet of cloth or a blanket and to protect his naked body parts.

Under natural conditions, *Apis dorsata* generally make their nests higher than 10 meters, but some are seen at lesser heights, particularly in areas where there are only a limited number of tall trees

(**Plate 2**). When a swarm is smoked at night or a large population dies of fire the remnant demoralized swarm may shift temporarily on a bush, a crop or an object much nearer to the ground than their original nesting height.

In a colony the brood is reared in worker cells. The drones are also reared in worker cells and they are randomly scattered here and there on the comb surface. The cappings of the drone cells are raised whereas those of the worker brood are flat. The queen cells are made on the lower rim of the comb (Vishwanathan, 1950a, b; Millen, 1942). This honeybee like other *Apis* species also produces reproductive swarms. Once swarming starts it may continue for sometime and a new swarm may be issued every 3-4 days. The new swarms generally settle at some distance in the area. A unique feature of this honeybee is that a nucleus colony may sprout out of the mother colony and settle at a distance of one metre on the same limb of the tree. A new comb is built and at first the workers keep moving between the mother colony and the nucleus but after a few days the nucleus becomes independent. It appears this honeybee is capable of nocturnal orientation.

Apis dorsata like other honeybees also has geographical races or subspecies. The most distinct is *Apis dorsata laboriosa*. Sakagami *et al*. (1980) called it the largest honeybee of the world and they observed it distributed in Nepal, Bhutan, Sikkim and Yunnan between altitudes of 1200-1400 metres. It is distinctly larger than the common *Apis dorsata* and remains active at comparatively lower temperatures and greater heights. It has exposed combs and in its area of distribution the temperatures range between 10° and –5°C. The dense coat of long body hair and the dark hair appear to have good survival value.

The Giant honeybee has a great economic value in India, and before the introduction of *Apis mellifera* it contributed 60-70% to the honey crop and practically all the commercial wax (ghedda) in the country (Thakar & Tonapi, 1961; Singh, 1975). In quality, honey is like the hive honey provided extraction is done properly (Mann *et al*., 1970). In Nepal and the Eastern and Central India it plays an important role in the economy and tribal life of the people. Those who have learnt the art of honey harvesting climb a tree or a rock with bucket or a basket. A small fire and smoke is started underneath and they reach the comb, cut out the thick top corner full of honey and pour it in a basket or a bucket and come down. When the bees migrate in the hot summer before the monsoon, the entire comb is brought down for wax. Among the tribals that art is developed to varying degrees, and no doubt there is a wastage and unnecessary killing of the bees. The art, romance and dangers of honey harvesting in Nepal are well illustrated in "The Honey Harvesters of Nepal" by Valli and Summers (1988, 1998) and by Strickland (1982). From the cave paintings found in central India it is evident that honey harvesting has been practised for thousands of years (Mathpall 1984; Joshi *et al*., 1980).

This honeybee is an important pollinator of crops (Atwal, 1970; Kapil & Kumar, 1975; Mann & Gurdip Singh, 1983). On various varieties of Brassica crops its number was 11.3 per 100 sweeps as compared with 182.3 of *Apis florea*. It was the second most common insect visitor, close to *Andrena* and *Halictine* bees. This honeybee was also common on *Eurica sativa,* sunflower, and cucurbits in Punjab, Haryana and Himachal Pradesh.

The seasonal migration of *Apis dorsata* is quite spectacular (see also Koeniger & Koeniger, 1980). They descend from the mountains to the plains of India after the monsoons or from the highlands to the coastal areas in South India, and flourish on the winter flora of crops, trees and shrubs. They raise their brood, gather honey (some is harvested by man) and in the summer they are ready to migrate to altitudes less than 1500 m. It is almost certain that they perceive and forecast the oncoming events, and make preparations to move to safer areas before the crushing monsoons in South India, East India and North India.

With the coming of summer, the desert areas and the plains of North India burn with heat. The atmospheric pressures fall and the dust storms are ready to whirl around in the countryside and it appears the bees are set to avoid them and hence they must migrate to safer areas to the mountain sides where there is thick vegetation cover. The way the mighty swarms move from their abodes on tall trees, as if at a short notice, by some elaborate system; the scientific phenomena is wroth studying and recording further.

The migration of *Apis dorsata* from the plains of Kalka[1] and up into the valley reaching Kasauli is very interesting. Around 15th of May every year you hear over powering humming sounds almost all day and you see groups of swarms of *Apis dorsata* bees coming up the valley almost following the contours of the terrain, from 5 to 30 metres above the ground moving upwards and settling into the jungle up to 1500 m high. The swarms when they come are a sight to be seen and the buses and cars on the Old Bridle Road to Kasauli invariably stop and people draw up the windows. Out of fear more than discretion they may stop the vehicles. Many a swarm arrive late in the day and stay overnight, settling on a low tree. The next morning they are gone ! This commotion goes on day after day for weeks. It seems that the migrating swarms somehow excite others on the way and they also make quick preparations (using up honey reserves) to move along. In the empty combs left behind hardly any honey is left since it is used up by the bees before leaving.

C. *Apis cerana* Fabricius (1793)

Apis indica is a synonym of this species and was described in 1798. Earlier in 1793 the same author had named this species as *Apis cerana* Fabricius, hence the latter name is accepted under the law of priority. In literature both these names are used for the two races *viz*. *Apis cerana indica* and *Apis cerana cerana*. Commonly the species is called the Eastern or the Asian honeybee (**Plate 3**). It occurs naturally up to 46° N in Asia east of Iran, and in the valleys of Hindukush and Himalayas upto 2,000 or even 3,000 metres in the southern warm valleys of these mountain ranges. Eastwards and northwards it is distributed in India, China upto Siberia, Southern Islands of Japan and throughout South-Eastern Asia. It has been introduced into the Hawaii Islands, Java and New Guinea (Ruttner, 1988). It is claimed that two other allied species, namely, *Apis nuluensis* and *Apis koschevnikovi*, exist in North-East Borneo (Koeniger *et al*., 1999).

Feral colonies of *Apis cerana* are found in hollows of trees and holes in the rocks or the walls of houses. In the indigenous methods of beekeeping in India the nesting situations are simulated and bees are kept in log hives and in hollowed out house walls with entrance hole opening on the outside and window opening on the inside. Records of this method of beekeeping in Kashmir are available dating back 1470 A.D. (Shah, 1980). Wicker and earthen hives are also made for keeping bees. In nature a colony has 6-8 combs (Muttoo, 1956), and 6-7 thousand workers in subtropical areas but 10-20 thousand in temperate areas like Japan.

The diameter of a worker cell and consequently the size of the worker varies a great deal in various parts of Asia (Sammatoro, 1970; Sakai & Matsuka, 1982). The cell size is 4.87 mm in Peshawar, 4.78 in Japan, 4.67 in Peking, 4.25 in plains of India and Thailand and 4.2 in Sri Lanka (Ruttner, 1988). This honeybee has the habit of gnawing wax of old cells down to the middle sheets (Singh, 1940). Thus wax debris accumulates at the bottom and provides medium for the development of wax moth. The standard Langstroth hives by itself is not suited for this bee but if the bee space in between the combs and the cell size is suitably altered the hive will be accepted (**Plate 4**). For the tropical South India a smaller hive containing 6-8 movable frames has been devised and it is commonly known as the Villager hive; it has many variations and the dimensions of frames and hives

1. Kalka : The place where the East-West Himalayas turn North-West.

are described in publications of Indian Standards Institution (1970) and also by Kapil (1971), Mahindre (1983b).

Apis cerana and *Apis mellifera* are similar in many respects and to differentiate between the two the size and colour cannot be relied upon; the cubital index in the former is higher. The drone cells of *Apis cerana* have characteristic holes like pores in the cappings which is not the case in *Apis mellifera*. Henel and Ruttner (1985) have discussed the origin of the pores in drone cell cappings. The caste system, colony structure, the parallel comb arrangement and the biology of the two species is very similar but the fanning position at the hive entrance is the opposite; in *Apis cerana* the workers keep their head away from the entrance whereas in *Apis mellifera* the workers keep their head directed towards the entrance (Sakagami 1959, 1960; Atwal and Sharma, 1964, 1968; Dhaliwal and Atwal, 1969, 1970). The sequence of functions by the workers are the same in the two, but *Apis cerana* workers perform the various functions a little early in life *e.g.* cell cleaning starts before the third day; brood care, pressing the pollen in cells and comb construction starts on the third day; development of pharyngeal glands between 4th and 16th day; the wax glands start functioning on the 3rd day but the maximum secretion is between 12th and 16th day; the orientation flights are between 7th and 11th day (Ruttner, 1988).

As compared with the Western Honeybee, *Apis cerana* bees are more excitable when disturbed and are prone to sting more readily, particularly when the weather outside is chilly. The smell of body sweat or a perfume also excites them. When a hive is knocked the bees inside make a hissing sound produced by quick movements of wings; (Fuchs & Koeniger, 1974). The same sound is produced when one blows at an exposed frame; the bees move away exposing the cells which can then be examined. When the bees are attacked by enemies like a wasp or a large black ant, they raise the abdomen and collectively make violent lateral shakings. The hissing sound and the shaking movements frighten the enemies. If a giant wasp (*Vespa mandarina*) comes to the colony as a predator he has to face the wall of bees who may ball around him and even kill him without suffering a single casualty (Matsura & Sakagami, 1973). As a honey gatherer and possibly an exploiter of flora, *Apis cerana* seems to be frugal in her behaviour; she can sustain herself at a lower colony strength fewer resources and lower colony activity as compared with *Apis mellifera*. It starts its daily activity at temperatures as low as 6-8°C in winter, whereas *Apis mellifera* is active at warmer temperatures with a stronger colony strength. A further distinction of behaviour is that *Apis cerana* visits a greater variety of cultivated and wild plants as in a natural habitat of a forest area, while *Apis mellifera* sticks to fewer but rich resources like, *Trifolium, Brassica, Eucalyptus,* sunflower etc. It is further demonstrated by the pollen spectrum of honey of the two species (Atwal & Sharma, 1970, 1971; Adlakha & Sharma, 1975, Thakar *et al*., 1962).

Communication among *Apis cerana* bees is through the round and waggle dances in a vertical plane on the comb and it is associated with sound production and sun orientation. (Lindauer, 1971; Towne, 1985). The dances are similar in the two species of honeybee but the dance rhythm for a given distance is slower in *Apis cerana* than in *Apis mellifera*. The transition from round dance to waggle dance in *Apis cerana* occurs at a close distance of 7-8 m (Goyal & Atwal, 1971, 1976). When the two species are trained to feed at a common feeding dish *Apis mellifera* is found to be more aggressive and it soon excludes *Apis cerana* bees (Sakagami 1960; Dhaliwal & Atwal 1970; Ruttner 1988). Mated *Apis mellifera* queens are accepted by *Apis cerana* workers who raise the brood, and the workers of both the species live together for sometime and eventually *Apis cerana* die out of old age and the colony becomes pure *Apis mellifera* (Atwal 1964; Atwal & Sharma 1964, 1968). The reverse queen introduction is not likely to be a success owing to pheromone differences of the queens; chemically their pheromones are alike but *Apis cerana* lacks a component whereas that of *Apis mellifera* has a broader spectrum. However, Inoue (1962) was successful in grafting *A.c.* queen cell in *A.m.* colony and whether the emerged queen could raise its brood was a different matter. The two species need different techniques for beekeeping (Inoue, 1985).

The biology of the two species is similar; *Apis cerana* queens deposit fewer eggs, say 300-500 in South India and 700-800 in North India per day as compared with 1000-1500 for *Apis mellifera;* and the former produces half the brood area than the latter in a given time (Kapil 1957, 1962; Alpotov, 1938a). It has also been observed that *Apis cerana* queens have almost half the number of ovarioles than *Apis mellifera* but the difference may not be a direct indication of their egg laying capacity (Sakagami & Akahira 1958; Shah & Shah, 1980, 1981; Saraf Wali, 1972). Queen rearing, queen emergence, mating and reproductive swarming is similar in the two species. In *Apis cerana* an average of 9 queen cells has been observed in the swarming season.

Reproductive swarming is a regular feature of *Apis cerana* and it starts when the colony strength is around 20,000 bees. The drones fly between 11.15-15.15 h and the maximum number comes out between 12.15-14.15 pm. (Hoshiba *et al*., 1981). In an area where there are mixed colonies of the two species *Apis cerana* drones may also fly to the drone congregation are of *Apis mellifera* which indicates that sex attractants of the two species are most probably similar. And yet in body development *Apis cerana* drones are smaller and they contain around 1 million spermatozoa as compared with 10-11 million in *Apis mellifera* (Woyke, 1975b, 1976a). The queen of *Apis cerana* on emergence goes out for mating flight after 4 days (Sharma, 1960), between 13.30-15.30 h and most probably multiple mating take place till the spermatheca of the queen is filled with 3.5 million spermatazoa. The queen starts laying eggs within 2-3 days after mating. According to Ruttner (1988), 8-10 matings during the nuptional flight is an average in both species. He also observed that in one instance an *Apis mellifera* drone mated with an *Apis cerana* queen but it was not successful and sting of the queen was damaged. It was demonstrated cytologically that the gamates of the two species on fusion did not develop further. Efforts are being made to genetically improve *A.c.* as there is plenty of biodiversity in this species (Verma, 1974a, 1978a, 1983,1984, 1988).

As in other Apis species drones in *Apis cerana* develop from unfertilized eggs and the workers and queens from fertilized eggs. On the loss of a queen the colony becomes a laying worker and from the unfertilized eggs drones develop and ultimately the colony perishes. The cell size on the foundation sheets is smaller suiting this species. Based on morphometric observations Ruttner (1988) has described four major races of this honeybee (See also Kapil, 1956; Verma 1986b, c; Verma *et al*., 1984; Ruttner, 1986a). *Apis cerana indica* is the smallest in size and is distributed in south India, Sri Lanka, Bangladesh, Burma, Malaysia, Thailand, Vietnam, Indonesia and the Philippines. The standard size *Apis cerana himalaya* is distributed in Afghanistan, Jammu & Kashmir, North India, China, North Vietnam. *Apis cerana cerana* is distributed in most of China. There is a large size race *Apis cerana japonica* which is distributed in Korea and Japan.

Apis cerana is attacked by a number of diseases and parasites. The wax moths are particularly damaging and they start breeding on wax debris on the bottom board and then attack the combs. They are so damaging that they may cause absconding of the colony. The ectoparasitic mite *Varroa jacobsoni* Oudemans, can also live as a serious parasite (Phadke, *et al*., 1966). The endoparasitic mite *Acarapis woodie* which lives in the thoracic tracheae causes a devastating Acarine disease. Even more devastating is the Thai Sac-Brood Virus (TSBV). It entered India from the East, spread to the Himalayan region and then to South India wiping out 95-97% of the colonies (Mishra 1995). The protozoan *Nosema apis* Zander, can also cause serious damage in areas where bees have to drink infected stagnant water.

Apis cerana has the habit of absconding during the dearth period, and the attack of enemies like wasps and wax moth further aggravates absconding behaviour. In order to check asconding, the colonies should be shifted to areas where there is good nectar flow and there are comparatively fewer enemies (Dulta, 1988). Competition from other honeybees (*A.m.*) may also cause commotion and absconding (Atwal and Sharma, 1971).

The wax debris on the bottom board should also be cleaned regularly so that wax moths do not multiply. This honeybee exhibits seasonal migration following the winter bloom in the plains; in the hot season they migrate to the cooler climate in the mountainous area. Making use of this behaviour beekeepers migrate their colonies from Himalayan slopes to the Indo-Gangetic plains, and between the Ghats and the coastal areas in south India.

The real trouble started with *Apis cerana* with the appearance of Thai Sacbrood Virus (TSBV) in Thailand in 1976 and in Shillong (East India) in 1978. Up to 1985-86 it had spread throughout the Himalayan region and was also noticed in Bihar and other places in the plains. The disease reached Tamil Nadu, Karnataka and Kerala by 1991-92 (Kashirsager, 1982, 1983; Abrol, 1993) and within 7-8 years wiped out 95-97 per cent colonies in the northern belt of India from the East to West (Mishra, 1995). The virus is specific to *Apis cerana* and it does not affect *Apis mellifera* colonies even though kept in the same apiary (Rana *et al*., 1986, 1987).

The gap produced in *Apis cerana* population was filled by the introduction of *Apis mellifera* in North India (Atwal & Goyal, 1973). Some traditional beekeepers in India showed concern at first, fearing that *Apis cerana* might be elimated ecologically. It has been observed both before and after the appearance of TSBV that *Apis mellifera* was superior to *Apis cerana* in honey production and gave 3-4 times more honey yield (Goyal and Atwal, 1979; Mishra, 1995).

In China, Japan and Korea *Apis mellifera* is the established honeybee for commercial beekeeping where *Apis cerana,* is only one-fifth or less in numbers and is confined to the mountains and remote areas. In the Indo-Gangetic plains of North India where there is plenty of crop and tree flora, *Apis cerana* could not be a success for commercial beekeeping but since then *Apis mellifera* has established itself remarkably well and apparently it has found an empty niche. It was proven experimentally that *Apis mellifera* could tolerate higher temperature as compared with *Apis cerana* (Verma & Edwards, 1971). It has been observed in the Himalayas that the latter is a frugal species and can live in cold marginal and poor beekeeping areas. Temperature alone may not be the sole determining factor, however, and inherent reproductive and behavioural characteristics may play even a more significant role in the comparative performance of the two species.

In its natural area of distribution *Apis cerana* is a common visitor of a number of crops like *Brassica,* clovers, sunflower, *Eucalyptus,* pome fruits, coconut and other palms, rubber etc. It is a gatherer of honey on most of these plants, and is an important pollinator of pome fruits. Studies on insect pollinators were made in North-Western parts of India during 1965-70, and the abundance of this honey bee was recorded on various flowers (Atwal, 1970). In the plains where *Apis cerana* does not occur naturally it was not recorded on the crops mentioned above (see chapter 15; Pollination of Crops by Bees).

The Indian hive honey bee (*Apis cerana*) has also been kept in skeps and wall hives for the last many hundreds of years and written records are available in Kashmir since the 15th century (Shah, 1976; Shah and Shah, 1980). It is described that successful beekeepers kept up to 50 colonies in wall hives and obtained pure and clear honey, probably by squeezing the honey combs. Honey was used as food by the royalty and upper social classes. The common man ate cane sugar that was brought from the plains of India.

With the advent of British, French and Portuguese influences from Europe, modern beekeeping with moveable frames was started in India towards the end of the 19th century. Scientific researches were also started in various parts of India and work on the behaviour, biology, brood rearing, and foraging by bees was published in Indian and foreign journals and some books were also written (Douglas, 1884; Ghosh, 1939; publications of Khadi and Village Industries Commission, Mumbai).

In the 20th century extensive researches were carried out in India at various centers on applied aspects of beekeeping; *e.g*. in Kashmir (Singh, 1972; Shah and Shah, 1976, 1980a, 1982a; Saraf, 1972, 1985); in Himachal Pradesh, (Sharma, 1958, 1960; Dhaliwal and Sharma, 1972; Salvi, 1975; Mishra & Dogra 1980; Mattu *et al*., 1985; Dulta & Verma 1987); in the Punjab plains, (Sharma, 1945; Rahman and Singh, 1946; Dhaliwal and Sharma, 1972, 1973, 1974; Dhaliwal and Atwal, 1974; Ahmad *et al*., 1983); in Uttar Pradesh, Bihar, Orissa, (Mutto, 1956; Chonduri, 1940; Naim and Phadke, 1976; Srivastiva and Tripathi, 1980, Singh, 1983); in Western Ghats, (Deodikar *et al*., 1957, 1958; Thakar, 1976a, b; Phadke, 1961, 1962; Kshirsagar *et al*., 1975, 1980; Kevan, 1989); in Karnataka (Diwan and Vartak, Reddy, Rajan, Chaubal and Kotmire; 1980); in Tamil Nadu (Narayanan and Sharma, 1960; Chandran and Shaw, 1974); in Kerala (Devadasan, 1971, 1972 ; Chandran *et al*., 1989; Shakuntala and Wakle, 1983; Nehru *et al*., 1984; Mathew, 1975). Among the various studies on the biology of honey bees, two queen system and multiple brood chamber systems were also tried with success. (Latif *et al*., 1960; Pandey, 1977). These are only a few of the references selected at random to show the wide spread interest for research on Indian honey bee.

Apiculturists from the West showed continued interest in *Apis cerana* and made valuable contributions to the study of its biology. Koeniger and co-workers studied the time of drone flights (1976) and the sympatric foraging activity of the four Apis species (1979, 1982). Working in Sri Lanka, he decoded the dance communication in Indian Apini (1980, 1982) which Lindauer had reviewed earlier (1957). Shearer and co-workers (1976) demonstrated the presence of queen pheromone in the Indian honeybees. Brood rearing, queen rearing, artificial insemination, colony management and absconding of colonies were studied by various scientists (Free 1977; Ruttner *et al*., 1972, 1973; Woyke, 1975a, b, 1976, 1979, 1980; Oleson, 1988). Morse, (1982) outlined the beekeeping program for Burma, Wongsiri for Thailand and Southern China.

Extensive literature on *Apis cerana* has been generated in Japan, China and other adjoining countries. Some of it has been reviewed in India in the form of books and monographs (Singh, 1962, Shah, 1976, 1983; Rawat, 1982; Verma, 1991; Mishra, 1995). Continued scientific interest is shown in this species throughout Asia.

2.6.4 *Apis mellifera* Linnaeus (1756)

This honeybee belongs to the Afro-European region (Fig. 5) and in its natural state it is distributed in the Middle East, Russia, Europe and Africa. It is adapted to diverse climates ranging from cold temperature to the tropical, from permanently humid areas to semi-deserts. Among animals it is the

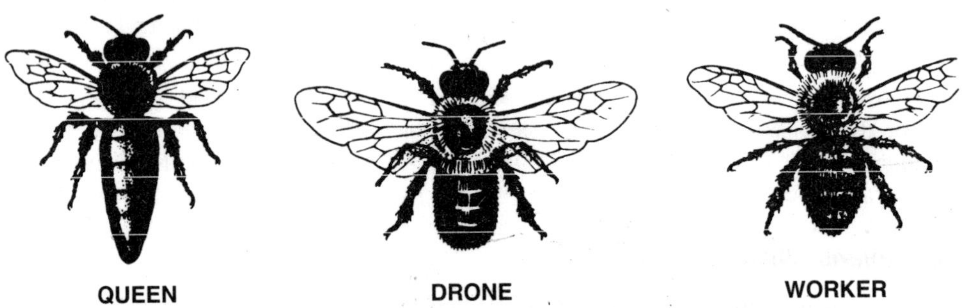

Fig. 5. *Apis mellifera* L. ; queen, drone, and worker

most widely studied species; more literature is published on it and more men have observed and studied it than any other species of the animal kingdom (except man). The early Aryans made it's mention in Vedas (2500-3000 BC) and since they came from the East European or Central Asian region it can only be conjectured whether they observed *Apis mellifera* or *Apis cerana* or both the species living together.

Apis mellifera L. has a great variation in size, body color, hairiness, wing span, lengths of tongue and legs. Based on morphometric characteristics, Ruttner (1988) has described the geographical races with separate names. Some authors refer to the races as subspecies although they interbreed easily. In general, the African races are smaller, have smaller comb cells, and in proportion to their body have longer legs, wings and tongues (Ruttner, 1988). No less than 24 races have been described but only the better known ones which have been exported for commercial beekeeping will be mentioned here. For the sake of brevity only the last race name is used rather than the full name-hence bees will be referred to as Mellifera, Ligustica, Carnica, Caucasica etc.

The geographical races have been evolved through the process of natural selection and they are well suited to the regional climatic conditions. Such characters as behaviour, ability to pass winter successfully, capacity to live under extremes of temperatures and synchronization of brood rearing with the availability of natural flora etc. make one race distinctive from another. Body color and pilosity must also have some survival value so also the tongue length. The account of four races from different parts of Europe is given here.

A. *Dark Bees (Apis mellifera mellifera* L.)

They are uniformly dark without yellow bands on their abdomen and are found in most of Europe, west of the Alps and in Russia. They are industrious, frequently aggressive and have weak disposition to swarming. In severe weather they winter well and in the spring they make a slow start in brood rearing. They are quite large in body size but their tongue is comparatively short. They are thrifty by nature and always produce some surplus honey. Their tendency to sting is considered a bad habit. Originally this bee was taken to the USA.

B. *Italian Bees (Apis mellifera ligustica* Spin.)

They are also known as yellow bees and Italy is their homeland. They have wide yellow bands across their abdomen. The hair particularly those of the drones are yellowish. The queens of this bee were sent from Italy to USA around 1859 and since then cross-breeding has been going on. They are generally gentle and calm on the combs and are somewhat smaller than Mellifera, with a slender abdomen and long tongue. The queens are very prolific and they have a strong disposition to breeding. They build the colony strength fast and maintain large brood area even when the honey flow declines. In spite of this they have a tendency not to swarm. They make beautiful white cappings on the honey cells. They do not do well in areas where there are prolonged winters whereas they seem to be quite happy in warm and even hot weather of Texas, California and Indo-Gangetic plains. Queens of this Honeybee have been bred near Bologna for the last 100 years and sent to USA and other countries. It is the most popular commercial bee of the world, and commonly it is referred to as the Italian honeybee.

C. *Carniolans (Apis mellifera carnica* Pollman)

They are quite similar to Ligustica, with slender body and long tongue. They have prominent dark and leather-brown bands on their body. They have short dense gray hair and drones are grayish brown.

Their homeland is the southern slopes of the Austrian Alps, Yugoslavia, Bulgaria, Rumania and Hungary. It is the quietest and most gentle bee and is not disturbed on the comb. Carnica can over winter even in small strength and in summer they build up fast if pollen is available. They are adapted to long cold winters and hot summers, hence they react fast and adjust brood rearing accordingly. They do not have robbing behaviour and collect only small quantities of propolis. These bees seem to be quite resistant to diseases. They work extremely well on early spring honey flow. After the Italian bee this race is popular commercially throughout the world. If the high brood rearing quality of the Italians is mixed by crossing with mild temperament of the carnica, we may be able to develop a breed "par excellence", suitable for various regions of India.

D. *Caucasians (Apis mellifera caucasica* Gorb.)

It is a black bee having very long tongue. The drones are dark with black hair on their thorax. The workers have a brown band on the first abdominal segment. They are calm and gentle on the comb and build colony strength rather slowly by about mid-summer. Their original home is the Central Caucasus but are commonly kept by the beekeepers of Russia. Their swarming instinct is weak and they are great users of propolis. This bee is readily susceptible to Nosema disease and is not a great gatherer of honey.

E. *Hybrids*

In USA and other countries where *Apis mellifera L.* has been introduced, there has been extensive crossing between the races. New breeds like Midnite and Starline have been developed by the commercial beekeepers in the USA. In Australia also hybrids have been produced. As will be narrated further under "Mellifera comes to India", an admixture of many breeds from various sources exists in India today. Judging from the color of workers, the yellow blood seems to be predominating in the process of natural mating.

In the Western States of USA the yellow blood is predominating although the "American Bee" is a mixture of many races. The prime qualities of a good honeybee are the egg laying capacity of the queen, ability to build up colony, gentleness and quietness on the combs, and good honey gathering capacity. Suiting the local climatic conditions the honeybees are to be selected based on their overall performance. It is most important to avoid intensive in-breeding, therefore, as a practice colonies should be purchased from areas far apart. In fact using the heterosis effect hybrid vigour can be achieved in the queens so produced. Indiscriminate import of Mellifera queens from USA and Europe into the Asian countries should be avoided for fear of accidental introduction of brood diseases. Now there is a threat of the Africanized blood within the *Apis mellifera* bees which may make a colony change to the habits of "killer bees". The danger is real particularly in the tropical parts of Asia if queens are imported indiscriminately without determining their genetic constitution.

2.7 *APIS MELLIFERA* HONEYBEE OF THE WORLD

This species was introduced both in areas of it's natural distribution and to places where it did not exist before. In the former case better strains or races were brought in to improve the honey yield. In the early times it was introduced into Egypt and Israel, and the standard method used was to introduce the queen of a desirable race into the dequeened colony or a nucleus of the local race. The colonies transformed into the new type within one season. Exhaustive history of beekeeping practices in the last 2000-3000 years has now been compiled by Crane (1999). The introduction of *A.m.* to various continents has been well recorded.

2.7.1 Spread of *Apis mellifera* to Various Parts of the World

According to records available *Apis mellifera* bees were shipped to new areas and countries of the new world from 1600 AD onwards (Eckert & Shaw; 1960; Ruttner, 1985; Crane, 1990). In a letter dated 5th December 1621, bee hives were included in a list of livestock sent from London to Virginia (USA). The bees were taken from Europe to Leeward Island in 1688 and to the Guadolupe in 1689. In 1784 they were taken from Florida to Cuba. This honeybee was quite well spread in the Eastern States of North America by 1800 A.D. According to one record they were taken from Spain and Portugal to South America in 1839 and further they reached Chile and Peru around 1857.

The honeybees were shipped successfully from England to Australia in 1810 and to New Zealand in 1839. Since it took 5 months for the ships to reach these countries detailed instructions were given as to how to handle the bees on the voyage.

In the second half of the 20th century *Apis mellifera* has been introduced into many Asian countries including Japan, Corea, Philippines, China, Thailand, India and Pakistan. Its introduction into India is discussed in some detail because efforts to introduce the species are spread over a greater part of the century and the final success was achieved through an interesting experiment of *Apis mellifera* queen introduction into a de-queened colony of *Apis cerana indica*.

2.7.2 *Apis mellifera* in India–The Background

Before the successful introduction of *Apis mellifera* into India since 1962-64 (Atwal & Sharma, 1964, 1968) many attempts were made by a number of beekeepers in various parts of India to import and multiply the exotic honey bee. The superiority of this honey bee must have been appreciated at an early stage. Douglas (1884) in his book mentions that the Italian stocks were imported in Bengal in 1882 and he was hopeful that the bees would be multiplied and given to the public. He felt that the monsoon was a difficult period for the bees and as soon as the rains were over they resumed activity. Winter was not the problem there or in South India but it was subsequently noticed to be so in the Himalayas.

Thompson (1940, 1944) and Sharma (1960a, b), have mentioned the earlier attempts at introducing *Apis mellifera* into India. The bees were brought from Europe and Australia by ships and apiaries were established in various parts of India: in 1919 at Pusa (Ghosh, 1920, 1939). Hatch in 1928 brought 6 colonies from Italy and kept them at Trivandrum; at about the same time J.S. Baldery maintained up to 30 colonies at Mahabaleshwar (Maharashtra); Beadnell (1939) imported several consignments of Caucasian bees and kept them in the Nilgiri Hills; bees from Italy reached Kashmir in 1951 and there were some unconfirmed reports of Intercrossing between *Apis mellifera* and Kashmir variety of *Apis cerana* which probably was not a correct observation. All these attempts were not successful in the sense that there were no reports of the continued presence of *Apis mellifera* in India at any of the places of introduction. When renewed attempts were started in 1962 to introduce *A.m.* in the Himalayan or the plains of Punjab there was no record of the presence of this honeybee anywhere in India or Kashmir.

With the coming of the Air Transport services by propellar aeroplanes and the renewed development activities after the second World War consignments of honeybee colonies were received in Kashmir (Vats, 1953) United Province (Mutto, 1957) and in Punjab in 1938 (Rahman and Singh, 1940). The colonies received in Punjab in April made a reasonably good start at Katrain but perished during the summer. Again 3 nucleus colonies were imported in Punjab through a RAF plane in April 1945 (Rahman & Singh, 1945). Two died on the way and the third started laying eggs at Nagrota but died in winter (Rahman and Singh, 1946). It seems the impression was that bees which came from the cold climate of Europe should be kept at cool places like the hill stations. However the bee brood which gets restricted during the rainy season does not get sufficient time to make a good start before the onset of winter in the colder parts of the Himalayas.If there is an autumn flower like *Plectranthus*, the bees further exhaust themselves, leaving the colony poorer of strength. Probably the bees received from the USA in April, 1945 also experienced the same climatic constraints of foraging on wild cherry and they did not survive the winter.

Them followed a long gap when there was no attempt made to introduce the exotic honeybee into India. It was rumoured, however, that an Indian pilot who flew between Kenya and Mumbai in early 1960's brought a few colonies which did not survive. It was a close escape for India because the African bees could have turned feral and subsequently created havoc in South India as they did in Brazil (Carr, 1957; Winston, 1987, 1992a, b).

2.7.3 *Apis mellifera* in India–New Attempts

The middle of the twentieth century brought India independence but it was also a period of food shortages owing to successive crops failures, as agriculture was by and large dependent on rains. Under the US–AID programme among other things, came the concept of establishing Agricultural Universities in various states of India, with new ideas of education, research and extension services for the benefit of the farmers and the people. In these Universities the system of internal assessment in the formal education of students was introduced. The concept was new and novel and students had to be persuaded and convinced by persons of stature that it was for their own benefit. Through the extension services of the Universities gates were opened to the farmers and they could approach any Faculty member for getting agricultural problems solved or for conducting meaningful researches on the unsolved. New concepts of research based on solving problems was conceived, organized, and executed both by borrowing technologies from abroad and by developing new ones suiting the Indian conditions. In this process of Institution Building with new ideas was an inbuilt system of change quite different from that observed in traditional Universities.

There was pressure of work but freedom of ideas also flourished under a non-bureaucratic system directed towards a personalised and understanding approach to the working staff. Consequently new wheat and rice varieties were brought from CIMMYT (Central International de Mejoramiento de Maize Y Trigo) in Mexico and IRRI (International Rice Research Institute) in the Philippines. Those faculty members who traveled abroad saw the giant businesses of food processing, flori-culture, fruit and vegetable export. The need for replacement of depleting timber, the value of the timely information on the coming weather, the necessity of farm machinery to cut out human drudgery, and the role of scientific nutrition in combating human diseases and sufferings in the West were appreciated by Indian scientists.

Under such a new climate of thought it was not a matter of coincidence that new dimensions of research in the field of apiculture were also envisioned. Since this topic is the theme of the book, it is discussed in some detail, partly because of the subject matter and partly as an example of persistence which is a rare quality among insects, including the honeybees. Moreover, there were so many interesting happenings during the course of research which beekeepers would like to know.

In 1962 efforts were renewed to introduce *Apis mellifera* into India and experiments were started in Punjab Agricultural University Ludhiana. As a first step fertile queens were imported and introduced into de-queened colonies of *Apis indica*. Since there was a fear of accidental introduction of bacterial brood disease only the fertilized queens of *Apis mellifera* were brought in and the worker attendants were killed. Permission was sought from the Directorate of Plant Protection and Quarantive Govt. of India New Delhi, for this experimental importation of the queens. I wrote to Mr Harry

Whitcombe, a well known beekeeper of Davis, California, making a request to send some queens by air.

He did not write back or ask for money and simply sent a packet of queens by air. I was surprised when a postman delivered the parcel of 8 queens on my office desk on 21st Aug., 1962; the queens and attendants had been on their journey for 9 days in temperatures ranging from 25-38° Celsius. The queens were taken to Nagrota Research Station of the University the next day and from there to Kulu (Katrain) Research Station on 23rd Aug., 1962 at an altitude of 5,000 ft. I followed in my old jeep that had given untiring service for locust control (there were no official cars in those days). There were breaches on mountain roads and after hours of wait, debris was cleared and avoiding the falling stones the driver and I made it to Katrain. It was cooller there and the research station was located on the mountain side where there was bright sunshine and cool breeze. My co-worker Om Parkash Sharma and I took up the task of introducing *Apis mellifera* queens into the dequeened colonies of *Apis indica*.

The two uppermost queens in the parcel had died on the way to Kulu. Six colonies of *Apis indica* were dequeened hurriedly and the queen cages were hung in between the frames as it was the normal procedure for hived colony of honeybees. One of the second uppermost cages also contained a dead queen. After 18 hours interval the fourth queen was released in a nucleus colony of young indica workers from which field workers were isolated. She was seen to fly out of the hive and after making a few circles around would tend to return to the hive. The indica workers did not follow her on her solitary flights. Inside the colony she moved rather excitedly. Some of the workers tried to feed her but she was too brisk and was not inclined to accept food. Some bees were even seen trying to groom her. The strange behavior of the queen seemed to excite the bees and they would try to catch her at legs or wings, but there was no balling. The hive was closed and observations were made every hour or so. The queen was seen moving from frame to frame or she would move away to the side frames where bees would follow her to tease. She would then fly out and after being on the wings for sometime, would come back to the hive. It was noticed that the field bees and the older hive bees were the most troublesome. The queen was then clipped and observed more closely. When the unfriendly bees followed her she tried to avoid them and was once seen sitting on the alighting board. Another time she was at the bottom board and bees were seen encircling her and offering food which she accepted occasionally. She was placed on the frames and queen gate was fixed at the entrance. On the next observation, she was again seen on the bottom board and a few bees were closeby. She seemed to be exhausted by now and was caged and hung between the frames. Next morning when the queen was released the same behaviour continued and as bees teased her a wailing hum (zeep or piping) was heard at a distance of 10-15 feet from the hive. Considering that it might be the rust red colour of the queen that was not being accepted, she was even painted black but the bees licked her clean. Some queen cells had been started in the colony and those were broken down. With this persistent intolerance and teasing the queen became exhausted and weak and died during the second night of release.

In the meantime the candy of the remaining four queens was being eaten up to release the queens (with *Apis mellifera* workers inside). They were to be released by the bees themselves and we hoped there would be a better acceptance by the bees by this delayed release. Another precaution taken was that the newly formed nucleus colonies were placed at a distance of 40 meters from the main apiary. The older house bees and the field bees were supposed to have gone back to the original site where a hive with frames and the old queen was placed. The strategy of alternate caging and releasing of the queens was followed and the queens were also clipped.

Two queens were released earlier than the others. As the Mellifera workers from one cage came out they tried to intermix with the indica workers, but were soon detected as intruders and were killed in the colony. From then on the Mellifera workers were killed by us as a precaution against brood diseases. These queens also behaved like the one mentioned earlier. They did not move on the frames with slow and steady movements, instead they moved fast and the indica workers chased them and pulled at their legs and wings. As a result the queens moved away from the frames onto the sides or on the bottom boards. Apparently all the older bees had not been separated in the colony and teasing continued with mutual unrest. When the queens appeared to be exhausted they were caged and released after some time. The remaining two queens were also released by the indica workers and their behaviour in the respective colonies was the same. By alternate caging and releasing, life of the queens was prolonged but they were not accepted by the bees. One of them even lost one leg and it was apparent that the queens were destined to die of exhaustion. The last remaining queen along with the colony was then moved to Nogrota (3000 ft) where it survived upto 2nd Sept. 1962, and in the end was seen dead in her cage. Considering that the queens found indica worker cells too small for laying eggs, frames of drone cells, being bigger, were offered but that made no difference.

The last queen that lived for 8 days in the colony was seen to live comparatively calmer life at times and once it was left overnight uncaged. It was observed that the field bees and the older hive bees were most active in teasing and harassing the queens; the youngest bees or novices did not participate in this mallee and they were either more receptive or unconcerned. It was felt that too long a confinement in a cage during journey could also be the cause of excitement of a queen. Even though disappointed by the unsuccessful experiments we were not discouraged.

In my next letter I explained to Mr. Whitcombe the results of the experiments. He was most surprised to know what we were trying to do ! He thought we had *Apis mellifera* bees and he was only supplying better queens. He immediately contacted Professor H.H. Laidlaw of the University of California, at Davis, for his advice. He told him, although no such experiments were reported in literature, there could be a possibility of introducing mellifera queens into indica colonies. Before he wrote me that he did not realize the nature of experiments, Whitcombe had already dispatched a parcel of 2 queens, sending the forward information through a cable. He said he had consulted US Post Office and was going to take precautions so that the bees reached India in the shortest possible time. He was surprised that the last parcel took so many days enroute and it was fortunate that any of the queens were left alive. Now he was going to send the queens on a commercial flight from San Francisco to New Delhi by a special delivery system.

The parcel was delivered to our special messenger at New Delhi on 24th September, 1962 at 1:40 a.m. which was brought to Ludhiana by jeep and was taken further to Nagrota the next day, thus cutting the journey time to 4 days. The nucleus colonies of young workers were prepared before hand and all field bees and older hive bees were isolated. Colonies of mild temperament were selected and were allowed to remain without a queen for a few hours. Then the mailing cages were suspended after removing the corks. Within 24 hours the indica bees cut into the candies and reached the queens but before the release they were clipped.

One of the queens showed erratic behaviour like the earlier ones and to save her from the teasing workers she was caged and released alternately and after two days of struggle she died in her cage. The nucleus colony for the second queen had been prepared with extra precautions. All the eggs had been removed and its position in apiary was shifted twice to get rid of the field bees. The workers in this colony were also seen to be quite anxious to release the queen which had a long trailing abdomen as if heavy with eggs. When released she accepted food from the bees and walked with easy movements on the frames. The bees made ways for her and groomed her. There was no teasing and pulling at legs and it appeared that apparently she was accepted. She had to lay eggs yet which was our main purpose. On the third day of release she was seen putting her head in one cell and then take the position of thrusting her abdomen into it. No egg was deposited in the cell, but one was seen sticking to the ovipositor and it was removed by the attendant bees. By the next day six eggs were deposited in indica size worker cells (there are 22 indica cells per 4 linear inches compared with 19.3 cells in Mellifera). Within two days a compact area of eggs was noticed and it was 60 sq. inches; it increased to 120 sq. inches within a week. By that time PAU development was also picking up and transportation through University vehicles was facilitated. During the project of *A.m.* introduction into India I must have made over 100 trips to the mountains and other places.

It was noticed that in the beginning multiple eggs (2-3) were deposited in some cells which condition was corrected by the indica workers, but subsequently the queen started laying steadily. The eggs hatched normally and the indica workers cared for the larvae as if they belonged to their own species. As the larvae were nearing full growth they were seen protruding out of the cells. When the cells were sealed the cappings were domical and the indica bees resorted to this adaptation to compensate for the larger sized mellifera larvae. "When the first adults emerged they were all drones but by the middle of November, worker brood of mellifera also emerged from the domical cappings. The workers of both the species lived together in the colony without any signs of hostility among themselves even though their body colour varied a great deal.

The mellifera workers raised in indica cells were slightly larger than the indica workers but were smaller than the normal size of their species. It was noticed that they performed no field duties and instead went out for play flights and simply returned to the hive. We wondered how they were to take over the colony duties when the Indica died in due course. Whether they performed any indoor duties we could not find out as the precious colony could not be confined to an observation hive.

At Nagrota September to January is considered dearth period when colonies cut down brood rearing. In September the average brood area of an *Apis indica* colony was 140 sq inches. Towards the end of November the brood area of colony with mellifera queen was only 24 sq in. More mellifera workers emerged from the sealed brood and the oldest ones though one month in age did not start field work and they flew around between 1-2 pm in and out of the hive, within a radius of 3 meters. Their number increased to 300. The colony strength was dwindling and the new young bees of Indica were given artificial feeding. The queen stopped laying eggs and when taken to the fields of *Brassica* flowers at a lower altitude the queen laid some sterile eggs in December. The indica workers started ignoring her and even tried to bump her off the frames and she died on 15 January 1963. In the meantime, bees had built some queen cells over the sterile eggs but stopped feeding the larvae after some days. Efforts were made to requeen the colony with indica queen but bees did not accept her. Later the colony was saved by grafting an Indica queen cell and the Mellifera workers continued to live in it along with Indica workers.

Once again the experiment seemed to have failed but we gained so much knowledge. Many questions remained unanswered, however. Why the Mellifera workers did not perform field duties ? Was their communication language different from those of Indica workers ? Looking back we wonder whether Mellifera queen and Indica workers exchanged the pheromones to maintain colony rhythm !

2.7.4 Next Summer and 1964

When I wrote to Mr. Whitcombe next he was excited and offered to send more queens in the spring. He had already sent mellifera size comb foundation sheets. We knew the technique of queen introduction and were well prepared. In the meantime we were also trying to procure a comb foundation mill for making the sheets ourselves. When the next parcel of 4 queens arrived in May 1963 it was kept overnight in the safe room at Delhi Airport where foreign consignments were always fumigated. All the queens died. When the next parcel of 4 queens arrived on June 25th 1963, we went to receive it in a borrowed station wagon. On return journey it broke down at 4 am on the other side of

Karnal. Rather than wait till it was day light we asked a man on a bullock cart to tow us to the next town. The people who passed by and saw the procession found it funny and remarked that it was the first time they saw a cart pull a car !

The queens reached Nagrota safely, everything was in readiness and all of them were introduced successfully. An incubator had also been sent from Ludhiana to get a constant supply of newly emerged Indica workers to strengthen the nucleus colonies. Brood rearing commenced in the second week of July when two of the queens started egg laying. The third queen took one week longer and the fourth took 3 weeks longer. It was again noticed that the mellifera workers did not perform field duties at the expected age. They waited for this task until their number reached 50 per cent of the colony. Why this critical concentration was required, we could not understand ?

Once the Mellifera workers took upon themselves the field duties they worked in harmony with indica. Except for their body size and colour the two species could not be distinguished in the performance of field duties (Fig. 6). But it was most interesting to watch mellifera workers fanning with their face towards the entrance and the Indica workers, with their face away from the entrance, both sitting side by side.

All the four colonies became mellifera by October 1963 and at that time they were spread on an average of 4 frames. Their brood area was double that of indica colonies of the same strength; it was 224-354 sq. inches in Mellifera and 108-120 sq. inches in indica. The mellifera colonies almost simultaneously suspended brood rearing in the end of October, as if they worked by some instinctive clock ! True, bee forage had dwindled very much, but how was it conveyed to the queens ! It was fascinating to see how nature worked !

When the colonies were moved on wild cherry flowers in October, queens did not resume egg laying and many Mellifera workers foraging, died of chilling and exhaustion. The Indica colonies lying beside, continued foraging and brood rearing and it appeared they were better adapted to work under frugal conditions, low temperature and low colony strength. The Mellifera colonies needed good strength (7-8 frames) to maintain the right environment inside and under conditions of low strength, they dwindled very fast. Two colonies perished and their queens were released in dequeened indica nuclei in January 1964. They were accepted, started egg laying in mid February, and the worker bees started emerging in March. The rate of brood rearing in one of these colonies was slow and it started building supersedure cells and by mid April, six ripe queen cells were noticed. When the queens emerged, they failed to mate although mellifera drones were present in other colonies. By the end of May, the requeened colonies also became pure mellifera. One of the queens seemed to be weak and in the colony Mellifera queen cells were grafted. The queens on emergence mated successfully as there were plenty of ripe Mellifera drones.

Subsequently, Mellifera queens from other sources were also obtained from abroad. The fifth batch of 4 queens arrived from Bologna, Italy, in June 1964, of which three were dead and one was introduced successfully. The sixth batch of 4 queens from England[1] was received in August 1964, of which two were dead and two were introduced successfully. The seventh batch of 10 queens arrived from California in June 1965 and all were introduced into Indica colonies.

2.7.5 The Technique of Interspecific Queen Introduction

The successful introduction of all the later 13 Mellifera queens received was a sufficient indication that the technique of Interspecific Queen Introduction had been perfected. The queens were sure to be accepted if the age of indica workers was between 24-72 hours and in no case they should be

1. Courtesy, Rothamstead Experiment Station, Harpenden, Herts, England.

Fig. 6. *Apis mellifera* queen introduced in the *Apis cerena* colony.

older than one week. In other words the queens were accepted by the young bees; the field bees and those who had started orientation flights were detractors and trouble shooter and they teased the queens and exhausted them to death. Moreover the queens should be introduced in spring season when a colony would become pure mellifera in 8-10 weeks. In warm weather there are less chances of brood being chilled.

Nucleus colonies of *Apis cerana*[1] can be prepared by placing sealed brood above a colony separated by screen or by incubating sealed frames at 86-92°F; ten frames of sealed brood will give the required number of workers of the required age groups at a given time. The nuclei should be kept in a protected place and not in the open. The entrance of the colony should be kept closed and inspections carried out twice a day. After 4 days the colony should be taken in the open for an hour and entrance opened so that bees can fly around and void faeces. In about a week the colonies are ready to be moved out to a permanent site and the entrance opened so that bees can start orientation flights and even start foraging at an early age. Sufficient amounts of honey and pollen should be present in the nucleus colonies and subsequently feeding should be given for stimulation. The queens should not be introduced in very strong colonies because noticing the slow start of a foreign queen the workers get demoralized; a colony on 3-4 frames is of the right strength, including one frame of honey and one of pollen. The site of colony raising should be 1-2 miles away from the apiary so that there are no Cerana colonies in the vicinity. In the early stages of the formation of a colony, more young bees can be given to increase the strength.

Introduction of queens should not be attempted when the number of Mellifera workers in a mixed colony are more than 50% because in that case they would be rejected. The mellifera size sheets should also be inserted when the mellifera workers are in preponderance because Cerana bees would not raise them. If the Cerana bees are seen removing the larvae, or they are not feeding them well and the brood does not enter pupal stage, even after 45 days of the queen introduction, it would mean that the colony had been demoralized. In that case it would be better to remove the queen and to release it again in a new Cerana nucleus (Fig. 6).

In the earlier trials when the queens were not accepted it was noticed that they kept coming back to the hive after flying around. It indicated they had a strong instinct to be housed in a colony and the *Cerana* bees, particularly the field workers, rejected them. When the experiment finally succeeded the very young indica workers were as receptive as the queen's instinct to be accepted. Success was achieved when the behaviour of the two coincided and both species were compatible.

When the first queen was accepted by the Cerana bees producing Mellifera bees as foster progeny, the queen stopped laying eggs after sometime. It was the Cerana bees who started building supersedure cells. The instinct was to preserve the indica race or the Mellifera race ? The most interesting observation was that in a mixed colony the Mellifera workers started field duties when their number reached near about 50% level. It indicated that there was a critical level of population when the Mellifera instinct for foraging became operative. What was the first stimulus that drove the earliest Mellifera scouts to go out and gather nectar and then through their dance communication recruited other members of their species ? It is now known that the two species have a different dance tempo transmitting different distances. Are there barriers of some kind that the dance rhythm of one species cannot be translated by the other ? Then again, the Mellifera queen in a Cerana nucleus rejected the larger drone cells for laying fertilized eggs but they managed to squeeze their abdomen into the smaller indica workers cells and laid fertilized eggs. Obviously, cell size was not the stimulus to decide whether the queen would lay fertilized or unfertilized eggs ! Then what was it ? Are there pheromones of male and female larvae and pupae left behind in the cell ?

1. Around that time the name *Apis indica* was changed to *Apis cerana*.

In the apiary at Nagrota where most of the earlier experiments were conducted there was always a mixture of Cerana and Mellifera drones flying around but no intermating was observed. Were there some kind of barriers of pheromones, speed of flight, general behaviour of flying queens or simply the structures of genitalia that prevented the interspecific mating ? All these and many other questions remained unanswered and there was a whole lot of knowledge to be gained at that time.

2.7.6 Packages of *Apis mellifera* L.

The second method used for introducing *Apis mellifera* into India was by importing nucleus colonies from California and other places by the standard method of raising colonies from package bees. Fearing the introduction of bacterial, brood diseases we were reluctant at first but later on consented. An American Peace Corps Volunteer in Mysore read some newspaper accounts of our Interspecific Queen introduction and showed a desire to work with us by importing package bees. Noticing his enthusiasm and ambition we made a request to the United States Department of Agriculture, Washington, who made the funds available through our PL480 scheme on Acarine Disease Problems of Honeybees. Ten nucleus colonies (3 lb each) arrived in June 1964 and the Volunteer[1], flew with them from Delhi to Kulu. The colonies were kept at Kulu (Katrain) Bee Research Station and university beekeepers helped him. The colonies made a good start and 57 kg sugar was fed to them. Their strength increased to 6-7 frames per colony in the premonsoon period and they dwindled somewhat during the rains. *Plectranthus* (a shrub) bloomed as the main honey flow after the rains in September and October. There were further losses of bees and the colony strength was reduced to 5 frames. Bees collected 10-15 kg of honey per colony but it was not extracted. No packing had been given as yet and in November the first cold wave devastated the colonies, reducing the strength to 1-3 frames. The Volunteer left for USA in December and the colonies were then packed together in twos or threes by giving partitions and seven colonies were saved which were then shifted to Nagrota in march, 1965.

Another consignment of 10 nucleus colonies arrived from California in June 1965[2] and bees were kept at Kulu. They gained strength fast but dwindled during the rains and in the post-monsoon period, confirming the earlier observations. A combination of factors lead to this dwindling; the main was that the major honey flow was in cold weather when temperature was often below 70°F. Since the colonies were weakened they were shifted to Nagrota in the first week of October and from then on Mellifera work was concentrated there, considering Kulu site as unpredictable.

In April 1966[3], 48 packages of bees (2 lb each) of Starline (Mid-West Hybrid) and Midnite (Caucasian hybrid) were ordered from Florida but the plane overflew to Bangkok and bees were returned to Delhi three days later where I and W.B. Wood (US Aid) were waiting anxiously. At Bangkok bees were kept in an air-conditioned room direct in the air blast. When the bees arrived at Nagrota 10-90% mortality was noticed in the packages and only 42 queens could be saved. By combining some of the packages, 32 colonies were given a start. They gained strength in summer, but in autumn some had to be combined. The total number of colonies at the end of winter was 50 only. The bees from Florida were considered to be free from brood diseases. By now *Apis mellifera* seemed to have been established at Nagrota. Some effort was made to keep the various stocks separate but in the end all of them were allowed to breed freely. Their honey yield was 4 times that of *Apis cerana* colonies. In the earlier experiments at Kulu (Katrain) (4500 ft) it was observed that ambient temperature played an important role.

1 . Mr. Larry Brown; Bee World, 45(4) : 128.

2. Courtesy, Kapoor Singh Siddoo Foundation; Vancouver, Canada.

3. Technical Cooperation Mission (Ohio State University) of US-AID Programme.

The Mellifera bees remained most active between 70-95°F and their activity decreased significantly outside this temperature range. Below 70°F the activity remained poor and many of the field bees failed to join their colonies. Some were seen crawling before their hives or were found sitting on plant leaves. Around 100°F they became listless and below 58°F totally inactive. *Apis cerana* on the other hand remained most active between 60-95°F became listless around 105°F and inactive below 50°F. These were preliminary observations and apart from temperature there could have been other influencing factors.

It was observed that Mellifera queens laid eggs in excess of the brood area that could be covered by bees in small colonies. In a pure Mellifera colony the workers in autumn could not cover the brood effectively so that in case of a sudden change in temperature they would throw out the outlying brood. Cerana colonies during that weather would have bee strength just enough to cover the brood. In other words, brood rearing was regulated according to weather and colony strength. They expanded and contracted their spread according to weather, thus saving the brood from chilling. With a smaller strength it is quite reactive to outside changes whereas the strong point of Mellifera is its greater colony strength but less reactive to sudden changes. It explains partly the autumn losses in the latter particularly when colony strength is weak. In the months of August, September and October (1963), the average brood of *Apis cerana* was 58, 82, 61 and that of *Apis mellifera* was 110, 110, 135 sq inches, respectively.

In June 1964, Punjab Agricultural University was to be inaugurated by the Prime Minister Pt. Jawahar Lal Nehru. *Apis mellifera* was one of the prime exhibits. We took the risk and brought one colony to Ludhiana to be placed in an observation hive. Outside the entrance hall of the exhibition was placed this hive. Mr. Nehru took out his glasses and tried to see the queen but she eluded him. He persisted for quite a few seconds but the queen did not appear on the broad side of the frame. When the exhibition was over and lunch finished, I had brought the observation hive on the exit side as the queen was now moving normally. Seeing us again Mr. Partap Singh Kairon, the Chief Minister Said "No, no, that is enough". Seeing my intent face, Mr. Nehru walked up to the hive, took out his glasses and had a good look at the foreign queen. He gave an imperceptible nod and with his magnetic eyes he conveyed that he had seen her ! In subsequent years up to the early eighties, honey from *Apis mellifera* colonies was sent to Mrs. Indira Gandhi, Prime Minister, and her family regularly by special deliveries.

2.7.7 Comb Foundation Mill and Other Supports to the Project

In preparation for the next experiments, comb foundation sheets of Mellifera size were required. One packet of the sheets was sent by Mr. Whitcombe and another was obtained in January 1963 from Ohio State University under US-AID programme. Our real need was that of a Comb Foundation Mill and for importing it from USA dollar funds were required. The AID-Technical Cooperation Mission at PAU had already committed the funds for that year but they asked us to explore the possibility of its manufacture in India; if available it could be purchased with rupee funds. We consulted G.B. Deodiker in Khadi Gram Commission, Bombay, who gave the name of a manufacturer for Indica size mill. I asked M/s Lotlikar of Vile Parle, Bombay, if he could also manufacture a mill for Mellifera size sheets ($19\frac{1}{3}$ cells per 4 linear inches). He was confident he could do it and an order was placed in August 1963. He manufactured it in three months. After negotiating the mode of payment and delivery, I went to attend meetings of the Indian Central Cotton Committee in Bombay, and visited his workshop to test the working of the machine which was found satisfactory. The mill was received at Ludhiana on 26th of May, 1964 and was sent to Nagrota the next day. This was another step forward in our quest for *Apis mellifera*.

(1) In the summer of 1965, I was on leave and went for a visit to the West Coast of Canada, and the University of California Davis near where Mr. Harry Whitcombe lived. I met Prof. H.H. Laidlaw at

the University and discussed with him our research project. He was most impressed and he demonstrated to me the working of his artificial insemination apparatus which he said would be required by us one day. I also held discussions with Prof. Gary.

In Canada, I had procured some dollar funds which I deposited with Mr. Whitcombe when I went to see him. I was impressed by his friendliness and enthusiastic nature. It was fascinating to see him pour a bulk of bees en masse through a funnel on to a weighing balance ! He promised to correspond with my colleagues at Punjab Agricultural University and have sent 10 nucleus colonies with spare queens. While I was still on holidays, the consignment was received at Delhi Airport in June 1965 and nucleus colonies were forwarded to Kulu. They gained strength in summer but during the monsoon and in autumn they dwindled and were then shifted to Nagrota in October 1965, for further care.

(2) On my return from holidays, I used to sit in my office and continue work in the afternoon. One day, I heard my attendant open the door and announce "Some sahib has come !". There walked in slowly an old gentleman of 75. I greeted him at the door and said "Mian Afzal Hussain ?" "Yes", he said. How remarkable that he with his age and unsteady legs could come up to the third floor when seemingly no one was in the building ! He explained he was passing by the university, he felt a great desire to see the daughter institution, a sion of Lyallpur College. He said he was sure he would meet someone in the Entomology Department even though the laboratories were closed in the afternoon hours. We sat together and I told him of my *Apis mellifera* project which he approved enthusiastically and said Pakistan should also introduce this species and he would suggest it to the entomologists there. He asked what were the new approaches to entomology ? I told him nematodes were an emerging problem particularly in vegetable areas and in citrus orchards. He agreed and said he was also pointing out to the agriculturists of Pakistan that nematodes would limit the production of agriculture. He was happy he visited PAU and he said with his age he did not know when he would get a chance to visit this place again. Within a few months the 1965 Indo-Pakistan war started and communication between the two countries was cut off for many years to come.[1]

(3) The year 1966-67 were trying for newly formed Agricultural University at Ludhiana. Not only that the new system of education was being introduced, there were many changes in Administration. In addition to being Head of Department of Zoology-Entomology, I was asked to take over the charge of Dean, College of Agriculture, the premier institution. I took over the responsibilities with fortitude but at the same time, I could not ignore my work on honeybees. I gave prolonged summer assignments to my graduate students, who finding all the four *Apis* spp. on the campus for the first time took up work on "Bee Flora For Four Species of *Apis* at Ludhiana" recording the bee populations and the forage available (Atwal *et al*., 1970). The work was spread over the period from 2nd week of November 1966 to the 3rd week of January, 1968. It was exhilarating that all the four species of *Apis* in the world were working side by side ! Ms. Eva Crane picked up our article and wrote an editorial in the Bee World (Vol. 54 : 1973).

She commented that "in the past there has been no opportunity to compare the foraging of all the different honeybee species because there was no overlap in the world distribution of two of them, the European *Apis mellifera* and the *Asiatic Apis cerana*. In recent years Dr. Atwal has successfully reared *Apis mellifera* in the natural territory of the Indian bee *Apis cerana indica*. The other two species, the large *Apis dorsata* and small *Apis florea,* build a single comb nest in the open, and these also occur around Ludhiana in the wild. They have recorded details of flowers in spring, summer, autumn and winter from which some interesting comparisons can be made about the four species of bees.

Within 1.6 km of the Agricultural University of Ludhiana there were 6-15 colonies of *Apis mellifera,* 2-5 of *Apis cerana indica,* 11-16 of *dorsata* (up to June-July only) and 15-20 of florea.

1. His son was the Pakistan High Commissioner in India.

Observations were made on several dozen plant species; nearly all were visited by all four bee species, although three were not visited by *Apis florea* (*Eranthymum*, bottle brush, *Bouhinia*) and three others were visited by *Apis dorsata* alone (jamun, siris, amaltas). Lucern presents pollination problems and it was visited more by *Apis florea* than any other species. *Apis cerana* and *Apis mellifera* overwhelmingly preferred barseem. All four species started to collect nectar from *Eucalyptus* around 8-8.30 and from lucern at 10-10.15 am.

Foraging activity during the course of the day showed one interesting difference., The two single-comb species, dorsata and florea kept a fairly steady number of bees in the field in the middle of the day. But both species of hive bees showed two peaks of activity, one at 1130-1200 and the other at 1630-1900 h. Around 1300 the number counted was less than half the morning and afternoon maxima. The mid-day breaks were shown earlier for Mellifera (Hambleton, 1925, 1944) and for *Apis cerana indica* by Lindauer in Ceylon (1957). The diminishing of activity at mid-day was not owing to temperature because it was even hotter in the afternoon."

(4) The 21st International Apiculture Congress was held in August, 1967 at Beltsville USA. A paper on Interspecific Queen Introduction was presented there by my wife and I was present there also. A colour positive photograph of a *Mellifera* queen surrounded by *Cerana* and *Mellifera* workers was projected by the old type magic lantern. It was not a very good projection but the scientists declared loudly they could distinguish the bees and gave a loud applause by clapping.

At the conference Otto Mackensen demonstrated to me artificial insemination of virgin queens. I discussed with many apiculturists that our queens at Ludhiana were not mating although there was no such difficulty at Nagrota. They said sometimes it was the case in California also particularly in dry years. Dr. Kerr from his own experience with African bees in Brazil wondered what were the chances of intermating between *Apis indica* and *Apis mellifera*. That made me cautious and I told him we would keep a watch on possible intermating. Prof. J.B. Free and Ms. Eva Crane were very friendly and they showed a desire to visit India, which they did many years later on, and so did Roger Morse. Some months later we made a request to Dr. Walter Rothenbuhler of Columbus and he sent us a sample of pollen trap. Many scientists were interested in our work and were so helpful and cooperative. It was a great encouragement even though the multiplication of Mellifera colonies with us was rather slow.

(5) In November 1967, Punjab was divided to carve out Haryana on the East and South and Himachal Pradesh in the North. The Punjab Agricultural University Ludhiana worked at first in the three states through South Campus at Hissar and North Campus at Palampur. Within a year the Hissar campus separated but Palampur campus remained with PAU up to March, 1970. Thus beekeeping collaborative research between Nagrota and Ludhiana continued for another three to four years and we continued to exchange materials and ideas received from abroad. A great deal of interest was generated in the public also and many beekeepers throughout India wanted to obtain Italian bees. Reluctantly we had to tell them that the work with the limited number of colonies we were doing was still in the experimental stage.

(6) Planning in time, the work at Ludhiana was also started in 1966 with 6 colonies obtained from Nagrota, and the number was increased to 15 later on. With the abundance of bee flora at the campus the colonies made an excellent start. Within three years their number increased to 27 but we observed that queens did not mate easily and mainly remained unmated. Perhaps the weather was too dry in April-May and sufficient number of mature Mellifera drones were not available at that time. Their number was quite small. To overcome this difficulty, fertilized queens from Nagrota were brought in. The old queens were replaced and new ones were given to nucleus colonies.

The problem of shyness of mating was discussed with scientists in the International Apiculture Congress in Beltsville in 1967. We were also in touch with Prof. Butler in England who was working on queen pheromones and he asked for some indica queens to carry on his experiments. Dr. Bailey was always helpful in diagnosing diseases of our bees. Dr. F. Sakagami wanted collaborative research but my students were still completing their degrees and were not ready yet to take up collaborative research at an international level.

(7) During advanced training in apiculture, one of my graduate students was so impressed by Wenner's work that he would ignore von Frisch's theory in our discussions. It bothered me and became embarrassing because I knew that there has got to be an intermediary approach to the whole system of bee communication. I was relieved when I finally convinced him of this. Later on he became a valuable asset and made good contributions in understanding the communication behaviour of *Apis cerana* and *Apis mellifera* (Goyal & Atwal 1971, 1976, 1979). He joined my project on the introduction of *Apis mellifera* in the Punjab plains from 1970 onwards.

(8) From 1960-1970 two PL480 research projects were completed under my supervision as principal investigator. The projects were "Acarine disease & Problems of honeybee (*Apis cerana*)" and "Biology, ecology and utilization of insects other than honeybees in the pollination of crops". Profs. R.A. Morse and G.E. Bohart were the collaborating scientists. They visited India and gave many valuable suggestions through correspondence. Many scientists at Ludhiana worked in these projects, earned higher degrees and later on established their own centres of research in Haryana and Himachal Pradesh.

Many of the standards on honeybee appliances, bee products and bee flora adopted by the Indian Standards Institution, New Delhi, were finalised under the chairmanship of Dr. Sardar Singh and, subsequently, under mine and that of others. It gave credibility to beekeeping industry and bee products.

2.8 QUEENS FROM AUSTRALIA

By 1970 there were signs that Palampur campus would also separate and we at Ludhiana started the work of multiplying *Apis mellifera* colonies in right earnest. There were over 150 colonies at Nagrota now but there were only 27 colonies at Ludhiana. Twenty more colonies came from Nagrota for strengthening work at Ludhiana.

Help also came from another part of the world. There was an exchange of fellow Rotarians from Albury, NSW and I had some Rotarians as house guests. When they returned home, Alan M. Frost of Albury made an offer of sending 100 Caucasian hybrid queens raised in the foot hills of Australian Alps. He said mild bees would be selected with qualities of wintering over more successfully. I was not sure whether we could handle that many queens but my student and collaborator, N.P. Goyal, was confident. Two air mail parcels of 50 queens each were received on November 23 and 28, 1970. Those were speedily cleared at the customs at New Delhi and were brought to Ludhiana.

In Australia it was spring but for us winter was approaching. The 45 colonies (by now) with us were divided and the queens (6 were dead) were released in nuclei. Winter packing was given to keep them warm and in the end 132 colonies were formed and some queens were kept in cages as reserves. With much effort the colonies passed the winter (even though mild) and artificial feeding was given.

Since winter days at Ludhiana are sunny and warm and there are Brassica and Eucalyptus flowers the queens started early activity in January-February and their strength in February, March, April, May and June increased to 4, 6, 9, 12, 11 frames and the maximum brood area to 1115, 2962, 5250, 3620, 3725 sq. cm. respectively. The pollen area ranged from 205-1615 sq. inches and the colonies stored 20 kg honey during the main honey flow in May, 1971.

We were most encouraged by the performance of *Apis mellifera* at Ludhiana and the speed of gaining strength by a colony was faster here than at Nagrota where winters are prolonged, are wet, and wild flowers appear late in the season. At Ludhiana on the other hand *Brassica* flowers were plentiful, which were followed by *Eucalyptus* planted all along the boundary of PAU. Then followed the Egyptian clover grown in over 70 acres for the dairy cattle in the campus and more in the surrounding areas.

When I wrote an account of our results to Albury Rotarians they were pleased and were excited. We were pleasantly surprised when another consignment of 100 Caucasian queens was offered to us in July 1971. During the hot summer and the monsoons there were some reverses and the weak colonies were combined. After the winter of 1971, 140 colonies in good strength were with us in February, 1972 and we were ready to receive the first parcel of 50 queens which arrived in April, 1972. Considering our climate the programme of queen rearing in NSW had been revised and thus those were produced in autumn which was our spring. The queens performed very well at Ludhiana and being the season of main honey flow the colony strength increased very fast.

In the next two years the performance of *Apis mellifera* colonies was studied at various places in Punjab and it was found that they gave even better performance than in the lower Himalayas. At the University campus, on reaching a strength of 8-10 frames during the spring and the summer, their growth slowed down somewhat, and then followed the dearth period and the monsoons. In October-November they picked up activity on toria (*Brassica campestris* var. toria). When the bees were in good strength in December-January they were placed in sarson fields (*Brassica campestris* var. sarson). The brood rearing activity continued and colonies became stronger. The queens stopped egg laying for a brief period of 7-10 days towards the end of December and the first week of January. At the campus the honey yield was 20 kg per colony but at farmers' fields it touched 35 kg. The number of *A.m.* colonies at research stations and with farmers in the initial years are given in Table 2.

It was interesting to observe mixed lots of mellifera drones at Nagrota as well as at Ludhiana. Their body colour at Nagrota was yellow as they were mostly from the California stock. At Ludhiana now most of the queens were of Caucasian stock from Australia and hence most of the drones also were black. The worker bees were mostly dark but yellow colour was still visible. It was observed that over the years as a result of natural mating yellow worker bees rather than dwindling were increasing in numbers steadily. It appeared that the yellow strain had some kind of genetic dominance over the Caucasian mixed blood.

2.9 TOGETHER LIVE FOUR SPECIES OF *APIS*

Apis mellifera was initially introduced into the Punjab Himalayas by interspecific queen introduction technique (Atwal and Sharma, 1964, 1968)., Nucleus colonies and queen bees were subsequently imported from California, Florida and New South Wales. The programme of multiplying the colonies was speeded up. Even though farmers in Himachal Pradesh and Punjab were anxious to have these exotic bees another ten years were spent at the research stations to study their behaviour, the potentials for beekeeping and their interrelations with the indigenous honeybees, *Apis florea* F., *Apis dorsata* F. and *Apis cerana* F.

The interspecific relations among the four species were first studied at Ludhiana campus of the Punjab Agricultural University which covered an area of 1300 acres and had abundant growth of crops, trees, shrubs and annual flowers. Their daily activity on various flowers was studied for over one year (Atwal *et al.*, 1970; Crane, 1973, 1976). It was a unique opportunity created because prior to this, there was no overlap in the world distribution of two of them, the European *Apis mellifera* and the Asiatic *Apis cerana*. Out of the several dozen plant species observed nearly all were visited by the

TABLE 2. ***The number of Apis mellifera L. colonies at the research stations and with private beekeepers from 1965-1987***

	Himachal Pradesh			*Punjab*		
Year	*Experi-mental colonies at Nagrota*	*No. of Private Bee-keepers*	*No. of Colonies with Private bee-keepers*	*Experi-mental Colonies at Ludhiana*	*No. of Private Bee-keepers*	*No. of Colonies with Private Bee-keepers*
1965-66	13	Not Given	Not Given	6	Not Given	Not Given
1966-67	45	”	”	10	”	”
1967-68	54	”	”	15	”	”
1968-69	85	”	”	20	”	”
1969-70	115	”	”	27	”	”
1970-71	163	”	”	45	”	”
1971-72	164	”	”	132	”	”
1972-73	150	”	”	140	”	”
1973-74	151	”	”	142	”	”
1974-75	156	6	25	136	”	”
1975-76	173	15	86	140	”	”
1976-77	190	24	220	120	5	28
1977-78	200	32	475	112	34	116
1978-79	202	40	1000	125	80	750
1979-80	217	60	2000	135	157	1500
1980-81	202	120	3000	150	250	3000
1981-82	216	200	3500	177	500	5000
1982-83	235	250	4500	200	950	10000
1983-84	225	350	5000	250	1941	20000
1984-85	184	450	5500	205	3000	30000
1985-86	186	650	7000	195	5200	55000
1986-87	185	750	7800	164	6500	68000

Note : In 1995-96 the estimated number of colonies with beekeepers in Punjab, Haryana & Himachal Pradesh and other states is around 200,000 with an average yield of 20 kg per colony.

four species of bees, although three were not visited by *Apis florea* (*Eranthymum,* bottle brush, *Bouhinia*) and three others were visited by *Apis dorsata* alone (jamun, siris, amaltas). Lucern presents pollination problems and it was visited more by *Apis florea* than by any other species. *Apis cerana* and *Apis mellifera* overwhelmingly preferred barseem (Egyptian clover). It was observed that the single comb species (*Apis florea* and *Apis dorsata*) maintained a steady population of workers in the field during the middle of the day whereas the two hive species showed two peaks of activity, one in the forenoon and one in the afternoon. There were indications that the four species in nature had the potential of adapting to their floral requirements. Akratanakul (1984) observed and discussed similar potentials. Koeniger and Vorwohl (1979) viewed the sympatric behaviour as a possible competition for food and Koeniger (1982) further discussed the biological significance of the interactions.

Close observations on the physical aggressiveness among the four species of *Apis* were made by confining them together in small spaces (500 cubic cm) in various combinations (Dhaliwal and Atwal

1969). Food was provided in one corner near where they had physical combats. *Apis dorsata* in the presence of *Apis mellifera* or *Apis cerana* was reluctant to fight at first but when encounter became inevitable it was victorious over the latter. Based on mortality the four species in terms of strength could be placed as *Apis mellifera* > *Apis dorsata* > *Apis cerana* > *Apis florea*. In a direct combat *Apis mellifera* was stronger than *Apis cerana* which was also observed in Japan (Sakagami, 1959). Overall success of a species under natural conditions would depend on complex factors, including aggressive behaviour of a species.

The aggressive behaviour of *Apis mellifera* over *Apis cerana* was further studied by introducing brood frames and adult bees of various ages of one species into the colony of the other. Whereas *Apis cerana* was more incriminating to the immature brood of the other, *Apis mellifera* was more aggressive in expelling or killing adult workers of *Apis cerana*. In exploiting food at the feeding stations *Apis mellifera* was dominant over *Apis cerana* which was also observed in Japan (Sakagami, 1959) while studying interspecific relations between the Japanese and the European honeybees. It was concluded that as a result of aggressive behaviour of *Apis mellifera* in commercial beekeeping areas the Japanese honeybee had receded to the mountainous areas.

Robbing among the two species was studied in the Himalayas, Punjab plains, and in Japan (Atwal and Dhaliwal, 1969a, b; Adlakha and Sharma, 1974; Sakagami, 1959). Both species in apiaries robbed each other but there were distinct patterns. *Apis cerana* tried to rob at any time of the year but bees were not successful as *Apis mellifera* guards were vigilant and sometimes they physically lifted the robbers and took them away from the hive and threw them at some distance. They were successful when the ambient temperatures were low and *Apis mellifera* guards were not so effective. Robbing was always mild. *Apis mellifera* bees during the dearth

period of monsoons and autumn resorted to severe robbing of the 'exterminating' type and were more successful as the *Apis cerana* guards could not defend effectively and they merely moved their abdomen from side to side. The robbed colony often perished unless moved to a distance and combined with a strong colony.

Apis dorsata robbers do not come near *Apis mellifera* colonies but occasionally they do rob *Apis mellifera* colonies; more often the individual robbers enter the hive and help themselves to honey. It is interesting that the drawf honeybee, *Apis florea,* can successfully rob a weak *Apis mellifera* colony (Chahal *et al.*, 1986) and it was observed in seven out of 109 apiaries visited. *Apis florea* behaved aggressively and they stung *Apis mellifera* bees but escaped being stung themselves by the latter. Casualties occur in both species. It is in contrast to their mutual behaviour in a confined space.

The mating habits of *Apis mellifera* and *Apis cerana* were studied based on available literature (Butler *et al.,* 1967) and peculiarities of the two species were noticed (Adlakha, 1971). The brood rearing activity and colony build-up were studied in Kulu valley (Atwal and Sharma, 1968, 1971; Hameed and Adlakha, 1973), in Kangra valley (Atwal and Sharma, 1970a, b), in Shimla Hills (Sharma *et al.,* 1980) in the Punjab plains (Atwal and Sharma, 1970b), in Pakistan (Ahmed, 1982). It was the general observation all over that *Apis mellifera* maintained a stronger colony strength and built it up faster than *Apis cerana.* It was owing to the fact that the former queen was more prolific and laid more eggs per unit time and owing to the larger size of the workers, they gathered more honey; the yield was 3-4 times higher than that of *Apis cerana*.

In spite of the excellent performance of *Apis mellifera* some people showed doubts whether its spread in India was in biological interest (Kshirsagar, 1968, 1981; Pant, 1980). The possibility of spread of bee diseases through this honeybee was cleared (Chahal and Gatoria, 1983) but some feared that there might be natural interspecific hybridization between *Apis mellifera* and *Apis cerana* eliminating the race of the latter on the analogy of intraspecific interbreeding between the European and the African races in South America. These fears seemed to be unfounded since Ruttner (1969, 1983, 1988) had demonstrated that there were physical natural barriers between the genitalia of the two species and moreover when ovum of one was artificially inseminated with sperm of the other, the gametes did not develop. The final verdict in favour of *Apis mellifera* was given by the Expert Consultation Committee on Beekeeping of the Food and Agriculture Organization, Rome (FAO, 1984; Akratanakul, 1986; Burgett and Krantz, 1984).

Apis mellifera is now living in a greater part of Asia alongside other three (or four) species of *Apis* and has already shown great promise for commercial beekeeping. Its greatest biological attribute is that it builds up the colony strength very fast and is quick to gather substantial quantities of honey during a major or even a minor flow.

Honeybee Structure

Honeybee is an insect. It has a segmented exo-skeleton made of tough chitin and on the sides of the body there are tiny breathing holes. It has three simple eyes, two compound eyes, one pair of sensory antennae, three pairs of jointed legs and two pair of wings which join together in flight. At the anterior end of the body around the mouth cavity there are arranged mouth parts which direct food into the alimentary canal.

As the other insects do, the honeybee also develops in stages; the four stages being, egg, larva, pupa and adult. Metamorphosis is also a characteristic feature in the biology of the honeybee. Unlike most other insects it has an elaborate caste system and the three castes, worker, drone and queen, have specific duties to perform in a colony, which comprises around 50,000 individuals.

The body structure of a worker bee is typical of the three castes (Fig. 7). There is a dense covering of hairs on the body which originate from epidermis, the outer most layer of the wall. The exoskeleton is segmented and, not coinciding with lines of segmentations there are transverse bands of yellow, gray and black pigmentation within the body walls. The chitin is quite hard at some places and rather flexible at others. The body has breathing holes on the sides, mouth opening in the front and excretory anus at the posterior end. The other organs of epidermal origin are wax glands, scent glands and salivary glands. The body is divided into three parts, head, thorax and abdomen. The detailed anatomy of the various body parts and some functional physiology of the honeybee has been described by (Snodgrass 1956; Dade, 1977). Only a brief account is given here.

3.1 HEAD

The head viewed from the front is a triangular dome made of six fused segments, but the fusion lines are not visible from the outside (Fig. 8 top). It has three simple eyes on the upper fronts and two large compound eyes on the sides. There are two segmented antennae in the front centre and they are so fixed that their bases lie close together. The antennae are sensory organs (for smell, taste, touch etc.) and the eyes are useful in seeing objects as well as the countryside. The three simple eyes arranged in a triangle help the bee perceive color and light intensity. The compound eyes have numerous facets which perceive multiple images of the objects. The perception of the various castes must be quite different as eyes of a worker bee have 4,000 to 5,000 facets each, those of a queen have 3,000 to 4,000, and those of a drone are around 8,000. Since the drone's primary job is to mate with a queen the large eyes must be a good aid in locating a queen while on the wings.

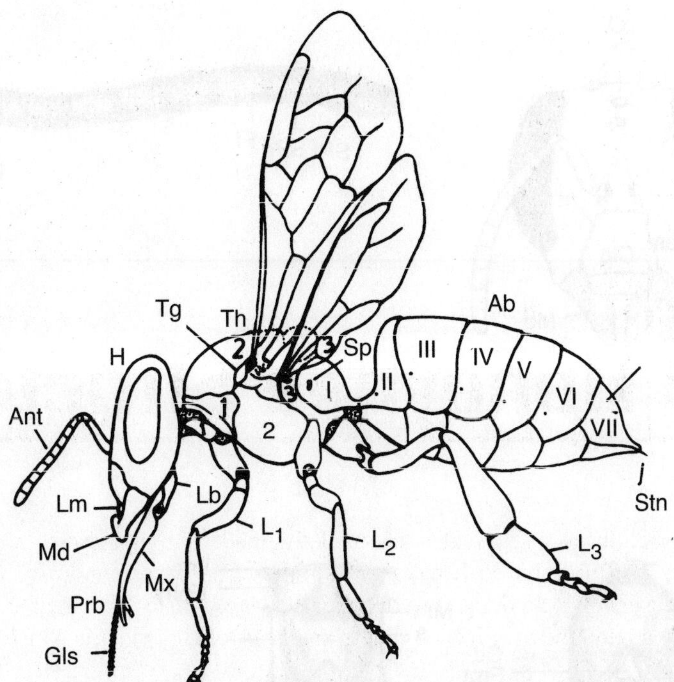

Fig. 7. Lateral view of honeybee worker : H—head, Th—thorax, Ab—abdomen, I—propodeum, II-VII—abdominal segments, Tg—tegula, Sp—spiracle, A—antenna, E—compound eye, Stn—sting, L—1-2-3, legs, Lm—labrum, Lb—labium, Md—mandible, Mx—maxilla, Prb—probosis, Gls–glossa.

The mouth parts are located on the lower side of the head; those are in pairs but some are joined together to form the probisis or the tongue, which is a complete structure (Fig. 8 bottom). In the front is the upper lip which is a single structure and on its sides are two hard biting mandibles. The mandibles are used for collecting and eating pollen; building and cleaning comb cells; gathering and manipulating propolis, the gum, used for plugging holes in the hive; fighting enemies; but not for eating fruits as some people believed earlier.

The tongue is elongated and is jointly made of maxillae on the sides and various parts of labium (lower lip) in the middle and the back. When not in use tongue is held backwards on the lower side of the head. When in use the bee makes its back-and- forth movement by extending it in the front and sucks liquids into the mouth through a food canal made by the foldings of maxillae and labium. The food canal is a functional tube who's sides are formed by the elongated maxillae and the bottom by the labium. The lobes or glossa of the labium at the end make labellum which has a separate salivary tube within formed by the modified parts (Galea) of the maxillae and the palpi of the labium (see the cross section).

The mouth leads to a funnel shaped pharynx inside the head and it narrows into a tube like esophagus, which passes through the neck and thorax into the abdominal region. It is capable of tremendous expansion in the abdomen and enlarges into crop which is used by the honeybee as honey stomach or the storage house of nectar from flowers and honey from the hive. The various appendages of the head described are moved by different sets of muscles located in the head. The muscles also control the functioning of salivary glands, and food glands (producing Royal Jelly).

The head is joined to the thorax by a slender neck through which pass the longitudinal internal organs like the esophagus, glandular ducts, nerves, trachea etc.

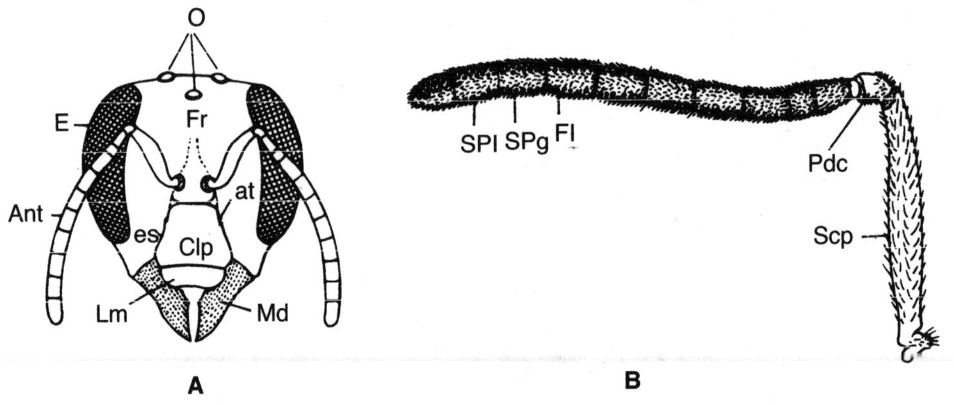

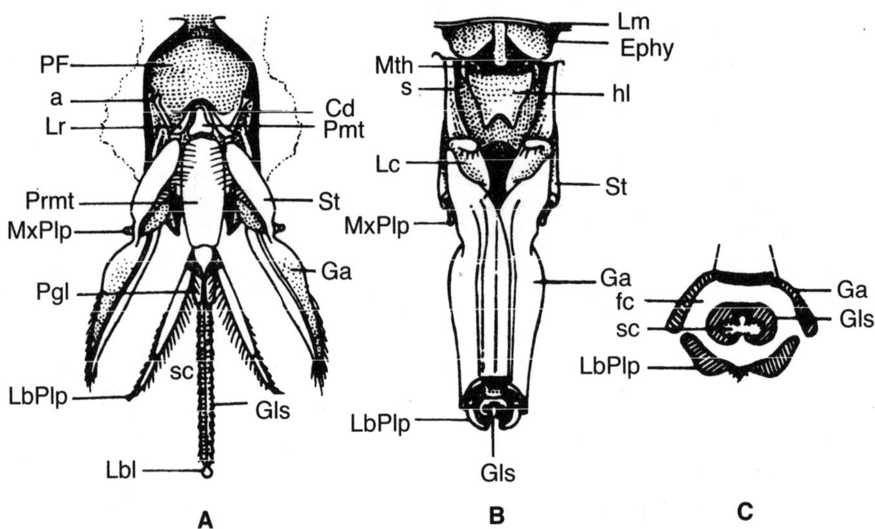

Fig. 8. **Top** : **A**—Front view of head of a worker bee : Fr—frons, Clp—clypeus, O—ocelli, E—compound eye, Ant—antenna, at—anterior tentorial arm, es—epistomal sulcus, Lm—labrum, Md—mandible, **B**—Antenna enlarged; Scp—scape, Pdc—pedicel, Fl—flagellum, Spg—sensory peg, Spl—sensory plate.

Bottom : **A**—Mouth parts (tongue) of a honey bee worker; **B**—Base of the tongue or proboscis with anterior food canal open; **C**—Cross-section through distal part of proboscis.

Cd—cardo, a—articulation of cardo, PF—fossa of proboscis, Lr—lorum, Pmt—prementum, St—stipes, Prmt—prementum, Mxplp—maxillary palpus, Ga—galea, Pgl—paraglossa, LbPlp—labial palpus, Sc—salivary canal, Gls—glossa, Lbl—labellum, Lm—Labrum, Mth—mouth, Ephy—epipharynyx, hi—hypopharnyx lobe, Lc—lacinia, fc—food canal, Sc—salivary canal.

3.2 THORAX

The middle strong, rounded part of the body is made of four fused segments, namely prothorax, mesothorax, metathorax and, propodeum which is the first abdominal segment but forms part of thorax. The prothorax is small, collar like and carries the first pair of jointed legs (Fig. 7). The mesothorax is the largest and carries the middle legs and forewings. The metathorax is narrow band and it carries highly modified third pair of legs and smaller hind wings. There are spiracles one on each side of the segments. The wings are membranous and transparent and on each side of the body the two wings are joined together with the help of hooks or hamuli which are on the anterior margin of the hind wings; they fit into a fold on the posterior margin of the forewing and thus the wings are paired in healthy bees. They get separated in bees suffering from acarine and other diseases and acquire the shape of letter *K*.

The three legs on either side are variously modified to perform different functions, but their basic structure and parts are similar (Fig. 9). The leg is made up of five segments; it is attached to the body with coxa, has a small bend called trochanter, the next robust part is femur, followed by a flat hairy part tibia, and the last part is tarsus which in itself has five segments. The first segment of tarsus is known as basitarsus which is followed by four pretarsi (foot), the last one having two terminal claws, serving as heel of the foot. In the basitarsus of the foreleg there is a notch on the inside aspect through which an antenna can pass; the two notches on legs are used for cleaning the two antennae. The middle leg has the typical parts except that the basitarsus is somewhat large and hairy and the whole leg is used for cleaning and brushing off pollen from the thorax of her body. The hind leg is highly modified; the tibia is swollen and on the outer side there is a concave basket (corbiculum) which is fringed with long hair to retain pollen and propolis; the basitarsus is very large and hairy and on the inner aspect has a pollen comb used for brushing together pollen from the body. The brushed pollen is scraped by the

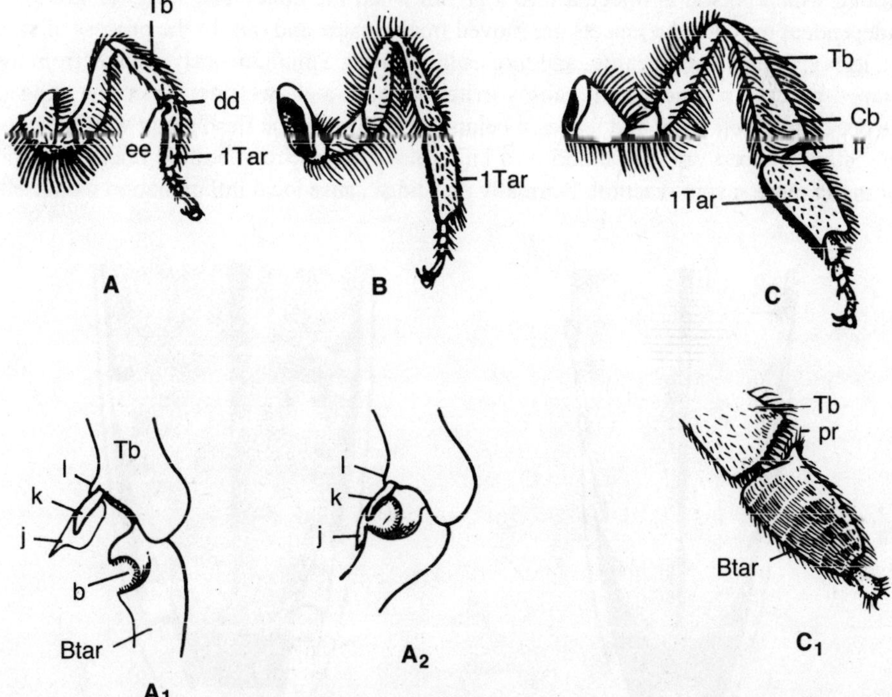

Fig. 9. **A**—Left front leg of worker; **B**—Left middle leg of worker; **C**—Left hind leg of worker; **A**$_1$—details of antenna cleaner in open position; **A**$_2$—antenna cleaner closed; **C**$_1$—details of pollen press and pollen brush.

Tb—tibia, 1tar—first segment of tarsus, Btar—basitarsus, Cb—corbicula or pollen basket, ff—wax pincers, px—pollen press, j—clasp of antenna cleaner, k—lobe of clasp, l—stop point for clasp.

brush–comb against a row of spines on the tibia of the opposite leg and is forced up into the pollen basket. The middle legs are used for unloading pollen and packing it into the cells in a hive.

3.3 ABDOMEN

The abdomen is attached to the thorax with a narrow short stalk or peduncle. There are ten segments of the abdomen which on the sides are joined by flexible and soft intersegmental membranes. Each segment has a dorsal plate known as tergite which overlaps a ventral sternite. The intersegmental membranes are weak points and it is there that the bees sting each other or the parasitic mites attach themselves to their body. On the lateral aspects of the tergites there are spiracles or the breathing pores one on either side. The segments II to VII are easily distinguished and, concealed under the last one, are segments VIII to X which are fused together and are modified to form ovipositor and the sting. Anus is located on segment X.

A number of glands open in the abdominal region. The wax glands open on the IV to VII sternites on each side, over the transparent polished surfaces known as the mirrors. Wax is secreted as a liquid which on solidification turns into thin flakes. The production of wax is regulated by bees according to the need of the colony. The scent gland or Nasonov (Nassanoff) gland is located under the mid tergite VII. The poison glands are located inside the body at the posterior end; the venom is excreted through the sting.

The sting consists of a dorsal stylet or sheath and two ventral lancets and together they form a cavity through which poison is injected into a victim when the honey bee stings (Fig. 10). With the help of independent muscles the lancets are moved freely inside and out. In the process of stinging the bee curves its body, pulls out the sting, and thrusts the lancets. Simultaneously, poison from the venom sac is pumped into the wound which causes irritation, pain and swelling. Since the sting is highly barbed the bee cannot pull it out and leaves it behind embedded in the flesh along with the poison sacs. If the sting is not scraped off with a nail or a knife blade the entire venom is injected in the animal flesh resulting in more severe reaction. Normally the stings cause local inflammation which subsides in 3-4 days.

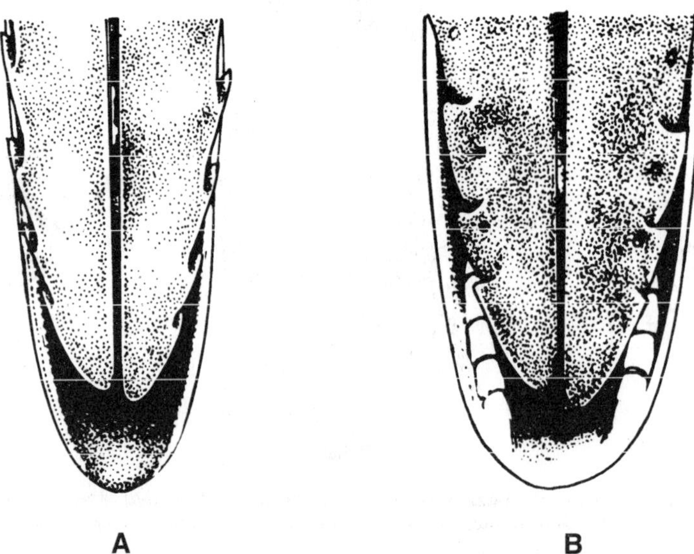

Fig. 10. Lancets of sting : **A**—barbed lancets of worker; **B**—sting of queen less barbed.

3.4 FUNCTIONAL ANATOMY

Most of the internal organs lie in abdomen along its length, and collectively they form the viscera (Fig. 11). The anatomical systems involve the physiological functions of the body. These functions are; food digestion, excretion, blood circulation, respiration, nerve transmission and reproduction. In addition there are certain glands that are involved in specific functions and control muscles that aid in movement of body parts (Snodgrass & Erickson, 1992; Wigglesworth, 1984).

3.4.1 Digestive System

The digestive system consists of alimentary canal or the food tract (intestine). The mouth leads to pharynx which is sort of sucking pump. It narrows into esophagus which passes through neck, thorax, and peduncle. As it reaches the abdomen it swells into crop or the honey stomach. When expanded fully it may cover a greater part of the abdomen. The honeybee occasionally fills it with nectar from flowers or fresh water for drinking and sometimes with honey also. At the end of the honey stomach there is a narrow one-way valve which controls the onward passage of liquid food and pollen into proventriculus the (fore-gut). The honey stomach acts as a temporary store room for food and water most of which are regurgitated for processing in the hive; some is also passed on for digestion and personal use of the bee. The valve can not only retain food but also spores of Nosema protozoa, of

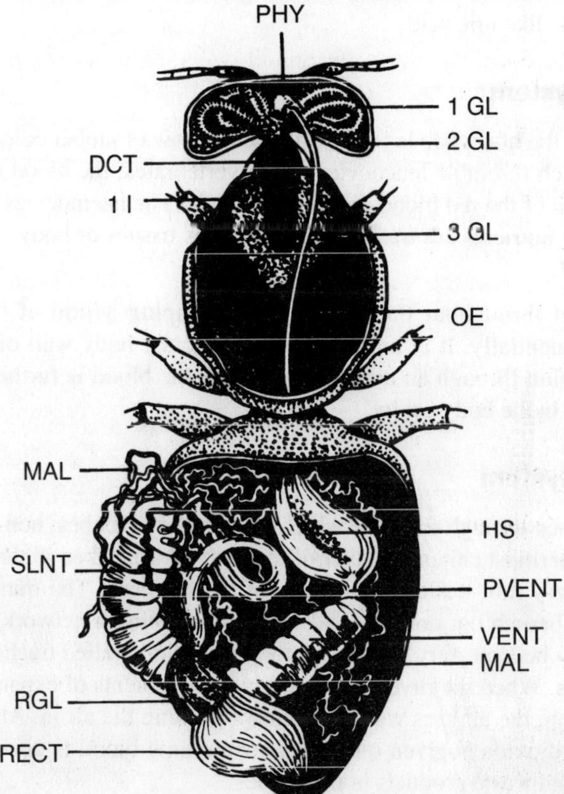

Fig. 11. Dissected honeybee worker exposing alimentary tract and associated organs: Phy—pharynx, 1GL—pharyngeal gland, 2GL—lobe of salivary glands in head, 3GL—lobe of salivary gland in thorax, DCT—salivary duct, OE—oesophagus, HS—honey stomach, PVENT—proventriculus, VENT—ventriculus, SLNT—small intestine, MAL—malpighian tubules, RGL—rectal glands, RECT—rectum.

Foulbrood bacteria and yeasts. After the proventriculus, follows the mid-intestine or the venticulus, where most of the digestion takes place. It is of thick tubular shape, having muscular layers on the outside and numerous folds on the inside, which greatly increase the digestive surface. On the inside it is lined with a thin layer of epithelial cells which multiply fast and provide the digestive fluids and enzymes. They act on carbohydrates and proteins of the pollen grain without affecting the spores of disease causing micro-organisms that might reach there. The substances of digestion pass through various outer layers and reach the blood that circulates in the body cavity.

Some of the digested food is passed on to the narrow hind gut or the small intestine. In the anterior part there is further absorption of food and in the posterior part where it enlarges somewhat is the rectum. In rectum the waste or toxic substances are absorbed and are passed outside through the anus. During winter or prolonged bad weather when bees cannot fly out faeces are retained in the rectum for prolonged periods. At the junction of the midgut and the hind gut, protruding outwards, there are about one hundred thread like malpighian tubules which float around in the blood. They are excretory in function and on absorbing waste chemicals from the blood throw these into the hind-gut to be excreted outside.

The surplus food is stored by honeybees in the form of fat body and it is meant for emergency use during their long flights. The fat bodies also contain glycogen, carbohydrates, hydrous iron oxide and even poisonous materials like uric acid.

3.4.2 Circulatory System

The body cavity of the honeybee is filled with pale yellow or amber coloured blood. The colour is due to haemocytes, which resemble leucocytes of the vertebrates; the blood is not red because it does not contain haemoglobin of the red blood cells of mammals. The haemocytes are suspended in plasmic fluid through which the nutrients are transported to various tissues of body. The metabolic wastes are also carried in the blood.

Blood is circulated throughout the body by the pumping action of the heart which has five chambers arranged sequentially. It is located along the dorsal body wall of abdomen and is carried forward to the head region through an aorta. The flow of the blood is further regulated by the dorsal and ventral diaphragms in the body cavity.

3.4.3 Respiratory System

Breathing takes place through tracheae which are tubular branches, non-collapsible invaginations of the ectoderm, the outermost chitenous layer of the body. Air is taken in through spiracles and enters air sacs which are silvery thin walled expansions of the tracheae. The transverse tracheae originate from the spiracles and branch out dorsally and ventrally forming a network throughout the body. As they branch further they become very narrow till they form thin-walled tracheoles reaching in between the cells of body tissues. When the rhythmic abdominal movements of expansion and contraction take place by muscular action, the air sacs work as bellows pushing the air in. At the cellular level oxygen is taken in and carbon dioxide is given out. At the same time blood is circulated in the body for the exchange of nutrients and waste products in the tissues.

In the honeybee there are ten pairs of spiracles, four in the thorax and six in the abdomen. The air sacs are quite large and extensive and they aid the bee in two ways; to supply large quantities of oxygen required in long flights, and to keep them air borne when laden with huge quantities of pollen and nectar.

3.4.4 Nervous System and Sense Organs

The nervous system consists of three parts;

1. the *central* nervous system
2. the *sympathetic* nervous system connected with digestion, blood circulation, reproduction etc.
3. the *peripheral* nervous system connected with sensory cells in the integument.

The central nervous system comprises brain in the head, a double ventral nerve cord along the floor of body wall and seven prominent ganglia on it, two in the thorax five in the abdomen. The ganglia are centres of nerve fibres connected with body appendages, glands and other organs establishing a neural circuit. The brain in the head is also a prominent cerebral ganglion, with many lobes, internally connected with mouth parts, antennae, eyes etc. Since drones have very large compound eyes, the eye lobes are highly developed. As a result of external stimuli the sensory impulses are transmitted to the motor nerves which in turn produce appropriate movements. Impulses are carried by the association nerves to brain and to other ganglia, thus setting up a coordinated behaviour, even independent of the role of brain. A honeybee with its head cut off can still walk, fly and sting, but these movements are not motivated.

Honeybees have a complex life, ranging from social living to solitary foraging, and home coming in all kinds of weather. They achieve this with the help of sensory hair (sensilla) and receptors on the body surface. The receptors perceive stimuli of various kinds such as mechanical, chemical, gravitational, thermal, photosensitive, electric, magnetic etc. They even have a biological clock like the higher animals do. The location of various stimuli on the body are summarized in Table 3.

3.4.5 Reproductive System

The honeybee that we see every day is a sterile female and normally does not lay eggs. The queen bee that lays abundance of eggs lives inside a hive of the bee colony and we do not see her. The males or the drones live in hives too but are often seen flying around outside. The complexity of colony life will be described later on. Here the male and female reproductive systems are discussed (Fig. 12).

The male internal organs consist of two flattened, kidney shaped testes that lie below the coiled alimentary canal. Testes are made of many tube-like testicular follicles which open into vasa deferentia and the sperms are carried further into the tubular seminal vesicles where they are retained. The vesicles join mucus glands and they all open into a single ejaculatory duct. The duct opens into penis which is a large complex structure. The penis is internal (endophallus) and is not seen from the outside. At the time of mating it is pushed out just as the closed finger of a rubber glove when inflated becomes full length. The penis aperture is located just below the anus, above the sternite IX. If a drone is held from the front and is given a slight touch on the ventral side of abdomen, the penis is everted and semen oozes out which can be collected in a syringe for artificial insemination of the queen. In order to gather a full store of the semen, a drone should be at least thirteen days old.

The queen bee has the complete female reproductive system. It has two large pear shaped ovaries made up of up to 160-180 tubular ovarioles, held together by terminal filaments. Ova start developing at the terminal end of the ovarioles and as they grow they travel down and absorb nurse cells. Each ovum with the nourishment cells around it is then coated with a thin shell (chorion) and an egg is formed. The chorion has a minute pore at one end (micropyle) for the entry of a sperm. The ovarioles at the thick lower end open into an oviduct and the two oviducts join together to form the median oviduct. It opens into vagina which is guarded at the end by bursa copulatrix. On the dorsal side of vagina joins a fine tubular spermathecal duct at the distal end of which is the sac like spermatheca, the warehouse of sperms. The sperms are gathered by the queen by sequential mating with a number of

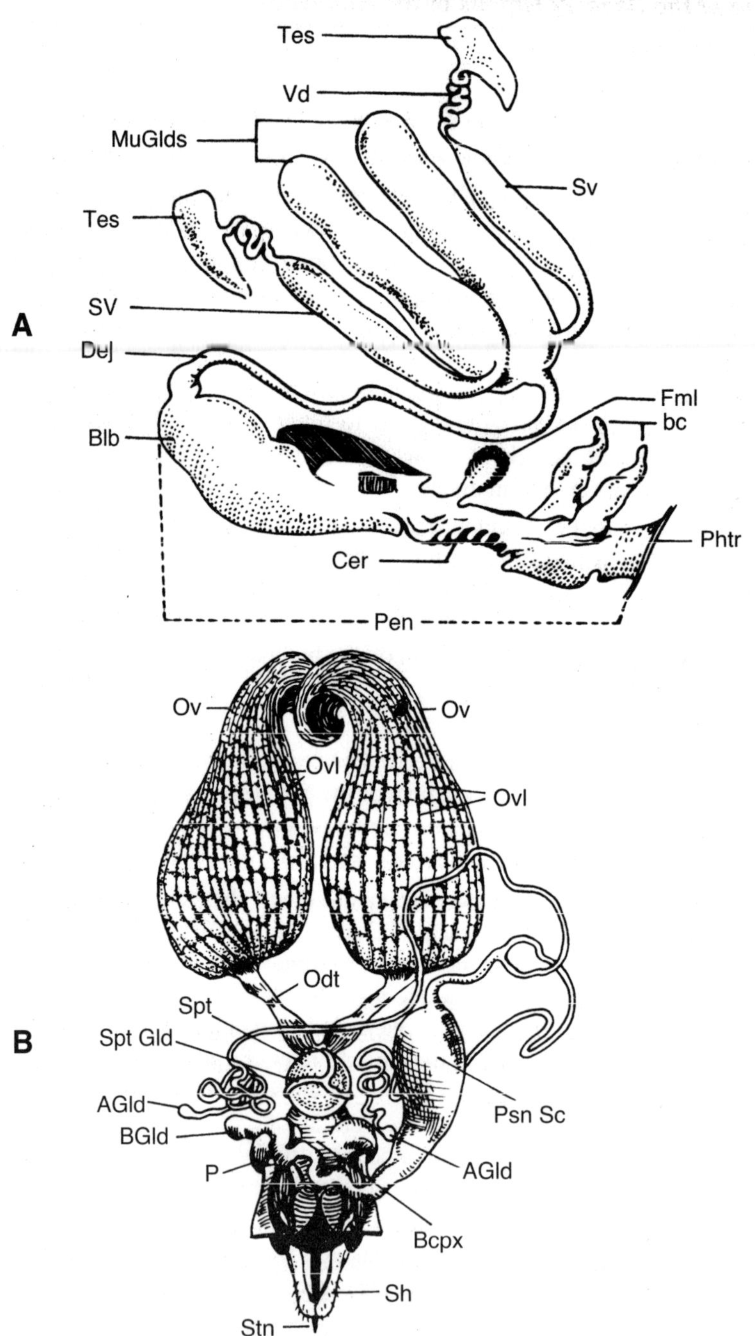

Fig. 12. **A**—Reproductive organs of drone : Tes—testis, Vd—vas deferens, SV—seminal vesicle, MuGlds—mucous glands, Dej—ejaculatory duct, Fml—fimbriate lobe, Cer—cervix (neck of penis), bc—bursal cornua, Phtr—external opening of inverted penis, Blb—bulb of penis, Pen—penis.

B—Reproductive organs of queen, Ov—ovary, Ovl—ovariole, Odt—oviduct, Spt—spermatheca, SptGld—spermathecal gland, AGld and BGld—accessory glands of sting, PsnSc—poison sac, Bcpx—bursa copulatrix, P—lateral pouch of bursa copulatrix, Sh—sheath lobes of sting, Stn—shaft of sting (adapted from Snodgrass).

TABLE 3. *Some of the Sensory Organs of the Honeybee*

Type of Stimulus	*Sensory Organ*	*Location*
Mechanical		
touch	s. trichodea	many sites
stress/strain	s. scolopophora	legs and head
	s. campaniformia	lancet of sting
gravity	s. trichodea	hair plates on neck and petiole; coxa; trochanter
movement	s. chaltica	antenna
flight speed (relative to food)	interneurons	compound eye
flight speed (relative to air)	long hairs	head; compound eye wing
temperature	s. ampullaceae	last 5 antennal segments
	s. coeloconica	"
Light		
intensity	ocelli	head
intensity; colour, flicker, shape	retinula cells	compound eye
Polarized light	un-sensitive retinula cells	compound eye
Chemical		
taste	s. chaetica and others	mouthparts; antenna; tarsus
carbon dioxide	s. ampillaceae and s. coeloconica	antenna
humidity	"	"
Time	interneurons	corpora pedunculata
Magnetic	megnetite 'granules'	fat body

After Crane (1990)
s = sensilla

drones during the mating flights. As the eggs mature and pass down the genital pore those are coated with chorion. The ovipositors in workers and queen bees are further modified as the sting. A queen in her spermatheca can store upto 7 million sperms which last for 2-4 years (Dade, 1977).

3.5 SPECIAL ORGANS

There are a number of special organs distributed throughout the body which are not directly connected with any of the systems. These include muscular sucking pumps, glands, oenocytes, rectal pads, Johnston organ, scolophorous organs and endocrine organs.

3.5.1 Sucking Pump

It is elongated organ in the head made of muscles. It is rounded sac lie structure which starts at the mouth and extends up to the neck where it becomes continuous with the esophagus. It is used by the honeybee for sucking and pumping in its esophagus water and other liquid foods. Perhaps, it also helps the bee to regurgitate nectar and honey stored in the honey stomach.

3.5.2 Salivary Glands

They are whitish and look like a bunch of grapes located in the posterior part of the head and the ventral part of the thorax. These glands secrete saliva which flows through tubular ducts from thorax and head into a common duct which open into salivary syringe at the base of the tongue. Saliva is pumped forward within the channel of the tongue (proboscis) to the tip where it is mixed with nectar or it is used as a solvent when solid food is available.

3.5.3 Food Glands

These paired structures are also called hypopharyngeal glands and look like long strings of small sacs arranged in many loops and coils. The two ducts that originate from the glands open separately in the floor of the mouth. The translucent secretion is called royal jelly and it is fed to the larvae for growth and to the queen and drones for maintenance. The nurse bees collect it from the mouth of the feeder bees and while carrying it they use their tongue and mandibles.

3.5.4 Rectal Pads

They are six elongated structures located at the anterior surface of the rectum. They are visible to the naked eye in a dissected honeybee. They absorb excess moisture in the blood and maintain water balance.

3.5.5 Oenocytes

These are large cells found in association with fat cells. Their function is to secrete certain enzymes required for various physiological functions in the body.

3.5.6 Johanson's Organ

It is located on pedicel or the bend of the antenna. The sensory hairs on the organ aid the honeybee in perceiving oscillations in air particles. Other scolopophorous organs elsewhere in the body of honeybee are involved in perceiving muscle tensions, vibration in substrate, stress and strain.

3.5.7 Endocrine Organs

Such body functions as growth, molting, development etc. are regulated by chemicals produced by various organs collectively termed as endocrine glands. Some groups of neurosecrtary cells have also similar functions. The glands include corpora cardica located at the junction of pharynx and esophagus; the glands secrete enzymes. In addition there are thoracic glands who's function is not known.

3.5.8 Mandibular Glands

These glands are located at the base of the mandibles and open into them. The secretion of the glands is milky white and forms a part of the royal jelly ; the other part being clear or translucent secretion of the hypopharyngeal glands. The nutritious quality of the white secretion is apparently superior and a greater quantity of this is fed to the queen larva as compared with that to the worker larva.

Biology of the Honeybee

In the life history of a honeybee there are four stages of development; these are egg, larva, pupa and adult. The eggs are of two types, fertilized and unfertilized. From the fertilized eggs develop workers and queens and from the unfertilized eggs develop males or the drones. As the egg hatches a larva emerges and it grows in size by discarding its skin five times and thus developing in stages.

When full grown, a larva becomes inert and then transforms into a pupa which is a quiescent stage. At the end of that stage the adult bee comes out of the puparium. The adults are of three types; the male is large and is called a drone; the worker is an imperfect or sterile female; the large queen is a fertile female. The process of growth and development in stages is called metamorphosis which is characteristic of all insects, including honeybees.

The biology and behaviour of the honeybee has been studied in much detail in the last one hundred years or so every since the invention of modern hives with movable frames. A frame can be taken out, the activity of the bees can be watched and it can be placed back in the hive without disturbing the bees unduly. The walls of the hive can even be made of glass and the bees can be watched under infra-red light which they cannot perceive and they carry on their work as if they were in the dark. The world literature has been complied by a number authors in their works which have served the cause of apiculture for the last many decades; they include Smith (1960), Eckert and Shaw (1960), Morse (1973,1975), Gojmerac (1980), Butler (1974), Winston (1987), Crane (1990), Graham ed. (1992). The biology and life history of honeybee is given here in a summarized form.

4.1 METAMORPHOSIS

As compared to a bird-egg the eggs in insects have less food, hence the young ones emerge at an early stage of development; the young are quite small and, much different from the adults in body shape. The immature stages or the larvae, as the young ones are called, have a covering of inflexible skin made of chitin which they have to shed five times and only then they can grow to full size (Fig. 13). The skins are cast off by the process known as ecdysis or moulting. Before moulting takes place the larva stops feeding and becomes quiescent for a few hours. The epidermal cells enlarge and divide mitotically. The old cuticle is detached and a layer of epidermal cells secrete a new cuticle. The space between the old and the new cuticle is filled with moulting fluid which contains two enzymes, protease and chitinase. These enzymes dissolve the old cuticle, not touching the new one; the dissolved fluids are absorbed by the epidermis. The old skin splits along the median dorsal line in the thorax region and the larva frees itself by muscular movements, leaving behind the old skin (exuviae). The interval

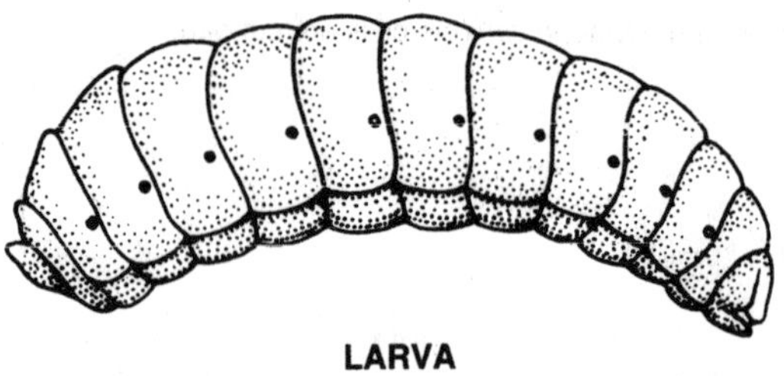

LARVA

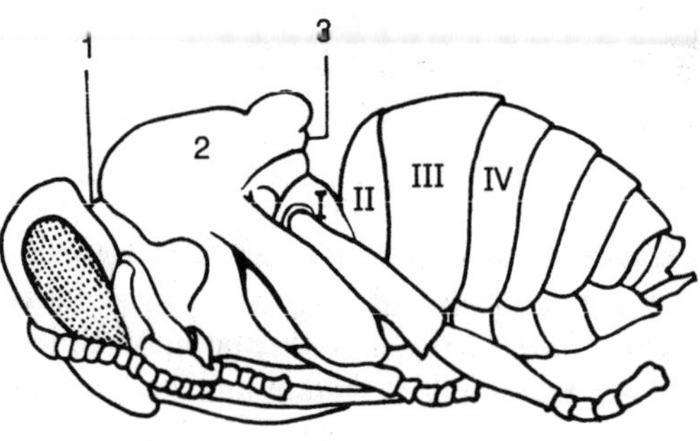

PUPA

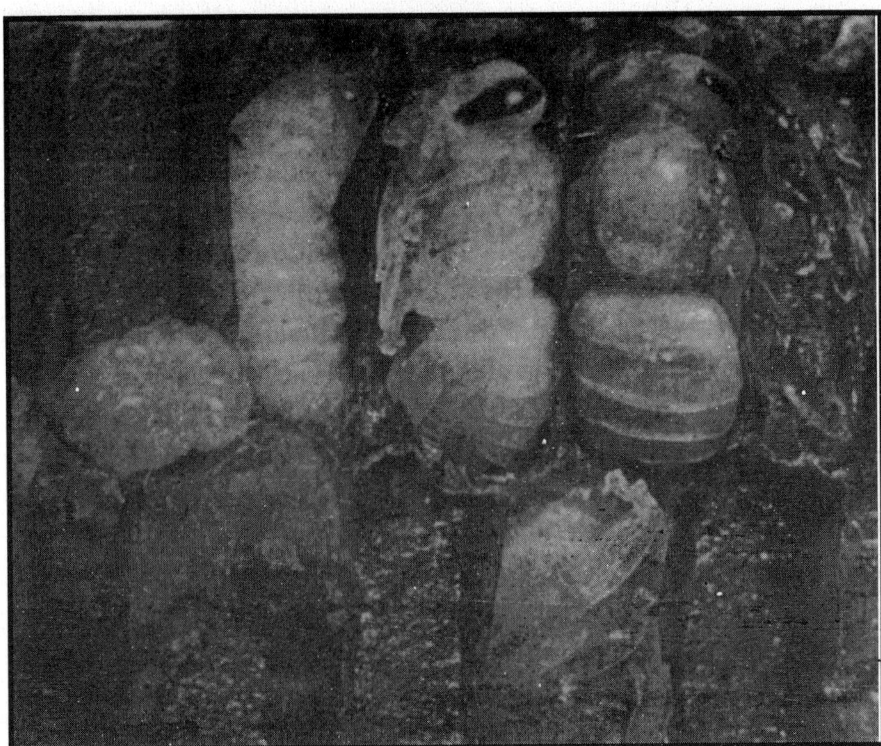

Fig. 13. Top and Middle : immature stages; larva showing spiracles; pupa marked as 1, 2, 3, the thoracic segments and I, II, III, IV etc. as the abdominal segments.

Bottom : larvae and pupae *in situ* (the cells are cut out).

between two moultings in a larva is known as stadium and the body forms in a particular stadium is called instar.

4.2 HOUSE OF THE BEES

Hive is a house of the honeybee where they live breed and raise their brood. Three castes of the honeybee, (worker bee, queen bee, drone) cannot live an independent life; they have been evolved to live in colonies. Queen is the mother of the colony, drones are males who mate with virgin queens and die, and the workers perform all sorts of duties in the hive. Those include food collection; feeding the brood, the queens, the drones; cleaning the hives; producing wax and combs; sealing cracks and crevices in hive; fighting the enemies and guarding the hive etc. All these duties are performed on the wax combs and in other parts of the hive.

The wax combs are made on wooden frames 1″ in width, 9″ in depth and 17″ in length (**Plate 4**). On their own bees can make brood and food combs on these frames but to save their energy and to have uniform attractive combs, a comb foundation sheet is fixed in the rectangle of the frame on supporting metal wires. The foundation sheets have on either side hexagonal impressions of cells. The sheet later on forms the midrib or base of the comb and on both sides worker bees raise horizontal long cells which have hexagonal periphery and are slightly raised at the open end, at an angle of 9–14°. A typical frame has both worker cells and drone cells on both sides of the frame. Their number is 55.3 per square inch for the former and 33.5 for the latter. *Apis cerana* has 80 worker cells per square inch and *Apis dorsata* has 50.8 cells. Frames having predominantly drone cells are raised at certain times of the year only. During the breeding season or sometimes earlier, if there is a need, bees make large queen cells raised and directed downwards generally on the bottom edge of the frame; they look like a groundnut pod. The worker and drone cells are used over and over again after the bees have cleaned them of the leftover pupal skins. It has been observed that if old combs are used for more than 3-4 years the cells size become significantly small and the resultant brood reared in them also becomes smaller in size which is not considered a desirable feature.

The beeswax required for comb building is secreted by the bees at a certain age when their stomach is full of honey. Therefore, in the hive comb building and honey gathering proceed simultaneously; as soon as the nectar flow slackens, the bees stop making combs. In other words bees make sufficient combs according to the need of the colony. Bees hived on new foundation sheets can be stimulated with sugar feeding.

4.3 LIFE HISTORY OF THE COLONY

It has been suggested that since no individual in a colony can live independently the whole colony should be considered a complex animal. The seasonal rise and fall of the colony population and its functional behaviour may be considered the performance of a whole animal. The changes that take place in the colony are determined by the time of the year, seasonal cycle, weather, and honey flow in nature. With lengthening of days in the spring and appearance of flora in the fields, more eggs are laid in the colony and the strength increases in preparation for the exploitation of honey flow. When hot summer comes, the nectar flow slackens and colony cuts down egg laying in order to reduce the number of mouths to be fed. In the heat of the summer the life of the workers is shortened, many of them die, and young workers are left behind to fend the colony during the wet monsoon when workers may be confined inside the hive for days on end; lots of food reserves are required for sustenance. In autumn and in winter in areas of mild climate when new flora and crops appear in the field the colony

larvae. Tolerance of drones depends upon the food stores available in the colony (Fukuda and Ohtani, 1977).

Sometimes very small drones are seen in colonies and they are sexually functional. If the colony is normal with a queen it means the drones have been raised in worker cells for want of drone cells. If there is no queen in the colony and drones are seen it means there are laying workers in it who are depositing eggs in worker cells. In general drones live 21–32 days in spring 14 days in summer, 90 days in autumn and do not survive temperate winter.

4.6 LIFE HISTORY OF QUEEN

Queen is the largest individual in a colony. Its wings are normal but the legs do not have special features for collecting pollen. Its antennae are longer than that of a worker although there are the same 12 segments of flagellum. It has no wax glands or brood food glands. Its tongue and stomach are smaller than those of a worker but the poison sac is much larger. The barbs of its sting are less dented and therefore it can be pulled out. A virgin queen held between the thumb and a finger can give a sting. A queen is a perfectly developed female and is the mother of the entire colony. On its survival depends the very life of the colony; without a queen sooner or later the colony perishes.

A colony rears a queen under any of the following circumstances :

1. A queen has been lost suddenly and it has to be replaced.
2. A queen is old, not performing well and a new one is required for supersedure.
3. A colony is at the stage of reproduction and swarming and extra one or more queens are required to go with swarms.

A queen can be reared only when workers are short of inhibitory substance (see pheromones Chap. 5). In the first situation queen cells are prepared from normal worker cells, containing an egg or a young larva two or at the most three days old. The queen cell goes deep in the comb up to the midrib. These larvae are fed rich brood food (royal jelly), which contains more of the white component as compared to the brood food given to worker larvae. The "royal jelly" has 34% sugar during the first three days and when larvae are 4–5 days old the sugar content is increased to 47%; thus sugar is the major feeding stimulant for queen larvae.

Starting from a 2 day old larva the virgin queen should emerge 11 days later. If the weather is favourable for mating the new queen should start laying 10 days later. Therefore, within three weeks of the loss of a queen the colony would have a new one laying eggs, thus saving the colony. If, however, the workers have not been able to raise a queen, or the virgin queen owing to bad weather is not able to mate, the colony becomes helpless, demoralized, and turns laying workers. Before that happens it is better to combine it with a weak colony having a fertile queen.

At the time of supersedure, or swarming, the workers raise different kinds of queen cells. In case of supersedure those are thick and bulbous and are attached to the face of the combs raised on wax so that they project outward and hang with their mouths downwards. The cells are about an inch long, have thick sculptured walls (somewhat like a groundnut pod) and taper slightly from base to top. The mouth or cap of the cell is at the lower end and a queen emerges by cutting it along the circular edge and raising it like the lid of a can; their head is towards the distal end.

The number of queens raised in case of swarming depends upon the size of the colony and the race of the bees involved. The Italian and Caucasian races produce only a few queens whereas Syrian and Egyptian races may at times produce as many as 100 queen cells. In case of supersedure two or three queen cells are generally raised and it may not be the first queen that emerges may take over the colony. In case of swarming many cells are raised at the bottom of a comb and they project downwards

(Fig. 14). In a colony getting ready to swarm there may be over one dozen cells. These cells are lighter in colour as compared with those of the supersedure cells.

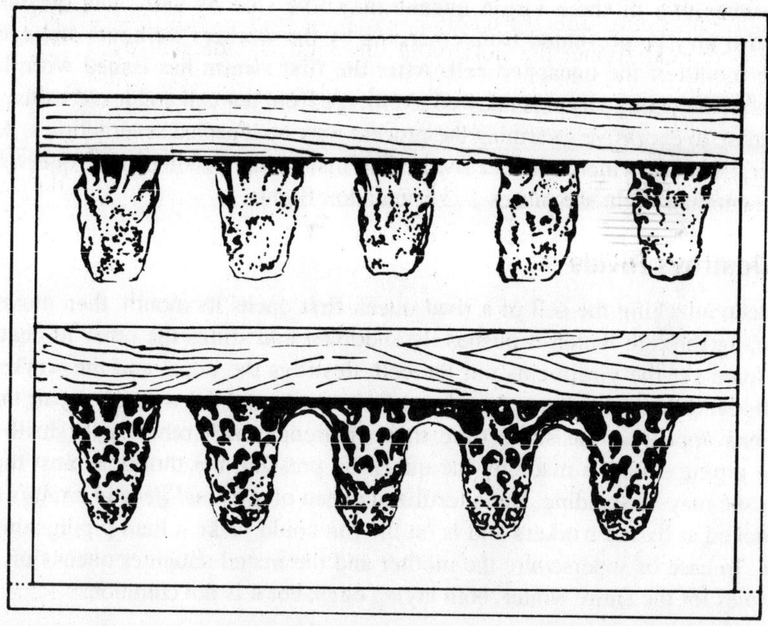

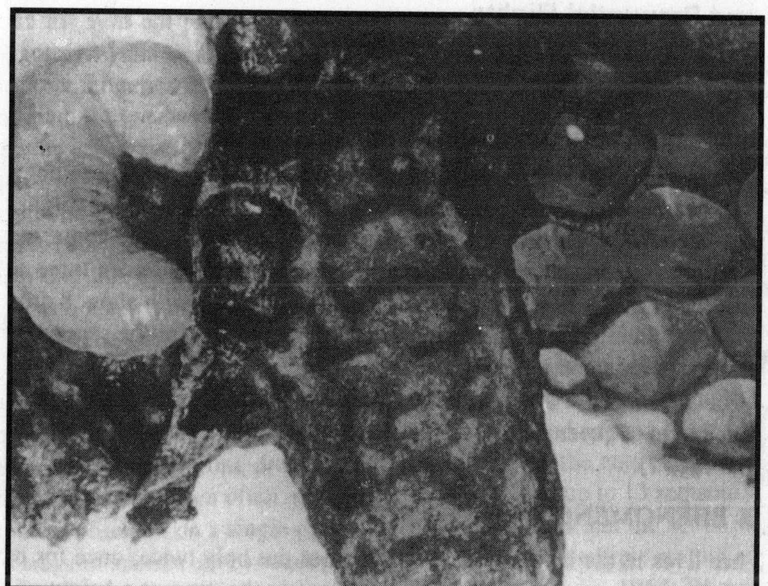

Fig. 14. Top—queen cells on wooden bar, Doolittle method of mass rearing of queens. **Bottom**—close-up of a queen cell and a cut out showing larva.

When a queen emerges its pheromone glands are not developed and workers do not notice her. Immediately, she feeds on pollen and nectar like the workers for 3–4 days and its internal organs mature on 5th or 6th day. The workers even treat her roughly and vibrate her 1200 times/h. She pipes frequently to protect herself. She searches other virgin queens that may have emerged which leads to a fight and one of them gets killed. If there is no other queen she turns to queen cells and destroys them. However, sometimes two or three virgin queens may live side by side. The queens in a colony preparing to swarm may be prevented from emerging by the workers for hours and even days. They are fed from the mouth of the uncapped cell. After the first swarm has issued with the old mother queen or a new one, the young virgin is allowed to emerge from her cell and leave with a swarm. In the strains that are prone to excessive swarming the process may be repeated several times. At last the bees allow the one virgin queen which kills her rivals. She mates, and becomes new queen of the colony. Before going on a mating flight she makes 1–2 orientation flights.

4.6.1 Virgin Destroys Rivals

A virgin queen attacking the cell of a rival queen first opens its mouth, then moves to the base, cuts out an opening through which it pushes its abdomen and stings the other to death. If the rival queen is not formed yet the virgin cuts out the cell, destroys the pupa, and the workers then throw away the dead body. Sometimes damage to the pupa is minor and the workers repair the cell. In case two virgin queens appear at the same time they challenge each other with shrill piping sound "ze.e.e.ep". The piping sound is made by the queen by pressing her thorax against the comb or the object on which she may be standing. But a fertilized queen of *Apis mellifera* in an *Apis cerana* colony that is being plucked at by the workers and is on the run could make a loud piping sound heard from outside the hive. In case of supersedure the mother and the mated daughter queens of *A.m.* may live together, sometimes for the entire winter, both laying eggs, but it is not common.

4.6.2 Nuptial and Prenuptial Flights

A virgin queen on emergence may be as large as a mated laying queen but after a few days it becomes smaller, perhaps in preparation for the mating flight. It is estimated that half of the queens mate on their first flight but some make a prenuptial flight for orientation. Mating takes place when virgins are three to five days old but they can mate up to 10 days of age. Queens mate with about 8 drones in one or two days, therefore, a mated queen may again go out for the second nuptial flight till she has obtained and stored 5.3–5.7 million sperms (Laidlaw and Page, 1984, 1989). Queens more than 3 weeks old are incapable of mating and it is no use waiting for this new queen to be productive in a queenless colony (Woyke, 1964; Adams *et al.*, 1977).

4.7 THE PHENOMENON OF MATING

A queen bee lives in the hive all her life. She goes out only twice; once for mating and second time with a swarm. Mating takes place in the open at a height of 30 to 50 meters from the ground. The abundance of drones is essential in the air space if successful mating is to take place. On warm sunny days drones fly out and saturate the air space near the apiaries. There are drone congregation areas which remain fixed over the years. The congregation generally takes place near a prominent land mark like a hillock, or a grove of trees. When flying around they make a loud hum and also emit male pheromone to which queens are attracted. Mating takes place, generally between 2–4 pm on sunny days, when queens are on their nuptial flight. The queen arrives with her genital aperture open and she gives out a strong pheromone smell. The chemical is 9–oxodec–trans–2–enoic acid to which males are attracted. Many drones give the chase but only one is successful; mating is rarely seen in nature. Only in recent years it has been possible to observe mating through new techniques and devices. A queen is tied to one end of a thin nylon thread and is tethered to a rubber balloon full of helium which is hoisted up in the air to an appropriate height. The queen is free to fly around and the drones chasing it can be watched. In another device a camera is fixed on an extended metal arm and a queen is harnessed to it at the end. In ideal weather, temperature should be above 20°C on a bright sunny day, and wind velocity should be less than 22 km/h.

In the phenomenon of mating and natural selection the drone congregation areas make a fascinating study. These areas are 90–120 m or even further away from an apiary, near a lake, a grove of trees, a hillock or some other landmark. Drones in hundreds and thousands fly around in these areas year after year. The queens come there from long distances and they as well as drones have been noticed to fly 2–3 km. These areas are 5–7 km apart. An area itself is 30–200 m in diameter full of flying drones 10–40 m above the ground.

Gary (1963) has made detailed observations on the mating behaviour at drone congregation areas. The drones are greatly stimulated by pheromones from the virgin queens. In the process of mating a number of drones chase the queen; and one of them who happens to be very agile rides the queen from behind holding her on the ventral side with his hind legs, and at the same time making a grip on her dorsal side with fore and meso-legs. He mates in a split second, is paralyzed, and falls backwards. The queen drags him along for a few more seconds and during that time the penis bulb reaches the vaginal pouch. Even though the drone is paralyzed it is due to the active participation of the queen that penis muscles ejaculate; the bursa is retained along with other membranes of the bulb and the organ is snapped at the cervix. The drone loses its grip and falls to the ground to die soon after. In the meantime sperms are released in vagina and they are sucked in by the queen's internal system and through the spermathecal duct they reach spermtheca. There they are mixed with a nutritious secretion and are stored for the life of the queen, that is, for 3 to 5 years. The secretion is essential for the survival of the spermatozoa. In a virgin queen the secretion is available for three weeks and if there is no mating and sperms do not enter the spermatheca it dries up, hence a queen cannot be impregnated. A fertilized queen has a store of about 5 million sperms and a drone has 4 to 7 million of them. Theoretically one drone should be enough for the entire collection, but in actual practice the queen often mates more than once and up to 13 sequential matings have been observed on a single day of her life. It indicates that the act of mating and the storage of sperms are two independent variables, and multiple mating is meant to have a full storage of the spermatozoa (Gary, 1963; Woyke, 1964; Winston, 1987; Ruttner, 1956, 1966; Lensky *et al*., 1985).

4.8 LIFE OF QUEEN IN A HIVE

After mating, the queen comes back to the colony, carrying a white membranous material on the dorsal side of her abdominal end which the experienced beekeepers have always noticed and called it the mating sign. It is a part of the penis bulb outside and the bursa inside the vagina. The worker bees try to remove it, but quite often the queen herself gets rid of it by scratching her abdomen on the comb cells. In case it is from the last mating. The question arises how queen gets rid of the material from earlier matings while on the wings. It is likely that is is removed automatically in the process of a subsequent mating.

In a hive the queen, surrounded by workers is cleaned up, licked, and fed by them. Within the next two or three days when the eggs mature in her ovaries, the queen also secretes many other chemicals which attract the workers and, the queen is now in full attendance. She starts laying eggs within three days and regular life starts in a colony.

After mating, queen starts laying eggs and continues for the entire life of three to four years; queens as old as eighth years have been reported in literature. The young queens are always surrounded by workers who keep her licking and also feed her. The older queens are less hairy and are shiny in appearance and are surrounded by fewer workers as if they are losing interest in her. Workers may even chew up her wings and give her the doom to die. Also the young queens lay eggs in compact areas but the old queens leave gaps while laying eggs in the cells. A brood of uniform age in a compact area on the entire frame indicates a vigorous young queen. In commercial beekeeping queens are replaced after one or two years even though they might be doing well.

A queen ready to lay eggs is fed by nurse bees secreting brood food mixing it with honey. Even though a queen can feed herself the nurse bees attend on her day and night and they control the food according to the number of eggs being laid by her (Allen, 1960).

5 Functional Behaviour of Honeybee

The functional behaviour of honeybees is aided by numerous body structures designed to carry out the life functions. Since the bees have been studied extensively the function of some of the structures are well known and we have a good picture of their role in the daily life of the bee. The external skeleton in most regions of the body, for example, is covered over with dense piles of hairs which have definite biological functions. The outside hardened plates on the skeleton are connected by flexible membranes and there are internal ridges throughout the body on which muscles are attached for the functioning of movable parts, including antennae, mouthparts, legs, wings, sting, etc. The skeleton, apart from giving attachment to muscles, also provides physical protection and prevents water loss through transpiration.

The functional behavior of the honeybees depends upon various body parts, sensory structures, chemical secretions and pheromones, and the instinctive qualities as expressed in communication and other dances, fanning, stinging etc. It is the combined effect of various functions that make the complex animal (honeybee) so efficient and successful. Since the honeybees are social beings and an individual is only a part of a large colony, their functions have to synchronize for the common good of the community so that the colony works as a single 'organism'. Thus the functional behavior reaches a level of perfection.

5.1 PHYSIOLOGICAL AGE AND HIVE DUTIES

Certain behaviour and duties among the worker bees are age related. During the first day of life sting is not developed and she can not sting. Young bees clean cells and feed the older larvae. When the brood-food glands develop they start feeding the young larvae who develop on the food. Then the wax glands develop and they secrete wax for building the comb. This activity is followed by guard duties and ultimately foraging in the field. All these jobs the bees take up instinctively and there is no master bee in the hive that gives directions. For that reason the term "division of labor" is misleading. There is no leader in the colony that allots them jobs. The sequence of work changes according to the physiological age and not the chronological age. There is also a great flexibility in the age related work according to needs of the colony and, bees of the same chronological age may change physiologically; the brood food glands in old bees may be activated if there are no young workers in a colony. In a new colony predominated by young workers, the work of older bees including foraging may be taken up by them. Thus bees instinctively feel the kind of work to be taken up at a given time regardless of the age, if it is so required to be done. However, by and large, bees stick to the age-specific work in a normal colony.

Another instinctive behaviour is that the bees share work. A single bee is capable of capping a cell but many of them will share the work and add a little bit of wax to it. The bees are commonly seen sharing food; honey is passed directly from one to another or many receiving bees (Fig. 18). This way, various kinds of nectars are mixed. The bees also exchange pheromones and other chemicals among themselves; those workers that lick and feed the queen bring along her chemical secretions which are shared by other workers. The activity produces many subtle changes in the colony but it does not mean that queen is actively guiding the colony. Even in the absence of a queen the workers can do the various duties on their own and maintain the tempo of colony life.

5.1.1 Genetics and Behaviour

The instinct among bees is an innate quality and they are born with it. A young larva coming out of an egg feeds itself on royal jelly supplied by workers. When full grown, it is sealed in a cell and on its own the larva spins a cocoon and with its head towards the cap so that on transformation into adult it cuts open the cap and emerges as a new bee. There is no intuitive, creative, thinking process but the behaviour depends upon genetic composition of the bee. Rothenbuhlar (1964) found that particular genes control house cleaning behaviour of workers; those that contain one of the genes uncap the cells but do not remove the dead brood and another group that contain another gene remove dead larvae or pupae from uncapped cells. Both genes must be present in a large number of workers in a colony for achieving housecleaning effectively; this behaviour is important for imparting colony resistance against diseases like the American Foul Brood, Varroa mite etc.

5.2 FUNCTION OF SENSORY STRUCTURES

5.2.1 Compound and Simple Eyes

The simple and compound eyes give a perception of light (and colours) and vision. The ocelli or the three simple eyes are arranged on top of head in an inverted triangle. Each of them has a dense lens and underneath there are sensory retinal cells which can perceive light intensity probably to orient the bee to diurnal hourly patterns. The compound eyes are capable of a number of photoreceptive functions like intensity, flickering, imagery (Fig. 15). Every facet on the large eye has a lens of its own to receive light, a pigmented cone to concentrate and focus it and sensory retinal cells which are sensitive to light. Each facet is capable of forming an image and it is separated from the next image by an angle of 1°. Through the compound eye the mosaic of images so formed gives a perception of

A

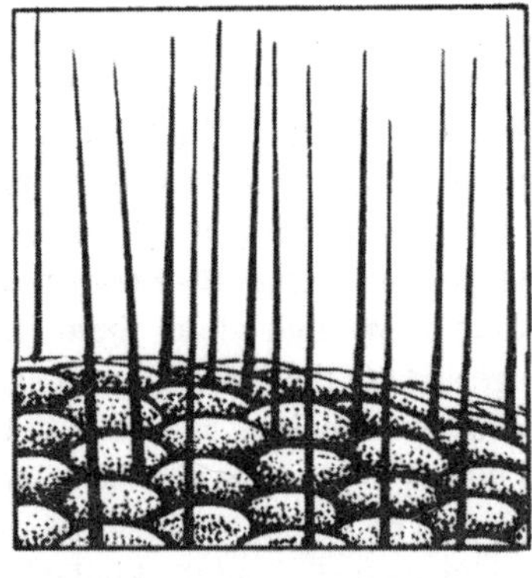

B

Fig. 15. A—Magnified (× 1000) surface of drone antenna; **B**—Magnified (× 600) part of compound eye of drone.

movement of the object being focused. The facets can respond to the light waves and von Frisch (1971) has proven that bees can "see" the shorter wave lengths (yellow, green, blue, violet) including the ultra violet which is invisible to the human eye. The honeybees cannot see red colour. A group of specialized facets can also perceive the plane of light polarization which quality is of immense help to the bee orientation in cloudy weather. At the junction of the facets there are sensory hairs which can feel the air flow and while flying bees can compensate for the speed of wind.

5.2.2 Antennae

Antennae in bees serve the same function as a nose in man and, their olfactory power for smelling essential oils is approximately the same. However, in general, honeybees are sensitive to smells of wax, flowers and other odours of biological significance, 10 to 100 times more than that of man. Moreover, they can perceive the direction from which food odour comes. On the antennae, particularly concentrated on the flagellum there are seven types of sensory structures that are in the form of pits, plates and hairs. The pore plates are important for olfactory sense and there are 3,000 of them on one antenna, mostly located on the last 8 segments of the flagellum (Fig. 15). The drones have 30,000 olfactory plate organs which indicates how important is their role in olfactory orientation in searching for a queen on wings, for that is their main occupation in life.

The function of all the structures on antennae is not known, but there is evidence that they are important for detecting, carbon dioxide, humidity, taste and temperature. Thus antennae play a significant role in detecting humidity and carbon dioxide which may induce fanning for circulation in the hive. In the bend of the antenna (pedicel) there is Johnston's organ having hairs connected internally with sensory cells. Their function is to measure air flow and adjust the speed of flight for bees to reach a specific spot in the field in varying weather and the winds. The drones and workers also have sensilla on their wings whose function is similar to that of the pore plates and hairs on antennae.

5.2.3 Physical Perception

In honeybees the appreciation of many objects is through mechanical touch. It is achieved by many hairs and plates distributed throughout the body. The structures that perceive stress and strain are mostly located in legs and head. The bees can also perceive and use gravitational force in their flight and orientation. It is of particular importance to the bees carrying heavy loads of nectar and pollen; compared to their body size the loads are sometimes quite heavy. The gravitational force is judged by the bees with the help of plates having hairs on them located on each side of the junction of head and prothorax, on junction of thorax and abdomen, and the joint between coxa and trochanter of a leg.

The honeybee learns the direction and location of foraging site through the communication dance but the measurement of the distance covered is achieved by keeping a tap on the energy spent by the bees. The source of energy is the nectar in honey stomach and no wonder the energy expenditure "account" is kept by sensitive cells in the walls of the stomach. Foraging is also connected with "taste buds" or the cells sensitive to chemicals which are connected with hairs on mouthparts, antennae and tarsi. The bees can also "hear" sound, and the organs or the sensitive cells, are located in the last 5 segments of the flagellum of the antenna.

Like mammals honeybees also have a sense of time which is regulated by corpora pedunculata. Within a given season the working hours in the field are attuned by this organ and the bees on their own start work and end the daily duties by a certain hour. They do not have to be "awakened" to start the work. The house bees inside are busy all the 24 hours and individuals have brief periods of rest. The field workers on the other hand have to spend lot of energy while foraging and they rest during the night on the combs.

During the field work the bees are exposed to environmental conditions, inclement weather, lightening and thunder and have to reach back home. They need to have sensitive body response by which they can "anticipate" the unfavorable events coming or to reach their hive in case they are caught in bad weather by a chance. The bees make such "decisions" with the help of a number of sensitive structures described above. They can perceive electrical charges in the air through various sensitive parts in the integument. They can also make use of magnetic fields for finding their way. This they achieve with the help of granules of iron salts deposited in the fat body.

5.3 FUNCTION OF PHEROMONES AND CHEMICAL SECRETIONS

Pheromones are chemicals secreted by workers, queens and drones used for communication among the members of a colony. There are perhaps 36 chemicals and 18 of them have been identified to control various functions associated with mating, alarm, defense, orientation, colony recognition and integration of colony activities. It has been demonstrated how these odours biologically influence the individuals and the colony as a whole. This field of science has just opened up with the availability of modern sophisticated instruments, and a great deal is yet to be learnt about the specific roles of various pheromones (Winston, 1987).

5.3.1 The Queen Pheromones

There are at least two queen substances produced in mandibular glands. Butler (1960) observed that workers given this chemical by the queen were so conditioned that they did not rear a new queen and felt the presence of their old queen in the colony. The chemical was identified as 9-keto-(E)-2 decenoic acid, also called 9-oxodecenoic acid, abbreviated as 9ODA (Barbier & Lederer, 1960). Subsequently another chemical identified was 9-hydroxy-(E)-2-decenoic acid (9HDA). Its function is similar to that of 9ODA but distinct in certain ways (Butler *et al.*, 1961).

Some other unidentified pheromones are perhaps produced by the abdominal glands by immature queens. Only a small quantity of 9ODA is secreted by newly emerged virgin queens but within 5-10 days the secretion becomes much greater. It is further increased in youthful queens laying eggs, but is much reduced in old queens. The biological function of 9ODA, 9HDA, and other unidentified pheromones have been demonstrated (Butler, 1974, Shearer *et al.*, 1976; Free, 1987; Winston, 1987; Crane, 1990; Blum, 1992). Their presence inhibit queen rearing in a colony and the consequent swarming and their reduction have the opposite effect. Under conditions of reduced secretion the workers starts ignoring the old queen and prepare for supersedure. In other words workers recognize their queen with the help of the odours. A young queen rich in pheromones leaving with a swarm attracts workers and stabilizes the swarm and induces the workers to forage around. So strong is the odour that a glove with which a queen has been held, in itself, becomes attractive to the workers. So long as the queen odour is present, the workers move around on that object. If a queen is caged and is taken away to be hived at some distance the bees will follow the cage and the person who carries it. While catching and controlling a swarm it is very helpful to search for the queen and cage it.

The queen pheromones are strong sex attractants. It is through their action that the young virgin queens as they fly out attract drones for mating. The strong odour is perceived by the drones with the help of antennae. So long as the queen keeps flying she may go on mating with drones one after the other, with an average of 8. The drones are driven to madness by the sex attractants up to a distance of 60 metres (Butler and Simpson, 1967, Butler *et al.*, 1967). Eventually the queen returns to the hive, leaving the other drones flying around.

The queen substance or the pheromones are known to inhibit ovary development in the worker bees. The queen smears mandibular secretions on her body probably with the help of her legs. The workers keep licking and grooming the queen and some of the queen substance is passed on to workers who further exchange food and these chemicals with other bees. The pheromones thus work as a pass word and bees spread the "news" that queen is well (Winston, 1987; Crane, 1990). If a queen dies accidentally or is removed and a new queen is not formed in the colony; some of the workers develop ovaries and they lay unfertilized eggs. It is indicated that bees become laying workers when they do not receive sufficient quantities of chemicals from a queen; probably the queen substance works on the hormonal system of the workers and suppresses the development of ovaries as long as the queen is vigorous.

More recent work indicates that the Queen Mandibular Phenomone (QMP) is the prime factor in queen substance that regulates the physiology and attraction behaviour of worker bees. However, when body wash of queen is added to it, the workers are attracted even more vigorously, indicating the occurrence of some other components also. At present it is known that QMP consists of five chemicals : 9-keto-2(E)-decenoic acid; (R)-and (S)-9 hydroxy 2(E)-decenoic acids; methyl p-hydroxybenzoate ; 4-hydroxy-3-methoxyphenylethnol. Although synthetic QMP results in the same pheromonal response it does not duplicate the attractiveness of a live queen. On the practical side the synthetic product in spray form has found its application in the field for attracting foraging *A.m.* honeybees on valuable crops that need thorough (cross) pollination (Kealing *et al.*, 1999).

5.3.2 Other Pheromones

The research is not conclusive but there are indications that certain other unidentified chemicals from the abdominal tergite glands and footprint pheromone (queen's tarsal glands) play important roles in conjunction with the queen substance. Their combined effect probably inhibits queen rearing and swarming.

The workers on their VI tergite of the abdomen have the Nasonov (Nassanoff) or the scent gland which plays a critical role in worker's behaviour. The Nasonov scent is a mixture of several chemicals identified as geraniol, nerolic acid, (E)—citral, (Z)—citral, (E, E)—farnesol, nerol (Pickett, *et al.*, 1980). The pheromone 9HDA stimulates the secretion from the gland and the workers are seen releasing it by fanning at the swarm cluster and at the nest entrance. The workers take up "orientation" or "scent" fanning by slanting their abdomen and the tip upwards (Fig. 18). The gland is opened, exposing a moist membrane that carries the volatile scent. The role of this chemical is to stabilize the swarm and to stimulate the foraging activity. Thus the queen and the workers jointly make an effort to settle down after swarming (Butler and Calam, 1969; Morse and Boch, 1971).

Odours from the comb are volatile in nature and they seem to stimulate the workers to increase the rate of nectar storage. Odours from the developing brood also affect the workers; they prevent ovary development and stimulate foraging. Studies have shown that both sealed and unsealed brood influence the workers in certain ways to collect more pollen which is the protein source for the colony.

The workers recognize their own queen and other workers of the same or another colony from the pheromones. It is known for a long time that a foreign queen if introduced into a colony is killed immediately hence a new queen is introduced gradually by confining it in a queen cage till the colony gets used to its odour. Similarly the guard bees at the entrance let their own bees in but check and stop bees from other colonies. Here also the pheromones play their role. The smell of the flowers on which bees of that colony are foraging is also instrumental in recognizing their own kin. In brief, bees recognize the inmates of their colony from the pheromones (mixture of chemicals), from the odour of combs and brood, and from the external odours acquired from flowers etc. It is altogether a new field of apiculture science and a great deal of work is required to be undertaken before a good understanding of the role of chemicals in bee behaviour is achieved.

5.4 OTHER FUNCTIONAL BEHAVIOURS

There are a number of tasks bees handle in the hive. Their behaviour is associated with specific functions such as cleaning, mending, polishing, fanning, gathering and storing pollen, gathering and ripening nectar, storing and sealing honey, propolis gathering, swarming and nesting etc. The elaborate system of foraging and swarming outside the hive will be discussed separately.

5.4.1 Fanning

Bees create circulating air currents inside the hive by fanning their wings (**Plate 5*a***). A group of bees numbering from a few individuals to a few hundred, stand outside the entrance facing their head towards the rear end of the hive (*Apis cerana indica* in the opposite direction). By fanning with their wings they push out air from one side of the entrance. Quite often another group sits on the inside of the entrance facing outwards and by fanning push out air from tne quiet end of the entrance thus circulating air throughout the hive; going in from half the entrance and going out from the other half. The current so created is very effective in evaporating water from the nectar. In hot weather bees place droplets of water all over the comb surface and the circulating air produce air-conditioning and lowers the temperature inside. In hot and humid season air humidity inside the hive is reduced by the same process. The air currents also circulate pheromones in the colony.

Vigorous fanning activity is observed in the summer and early autumn whenever it is hot. When ambient temperature is close to 40°C wax becomes soft and honey starts oozing out of the combs. Much effort of the bees is needed to lower temperature in the hive. Bees are needed for this work in large numbers and, therefore a strong colony is in a better position to control the environment. In cold weather a honeybee regulates its body temperature by shivering, heat conservation, foraging etc. (Heinrich, 1996).

"Orientation" or 'scent' fanning is another bee activity noticed at the entrance of hive or at a newly formed nest after swarming. A bee while fanning raises its abdomen upwards and exposes the Nasonov scent gland located on the dorsal side of segment VI (Fig. 18). The moist secretion volatilizes rapidly and the odour is very attractive to the bees. It helps in orientation of the bees to settle down after swarming. The pheromone has a mixture of chemicals, *viz.*, geraniol, citrol, nerol, farnesol, nerolic, geranic acid and perhaps some others.

5.4.2 Rocking Behaviour

This activity is seen in the evening on the front wall surface of the hive. The bees stand on meso- and meta-legs with their head bent downwards facing the entrance. With the help of their mandibles and bent tarsi of the front legs bees scrape the surface with quick movements, as if they are cleaning and polishing the surface. Their antennae constantly touch the surface. The material that accumulates on mandibles and tarsi is cleaned up by the bees. The scrubbing and polishing activity is generally performed by young bees both on the inside and outside surface of the hive.

5.4.3 Nest Cleaning Activity

More than 90 per cent of the old workers die in the field while foraging, but some do reach the hive exhausted and they die inside. Soon the bees remove the dead sisters and carry them more than hundred metres away. The dead brood inside the cells is also removed likewise. This way the area around the hive is kept clean and no scavengers gather there. In the post-monsoon period moldy growth is observed on combs, particularly in weak colonies. When such combs are given to strong colonies they try to clean them up. Any foreign matter or dead bodies of predator wasps are also removed and thrown outside at some distance. More than one bee may join hands.

5.4.4 Working Habits of Field Bees

The field bees once they start foraging for water, pollen, or nectar, stick to their jobs for many days and there is a restricted change over of duties (Figs. 18, 19). Some pollen collecting bees do shift to nectar collection but rarely it is the other way round. Once bees start foraging on a crop they continue it for a number of days, up to 3 weeks. Bees have memory for the timings of the day also; pollen and nectar available at a certain time of the day are visited by the bees during those hours only. In other words it is not a random search in the field for pollen and nectar. Although bees on a trip do visit several plants, differing in odour and structure of flowers, they generally forage on a single species of crop or tree and also confine their activity to a restricted area. However, Ribbands (1949) observed that, bees do not start or finish their trip on any fixed point.

The foraging behaviour is not fixed but is adaptable; bees exercise choice and they keep comparing the present availability with memory of the past and within this comparison they remain attached to a particular area in a given crop field. The size of the area will naturally depend upon the abundance of flowers; it may be 3-5 metre square or larger if the flowers are scanty towards the end of the crop. The speed at which bees visit flowers depends on the amount of nectar and pollen present; in general the pollen trips are shorter than the nectar trips.

Active foraging starts at a minimum of 13°C and activity becomes faster at higher temperatures and around 43°C it becomes too hot for them to forage. The collection of water continues even at higher temperature. Bee foraging in the field is directed by their communication dances which establishes a link between the hive and the field from where food comes. It is through this remarkable behaviour of the bees that von Frisch (1967a, b) solved the mystery of their communication and even called it their "language" which word has been criticized lately in the scientific circles (Wenner & Wells, 1990 ; Robinson, 1986; Gould, 1976, 1982).

5.5 COMMUNICATION DANCES

If honeybees were to remain confined in their hive, not unlike a housefly breeding in a dung heap, they would not have to fly long distances for food and then come back to their own house. Since bees are social insects and work together as a community they not only bring food for all but have to ensure they do not get lost. Therefore as individuals they cannot survive for long. They are evolved with a system of communication which means transfer of stimuli that induce behavioural and physiological responses in the individuals that receive them. It is not intelligence or human-like understanding of a language.

It was observed many decades ago that a dish full of sugar solution placed a kilometer away from a hive when visited and tasted by a honeybee, attracted dozens of followers from the same hive. In other words the scout bees were able to communicate information of the dish and its location to other bees in the hive. Subsequent studies proved that bees achieve this by an ingenious and elaborate method of communication. Since the environment inside a hive is dark it is perhaps not the vision but other stimuli like odour, touch, sound, etc. that play the crucial role. von Frisch (1967) described in his book how he solved the mystery of communication as also the colour perception and for this contribution he shared a Nobel Prize in 1973. He observed the behaviour through a glass walled hive and noticed bees dancing and communicating the information to others. The experiments were conducted by offering flavoured sugar solution at various distances and marking the bees that visited or those that were recruited later on. The literature on communication dances has been reviewed by many authors (von Frisch 1967a, b; Lindauer, 1971; Michener, 1974; Gould, 1976, Dyer and Gould, 1983; Gary, 1967, 1992).

The honeybee scouts fly around with hive as the centre, in search of flowers of plants rich in nectar and pollen. We as humans can know the seasonal cycle of various flowers in a locality but the bees do not know this and they have to search all the time, evermore for richer sources. A scout bee finding a rich source of nectar brings back a sample and gives its taste (odour) to other bees. They make three main type of dances with some intermediary formations to convey this information. The three main dances are the round dance, the wagtail dance and the dorsoventral abdominal vibrating dance (DVAV). In between the round dance and the wagtail dance there are transition or the sickle-shaped dances. These dances are performed on the surface of a comb inside a hive or on the surface of a newly emerged swarm temporarily settled on a tree or some other suitable place.

5.5.1 Round Dance

The round dance is performed to inform workers that there is a source of food (nectar) within a distance of 15 metres from the hive. In this simple dance the precise distance or direction of the food is not conveyed but some sounds are produced (Fig. 16). The bee that brings the food sample gives its taste to others and runs in quick short steps in narrow circles on the comb. She makes one or two circles clockwise and then anticlockwise and continues the dance for several seconds, may be up to one minute. After making about 20 circles she may again exchange food with more bees and resume the dance. She then leaves the hive for another foraging trip. During the dance the followers touch the performer with their antennae and obtain information on the odour of the nectar. They clean themselves, eat some honey for energy and leave the hive in search of food in their vicinity. These recruits fly around in widening circles and locate the source of nectar on which they start foraging.

The richness of the food source (*i.e.*, concentration of sugar) is also conveyed to the hive bees by making dance for a longer period and with a greater "vigour", with more pronounced abdominal vibrations. As a result of the vigorous dance a larger number of bees are recruited. The round dance is also performed by the pollen foragers who impart odour of the flower to the recruits. The hive bees reach the pollen the same way as they find the nectar.

5.5.2 Transition Dances

When the distance of food source ranges between 15 and 100 meters at first a simple crescent or sickle dance is performed which changes to some transitional forms and finally near 100 meters to a

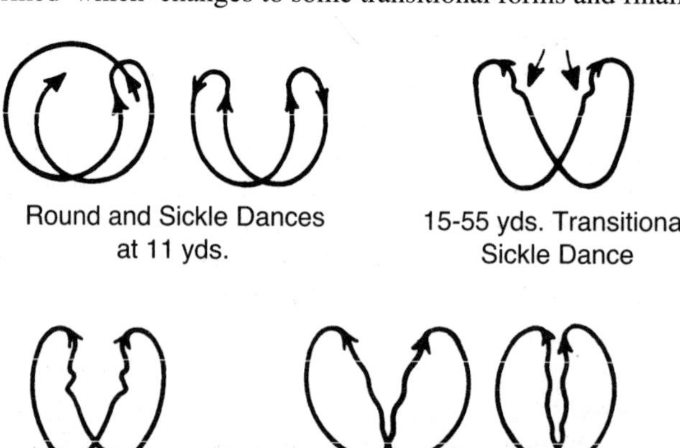

Fig. 16. Round, sickle and figure-of-eight dances (after Ribbands).

dance that resembles two half moons placed side by side. In the sickle or the crescent dance a bee starting from a point moves upwards and outwards along a convex course and reaching the top turns sharply and travels along a concave course reaching the starting point. Then she travels on the opposite side forming a mirror image of the first crescent, with the upper ends wide apart. The upper ends come closer as the distance of food source from the hive increases. Approaching a point 50 meters from the hive the crescents become quite broad, so that the entire course appears like a twisted figure eight. Near 100 meters the crescents become half moons and the upper ends lie close to each other. Now the concave runs are two parallel straight lines on which the dancer moves its abdomen from side to side. The transition from round dance to sickle or crescent dance and then figure 8 dance takes place gradually and the dancer is able to convey to the recruits the taste of the food and the distance at which it is located.

When the location of food is beyond 100 meters from the hive the dancer-changes its pattern further and wags its abdomen from side to side. It is wagtail dance and is perfect in conveying not only the distance, but also the direction, and the exact location, correct within a few meters.

5.5.3 Wag-tail Dance

The path of this dance is a circle made in two halves with the dividing line in the middle. The bee starting from a point runs in a small half circle turns sharply and goes in a straight-line reaching the original point, turns to the opposite direction making a half circle and then running along the original straight line. Thus a circle is complete as if two halves are joined together. Along the middle straight line portion of the dance a bee moves her body from side to side with a vigorous wiggling motion, hence this tailwagging dance is called a wag-tail dance. Nearer to 100 metres the transition or the figure 8 dances are performed with wiggling motion and at greater distances the dance adopts the shape of a complete circle, with a division in the middle (Fig. 16).

The wiggling is at the rate of 13-15 times per second and the forager bee stops now and then to distribute food from its honey stomach to the followers who extend their antennae to perceive the odour. A buzzing sound (inaudible to human ear) is given by the bee which is due to muscular and skeletal vibrations. The workers that follow the dancer also make a squeaking sound of 0.1-0.2 second duration. It is called "begging signal" and on hearing it the dancer halts and exchanges food with the bee that begs. Both the nectar and the pollen collectors perform this dance.

Waggle dance is performed to communicate information about the distance, direction and quality of food distances greater than 100 m from the hive (Fig. 17). These are conveyed in a number of ways, namely, the length of the straight run, the duration of wiggling and buzzing, the number of circles made per unit time, and the amount of energy (honey) spent to reach a certain distance. The distance of the food source is conveyed by the dance tempo which slows down with greater distance; the longer the time spent on the straight run, farther away is the distance of the source. The communication of the distance is not perfect and the recruits fly within 2-10% of the actual spot, and with that approximation the relation between speed and distance (von Frisch 1967) are shown (Table 4).

The comparative round and wag-tail dances of the two sister honeybees, *Apis mellifera* and *Apis cerana* have been studied by Lindaur (1956) and Goyal and Atwal (1976). In a round dance (including sickle variation), *A.m.* bee did not repeat turnings in the same direction more than twice but A.c. turned in the same direction a number of times; moreover the former performed round dance when the food was at a distance of 0.5-27.0 m and the latter between 0.3 to 7.0 m. (Goyal and Atwal, 1976). Also, more training acts were needed by *cerana* to remember a feeding station than *A.m.* In a round dance there is indication of direction at the turning points and *A.m.* conducts this dance mixed with intermediary versions upto 100 m, beyond which the waggle dance is performed (Lindaur, 1956). It

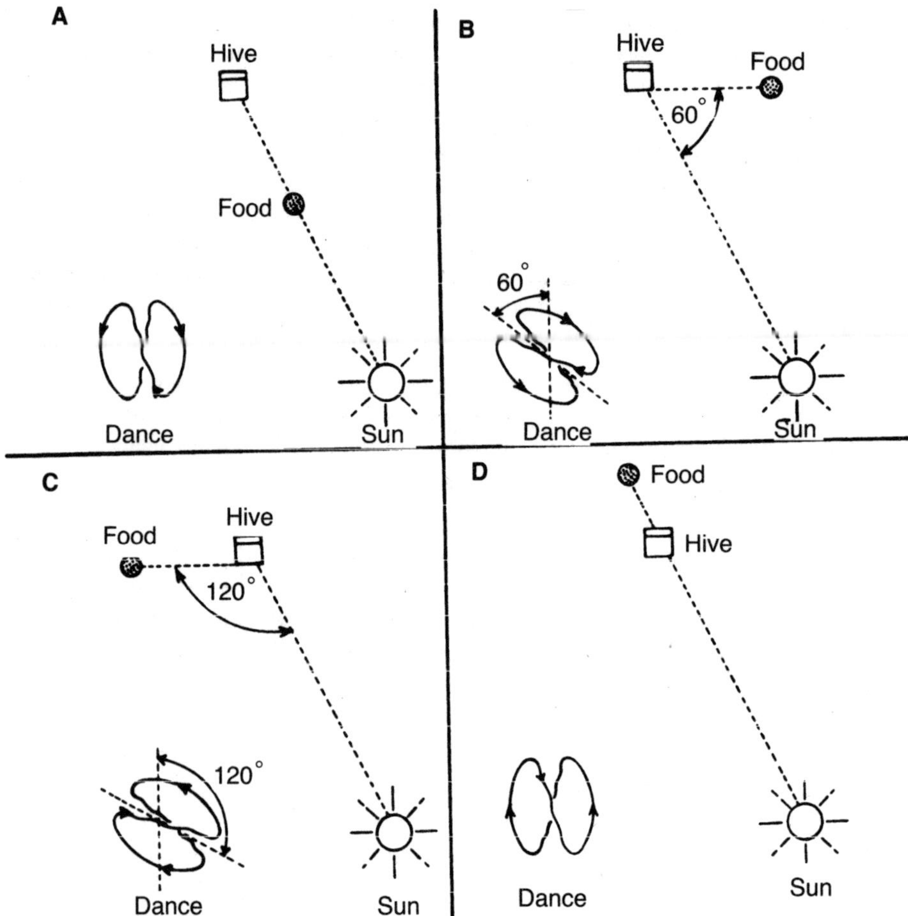

Fig. 17. Direction of dance on a vertical comb in relation to the direction of feeding place (after Ribbands).

TABLE 4. *Food Distance Communicated by Dancing Bees*

Distance (metres)	*Straight runs of dance per 15 seconds*
100	9-10
600	7
1000	4
6000	2

was, observed by Koltermann (1973) that speed of training by the two species varied with different scents and they had their own preferences.

It has been observed that within *A.m.* the dance tempo and distances conveyed vary in different races. According to Winston (1987) there are three basic dances : a round dance, a waggle dance, and a dorso-ventral abdominal vibrating dance (DVAV). In round dance bees recruit others to new resources. Waggle dance is for recruitment and communication of information on the location of new resources and DVAV dance is for stimulation of activities like foraging, controlling queen emergence, timing of swarming etc.

Apart from the distance the bee must also "know" the direction in which food is located. It is conveyed by the direction of the straight run portion of the dance in relation to the line of gravity on surface of the upright frame in a hive. If a frame is kept horizontal with upper bar facing the sun and bees are watched dancing, their straight run being directly towards the food source, the bees immediately fly in that direction. On a vertical frame, the horizontal direction is indicated by transposing the solar angle, into the gravitational angle. If the food is in exact line of the sun the straight run will be vertical, towards the top of the gravity line on the frame. On the other hand if food is in opposite direction to the position of sun the straight run will be towards the bottom of the frame. If food is 60 degree to the right or to the left of the sun, the straight run of the dance will be at 60 degree to the right or left of the vertical gravity line, respectively (Johanson, 1967; von Frisch, 1967a, b; Wenner, 1967).

As the bees go out for field work they can be delayed and in the meantime the position of the sun in the sky might have changed. For returning to their own hive the bees can compensate for this change and if clouds have appeared they can perceive the exact position of the sun. For the calculation of correct distance and direction in heavy wind or a mountain obstacle the bees can supply correction factor for their orientation. Even so there are some errors up to 10%, the older bees being more accurate than the younger ones. For that reason the young recruits attend six or seven dances before flying out and even then they have to make 2-5 flights to locate the foraging site.

Dances in the hive can also indicate the location of water and perhaps also of propolis. When the dances are performed on the surface of a new swarm the location of a potential new nest site can also be indicated.

5.5.4 DVAV or "Joy Dance"

Like the dances described earlier the Dorso-Ventral-Abdominal-Vibration (DVAV) dance is also of the field bees. In this dance a bee places front legs on some part of the body of another bee or the queen and makes five or six up and down shaking movements with the abdomen; there is also a slight swinging forward and backward. The performance is repeated with a series of bees on various locations in the hive. This dance may also be performed on a sealed queen cell; in a queenless colony when a virgin queen emerges a number of bees start performing. The total performers may number over 200 in one hour. The youngest worker seen to perform by an observer was 9 days old and the oldest was 151 days of age (Milum, 1955; Fletcher, 1975).

The dance may be performed at all hours of the day and night but there is greater activity during the peak foraging. The dance also takes place throughout the year and is more frequent during the abundance of flora, but performance has been seen even in queenless colonies or those nearing starvation. It is generally believed that the dance is associated with foraging, perhaps activating it. It is also connected with queen rearing and swarming. The dancers are more active prior to the queen rearing and the mated queen is vibrated quite frequently. The dance stops suddenly a few hours before swarming as if anticipating the event. The dancers in some way influence the queen left behind and prevent her from destroying the queen cells; but when the dance stops the queen is not inhibited and may even leave with a swarm. The exact role of DVAV dance is being studied and it is believed that it is performed to regulate the daily and seasonal foraging patterns according to food fluctuations and to regulate the queen activity in relation to swarming (Schneider *et al.*, 1986).

5.5.5 Extent of Information

In the communication dances it is difficult to evaluate as to how much information is given by the dancer to the followers. It is understandable that the younger bees who have not been to the field duties yet will gather less information than the bees who have been active foragers. Even among the active foragers the transfer of information may not be perfect. In experiments with marked bees it was recorded that approximately half the followers brought the same nectar or pollen which was distributed by the dancer. Furthermore, 25% of the learners not only brought the same nectar and pollen but also danced for the same source from which they brought the foraged food.

The old experienced bees are quite selective in their choice for foraging; they congregate in certain areas of the comb where that food source is being displayed to which they are already used. Probably they do this based on the odour of the flowers they are familiar with. As against random searching the dances do infuse saving of energy of the workers in a colony. The scout bees that search new sources of nectar and pollen are an important segment in a colony population and they have the ability to return to the hive on their own without prior information or directions from the dancers. If a certain sugar concentration is available near or far the bees prefer the nearer place.

5.5.6 "Cleaning" Dances

These dances are related to body cleaning when bees enter the hive contaminated with dust, hairs, or other foreign materials. The contaminated bee is seen stamping her legs rapidly and swinging her body side to side. It raises and lowers the body and cleans the bases of wings with a middle pair of legs. The bees close to the dancer touch her with their antennae and with mandibles wide open touch under the thorax of the dancer near the wing bases. The dancer stops, spreads her wings on one side, and curves the body somewhat upwards as if to accommodate the cleaner. With shearing motion of mandibles the cleaner bee bites around the wing base, holding her antennae close to the tip of her mandibles. While working, the cleaner stands on the mid and hind legs and the front pair is held in the air. The cleaning is also done in the grooves of the thorax from the front end and at the base of the other wings by climbing over the dancer. Then she stops and another cleaner may take over with or without another dance. At a time there are about 10% general cleaners in a colony, and they may also groom the drones and other bees who may not perform the dance. In one observation a bee groomed 26 bees in 25 minutes and they were of the hive age (Milum, 1947, 1956).

5.5.7 "Alarm" Dance

When a field bee brings nectar poisoned with insecticide, the hive bees and field bees inside become excited and perform an "alarm" dance. They run in spiral or zigzag path and vigorously shake their abdomen sideways. The neighbouring bees also start dancing and flight activity is suspended. With dinitrocresol the poisoned bees died within 1-2 hours and the colony resumed normal activity in another 2 or 3 hours.

5.5.8 "Message" Dance

The dance is started by a bee on a comb when it bends its head in a peculiar manner. One or more hive bees get excited and approach her, using the antennae and front legs; and from below they pull at the joints of the middle and hind legs, touching sides of the body. The dancer opens widely her mandibles and protrudes the upper part of her tongue. As soon as a hive bee, comes for examining, the dancer unfolds the entire tongue, extends the second pair of legs and constantly cleans the tongue with her front legs, stroking from above downwards. The tongue becomes distorted and its parts spread out. The massaging bees sometimes pull at the extended mandibles and the tongue. After some minutes the "sick" bee turns normal and moves away, and cleans its antennae, eyes, hind legs, wings and tongue. This activity is observed in winter and also in autumn or spring when chilled bees come after strenuous field duties.

5.6 ORIENTATION FLIGHTS AND DRIFTING

The worker bees when young neither participate in dances nor acquire the skills to search for food in the field. They leave the hive everyday on short orientation or play-flights when they defecate and get familiarized with the landmarks but return to their own hive; some do get lost and drift away to other colonies. Each successive flight increases in duration and the area covered. Such flights are particularly noticeable in the spring on a warm sunny noon after a few days of bad weather when they had remained confined to the hive. These flights can be confused with robbing but there are no fights and the bees are seen hovering around near the hive.

When there are no significant landmarks in the apiary and the hives are in a regular pattern there is more drifting and up to 50% of the young bees of a colony may go to others; it is common among bees up to 4-5 days of age and they keep drifting between 6-11 days of age. Painting the hives with various colours cuts down drifting to a considerable extent. Keeping the entrances in different directions also cuts down drifting. Drones drift even more than the workers do. It is observed happening more from a weak to a strong colony and from queen right to queenless one.

5.7 FIELD JOURNEY

The older workers on getting information through communication reach the vicinity of the foraging site but it is not known how they actually get there or how they return to their hive. For navigation bees use a combination of senses in the ever changing environment; the senses used are visual, olfactory, and magnetic. Two important factors are known to aid the orientation process; the bees can see colours and they use two sets of mechanisms *viz.,* the primary and the secondary. The colour vision of bees is very similar to that of humans except that they cannot see red but have the additional perception of ultra-violet. While using the visual senses they make use of the location of the sun, polarization of light waves and landmarks. The position of the sun in cloudy weather is found with the ultraviolet rays that can penetrate light clouds. The bees not only locate the sun but also compensate for its daily movement in the sky. When sun is hidden and the clouds are broken the bees orient themselves with the help of polarized sun rays. If the bees were to see through a filter of polarized rays they will be totally lost when sun is hidden. It has been proven experimentally that bees in reaching their food as communicated by a dancer they can compensate for light polarization.

When neither the sun's compass position nor the light polarization pattern is visible the bees make use of a back-up mechanism in that they follow the landmarks, like a tree, a group of bushes, a tall building, a river or a hillock. This is the secondary mechanism.

Thus foraging involves not only the communication stimuli but also other senses like vision, odour, and their past experience in avoiding poisonous food and water, which makes the honeybee a remarkable creature.

The wing frequencies of *A.m.* were lower than those of *A.c.*; wing beats of workers were 235 versus 306 per second, and those of drones were 225 versus 283 per second (Goyal and Atwal, 1977). Probably size of the wing played some role. The honeybees seem to have nervous control of the indirect flight muscles (Bastian and Esch, 1970). Among the two species of honeybees *A.c.* flies unpredictably in a zig-zag manner perhaps to avoid flying enemies but *A.m.* on the other hand has a steady clumsy flight (Lindauer, 1956; Sakagami, 1960). In a field experiment when both were realised 50 m away from the colony the former returned home in 192 s, and the latter in 295 s, (Atwal and Dhaliwal, 1969).

5.8 INTERSPECIFIC RELATIONS

Successful introduction of mated *Mellifera* queens into *A.c.* colonies was first achieved in 1962 (Atwal, 1964; Atwal and Sharma, 1964, 1968). When *Mellifera* workers emerged the workers of both the species worked together, the former were seen fanning outside the entrance facing their head towards the rear of the hive and the latter away from the hive entrance, each keeping its inherited position. When mature queen cells were grafted, *Mellifera* queens hatched among Cerana workers, mated with drones of their own kind, returned to the colony and laid eggs. The experiments were repeated again successfully (Adlakha and Sharma, 1971). In a reverse effort Inoue (1962) grafted 100 *A.c.* larvae into *A.m.* colonies and obtained 3 queens only which were not accepted by the colonies from which they emerged.

Based on forced combats in a limited space it was observed that *A.m.* dominated over the giant honeybee, *Apis dorsata* (Dhaliwal and Atwal, 1969). When brood frames of one species were introduced into colony of the other the host bees removed eggs and larvae but left the sealed brood out of which adults emerged and participated in indoor and guard duties for sometime; field duties were performed by the workers predominant in number. Young bees (1-2 day old) of one species were introduced into colony of the other and were accepted on examination but older bees were rejected outright. Once the young bees were accepted they participated in social work but *A.m.* was unpredictable and at a later date they could turn against *A.c.* whereas the latter could accept the quests for prolonged periods. (Dhaliwal and Atwal, 1969, 1970). Earlier in Japan, Sakagami (1959) also observed that only limited co-existence was possible between the two species.

It was observed at three experimental stations in northern India that *A.m.* always produced more brood and gained more colony strength as compared with that of *A.c.* At Nagrota (3000 ft) the latter produced 1/2 to 3/4 brood area of the former throughout the year and, at Ludhiana in the plains, comparative performance was the same (Atwal and Sharma, 1970a, b). At Kulu (4500 ft) the brood area of *A.c.* in August-September ranged between 2,271 and 2,415 cm^2, whereas that of *A.m.* ranged between 4,742 and 4,862 cm^2 (Hameed and Adlakha, 1973).

Apis mellifera behaved aggressively not only at experimental feeding dishes but also while foraging in the field. They robbed *A.c.*, caused absconding, and drove away wild and solitary bees from barseem upto a distance of 1000 m or more (Atwal and Sharma, 1971; Dhaliwal and Atwal, 1974). There was reverse robbing in cold and in inclement weather when *A.m.* was robbed by *A.c.* and even by the dwarf bee *Apis florea* (Adlakha and Sharma; 1974; Chahal *et al.*, 1986). Interspecific relations are also discussed in Chapter 2 and under Pollination of Crops by Bees in Chapter 15.

5.9 COMMUNICATION AND SOCIAL BEHAVIOUR

5.9.1 General Account

Quite a few thousand papers have been published on the activities and behaviour of honeybees yet less than 1% has been discovered and reported by scientists so far. Such studies were difficult to make as both the European honey bee (*Apis mellifera*) and the Asian honey bee (*Apis cerana*) lived in dark hollows of rocks, trees and man made walls, or other structures. About 200 years back, the practicing beekeepers were driven by a desire to know more about them and they made peep holes to watch what the bees were doing and how they made and stored surplus honey, living in a colony where there were many thousands (upto 50,000) of mouths to feed. It was in the middle of 19th century that a proper wooden live with 9-10 vertical frames was constructed with both sides made of glass. With the help of red light focussed inside the hive, activities and functions of bees of various ages and that of the King (of Vedas) were observed. A great deal of literature has become available and on dissection and watching inside of the large mellifera inmate was found to be a female and was called a queen. Hence most of the scientific progress on the bee activities and their functions have been made on *A.m*, although some studies on Asian honey bee (*A.c*) have been made in the last 50 years in Japan, China, India and other countries using the modern techniques of observing and hearing. The two other world honey bees of Aisa, *Apis florea* (the little) and *Apis dorsata* (the Giant honeybee) lived on single combs exposed to the elements but at protected places. Their life behaviour and activities have been studied to a limited extent and that too in recent decades only. Most of the work on Asian species, was initiated in the beginning of 20th century in connection with their role in pollination of crops. In this book most of the observations and scientific accounts are based on our knowledge of *Apis mellifera*.

Inside a hive what happens is a justification of the often used expression "busy as a bee". Outside a hive bees were observed near the vicinity, or out in the field on flowers or other sources of food to feed on, water to drink, and resins (propolis) of trees to give support to their dangling combs. With the invention of many new electric and electronic instruments it became possible to study the influence of wind, temperature, gravity, ultraviolet rays and even sound on their activity in the field. The synchronization of the opening of flowers at various times (under clouds and in the sun), in a day or in various seasons were all important for the success of biological existence of bees. The human observations and electronic measurements aided the beekeepers in helping them to study the survival of bees to make them produce more and store honey.

More recently, many biochemical techniques have been developed which revolutionized the entire thinking of scientists and beekeepers as to how honey bees lived together in thousands in a confined space, lived daily life and the entire life cycle in a regularized manner.

Honeybees do have eyes to see, mouth parts to chew food, antennae to feel and smell, wings to fly out of danger. They have spiracles to breath, blood that circulate inside the body, muscles to activitate the body and a well spread nervous system inside for coordination but they do not have speech to communicate among themselves in a colony or to listen to the orders of their only queen. Man for centuries has surmised according to his own thinking that there was some kind of plan for "assigning" "duties" to various workers.

While making scientific studies on the behaviour of honey bee (*Apis mellifera*) it was found that there were two distinct sets of causes i.e. the internal and the external. The internal causes consist of age or the stage of development of the body. For example, wax for making combs is secreted when the wax glands mature. They are located on the ventral side of abdomen and open when bees are ready to collect wax. Royal jelly, a superior food is secreted by bees when salivary and other supplementary glands are fully developed. On emergence the young bees cannot fly as muscles are not properly developed. The nervous system is not developed for good coordination of body movements.

The genetic constitution of individual bees plays an important role. For example, bees, as controlled by their specific genes, would handle inside the hive certain jobs only; some would remove the cell caps, others would remove the diseased and dying larvae. The colony is thus cleaned of diseases like the American and European foulbroods. Genes determine the behaviour of bees in foraging for pollen or nectar. However, in case of a catastrophic deficiency in a hive the bees can switch to the required job or food item. Such invisible causes are sometimes mistaken by beekeepers as shades of human like intelligence.

The external factors also stimulate various responses. They include sound, light, touch, magnetic fields, etc and are perceived by thousands of specialized cells activated by sensory hairs. The rhythm of behaviour or evidence of the "internal clock" is triggered by external factors like day-light (photoperiod), temperature, humidity, weather, learning from experience, etc. Apart from the subtle effects of genes (internal) the strong effect of certain chemicals is spectacular and instantaneous. The chemicals include smoke, certain perfumes used by humans or perspiration particularly induced by fear (adrenalin effect). In studying the external factors the most important to remember is that bees do not 'know' or 'anticipate' that a man was approaching their hive with a smoker in his hand.

The importance of internal and external factors is that they govern (not the queen) the social life of the colony through communication. Some of the working systems have been studied already and new clues are being found by scientists every day.

5.9.2 Communication Through Dances

It is quite apparent that some kind of communication among the honey bees is essential for their successful social behaviour. Since food collection and feeding their young form the core of their success as a species most scientists have devoted their time on such studies. Feeding is an indoor activity whereas gathering food is an outdoor activity The former is influenced by internal factors and are discussed here.

In this context it implies that individuals in a species transfer physical and physiological stimuli that are perceived by others under specific situations. It does not imply that they are human like messages of intelligence to solicit awareness by the receiver bee as desired by a sender. However, like man honey bees are elicited by the basic stimuli such as light, chemical smell or physical stimuli. Since men live in day light, their eyesight and cognitive power is well developed. The honey bees (*A.m* and *A.c*), on the other hand, live in a dark space in the hive and they have to depend more on other senses like odour, touch, sound, etc. These senses in bees are highly developed and have a great sensitivity, including certain wave lengths of light and sound not perceived by man. Moreover, in a dark space bees have to use a combination of stimuli, such as odour, sound and tactical signals.

Food location by honeybees is fundamental to their survival, particularly due to the availability of pollen and nectar produced outside their hive and their production depending on such powerful factors as day-light, water, temperature and wind. When food becomes available in abundance time is short, so the bees have to rush to the source and bring it to their hive and feed their young. Timing of the availability and population of the mouths to feed had to coincide. One of the most remarkable example to evolution was the appearance of flowering plants (Angiosperms) and their pollinators (seed produces) appeared on the earth at about the same time. The flowers and the pollinating honeybees (or relatives) have been found in similar ancient.

The presence of flowers is one thing but the actual gathering of nectar and pollen during that particular hours of the day in a short time is another. The bees have to reach there quick as there are other competitors and, survival of the fittest was a very ancient law that worked. So how were bees to know? Who was to ring the bell? As it is known now the honey bees developed a language for communication which is generally called the "Dance Language" of foraging. The activity was unfolded by scientists by ingenues ways, experimenting for a number of years.

In earlier experiments sugar syrup flavoured with mint, lavender, clove oil, etc. was placed in dishes within 10 feet of the hive. Honey bees were attracted to the dishes and they had a good meal of sugar. As they went back to their hive, other bees were excited to know of the food with a specific flavour. On the next trip they also followed the original scouts. Noticing a regular traffic scientists marked bees at the dish and applied indelible colour (nail polish). It was noticed that the bees did not come from one colony only but from others also. The dishes were then placed in all directions at various distances from a hive. The greater the distance fewer the marked bees returned to the dishes. In other words they lost their track. Von K. Frisch an Austrian in Switzerland was the pioneer (won

Noble Prize in 1973) who realized soon, that there must be some other way of communication and that a particular aroma, though important stimulant, alone could not keep bees on the right path, particularly when the food source would be a kilometre or more distance away. He worked to solve this riddle for many years and explained the details of Dance Language in his book published in 1967. It took some time for the book to be well known in far off countries.

There are basically two dances, the Round dance and the Waggle dance, the former relates to the food sources within 10 meters of the hive and the latter is performed by bees when the food source is beyond 100 meters. There are transient sickle-shape or crescent-shape dances in between the two forms and depicting the distance of food source. In between, there are small variations depicting the distance of food sources. The round dance is in a complete circle, not like a perfect geometrical line—the bees while trying to walk on an imaginary line swerve a little while turning around to reach the spot from where they started. Since the bees following the dancer cannot see in the darkness of hive there are some sounds created associated with the disturbance of air particles and not air pressure as in the case of human sounds. Bees also touch the dancer with their antennae to smell the flavour. The dancer performs the dance for a few seconds (less than a minute) and repeats it at 2-3 places in the hive. It then goes to the entrance and flies away.

The follower bees do not exactly fly after the dancer but fly around above the hive and with the help of antennae perceive the smell of food and settle on the desired flowers or a dish to feed. They invariably go back to their own hive and in case an odd bee by a mistake goes to a nearby hive or a strangers' colony, she is welcomed and is allowed to unload the harvest from the dish. If there were rich flowers growing or cane juice in a vessel within that distance honey bees will follow the same procedure.

Earlier it was believed that the dancer does not convey any information on the direction or location of food within the short distance of 10 meters. Some recent observations indicate that sound signals and even some sort of information on direction and distance is probably given by the dancer. Similar information for longer distance is elaborated into sickle or crescent and to the wag-dance. With the invention of some suitable instrument it might be possible to trace the path of the followers as a proof of this faculty. For this reason it is not clear whether upto a distance of 100 meters the followers learn the direction or distance of food source or they search at random and matching flavours reach the food. If that be so the foraging bees must acquire proficiency in recognizing the land marks. In case of the wag dance, such aspects have been studied more thoroughly with sufficiently convincing evidence.

The names and distances depicted by various dances are based on human observations. The honey bee most probably makes these activities instinctively in a continuous process under the stimulation of external factors *viz*. light, ultra violet light, wind and the air particles, acoustics and sound waves both audible or inaudible to man, humidity and rain, gravity etc.

The waggle or wag-dance in which an informer bee exhibits a certain pattern of activity is on a comb inside the hive. The imaginary line along which a dancing bee moves forms left half of the moon (Fig. 15) and then goes right to complete the other half forming a complete circle. She dances along a straight line on which she had walked twice before, while completing the flat lines of the two half moons. On this straight line she waggles (shakes) her abdomen and other parts of body. The run relates to gravity (not the compass direction of the sun) and it corresponds to the compass direction of the food.

Apart from the direction, distance is also conveyed by the dancer by two indications; one, how vigorously or slowly the abdomen is waggled; two, the number of air borne low frequency (250-300 heuatz) sounds made by the dancer. It is postulated that the follower bees on picking these indications fly out to the fields in search of the tasted food and can locate it. If the distance is too long bees reach that vicinity and locate the source from its flavour as indicated by the dancing bees.

The number of straight runs per 15 seconds is indirectly proportionate to the distance. For example, when the distance is 100 meters the dancer makes 9-10 runs in a straight line in the middle of the circle; at 600 meter distance 7 runs are made; at distances of 1000 and 6000 meters only 4 and 2 runs respectively, are observed. In other words, greater distances of food are shown by the dancer by running very slow. By making repeated observations, the data can be tabulated as a ready recover for making field observations on bees marked in the field.

The direction in which food is located is also communicated by the dancing bee. It is performed during the straight run of the bee. The direction is shown not directly like the needle of a compass but on the vertical frame the bee runs towards the ground or away from gravity, and the intermediary directions depending upon the location of food. In doing so the bees take help from the position of the sun or, on a cloudy day, the supposed position of sun. It is known now that bees can perceive certain rays (ultraviolet) which the human eye cannot detect. Drawings in Fig. 16 show some of the locations of food as seen from the hive. Bees make use of their faculties to locate food anywhere at any time of the day. A simple situation may be explained first. Seen from a hive (and the concerned bee dancer) if food source is directly towards the sun, the bee makes a straight run upwards towards the top of the frame while making a wag line dance and reaching the desired length turns left. Making the curve of a half circle it reaches the starting point. She again goes upwards and at the end turns to the right and making a half circle comes to the starting point. It takes her a few seconds but less than a minute, and she makes 2-3 such runs. She then moves to another group of bees in other parts of the hive. It is during the straight runs of the dance that she passes all the communications' information to the field age bees that follow her and touch her with their antennae now and then. She is the single actor and less than a dozen bees follow her. The tactical information picked up by the followers has been explained above. When food is away from the sun line clockwise or anticlockwise, the dancer shifts its straight run in either direction by the same degrees of the angle. For example, seen from the hive if food is at 90° angle to the left the bee makes the run at that degree on the left of the (now imaginary) original straight run described above. Thus ultimately, when the food as seen from the hive, is located on the opposite side of the sun the straight run will be towards the bottom of the frame (Von Frisch, 1971).

Communication for pollen is similar to that of nectar and both odour and flavour play an important role. Some scientists have even tried to prove experimentally that odour and not the dances play a crucial role. Both schools of thought have expressed points in their favour. In literature it is called "the Dance Language Controversy". (Gould and Gould, 1976, 1988 ; Wenner and Wells, 1990).

It is not clear so far how such complex information as altitude (a tall tree vs. a field crop), colour of flower (so important), and fragrance (of nectar and pollen) can be conveyed so effectively in the darkness of a hive. Probably Johanstons' organ (located on two sides of the 2nd segment of an antenna) is sensitive to movement of air particles in the hive. For learning about the colour of flowers they have to have a sort of computerized prism with memory, to record and later match the wavelengths in a field. Owing to the gaps in our knowledge such complicated explanations are thought of which may or may not turn out to be right. However, it is certain that followers do acquire certain information skills from the dancers with the help of which they become far better gatherers of nectar and pollen than the individual bee which might have to scout around in fields, to determine the quality and then bring that food. There must be a system of communication by which a large number of bees in a short time gather the same quality of food in their hive.

5.9.3 *Apis melligera* Queen's Introduction in a Colony of *Apis cerana*

Here a successful experiment carried out in mid-sixties in Punjab is of relevance. Fertilized *Apis mellifera* queens were imported from abroad and were released by a new technique into dequeened colonies of *Apis cerana*. The eggs laid by them in the Cerana size cells were cared for by the host workers, and were brought up as foster adults. They were fed, moved around in the colony, as if they were the same species but their colour was distinct. They went out for play flight and returned to the hive. Even after they were more than 20 days old they did not bring back food and merely fed on reserves of the hive. As the number of mellifera neared 50% or more they started foraging. Probably their scouts while flying around brought food samples and passed on information to other workers of their kind who started performing field duties. It is indicated that mellifera workers eventually by chance, detected in a field the same fragrance on which they were fed and amongst which they grew. On returning to the hive they instinctively or due to the effect of their own genes, they performed the dance of mellifera and communicated with all others of the same species. The observations indicate the following (Atwal and Sharma 1967, 1969).

1. *Apis cerana* workers were tolerant to the continued presence of *A.m*. worker till they themselves died of old age.
2. They could not teach the mellifera workers their own dialect of the wag tail dance although they had been hatched and matured in the same colony.
3. The *Apis mellifera* workers reproduced at maturity the genetic memory of their own way of dances, probably stimulated by fragrance of the food on flowers now visited by them.
4. Mellifera workers did not tolerate Cerana workers living among them and killed them, sooner or later if introduced physically. In an experiment they survived no more than 12 days. The complexity of inter specific exchange of language is yet to be solved even though some ideas have been floated with or without sufficient scientific proof (Su *et al*., 2008).

There is another difficulty in understanding the immediate value of communication to the followers. Even if there is a surety that a message has been received by a follower, how soon the response must occur? In the naive bee the response is generally not immediate. It may take hours or a day or more before sufficient stimulation "accumulates" in the nervous system of the receiving bee to trigger the first foraging trip (Von Frisch, 1967; Lindaur, 1971, 1978; Wenner, 1964). The richest food sources cause the greatest stimulation of the forager and is expressed by higher number, greater vigour and longer duration of dances, but they may take time to make a decision (Gary 1962). However, in the above experiment with a colony with two species of workers the duration was so prolonged and persistently observed in six colonies sequentially that such subtle differences in delayed action cannot be compared. The mixed species phenomena of differential communication has to be explained by some factors. More recent researches indicate that octopamine neurotransmitter integrates communication system of bee dances. Thus there is neurochemical effect on the bee dances and communication for foraging (Barron *et al*., 2007; Michelson *et al*., 1986).

Generally it is accepted that honey bees do have a very effective communication system. Most apiculturists agree that a multiple system has evolved and various stimuli like, chemical, acoustic, factual have utility according to the variable conditions of foraging in the field. They might have exerted selective pressure during the evolution of bees, Genetic variation probably predisposes, individual bees to have differential sensitivity to the difference as proven by the finding that honey bee workers (*A.m.*) are born with the instinct to perform, of themselves, specific cleaning jobs in a colony (Rothenbuhler 1964). It is believed that the instinct of doing a specific job is the expression of a gene with which the worker is born.

In addition to dances for the communication of forage location and gathering, there are other dances which fulfil needs of the colony, such as, cleaning dance, joy dance, message dance, etc.

5.9.4 Cleaning Dance

When foraging honey bees fly out they collect dust and other particles by the flying movements and even by the static effects of the atmosphere. On return to the hive they help each other in pairs for cleaning and being cleaned. They act as cleaners and dancers. The dust particles on the body of worker bees basically stimulate the 'cleaning dance'. A bee in need rapidly stamps her legs and swings her body from side to side and also raises or lowers the body and cleans it near the base of wings using the middle pair of legs. Such shaking of the body may be observed at any time during the year. A bee close to the dancer touches her with antennae and begins to clean her. The cleaner spreads its mandibles open and touches base of the wings in thorax region. On this touch the dancer bee stops the general dance and cooperates with the cleaner, bends its abdomen and curves the body sideways and upwards to accommodate the cleaner and let her do the job. The cleaner works vigorously, keeping antennae close to the mandibles and working as if cutting something. The bases of the wings are then made to look clean. In this process the cleaner takes moments of rest by standing on the hind pair of legs and keeping the front pair in the air. In this position she works the mandibles in air as if chewing something she found from the body of the other. The tip of antennae keeps touching the mandibles as if to identify the odour of the dirty material. With the mandibles working the same way cleaner moves from hind end of thorax towards head, and continuous the cleaning process on all sides including the other pair of wings. In order to do a thorough job she may even climb on the other bee and clean her up well before stopping and quitting that work.

Sometimes there is a general cleaning activity in the hive without a dancing solicitation. It takes about one minute to clean a bee or a drone if he is in the crowd (Milum, 1956).

5.9.5 DVAV or Joy Dance

The Dorso-Ventral-Abdominal-Vibration or the Joy dance is performed in a colony by workers who are older than 10 days and go on upto their maturity of life, or even at the age of 20 weeks (Haydak, 1929; Milum, 1955). A dancing bee places front legs on any part of the body of another bee and makes characteristic body movements. It makes five or six shaking movements of abdomen up and downwards, while swinging forward and backwards. The dancer then moves away to other parts of the hive touching many bees she meets on her way. Some bees are seen dancing on a sealed queen cell. In a queenless colony when a virgin queen emerges from her cell this dance is performed simultaneously by a number of workers, expressing a joy all over.

The dance is performed at all hours of day or night and practically all seasons, even in queenless and starving colonies. The dance may be performed on a queen or an empty cell, a virgin queen before she goes out for copulation or even if the colony is ready for swarming. It is believed that dancers in some way exert influence on foraging behaviour. There is no indication as yet that any pheromones are involved in this strange activity.

5.9.6 Massage Dance

The purpose of this dance is also not well understood. The activity starts when a bee on a comb holds its head characteristically. Many bees near her get excited and as if to investigate with heir antennae and forelegs. The bee shows so much interest, has her mandibles extended and tongue protruded out. The bees surround to her all over, above, underneath, on the sides. They touch her from underneath with the antennae, mandibles and forelegs. As if polluted with something they clean their own antennae. The subject bee when approached from the front side turns her head towards investigator. With that action she takes out the entire tongue, sits on her middle legs and the front legs are used to clean the tongue herself.

5.9.7 Alarm Dances

Certain disturbances have been observed among the hive bees when some kind of fear or alarm is spread among them. There may be some kind of poison in the atmosphere which tints the nectar of flowers brought in by foragers or a newly sprayed insecticide gives an alarm. Von Frisch (1923) observed that disturbed bees moved around in circles or zig-zags and vigorously shake their abdomen sideways. Schneider (1949) called and described these disturbances as 'Alarm Dances'. He mentioned that flight activity of the bees stopped completely and other bees nearby began to respond to the dancers. Some scientists have also observed similar dances performed by the ordinary nectar gatherers. It is likely some nectar odours illicit the same response. In any case, these dances and behaviour studies need more precise observations backed by statistical data. It has now been determined that alarm dances are also a significant expression of a string pheromone. It will be discussed in detail further on.

5.10 COMMUNICATION THROUGH PHEROMONES

5.10.1 Introduction

In the animal kingdom a large number within a species communicate with one another in one way or the other. Daytime they can see and meet of their own kind and night time they glow (cold light). Some use acoustics and make specific animal calls to meet in a group or in locating a mate. The species that live in groups or have a social life go by their body smells. It is most highly developed among those who live in societies or colonies such as honeybees, termites and ants. Invariably they secrete the chemical smells from a number of glands spread in various parts of the body. Many intra-specific behaviours and that of individuals of the same species are the results of chemical stimuli. These activities include sexual attraction, dispersion, aggregation, trail following are triggered by volatile compounds. The individual that secretes a chemical is called an emitter and the one stimulated is called receiver and the chemicals as a whole are called Pheromones which are a long-chain fatty acids. Among the highly social honeybees the pheromone glands are most developed and their effects on other honeybees have been studied quite thoroughly. The characteristic behaviour of queen, workers, drones are the result of specific stimuli of these chemicals. Some signals regulate the social behaviour of bees in a colony so that they work as a group for their common good and hence biological survival. The honeybees use acoustic and visual signals also but it has been found that pheromones play most crucial role and the species have evolved practically a "Chemical Language".

Pheromones are distinct from hormones which are the internal chemicals affecting the individual that secretes them. Pheromones and the individuals releasing them affect others of the same species. A pheromone is secreted by an internal gland and it may consist of a single chemical or a mixture of more than one. All glandular secretions that open on the outside are called semiochemicals, and are utilized for communication between individuals of the same species. The pheromones as a group are such secretions. The pheromone chemicals are used by individuals with intent and with a message as a language which the receiver individuals of the same species only can interpret or decipher the code rapidly. The queen can make known quickly "I am the queen", "Come to me", etc. There are some chemicals that are used for communication among individuals of different species and are called allomones, but those are not used as directed messages, hence are distinct from pheromones, the true components of a language.

In honey bees, as in other insects, antennae are the primary olfactory organs and on their surface there are specialized cells, the chemoreceptors that are sensitive to smell.

Recent advances in the laboratory techniques of analytical chemistry have made it possible to identify chemicals by which the honey bees communicate among themselves and consequently regulate their social behaviour. The various pheromone mixtures are categorized according to their action and the behaviour elicited. The odour pheromones are often volatile and their effect is perceived by chemoreceptors on antenna with a fast speed. The trigger is caused immediately through nervous system by the released pheromones. Other pheromones are much less volatile and are generally contacted through oral responses by chemoreceptors located on the mouth parts. They belong to the category of Contact Pheromones. The rapid behaviour responses are made by the Release Pheromones and they form a greater portion of the honey bee pheromones and cause immediate responses like sex attraction.

Another group of pheromones are the Primer Pheromones. They prime a condition in individuals to exhibit an altered behaviour at some future time, not showing any immediate altercation of behaviour. The primer pheromones physiologically alter the reproductive or hormonal (endocrine) behaviour and at a later date certain stimuli may so affect the bee that its behaviour is changed entirely. For example, a normal queen with the help of mandibular hormone can influence the worker bees to turn the female reproductive organs non-functional. A foraging bee while working in a clover flower trips the stamen and gets hit. She learns to avoid this unpleasant experience by collecting nectar from the sides, rather than from the front. All the pheromones, whether releasers or primers, are made available in a mixture of component chemicals, hence a particular pheromone is seen to stimulate to perform multiple functions as if those are functionally related. Apart from pheromones and allomones there are some other glandular secretions.

5.10.2 Pheromone Glands

Only those pheromone glands are discussed here which open on the outside and are exocrine functionally. The chemical behaviour of honeybees is mostly influenced by abdominal glands although some (though important) are also found in head and thorax. From the point of view of bohaviour functions the most significant are found either in queen or in workers.

5.10.2(a) The Mandibular Gland Pheromones

They are the dominant pheromones of the queen bee and also of the worker bees. The glands are found at the base of a pair of hard cutting mandibles of workers and the queen. Their ducts that carry secretions of the glands open outside in the head region.

Mandibular Queen Pheromone

It was identified a number of years ago and was originally called "Queen Substance or Queen Retinue Pheromone". A queen bee produces both the primer and the releaser pheromones. They have a variety of functions, some of which have been scaled down to single compounds. The queen pheromone is the primary source of queen's authority. It is made up of eight components. One of the main component is 9-oxo-2-decenoic acid (9-ODA) which has a powerful odour. It attracts bee drones during the mating flights of the queen. It attracts workers who form a retinue of the queen. The reproductive role of the workers is retarded and in behaviour they submit to the needs of the queen and the entire colony.

In the first studies it was determined at what age virgin queens mated. It was also determined that newly emerged virgin queens were ignored by the workers but after eight days they became strongly attractive to workers, even though not mated as yet. It was then established that after mating, the queens had to wait for one year, before they really attracted the workers. More recent studies established that there are four volatile components of the queen pheromone 9-ODA which are most attractive. (Butler and Free, 1952, 1961; Collins and Blum, 1982)

The queen mandibular pheromone also plays an important role in regulating the movement, cohesion, and stability of the swarm. With the help of 9-ODA workers in a swarm ensure that it is queen right. This pheromone works more effectively in the presence of Nasonov Gland pheromone. It is a very strong sex attractant and drones go mad after a flying virgin queen. The virgin queens generally fly around with their retinue of workers. The attractant is not required for a lonely flight for mating but is a general attractant for all mating flights. It has been proven so far that a experienced queen may mate 13 times (Gaury, 1962; Kaminski *et al.*, 1990; Katzavs Gozansky, 2006).

QMP (queen substance) in general maintains stability and regulates the functioning of workers for cleaning, feeding, foraging etc, for the common good of the colony (Vaitkviciene *et al.*, 2006). The mandibular gland secretion is more active in mated queens than in virgin or superceded queens. It has been determined that 9-ODA inhibits ovarian development in workers and restricts the rearing of more queens by workers and operates in conjunction with another component (E) – 9-hydroxy-2-decenoic acid (9-HAD).

Although queen pheromone suppresses the development of ovaries in workers they keep the inherent power genetically. Thus in a queenless colony the laying workers develop the power to secrete the same pheromone with the same power as possessed by queen mandibular glands. Hence the secretion powers of the two castes are not fixed class specifically but are inherently interchangeable physiologically. Glands of workers in a queen right colony show only synthesized hydrocarbons which are a typical worker compounds. In a queenless colony the workers produce de-novo esters which are typical queen compounds. Such workers also lay eggs, typical of the queen function (Hafetz; *et al.*, 2009).

Independent researches have shown that in the evolution of caste system in social insects there is an indication of 'Arms Race' between queen and workers. Queens have a multi-pheromone multi-glandular system by which workers are normally inhibited to produce progeny but have the plasticity to produce multiple pheromones when necessity arises i.e. to bypass queen's desire to produce sons only. However, workers can produce nephews only as they are born fatherless. This is the biological conflict. As a result of this phenomenon queens can produce more homogeneous progeny whereas the male nephews of workers could produce diversified progeny (Katzavs Gozansky, 2006).

5.10.2(b) Worker Gland Pheromones

These glands become active with the production of brood food, or commonly known as royal jelly, and is always stored in cells as a food reserve, particularly when brood rearing is in progress in a colony. The main component of brood food is 10-hydroxy-(E)-2-decenoic acid (a protein). Worker bees when newly emerged have small quantities of this component but in the nursing stage secretion becomes the maximum. Some simple fatty acids like hexanoic, octanoic and others are secreted by the same glands and act as preservatives in the stored royal jelly. When bees become more than two weeks old they take up duties of guards or foraging in the field and, they produce in the same glands a smelly compound, 2-heptanone (2-HP). The secretion has the smell of strong cheese (like Blue Cheese). It acts as a repellent, that is, a releaser of alarm behaviour. The excited workers sense danger and rush to the emission source. They also secrete the alarm pheromone, thus magnifying the level of danger. The alarm behaviour signifies agitated bees, their forelegs raised off the working surface, antennae projected forward and wings and mandibles spread. They move towards the source of the odour, whether emanating from the hive or outside. When the target is located some of the bees intensify odour and the stingers among them attack the object with their sting, indicating a defensive behaviour. The alarm and defense are two separate behaviours elicited by two pheromones, the former by 2-HP and the latter by (Z)-11-eicosenol isopoenyl acetate (IPA).

The stinger bee performs the ultimate act and in this process gives her life, also spreading the news around. The stinger apparatus has glands at the base and about 40 compounds have been isolated. Of all the compounds IPA is the most powerful and acts as a target-marking pheromone for other stinger bees to reach the target. Among the stinger pheromones at least 15 are alarm eliciting and they mobilize stinger bees for effectively defending their colony. The IPA and Z-11-eicosenol together act as releaser primer pheromones and hence their effect lasts for a longer period. In brief, alarm defensive behaviour is elicited by multiple component pheromones and ten or more may act together.

The venom of the honey bees is secreted through the sting. The gland that produces venom has a reservoir or a sac to store it. The queen at the time of emergence has three times the poison compared with that of a worker. Queen is always ready at hand with the venom, because on emergence she has a compulsive instinct to kill sister queens yet to emerge.

Till *A.m.* was taken to China in 1959 the Eastern honeybee *A.c* was indigenous there and was kept in skips or hollows in trees and rocks. *Apis mellifera* was on movable frames and were brought in 1896; the dark bees were brought from Russia after N.E. China and Xingjian were invaded in 1900. The exotic species were known in Manchuria in 1923. Descriptions of beekeeping with *A.m* in Vietnam was given in 1909 and in Philippines in 1913. A Chinese born in Singapore Lim Choo Kiat and living in Malay Peninsula imported *A.m* honeybees from Australia in 1930 and maintained 22 colonies. The honeybee was kept by some selected Chinese in Johore and other places in the South and one of them raised 250 colonies in 1988. Yields from 30 to 70 kg per colony were obtained by them. *Apis mellifera* was introduced in stages in various countries of Asia throughout the twentieth century.

The history of introduction of movable frame hives for *Apis cerona indica* and the introduction of *Apis mellifera* into India (partly including Pakistan) is equally interesting. Only scientific and some administrative aspects were mentioned in my earlier book "Essentials of Beekeeping and Pollination" (2000). As they say truth is stronger than fiction, and in real life one is wonderstruck by the temerity and the generosity of some blessed individuals.

5.10.2(c) Drone Pheromones

Although mandibles in drones are vestigial yet there are sex glands in the head region equivalent to the mandibular glands of the queen. Thus drones exude a sex pheromone that attracts other flying males to certain areas in the air, commonly known as Drone Congregation Areas. An enormous number of drones fly around there during mid-day generally when there is bright sunshine. The drones gather in the region at the same place near where beekeeping is practiced. These congregations generally are above a land mark like a lake, a hillock or a prominent land structure. Virgin queens are driven there followed by a retinue of workers. It is observed that such sites are very suitable for mating with high success.

It was determined that in flight drones were attracted to the queen sex pheromone (9-ODA). As mentioned above 9-ODA has four effective components which are perceived by the drone antennae. It was proven recently that one of the four components responded to drone odourant receptors. It was the only functional component, the other three being ignored. With the help of this receptor a drone could detect an available queen in the air upto a distance of 60 meters (Robertson, Loeute *et al*., 2009). It had been reported earlier (Renner and Baumann, 1964) that there were 170 odourant Chemo-receptors on the body of a honey bee worker. In 2006, Robertson's team also mapped these Chemo receptors (genes) for smell and taste. It is believed that it is through smell and taste the older bees maintain order in a colony and direct the foraging bees to the location of specific flowers for the collection of nectar and pollen.

5.10.3 Salivary Glands

There is one pair in the head and is called cephalic labial gland and, another pair in the thorax called thoracic labial glands. Their secretions flow through separate ducts which join together in the head region and the medial duct opens outside at the base of labium (the lower lip). The saliva (the secretions) is a mixture of many compounds and their functions are still being studied. The enzyme sucrose metabolizes sugars in nectar of the flowers on which the honeybees forage. Saliva is probably used by workers in cleaning body of the queen. It is also used in processing wax which is secreted by the bees. It is considered that saliva along with other secretions of hypostomal (near gena), and hypopharyngeal (maxillary) glands are used in processing and preparing food of the larvae and possibly that of adults also.

These and other exocrine glands (e.g. Dettour gland; Sting associated glands) whose secretions open outside are non-pheromone products (in both female castes) but their status may change with further studies. In combination with other compounds they might be eliciting influence in the social life of the honeybees. Hypopharyngeal glands are highly developed in young workers and produced compounds (non-pheromones) that fortify food fed to larvae. The enzyme sucrose is also secreted by those glands. Recent findings show that chemicals like ethyl oleate (EO) produced by salivary glands of larvae have releaser effects in the recognition of the larvae by adult bees.

5.10.4 Nasonov Gland Pheromone

It is located in the back end of abdomen and opens on the last tergite of a worker (Plate 6). Its secretion is a strong attractant used by bees in a variety of situations. Honeybees are after seen at the entrance of a hive with their body roused upwards at an angle and fanning the pheromone inwards as if intimating their presence through the powerful attractant. The major odorous compounds are oxygen containing monoterpenes that have the odour of flowers; Geraniol (from geranium) and rose oil are the major compounds in the pheromone (Boch, Shearer, 1966). In a way the odorous alcohols mimic perfume of flowers which act as attractants for other bees whose presence is required for some social activity. There are some other alcohols (nerol, farnesol, citral) which are produced on oxidation of the major compound geraniol. Those are found in small amounts in the Nasonov pheromones but can be very strong attractants e.g. citral.

The Nasonov secretion is used by bees in the field while foraging on flowers. The bees that reach the rich flowers first issue the pheromone and, other bees perceiving the odour rush to those flowers for foraging. In turn they also open their glands and increase the intensity of perfume.

The Nasonov gland pheromone plays a key role in regulating the movement and formation of swarms. The secretions work in conjunction with queen substance (9-ODA; Morse and Boch, 1971). Let us take up the situation of a colony part of which has swarmed taking the queen along with. It may be a seasonal multiplication swarm still under the influence of queen pheromone. Since a good part of the colony is still intact, the flying swarm may be considered undecided to fly away and instead of being homogeneous it breaks up. The queen dissuades the swarm from flying away and in the meantime a group of honeybees elicit the Nasonov pheromone tilting the balance towards returning to the hive and the remaining swarm also comes back to the mother colony.

In a reverse situation if the bee scouts have already located a new nesting site they dominate the split swarm and with the strong concentration of Nasonov pheromone they attract the swarm with queen to fly to the new site. It is rarely a permanent site to settle, but only a temporary stop over until the scout bees have located a suitable well protected site. It is in the temporary phase that swarms are generally caught and hived. The queen is located and confined to a wire-gueze queen cage and placed in a box where come the swarm. Nasonov pheromone is fanned in abundance and the swarm settles down. The box is then taken to the apiary and shaken into a topless proper hive. In case the original queen is not treble a new one in a cage is kept at hand, placed amidst the swarm, hoping that honey bees with Nasonov pheromone settle the swarm till the released queen secretes the queen substance (9-ODA) in abundance and gives a stable social life to the colony and the workers take over duties, rear new brood, and give stability to the colony.

Under field conditions, the honeybees fly around with the swarm only if the queen is free, not confined. Under management of colonies queen-confining is practiced by beekeepers to make life easy and to multiply colonies as desired.

In scientific studies, various pheromones are extracted from the desired body parts of honeybees and the queen. Some of the pheromones have also been synthesized artificially by mixing certain chemicals in right proportions. The synthesized and the naturally extracted pheromones mixed can further stimulate bees to open their Nasonov glands and enhance various social functions. According to Ferguson and Free (1981), the gland is induced by many stimuli like, hive queen, pollen, propolis and mandibular gland pheromone of the queen (9-11 DA).

5.10.5 Alarm Pheromone of the Sting

The sting of *Apis mellifera* is more barbed in the worker than that of the queen. A worker when stings the barbs get embedded in muscles of the victim or the threatening attacker and cannot pull it out. The sting alongwith some muscles and a bladder full of poison stays there embedded and the muscles on their own keep pumping poison. The worker that stings flies away but dies in an hour. The queens use their sting for killing the rival queens who might still be sealed in the cells. Thus a queen can kill more than one rival by pulling out the barbed sting without herself being hurt. A beekeeper, in case he handles a queen carelessly, can also be stung.

The sting pheromone has been identified as isopentyl acetate (ipd) by Boch *et al.* (1962). As soon as an enemy of the colony (their house) is pin-pointed, the pheromone elicits other workers as recruits so as to defend their colony with a greater force. A bear seeking honey or an unaware man are big objects and dozens or hundreds of defending workers are required. However, their defensive attack is generally not as vicious as that of the wild honey bees that make their colonies outside on single combs hanging prominently downwards.

The sting pheromones of the other three honeybees viz. *Apis cerana, Apis florea, Apis dorsata* have also been studied and their pheromones compared. Morse *et al.* (1967) found that IPA was the main constituent in the sting extract of all the four species, but its quantity varied, and so did the pain inflicted to the intruder enemy. The alarm system of the Asiatic honey bee colonies was similar. There was one significant difference in the prolonged effect of sting pheromones in the two wild species, *Apis florea* and *Apis dorsata* which were observed closely in flight rooms. Their stings were removed artificially and were stored at 10°C for chemical analysis within an hour or two to avoid any changes by oxidation. Yet the lasting effect of those two species was longer. It was determined by mass spectrography that their pheromones contained another component 2-decen-1-yl acetate (2da) (Verth *et al*., 1978). Quantification studies and response experiments proved this phenomenon. Moreover, it was found that *Apis dorsata* contained 10 times and 15 times more IPA than that found in *Apis mellifera* and *Apis cerana*, respectively. It shows two facts: one, that *Apis dorsata* sting is more painful to man; two, that pheromone remains in the atmosphere for a much longer time and their defensive behaviour persists for 24-48 hours, bees slowly coming down in their frequency and intensity of flights in the territory (over 1 km) of their initial reaction. Also, 2da is less volatile and hence the two wild honeybees can recognize a stung object for a longer time. Their stings are so barbed that they can be embedded deeper in animal tissues. For this double advantage, *Apis dorsata*—the largest honeybee, is so feared by man.

Apart from the above mentioned two alarm pheromones, there are others that have similar activity but exhibit much less potency. Further studies may indicate that their influence on social behaviour of honey bees could be quite important.

5.10.6 Koschevnikov Gland Pheromone

At the rear end of the abdomen of queen, there is Koschevnikov gland that produces a powerful pheromone in a mated queen to which workers are attracted. Not much is known about the chemistry of this gland and it degenerates in a mated queen within one year. The workers produce in their sting a complex strong alarm pheromone identified as isopentyl (= isodmyl) acetate (IPA). It is a volatile component of sting pheromone and is over 20 times more active than 2-heptanone, the mandibular gland secretion. The foraging and guard workers produce the maximum IPA and give a banana like odour. The foraging workers produce at least 40 pheromones some of which have been identified as initiators of alarm. Many of them resemble floral perfumes like lavender (2-nonyl acetate) and jasmin (2-monanol).

IPA is an alarm pheromone which is antagonist to Nasonov pheromone which is for making the swarming workers quiet down and settle to their work. So, depending upon the conditions of a colony and presence or absence of a queen, both these pheromones and their components play an important role.

5.10.7 Queen Retinue Pheromone

The honey bee queens have sophisticated systems of chemical signals (pheromones) that modulate the behaviour and physiology of colony population. Out of these the most stinking commands of the queen are through mandibular pheromones (QMP). The behaviour elicited is quite the opposite to the alarm behaviour. A combination of three releaser pheromones forming the queen mandibular pheromone (QMP) elicit a group of workers to gather around the queen, perform functions not only for the health of the queen, but also for the well being of the entire colony. According to Slessor *et al*. (1990), the components include (9-keto-2-CE)-decenoic acid, 9 hydroxy –(E)-decenoic acid (88% R-(–) and 12% 5-(+), methyl p- hydroxybenzoate, and 4-hydroxy-3-methoxy-enylethenol. QMP elicits a compelling behaviour for a certain period of time like a gang of workers. They touch queen's body with antennae, lick and groom her and at the same time remove pheromone from her body. The pheromone is passed on from worker to worker as if to transmit some message (Neumann, 1998; Keeling *et al*., 2003). Since the components are non-volatile, the effect is meant to last for some time; it is a direction to regulate the timing of swarming (colony reproduction). As long as the concentration of QMP is maintained at an optimum level, the workers are inhibited to form queen cells and rear new queens. As soon as the concentration falls below a certain threshold, the workers' population having increased causing crowding and dilution of QMP, the colony workers become excited and are induced to make queen cells. This is the mechanism by which swarming and division of the colony takes place. Strangely it is the mother queen that goes with the swarm, naturally taking alongwith her experience to organize a colony. The role of chemicals in life behaviour is not understood which until recently was considered an instinct.

Further studies have shown that perhaps heredity (gene constitution) also plays a significance role (Pankiw *et al.*, 1994). Irrespective of the concentration (dose) of QMP prevalent, certain workers are more responsive to the pheromone (9-components in all) and are also elicited to retinue behaviour more readily than the low response workers.

In the antennal lobe of the honey bee brain are odour evoked responses to some components of queen mandibular pheromone (QMP) and these responses coincide with the plant odour establishing a direct line between the two.

Furthermore, it is to be understood how the chemical reactions in brain are to be conveyed on neural pathways to the peripheral and then transmitted to the motor system and body movements. It indicates how valuable a subject honey bee has been for taking scientists to the frontiers of science. Genes have been identified whose brain expression levels were altered by exposure to QMP; and RNA levels were significantly down regulated. However, the foraging activity of field bees is not altered significantly (Growzanger *et al.*, 2007). It had also been shown that QMP pheromone elicited multiple distinct behavioural responses and that releaser and primer responses were uncoupled. The team of scientists also experimented with artificial insemination of virgin queens with semen of a number of drones. It was interesting that the latter queen elicited a larger retinue of worker bees in hive, and also their brain activity. In other words, a queen that mates with many males attracts more workers to attend on her. Possibly the over all performance of such a queen in egg laying, etc could be superior. The "stronger the colony higher the honey yield" depends on esters (pleasant smells) found both on the body of queens and workers, and are very important in regulating foraging. The effect is dose dependent; low amount of esters induces early foraging whereas the high dose delays it. It indicates that high concentration of pheromone imparts a sort stability in a colony with the result that foraging and broad rearing is delayed as also crowding and queen cell construction. According to the recent researches (Beggs *et al.*, 2007), this pheromone in the brain is at cellular level through dopamine pathways. These pathways play key role in the regulation of behaviour and motor control. In worker bees the levels of dopamine receptor gene expression and response to this amine are affected by QMP. Now ovamillyl alcohol is the main contributor to QMP-induced changes in the brain that influence the behaviour of the bee. It is through this mechanism that queen regulates the life and behaviour of her offspring. Further work is going on to study how the odours are processed in honey bee brain making use of a combination of pheromones and ensure effecting intraspecific communication when there is a wide range of individual and social behaviors to account for (Sandoz *et al.*, 2007).

5.10.8 Brood Pheromones

On the body surface of larvae and pupae of honey bees there are found secretions that act as pheromones. At least 10 components have bees isolated. Those are methyl and ethyl esters of linoleate, linolenate, oleate, palmitate and stearate (Koeniger *et al.*, 1985; Le Conte *et al.*, 1990). All these compounds have more biological activity than any other known pheromone of the honey bees. Broadly speaking, they are of two types; Inhibitory pheromones and Brood-Recognition pheromones.

The development of ovaries in workers were inhibited by worker larvae and pupae in colonies that had no queens, showing the effect of brood pheromones. In the presence of larval brood pupal brood, and a queen the inhibitory effect was further accentuated and larval ovaries hardly developed at all. In final analysis, the inhibitory effect of larval and pupal brood was as great as that of a mated queen which is known to suppress development of ovaries in workers and thus proves supreme in a colony in this respect.

There is a nonvolatile pheromone on the body of worker larvae and drone larvae. The workers moving in a colony can distinguish between them by making contacts with them. It is done with the help of recognition pheromones and thus gives different quality of food—the richer and more to the developing drones than the workers. Of course the queen larvae are given the richest food of all.

Foraging is stimulated by the presence of brood; the greater the stimulating pheromone the higher the quantity of pollen collected by workers (Free, 1967). The pheromone is effective on contact by the workers. The odour of brood (volatile component) is less effective in stimulation foraging. It is not fully known as yet whether he foraging stimulation, ovarian inhibition and brood recognition are all functions of the same pheromone or those are by different component pheromones.

5.10.9 Tergite Gland Pheromones

On the dossal surface of queen's abdomen, pheromones are secreted from the tergites. Their function is similar to that of queen's mandibular gland pheromones (9-ODA). They release signals to the workers to recognize their queen, to inhibit queen cell construction, to inhibit the development of ovaries in workers and display retinue behaviours. The tergite gland pheromone is not volatile and young workers are seen gathered around the queen to make contacts with her thereby recognizing her. Both these pheromones secreted at different parts of the body of queen enhance their effectiveness to elicit worker behavior, one functioning as a releaser and the other as a primer.

These two pheromones also help in the copulation of flying queens; the releaser is perceived by drones from a distance of 50 meters and the primer from as close as one foot. Thus their function is complimentary to each other. The chemical composition of tergite pheromone is not known as yet. Once these pheromones are synthesized it should be possible to achieve a greater success in queen mating and thus promote the multiplication of honeybee colonies.

The combined effect of pheromones has profound effect on the social organization of a colony. At least 50 of them have been identified and the synthesis of some of them has been achieved. Their doses can thus be regulated for obtaining the desired activity of worker bees.

5.10.10 Beeswax Pheromones

Young hive workers who do not fly out for foraging perform other functions producing brood food (royal jelly), cleaning and repairing comb cells. They are elicited by pheromones to secrete beeswax from paired wax glands on the ground side of their body. The wax is in the shape of small scales with which a sizeable force of workers makes combs. The newly built combs have a characteristic smell of many volatile compounds like octanal, nonanal, decanal, furfural, benzaldehyde, 1-decdnol and other aldehydies (Rinder, 1978; Blum *et al.,* 1988). These compounds with their odour affect the storage intensity of food (pollen, nectar, etc.) by increased or decreased foraging. The exact role of individual compounds is being studied.

Volatile esters on the body of queens and workers have a profound effect on the foraging activity of workers. A low amount of esters induces early foraging and a higher amount delays the start of foraging by workers. As we compare the effects of various pheromones it appears amazing how honeybees can perform inter-linked yet distinct jobs as regulated by pheromones. The overall effect seems to imply that the colony is a whole organism. Instead of the directions of a circus master, the bees work on their own for the common good of the colony and their biological survival.

5.10.11 Tarsal (Arnhart) Gland Pheromones

Some secretions exude from the tarsi of both queen and workers which are deposited on surface where they walk. Although their chemical nature has not been identified, it is noticed that these foot-print pheromones play some crucial role in the life of social insects.

The tarsal pads of queens produce a secretion which is deposited on the surface of combs. This pheromone together with the mandibular secretion of the queen when found deposited on the bottom edges of combs in a crowded colony inhibits queen cell construction. Thus the reigning queen does not let the workers raise a new queen required for swarming. When overcrowding reaches beyond a limit, it is not easy for the queen and workers to reach the bottom edge of comb and then workers take the liberty of raising new queens. In general, it was observed that young queens less than a year old deposit more pheromone than queens older than two years. A queen normally produces 10-15 times more pheromone than a worker produces her pheromone (Lensky., and Slabezki, 1981: Lensky *et al.*, 1984).

The workers deposit a trail pheromone through tars, but it is not sure whether it is secreted in the feet; its production could be in some other part of the body. Apart from the function of obstructing queen cell production in conjunction with queens, tarsal pheromone probably has other function too. Near the entrance workers keep landing constantly, therefore the concentration of trail pheromone becomes high and more effective (Butler *et al.*, 1967, 1969) in luring the lost bees which send signals for help by secreting and fanning around the odour of Nasonov pheromone. Thus the disoriented workers locate their own hive. It is also considered probable that honeybees leave signals of their trail pheromone at the foraging sites to make the place more attractive for foraging on the same analogy as crow calls for the clan when food is given in abundance (Ferguson and Free, 1981; Free, 1987 ; Butler and Free, 1952).

5.10.12 Worker Repellent Pheromone

This odour is given out by the young virgin queens for two weeks on emergence. It is discharged as a fecal matter from the rectum. As a result of its odour, workers leave the queen, start grooming themselves as if they have been tranquilized and to remain idle (or to clean themselves up). It has been identified as o-aminoacetophenone, a very small portion of the fecae which gives a characteristic smell of grapes. Its source of secretion or function is not known.

5.11 UTILIZATION OF PHEROMONES IN AGRICULTURE

Identification of a large number of natural pheromones and synthetic compounds leads to the possibility of their utilization in manipulating the behaviors of all castes in apiculture. Already their initialization in the field has been demonstrated. Synthetic Nasonov is highly effective in bringing feral swarms and hiving them (Free *et al.*, 1984). A trap box for containing the pheromone has been devised for this purpose (Schmidt and Thoenes, 1987). The trap is already being utilized to monitor northward movement of Africanized swarms of *A.m*. The same pheromone can be utilized for attracting foragers to crops that need pollination (Williams *et al.*, 1981).

The queen substance (9-ODA) used in hives can prevent rearing of new queens and stop excessive swarming. It can also be used for attracting virgin queens in the mating areas and increase reproduction of colonies. The commonly produced mandibular gland pheromone can be made available for rippling workers bees.

The science of pheromone is new and there is a great potential for its utilization in commercial beekeeping.

5.12 RECENT ADVANCES IN BEE BEHAVIOUR

Recent advances on honeybee (*A.m*) behaviour indicate that workers through their antennae pick up responses to pheromones (QMP) of virgin mature queen (Vaitkeviciene *et al.,* 2006). They have an odorant receptor for the pheromone (Wanner *et al.,* 2007) and the chemical blend induces young workers to feed and groom their queen and to take up other tasks in the colony. The response to chemicals reaches the brain and through dopamine pathways regulate the beahviour of young workers and their motor control. Thus the ability of queen to regulate behaviour of the workers is enhanced (Beggs *et al.,* 2007).

It has been reported earlier that QMP is a complex chemical compound of which five components have been identified and the some other have yet to be discovered. A natural compound functions both as a releaser and a primer. Its role includes, attracting a retinue of workers, attracting drones on mating flights, preventing workers to lay eggs, and other functions in the colony living exhibiting the social behaviour. The queen produces a synergistic, multiglandular pheromone blend of at least nine compounds (Keeling *et al.*, 2003).

Since the early times of observations on honeybee, pheromones such as secretions of queen mandibular gland, the sting and defensive weapons, the lure of Nasanov, etc., emphasis has been on the chemical identification of their components for instance, five of long chain components of QMP have so far been identified. It is now known that isoamyl acetate (plus some complementary compounds) are the active complements in the sting pheromone (Collins *et al.,* 1982; Boch and Shearer 1962; Beggs *et al.,* 2007). Geraniol, having the smell of the flowers, is the active component of Nasanav pheromone that lures the honey bees (Free *et al.,* 1984, 1987). New discoveries are being witnessed every year.

5.13 HOW HONEYBEES GIVE AND TAKE HONEY-MAKING MESSAGES

There are two kinds of semi-domesticated honeybees that live in wooden hives (with 9-parallel frames inside). Closely related, they are often called sister species. The bigger of the two, commonly known as the Western or Afro-European honeybee, lived in the western part of the old world, it has many races of which *Apis mellifera ligustica* is the most popular. The other, native to most of Asia and the adjoining islands is known as the Eastern or Asian honeybee – *Apis cerana*. It also has many races, of which *Apis cerana indica,* the Indian sub-species is one.

Mellifera has been worked on extensively by the Europeans, since the time of the pharaohs, who for close to 20 or 30-centuries, exploited them for domestic and commercial honey production Over time, it proved to be a more productive honeybee compared to the Asian species That is why, it has been taken to practically all the continents of the world, and lately to the Indian sub-continent, as well.

In Asia, while the Eastern honeybee has been kept and cared for since a very long time, yet its honey production does not match that of the mellifera. In fact, it was only at the beginning of the 20th century that it began to be studied scientifically, ever since the modern wooden hives with removable top covers and movable frames of combs became available, primarily due to the efforts of western colonisers. Since then, people in Asia became desirous of possessing colonies of the western honeybee.

In India, many British traders and officials brought mellifera colonies from Europe in the early part of the 20th century, and the colonies were kept at Calcutta (Douglas, 1882), at Pusa (Ghosh, 1920), Trivandrum (Hatch, 1928), Mahabaleshwar (Baldery, 1930), Niligiri Hills (Beadnell, 1939). After various durations of survival, the colonies perished due to one reason or the other. More efforts were made in the mid-fifties to bring colonies by special aeroplane service and to keep them at Kulu (Rehman, 1945), they also did not survive beyond one or two seasons. Renewed efforts were again made by the author in 1962-64. This time an altogether new method was used to try and introduce the Mellifera gravid queens into Indica colonies with the hope that their progeny would eventually take over the colony when the host workers died of old age. Requests were made to well-known bee keepers in USA and Australia and other places for mellifera queens. They all offered to supply gravid mellifera queens free of cost, sending them via Air Mail or Air Cargo Service.

It is relevant to mention here that Bee Science scientists in India resisted the import of exotic honeybees, fearing accidental introduction of diseases which could harm the Indian honeybee species (including *Apis florea* and *Apis dorsata,* the wild bees), so useful in the pollination of many valuable crops. In fact, Acarine, a new disease had been observed in Indian bees, which most probably came along with Mellifera bees imported from the Isle of Wight in UK. The disease had spread among the Indian bees up to Simla by 1953. This illegal act performed by a private beekeeper in Kashmir, made the scientists fear understandable.

Since Dr Atwal and his team were anxious to experiment, they dug out literature and found that all the known honeybee diseases, whether parasitic, bacterial or viral had already been recorded in South Asia. This helped them convince the Government of India Quarantine Authorities, New Delhi, who gave them permission to go ahead with the proposed experimentation. A lay out of the experiment was then planned. The idea of importing whole colonies of Mellifera was discarded for fear of accidentally introducing unknown strains of diseases into the country. Instead, it was considered safer to import fertilised Mellifera queens with four to five attendants by Air Mail service. The effort was to find out, whether de-queered Indica colonies were receptive to the Mellifera queens.

It was hoped Mellifera queens would, once at ease, lay eggs which along with the emerging larvae of the exotic queen be cared for and fed by the host species. Once that happened it was natural that in due course, the hosts would die of old age and the young workers with Mellifera blood, would take over the colony. Many a times, inspired ideas such as this had worked, giving birth to a new science. In the case of four species of honeybees, such cross-breeding (as produced mules, yak, etc.) was found to be impossible and all attempts had failed in nature. It was observed that if the embryo of one species was artificially inseminated under a microscope, the embryo would abort. Hence, the PAU scheme of using foster relationships seemed highly plausible.

Harry Whitcomb, a famous beekeeper from Davis, California, USA was requested to send us Mellifera queens with attendants, and he did so without cost by ordinary Air Mail Service. The little parcel with four queens was delivered at Ludhiana like ordinary air mail after 12-days of travel. From Ludhiana they were taken by road to Nagrota, where there was an established Indica bee farm of the University. Dr Atwal and his co-worker O.P. Sharma, hung the Mellifera queen cages among frames of the host bees. They noticed that the queens were happy with the Mellifera attendants who fed and groomed them inside the cages. The two alive and active Mellifera queens were released one by one (without attendants) among the Indica workers, who did not accept the exotic queens and instead killed them by balling (suffocating by smothering). It was noticed that the last queen with full stomach, who seemed full of eggs was allowed to live the longest, with the hosts even feeding her. This small hope kept up our spirit of introducing Mellifera species in India.

Once again, Dr Atwal got in touch with the beekeeper in Davis, who was surprised to learn of the crazy experiment the Indian scientists were trying to do. Immediately, he consulted Prof. H.H. Laidlaw, one of the world's senior most and highly respected bee specialists. Without discouraging Harry, he told him frankly that to his knowledge, no such experiment had been done before, though personally he did not see why it could not be a success. This opinion in a way committed Harry, who devotedly continued to supply the experimenters in India with Mellifera queens free of cost, whenever requested.

Observations of queens released among Indica (Cerana)workers revealed that the older bees were the trouble makers, pushing the queens, pulling at their legs and wings, refusing to feed them, so that they died of exhaustion. The ISQI (Inter-specific Queen Introduction) technique was improved by raising a queenless colony of Indica with young bees, only 3-4 days old. It was done by keeping frames of sealed brood in an electric incubator with temperature control, and as the bees emerged out of their cells, they were transferred to a new hive which contained frames with food and brood.

It was possible then to successfully release the next batch of four imported Mellifera queens into the de-queered Indica colonies. Thus, the social behaviour of the two species living together was observed. It was noticed that the workers of the two species lived together harmoniously, even when Mellifera worker bees were twenty or more days old. However, it was a shock to notice that instead of going out to gather nectar and pollen for feeding the colony, they merely helped themselves to the colony's food store, went out for play flights and came back to live a life of ease.

It was disappointing to find that the experiment had failed, since the main purpose of greater honey production seemed to remain unrealised. Then as a pleasant surprise, it was observed that as the population of Mellifera worker bees was nearly 50% of the entire colony, they started field activities, taking over the various indoor and outdoor jobs in the colony. Not only that, the Indica bees, by now had all become field workers and went out as before, working in total harmony with the Mellifera workers, jointly carrying out their indoor and guard duties. The language mystery remaining, it was also observed that whereas the Mellifera workers fanned for air circulation in the hive with their face towards the entrance, whereas the Indica workers fanned facing away from the entrance.

Whatever the mystery of the inter-specific behaviour of the workers, the main aim of the experiment was to produce large populations of disease free Mellifera workers, so that more queens could be imported and released into their nuclei to help multiply the exotic bees and give colonies to the farmers. Now it appeared the aim was likely to be achieved, due to the time, devotion and hard work of scientists and experienced beekeepers, who were anxious to multiply the colonies on their own and sell them to others close by, and even as far as Kerala. Eventually, this elite honeybee was introduced into India with great success in a large part of the country, even where there were extremes of temperature, rainfall, wind, etc. With time, the harvest of honey increased tremendously and in 1995-96 for the first time, India began to export honey for commerce to Europe and other countries. According to the figures of trade and commerce, 1,000-tonnes of honey were exported. Since then, the export of honey has increased every year, though the producers of honey are not getting a fair price for their product, posing a threat to this new rural industry.

The University with the help of ICAR went ahead multiplying the Mellifera colonies for extending modern beekeeping to all parts of India. However, in fundamental science, scholars are not generally interested in immediate production or gains, but they strive to unfold the Principles involved. Just when scientists at Ludhiana (and Nagrota) were working on inter-specific queen introduction, an Austrian, von Frisch, was working on the dance language of Mellifera workers to find out how bees are able to locate a floral source of food for nectar and flower pollen. He found that by dancing in the hive the scout bees conveyed the distance, location, direction and richness of food to their sisters. Thus, foraging skills were taught to all the bees, so that they could move out and work for common good of the colony. Completing his study after many years of work, he proved and described the *'waggle dance'* and its variations, as their tool of communication to teach the other bees, so that every individual did not have to needlessly spend time and energy in search of nectar of flowers for converting it into honey. He also discovered the colour perception of honeybees in nature, winning the Nobel Prize for his works.

Almost thirty years after his discovery, five Chinese scholars and one Australian got together and decided to decode the language of communication of the two species of worker bees living together in the same hive. It was an extension of the earlier observations made in Punjab in 1962-64 that Mellifera and Indica workers could live together in a colony.

In 2007-08, the Chinese scientists used the same technique, but devised a simplified method by placing Mellifera sealed frames amidst the Cerana frames full of brood and active workers. As the Mellifera adults emerged from the sealed cells, they were cared for by the Cerana colony, with about 5,000 of them living along with the same number of hosts, and were observed alive for 12-20 days. It was found that they lived together without intolerance or molestation.

The alternative inter-specific mixed colony was also prepared by them with Mellifera queen and workers in which the Cerana workers were introduced as described above. But, the foster Cerana workers were not tolerated peacefully and were killed sooner or later; no bee was allowed to live beyond day 12. This part of their experiment was not successful. However, the scientists continued their work on the first colony with the Cerana queen and workers and with Mellifera workers hatched as foster species. Like most other scientists of the world, the *Waggle Dance* of honeybees was accepted as the standard method for the comparison of variations in dances by individual species or two species of honeybees living together in a colony.

The dances in a mixed colony (of Mellifera and Cerana) workers were video-taped and their tempos (dance dialects) were analysed statistically. The dances of the two species were significantly different from each other. The scientists conducted a series of experiments observing the social behaviour of the two species living and foraging together, concluding that statistically it was proven that Mellifera workers could *'teach'* their own dance dialect or language to Cerana workers and make them forage for sugary food in dishes placed at distances of 300, 400, 500, 600 metres from the hive.

It is a part of science to express scientific doubts where due. The experiments of 2008 do not prove clearly that the two species of honeybees studied could learn each other's language. However, under a simulated situation (perhaps adequately copying nature) Mellifera worker scouts with the help of their own dance succeeded in making the workers follow them and reach the desired food dish. The experiments of 1962-64 indicated that Mellifera workers under appropriate situation of numbers of their own kind, instinctively *'rememberd'* their own dance language (after the experience of a stray worker visiting flowers?) and workers of their own kind followed the example and all of them started foraging. Or, could it be that Mellifera workers, even though in a mixed colony waited for foraging when they were 20 and plus days old, and instinctively learnt their own dance language without involving the other (Indica) species?

It is clear from 1962 – 64 and 2008 experiments that Indica workers in a Mellifera colony are not tolerated and are killed, so in a natural or a simulated situation, the former has no occasion to learn the language of the latter. We can, therefore only surmise on a hypothetical situation as to the significance of this recruitment of Cerana follower bees. It is most interesting to appreciate that with the help of modem technology, we can perceive and record even minor details of social behaviour of animals, both under natural and simulated situations.

SELECT BIBLIOGRAPHY

Abrol DP (2006). Factors influencing flight activity of *Apis florea* F, an important pollinator of *Dacus carota* L. J. Apic. Res.; 45(2).

Abrol DP (2006). Diversity of pollinating insects visiting litchi (*Litchi chimensis* Sonm) and path analysis of environmental factors influencing forage behaviour of four honey bee species. J. Apic. Res.; 45(4).

Adlakha RL, Sharma OP (1975). *Apis mellifera vs. Apis indica*. Glean. Bee Cult.; 103(5): 160.

Arias M, Sheppard W (2005). Phylogenetic relationship of honey bees (Hymn. Apinae, Apini) inferred from nuclear and mtochondrial DNA sequence data. Mol. Phylog. Evol.; 37: 25-35.

Atwal, AS (1964). Introductione de api regine de razza italiana (*A.m.*) in alveri di *Apis indica* 1. Apicobt. Ital.; 31(4): 100-101.

Atwal AS and Dhaliwal GS (1969). Robbing between *Apis indica* F and *Apis mellifera* L. Am. Bee J.; 109(12): 462-63.

Atwal AS and Goyal NP (1971). Infestation of honey bee colonies (*A.m.*) with *Trophilaelaps* and its control. J. Apic. Res.; 10(3): 137-42.

Atwal AS and Sharma OP (1967). Acclimation of *Apis mellifera* in the Punjab (India) – paper read and colour slide projected of *A.m.* queen being attended by *Apis indica* (cerana) workers. Proc. 21st International Apicultural Congress, Maryland, USA.

Atwal AS and Sharma OP (1967). Studies on the Acarine disease problems in *Apis indica,* (symptom also suspected in *Apis mellifera*). Proc. 21st Internat. Apic. Congress, Maryland, Aug. 14-17. pp 487.

Atwal AS and Sharma OP (1969). The introduction of *Apis mellifera* queens into *Apis indica* colonies and the associated behaviour of the two species. Indian Bee J.; 30(1): 41-56 (1962-64; 1969).

Atwal AS, Bains SS, Balraj S (1970). Bee flora for four species at Ludhiana. Indian J. Ent.; 32(4): 330-34.

Atwal AS (2000). Essentials of Beekeeping and Pollination. Kalyani Publishers, New Delhi: 394 pp.

Beggs KT, Glendining KA, Marechal NM, Vergos V, Nakamura I, Slessor KN, Mercer AR (2007). Queen pheromone modulates brain dopamine function in worker honey bees. National Academy of Sciences of USA; 104: 2460-64.

Blum MS, Jones TH, Rinderer TE, Sylvester HP (1988). Oxygenated compounds in beeswax: identification and possible significance. Comp. Biochem. Physiol. 91B: 581-830.

Boch R and Shearer DA (1962). Identification of geraniol as the active component in the Nasonov pheromone of the honey bee. Nature; 194: 704-706.

Boch R, Shearer DA, Stone BC (1962). Identification of iso-amyl acetate as an active component in the sting pheromone of the honey bee. Nature; 195: 1018-20.

Boch R, Rothenbuhler WC (1974). Defensive behaviour and production of alarm pheromone in honey bees. J. Apic. Res.; 13: 217-221.

Brar HS and Atwal AS (1985). Effect of some meteorological factors on the dance tempo of *Apis cerana indica* and *Apis mellifera* L. Indian J. Ecol.; 12(2).

Breed MD (1998) in Pheromone Communication in Social Insects, Ants, Wasps, Bees, and Termites eds. A review by Vander Meer RK, Breed MD, Espelie KE, Winston ML; Westview Press, Boulder, Co. pp 57-78

Butler CG and Free JB (1952). The behaviour of worker honey bees at the hive entrance. Behaviour; 4:262-292.

Butler CG, Callow RK, Johnston NC (1961). The isolation and synthesis of queen substance (pheromone) 9-oxodec-trans-2-enoic acid, a honey bee pheromone. Proc. Roy. Ent. Soc.; 155: 417-432.

Butler CG and Simpson J (1967). Pheromone of the queen honey bee (*Apis mellifera*) which enable her workers to follow her when swarming. Proc. R. Ent. Soc. London; A-42: 149154.

Butler CG and Calam DH (1969). Pheromones of the honey bee – the secretion of Nasonov gland of the worker. J. Insect Physiol.; 15: 237-244.

Callow RK, Chapman JR, Paton PN (1964). Pheromones of the honey bee: chemical studies of the mandibular gland secretion of the queen. J. Apic. Res.; 3: 77-89.

Collins AM and Blum MS (1982). Bioassay of compounds derived from the honey bee sting. J. Chem. Ecol.; 8: 463-70.

Dhaliwal GS and Atwal AS (1969). Comparative aggressiveness of the four species of *Apis* under controlled conditions. Indian Bee J.; 31(1): 17-21.

Dhaliwal GS and Atwal AS (1970). Interspecific relations between *Apis cerana* (*indica*) and *Apis mellifera*. J. Apic. Res.; 9(2): 53-59.

Dhaliwal JS and Atwal AS (1976). Competition among honey bee (*A.m.*), wild honey bees (*A. dorsata; A. florea*) and wild bees (solitary and social bees) for collecting nectar and pollen from *toria* (*Brassica campestris*). Indian Oilseeds J.; 6: 7-9.

Dyer FC, Seeley TD (1991). Dance dialect and foraging range in three Asian honey bee species. Behaviour.

Ferguson AW and Free JW (1981). Factors determining the release of Nasonov pheromone by honey bees at the nest entrance. Physiol. Ent.; 6: 15-19.

Free JB (1967). Factors determining the collection of pollen by honey bee foragers. Anim. Behav.; 15: 134-144.

Free JB and 4 others (1984). Honey bee Nasonov pheromone lure. Bee Wld.; 64: 22-24.

Free JB (1987). Pheromones of Social Bees. London: Chapman and Hall, 218 pp.

Gary NE (1962). Chemical mating attractants in the queen honey bee. Science; 136: 373-74.

Gary NE (2003). Activities and behaviour of honey bees in "Hive and the Honey Bee", 269-361.

Gould JL (1976). The dance language controversy. Quart. Rev. Biol.; 51: 211-244.

Gould JL and Gould CG (1988). The Honey Bee. Scientific American Library, New York: 239 pp.

Goyal NP and Atwal AS (1971). Comparative communication behaviour of *Apis indica* and *Apis mellifera*. Proc. Int. Apic. Congr. Moscow: 328.

Goyal NP and Atwal AS (1975). Studies on the relation of population of insect pollinators with seed production in lucern. Indian J. Ecol.; 2(1): 23-29.

Haydak MH (1945). Larval food and development of castes in the honey bee. J. Econ. Ent.; 36: 778-792.

Haydak MH (1960). Supercedure and swarming. Glean. Bee Cult.; 88: 265-269.

Hsu CY, Ko FY, Li CW, Fann K, Lue JT (2007). Magnatoreception system in honey bees (*Apis mellifera*). PLoS ONE; 2(4): e 395.

Johnson DL (1967). Honeybees: do they use the direction information contained in their dance maneuvers? Science.; 155: 844-847.

Johnson RN, Odlroyd BP, Barron AB, Crozier RH (2002). Genetic control of the honey bee (*Apis mellifera*) dance language: segregating dance forms in a backcrossed colony. J. Hered.; 93: 170-173.

Kaminski LA, Slessor KN, Winston ML, Hay NW, Borden JH (1990). Honey bee response to queen mandibular pheromone in laboratory bioassays. J. Chem. Ecol.; 16: 841-850.

Kapil RP, Jain KL (1980). Biology and Utilization of Insect Pollinators for Crop Production. Hissar (India). Haryana Agri. Univ. Press: 81 pp.

Keeling CI, Slessor KN, Hugo HA, Winston ML (2003). New components of the honey bee (*Apis mellifera*) queen retinue pheromone. Proc. Natl. Acad. Sci.; 100: 14486-91.

Koeniger N and Veith HJ (1985). Glyceryl-1, 2-dioleate-3 palmitate, a brood pheromone of the honey bee (*A.m.*). Experientia; 39: 1051-51.

Laidlaw HH (1987). Instrumental insemination of honey bee queens: its origin and development. Bee Wld.; 68-36, 71-88.

LeConte Y, Arnold G, Trouller J, Masson C (1990). Identification of a brood pheromone in honey bees. Naturwissenchaften; 81: 462-65.

LeConte Y, Mohammedi A, Robinson GE (2001). Primer effects of a brood pheromone on honey bees' behavioural development. Proc. R. Soc., London; B268: 1-6.

Lensky Y and Slabezki Y (1981). The inhibiting effect of the queen bee (*A.m.*) footprint pheromone on the construction of swarming queen cups. J. Insect Physiol.; 27: 313-323.

Lensky Y and 6 Others (1984). The tarsal gland of the *Apis mellifera* queens, workers and drones and the chemical characterization (in French). Am. Sci. Naturelle, Zool.; G: 167-175

Lensky Y and 4 Others (1985). Pheromones activity and fine structure of the mandibular glands of drones. Insect Physiol.; 31: 265-276.

Lindauer M (1953). Division of labour in the honey bee colony. Bee Wld.; 34: 63-73.

Lindauer M. (1954). The water economy and temperature regulation of the honey bee colony. Z. Vergl. Physiol.; 36: 391-432.

Lindauer M (1971, 1978). Communication among Sociai Bees. Cambridge, Mass., Harvard U. Press.

Mitchener CD (1974). The Social Behaviour of the Bees. Cambridge, Mass., Harvard University Press. 404 pp.

Michelson A, Kitchner WH, Lindauer M (1986). Sound and vibrational signals in the dance language of the honey bee *A.m.* Behav. Ecol. Sociobiol.; 18: 207-212.

Milum VG (1956). The significance of some inhibitory substances. Proc. Xth, Intl. Congress Entomology; pp 1085-108.

Mohammedi AP, Crauser D, LeConte Y (1998). Effect of aliphatic esters on ovary development of queenless bees (*Apis mellifera* L). Naturwissenchaften; 85: 455-458.

Moritz RFA, Southwick EE, Harbo JB (1987). Generic analysis of defensive behaviours of honey bee colonies in a field test. Apidologie; 18: 27-52.

Morse RA and Benton AW (1967). Venom collection from species of honey bees in south-east Asia. Bee Wld.; 48: 19-29.

Morse RA, Shearer DA, Boch R, Denion AW (1967). Observations on alarm substances in the genus *Apis*. J. Apic. Res.; 6: 113-118.

Morse RA and Boch R (1971). Pheromone concert in swarming honey bees. Ann. Ent. Soc. Am.; 64: 1414-1417.

Naumann K, Winston ML, Wyborn MH, Slessor KN (1990). Effects of synthetic honey bee queen mandibular gland pheromone on workers in packages. J. Econ. Entomol.; 83: 1271-75.

Naumann K, Winston ML, Slessor KN, Prestwich GD, Webster FX (1991). Production and transmission of honey bee queen (*Apis mellifera* L) mandibular pheromone. Behav. Ecol. Sociobiol.; 29: 321-332.

Naumann K (1991). Grooming behaviour and the translocation of queen mandibular gland pheromone on worker honey bees (*A.m.*). Apidologie.; 22: 523-531.

Pankiw T, Winston ML, Slessor KN (1994). Variation in worker response to honey bee (*A.m.*) queen mandibular pheromone. J. Insect Behav. 7: 1-15.

Pankiw, T. Winston ML, Slessor KN (1995). Queen attendance behaviour of worker honey bees (*A.m.*) that are high and low responders to queen mandibular pheromone. Insect. Soc.; 42: 371 -78.

Pankiw, Tanya (2004). Cited in: Honey bee pheromones as information flow and collective decision-making (a review). Apidologie; 35: 217-226.

Potichot S, Wongsiri Dietz A (1993). Attempts in queen rearing of *Apis cerana* larvae in *Apis mellifera* colonies. In: Conner LJ, Rinderer TE, Sylvester HA, Wongsiris, editors. Asian Apiculture Wiewas (Cheshire): Conneticut.

Ribbands CR (1952). Division of labour in the honey bee community. Proc. R. Soc. B.; 140: 32-43.

Rinderer TE (1982). Regulated nectar harvesting by the honey bee. J. Apic. Res.; 21: 74-87.

Rinderer TE ed. (1986). Bee Genetics and Breeding. Academic Press, Florida.

Rosin R (1980). Paradox of the honey bee dance language hypothesis. J. Theora. Biol.; 34: 775800.

Rothenbuhler WC (1964). Behaviour genetics of nest cleaning in honey bees. Ame. Zool.; 4: 111-123.

Sakagami SF (1954). Occurrence of an aggressive behaviour in queenless hives with consideration of the social organization of honey bees. Insect Sociau; 1: 330-43.

Sandoz JC, Deisig N, Sanchez MG de Bıito, Giurfa M (2007). Understanding the logics of pheromone processing in the honey bee brain: from fabled lines to cross-fiber patterns. Frontiers Behav.; Neurosci.; 1: 1-12.

Schmidt JO and Thoenes SC (1987). Swarm capture with pheromone-containing trap boxes. Amer.Bee J.; 127:435-38.

Schneider SS, Stamps JA, Gary NE (1986 a, b). The vibration dance of the honey bee, I Communication regulating foraging on two time scales. Anim. Behav.; 34: 377-385. II The effects of foraging success on daily patterns of vibration activity. Anim. Behav.34: 386-91.

Seeley TD (1979). Queen substance dispersal by messenger workers in honey bee colonies. Behav. Ecol. Sociobiol. 5: 391-415.

Seeley TD (1985). Honey Bee Ecology. Princeton Univ. Press, NJ.

Sheppard W, Berlocher S (1989). Allozyne variation and differentiation among four Apis species. Apidologie; 20: 419-431.

Slessor KN, Kaminski LA, King GGS, Borden JH, Winston ML (1988). Semiochemical basis of the retinue response to queen honey bee. Nature; 332: 354-56.

Slessor KN et al (1990). Semiochemicals of the honey bee queen mandibular glands. J. Chem. EcoL 16: 851-860.

Su SK, Albert S, Chen SL, Zhong BX (2005). Molecular cloning and analysis of four c DNA from the heads of *Apis cerana cerana* nurse honey bees coding for major royal jelly proteins. Apidologie; 36: 389-401.

Songkum Su, Fang Cal, Aung Si, Shaowu Zhang, Jurgen Tautz, Shenglu Chen (2008). East Learns from West: Asratic Honey bees can understand Dance Language of European Honey bees. Plos One/ *www.plosme.org*, june2008/vol3/issue61/e2365.

Thorn C, Gilly DC, Hooper J, Esch HE (2007). The scent of the waggle dance. Plo Biol. 5(9): e228.

Trouiller J, Arnold G, Chappe B, LeConte Y, Billion A, Masson C (1994). The hormonal esters attractive to the *Varroa jacobsoni* mite in the queen brood. Apidol.; 25: 314-321.

Vaitkeviciene G, Apsegaite V, Tamasauskiene D (2006). Antennal response of workers to pheromone of virgin mature queen in honey bee (*A.m.c*). Acta Zool. Lituanica; 16: 31-36.

Vaitkeviciene et al. (2006). Queen substance (QMP) affects all workers in a colony and they assign to themselves various jobs for common good – cleaning, feeding larvae, foraging, etc. Acta. Zool. Tituanica: 16.

Von Frisch K (1967 a). The Dance Language and Orientation of Bees. Cambridge, Mass. USA, Belknap Press: 566 pp.

Von Frisch K (1967 b). Honey bees: do they use direction and distance information provided by their dancers? Science.; 158: 1072.

Von Frisch K(1971). Bees, Their Vision, Chemical Senses and Language. Ithaca, Cornell University Press.

Wanner KW, Nichols AS, Walsen KKO, Brockmann A, Luetje CW, Robertson HM (2007). Honey bee odorant receptor. Proc. Nat. Acad. Sci.; 104: 14383-88.

Watmough J, Winston ML, Slessor KN (1998 b). Distinguishing the effects of dilution and restricted movement on the intra-nest transmission of honey bee queen pheromones. Bull. Math. Biol.; 60: 307-327.

Wells H, Wells PH (1983). Honey bee foraging ecology: Optimal diet, minimal uncertainty or individual constance? J. Anim. Ecol.; 52: 829-36.

Wenner AM (1964). Sound communication in honey bees. Sci. Ame.; 210(4): 116124.

Wenner AM, Johnson DL (1967). Honey bees: Do they use the direction and distances information provided by their dances. Science; 158: 1076-77.

Wenner AM, Wells PH (1990). Anatomy of a Controversy. Columbia Univ. Press, New York.

White JW Jr (1979 b). Spectrophotometric method for hydroxymethylfurfural in honey. Assn. Off. Anal. Chem.; 62: 509-14.

Williams IH, Picket JA, Martin AP (1981). I Nasonov pheromone of the honey bee (*Apis mellifera* L). II Bioassay of the components using foragers. J. Chem. Ecol.; 7: 223-237.

Wilson EO (1963). Pheromones. Sci. Am.; 208(5): 100-114.

Winston ML (1987). The Biology of the Honey Bee, Harvard Uni. Press, Cambridge, MA.

Winston ML, Higo HA, Slessor KN (1990). Effect of various dosages of queen mandibular gland pheromone on the inhibition of queen rearing in the honey bee. Entomol. Soc. Am.; 83: 234-38.

Yadava RPS and Smith MV (1971). Aggressive behavior of *Apis mellifera* L workers towards introduced queens. I Behavioural mechanisms involved in the release of worker aggression. Behaviour; 39: 212-226.

Swarming

A normal hived colony lives peacefully and works according to its own rhythm and does not like to be disturbed. When the number of colonies is small a beekeeper has the temptation to open them and to see that all is well inside, but a colony should not be opened more frequently than once a fortnight. In a large apiary, or with migratory beekeepers who keep hundreds or thousands of colonies, it is not possible to give individual attention to them. Their health and progress is watched at the entrance; if the rhythm of flow of the bees going in and coming out is good, there are no dead bees in the front in excess, there is no foul smell coming from the hive, and there is no commotion of bees fighting, the colony may be considered in proper health. The colonies are given extra frames and supers for space or honey collection according to a set time schedule which an experienced beekeeper has developed for that locality and season of the year.

6.1 WHAT IS SWARMING ?

There comes a time when the calm of the colony is disturbed by the commotion of bees inside as well as outside a hive, and they appear to be getting ready to leave their house and fly out in a swarm. It consists of 30% to 70% of the colony strength. Swarming takes place under special circumstances, and the literature has been reviewed extensively by a number of scientists (Snelgrove, 1972; Ribbands, 1953; Ruttner, 1956, 1966, 1985; Butler and Simpson, 1967; Johnson and Johnson, 1978).

Swarming takes place under the following circumstances :

1. When the seasonal conditions are favourable, the bees as a result of their natural instinct to swarm (in spring) and colonize for the purpose of multiplication. This is the ordinary swarm.
2. When it becomes imperative to supersede the old and exhausted queen, the bees with the new and mated queen return to the hive and join others that remain on the frames.
3. When the living conditions become so unfavourable that the bees are forced to desert or abscond from their house. These swarms are issued under conditions of food shortage, unhealthy atmosphere in a hive or a severe attack of vermin (Wax Moth) and predator enemies, etc.
4. When a virgin queen leaves the hive for mating it carries with it a large number of workers and drones most of whom return to the hive later on. These are mating swarms.

Since swarming of any type disrupts the colony life it is not considered good for honey production. It has not been possible to understand fully why bees swarm, but many general observations and detailed studies on their activity have been made. It has given some insight into the conditions under which swarming takes place, and it has been observed that the factors which govern swarming are complex. A strong colony that has been in existence for 2-3 years is pre-destined to swarm but the phenomenon is also influenced by genetics of the stock (race or species), physical crowding, and age of the queen.

What is a swarming fever ? Rev. L.L. Langstroth in USA in the middle of the nineteenth century, described the fever thus: "On the day fixed for departure, the queen is very restless and instead of depositing her eggs in the cells roams over the combs and communicates her agitation to the whole colony. The emigrating bees usually fill themselves with honey just before their departure; I saw them lay in their supplies more than two hours before they left. A short time before the swarm rises, a few bees may generally be seen sporting in the air with their heads turned always towards the hive; and they occasionally fly in and out, as though impatient for the important event to take place. At length a violent agitation commences in the hive, the bees appear almost frantic, whirling around in circles continually enlarging, like those made by a stone thrown into the still water, until, at last, the whole hive is in a state of the greatest ferment, and the bees, rushing impetuously to the entrance, pour forth in steady stream. Not a bee looks behind but each pushes straight ahead, as though flying for "dear life", or urged on by some invisible power, in its headlong career".

6.2 THEORIES OF SWARMING

It has been studied now that in swarming, sensory stimuli, instinct and dances are involved. The general bee activity under all the swarming situations is more or less similar. Those who have studied the swarming impulse under the natural instinct of colonizing have put forth their theories, explaining the causes of swarming. The most accepted theory is that of Gerstung (1930), who considered that the desire to swarm is occasioned by the presence in the hive of an excess of nurse bees and a consequent super-abundance of larval food. Such a condition is reached when the queen has passed her peak period of laying or when she is limited for space in which to lay, whilst at the same time young bees are reaching the nursing age in increasing numbers. The theory is supported by the observation that the addition of unsealed brood to a colony which gives increased employment to the nurse bees delays or prevents swarming.

Agreeing with this theory, Snelgrove (1972) has observed that : "As the spring advances the queen, under the stimuli of rising temperature and more liberal supplies of food presented to her by her attendants, gradually increases the rate of her laying until this reaches a peak of maximum. She may maintain this peak rate for a few days, after which her laying powers slowly decline. Since the time of emergence of the young bees from the cells lags behind that of the laying of the eggs by three weeks, it follows that a time comes when the bees are hatching in increasing numbers, while the queen is laying at a diminished rate. The result is that in the seasonal history of normal stock there is a period when large number of nurse bees, eager to feed larvae, and of wax-secretors ready to build new combs can find little employment, and are therefore in excess of the hive requirements. This period coincides with the time when the foragers are ready to bring in quantities of fresh nectar and pollen. Instinctively the bees seem to realize that this is the favourable time for colonizing; they quickly develop the swarming impulse and the bee-keeper often finds it extremely difficult or impossible to prevent them from fulfilling their purpose". In support of this he proved that by separating nurse bees from the field bees the swarming instinct could be regulated and swarming in the parent colony could be controlled or at least delayed considerably.

6.3 HIVE ACTIVITY BEFORE SWARMING

What happens inside a hive in connection with swarming has been studied with much interest. In temperate climate winter is severe and bees are in a compact cluster in the hive. As the season warms up, or during early spring in moderate climate, the bees spread out on the combs, where they have a tendency to remain in a certain number per unit area. The newly emerged bees remain on the combs since it is warm there. The older bees of 4-10 days age move outwards where they clean cells and feed the queen during her rest period of 10-15 minutes. The queen remains busy laying eggs and uses for this purpose as many cells as there are prepared and polished by the nurse bees. The number of attendants to the queen increases from 10-12 in normal season to double that number when egg laying is considerably more. The feeding of the queen is also stepped up and bees in the inner circle become excited, pushing their heads under the head and thorax of the queen. At that stage she lays more than one egg in a minute, approximately 1500 eggs in a day, and she is always in search of empty cells, including the queen cells. The workers feed her vigorously and sometimes jump on top of her and perform the "joy" or the DVAV dance, lasting 3-4 seconds at a time.

When eggs in the queen cups hatch, bees supply them abundance of food (royal jelly), at the same time curtailing the feeding of queen. The bees push her around, make her go and be always on the move. The queen may continue to lay a few eggs each day even to the day of swarming. Owing to the cut on her diet she diminishes in size in the abdomen and becomes lighter in weight by about one-third, and thus can be airborne more easily.

The decrease in egg laying by the queen results in a further rise in the number of inactive or "displaced" nurse bees. They fill all the available space in hive and even hang outside the hive entrance. They are the active swarm bees (4-23 days old) ready to leave with the swarm. By now the queen is light-weight and agile. One week before swarming and up to the last day she makes piping sounds, in the vicinity of the queen cells. At times the piping sound may be as frequent as one per minute. In the swarming season, if we find larvae two or more days old in queen cells, and the queen is seen loosing weight, these are sure signs of issuance of a swarm from that colony.

A few days before swarming, a comparatively large number of bees may be seen "resting" on the bottom of the combs. At that time searcher bees go out and begin to look for a new nesting place. On return to the hive the searchers perform wag-tail dance for hours and days, indicating the direction and distance of the possible new site. Since the dance goes on all day the dancers go on adjusting the run according to change in the position of the sun. It was experimentally observed that when artificial queen piping sound was produced in a colony honeybees were activated for swarming preparations (Simpson and Greenwood, 1974).

6.3.1 The Day of Swarming

On the day of departure the searchers perform another dance. They get very excited, force their way amongst the bees in a zigzag manner, vibrating their abdomen, and producing "whir" sound with their wings. Sometimes it is called the "Whir Dance". One or two bees start this dance but after a minute or so there come dozens, till 20-30 thousand bees are excited and the whole hive is in a ferment. At this stage the queen is not leading but acts like an ordinary worker and follows the swarm. If queen is prevented from following the swarm by caging it or it is lost in the turmoil the swarm will return to its hive.

Before actually leaving the hive, the bees fill their stomach with honey and are set to go on their journey to the new pre-selected place where to settle. Full stomach is also related to the loss of memory of their mother colony. Those who do not have full stomach often return to the hive. Forgetting the mother colony and its location is a very important factor in the colonizing effort. It has been noticed that bees have their own ways of deciding on the time of departure and the selection of a new site. Generally swarming takes place between 10 am and 2 pm on a sunny windless day (also see Morse *et al.*, 1969).

6.4 FATE OF THE SWARM

From a beekeeper's point of view swarming never takes place when he is around or he is expecting it and a swarm rarely settles on the most prominent place where he thinks it should have settled. He is often surprised to see a swarm settled on a low obscure branch, on the top of a tree, or even a wooden pole. Sometimes neighbours will call a beekeeper to help them remove the swarm from the oddest place inside a barn or a verandah. The multiplicity of the sites and the deftness with which bee-keepers capture and hive the swarms becomes a part of their beekeeping art (**Plate 6**). One thing beekeepers and the village elders must do; they should be always on the look out for ignorant boys who have an ever lingering childish impulse to burn straw under a swarm to smoke it out. The bees in a flying or a newly settled swarm rarely if ever sting. It holds true even for the wild ferocious giant bees *Apis dorsata* F. Once I watched it sitting on *arhar* (*Cajanus cajan* L.) plants, hardly two feet from the ground and 12-15 feet away, were men cutting the crop. The bees did not attack the men and finally they took off as the harvesters were closing in on them.

6.4.1 A Moving Swarm

A moving swarm of *Apis mellifera* is 15-30 meters in diameter, the queen in the centre and workers flying around it. It does not seem to be in a hurry and at a leisurely pace it moves up and away from the mother colony. Before it settles on a tree or a bush a dozen or more workers take a leading part in guiding it. They reach the proposed spot, expose their scent glands and attract the rest of the swarm bees and, the queen follows as one of them. Pheromones seem to play an important role in controlling the swarm.

A swarm when it settles temporarily consists of two distinct parts : an outside compact shell about three bees thick with a hole going inwards; and an inner loose core in which the bees in bands are connected with the shell. The queen is in the centre surrounded by a ball of bees. On the whole the swarm is quite loose and sometimes the queen may fall on the ground along with a lump of bees but the queen generally joins back the swarm. Actually, if the queen and the bees surrounding it on the ground is caged or is put in an empty hive, the entire swarm will come to the ground and try to enter the cage or the hive. It is also possible that the queen may be lost or picked up by a bird and in that case the swarm many go back to the mother colony. It shows the bees do not lose memory of the old site suddenly. If on the other hand, the fallen queen with the lump of bees is touched with a gloved hand and you move away with your hand in the dress pocket, bees from the lump will follow you, tracking the pheromone odour on the glove.

In a normal, cluster there is division of labour ; on the inside there are bees up to 18 days of age; bees forming the shell are 18-21 days old; and the searchers are older bees. The shell consists of bees that are always moving from the inside outwards and two third of them change places every 10 minutes (Seeley and Morse, 1976, 1977; Ambrose, 1976). Settling of a swarm has been studied in detail by Lindauer (1955b), Meyer (1956a, b), Seeley (1985a, b).

6.4.2 Search for Permanent Site

A swarm almost always first settles on a temporary site till the searchers have found a good permanent place. According to Lindauer (1955b) the searchers who fly around in the field in search of a site keep coming back to the settled swarm and they perform wag-tail dance all over the surface of the shell indicating more than one prospective sites; the better the place the more lively is the dance. The searchers seek open space, availability of sun and protection from the wind.

A distant place is preferred over a close one for settling. Seeing a vigorous dance, the searchers of the humble places visit the better place and on return they also perform the worthy dance. Out of this wide "advertising" the searchers agree upon the best site. If they cannot form a consensus and there is fear of swarm being divided into two they will rejoin the section where the queen is. Sometimes they discard all the sites selected and may build a nest right where they have settled.

When the searchers have "agreed" on the new site they start the "whir dance", boring their way into the cluster. Activity is generated all round and a loud humming noise is heard from inside the cluster. The activity is also controlled by pheromones. Bees recognize their queen from the odour of the queen substance (90DA) and they are stimulated by 9HDA to expose their scent glands (of workers). By fanning the odours are circulated and the workers are attracted to the queen and to the swarm. Thus, both pheromones and physical movements are involved in swarm orientation. If a cluster is queenless bees perform more of the breaking dances and eventually the cluster breaks up and returns to the mother colony. In calming and settling a swarm the Nasonov scent acts in synchronization with the effect of queen substance.

In case the searchers have agreed on a new site, have performed the "Whir dance", bees run to and fro and finally 5 to 10 bees fly out from the cluster, followed by about 100 leaders who reach the new site. The cluster disbands, a bulk of the bees go to the new site and at first a regular communication is maintained between the new site and the cluster site. When in a swarm, bees are gentle and can be handled with ease. Since a swarm is without a house there is no place to defend-hence they do not sting. A swarm unable to find a suitable place, and if bad weather exhausts their food reserves in the stomach, bees are stuck at the temporary site and start considering it their home. The bees defend it and may sting the intruder. Such a swarm is difficult to hive.

6.4.3 Secondary Swarms

When swarming fever starts in a colony it will last for sometime and a strong colony may issue 2-4 swarms in a season. Since the mother queen leaves with the primary swarm it is the agility and proloficity of the daughter queens that will determine the future behaviour, in issuing secondary or tertiary swarm. The availability of good forage and weather are also important determining factors. Some races swarm more than others and the Italian honeybees (Ligustica) does not swarm excessively.

In case more than one virgin queens emerge at the same time all of them may accompany the secondary swarm but it is a rare phenomenon. Normally one queen accompanies the secondary swarm or such a swarm may not issue at all. If a secondary swarm does not issue, the virgin queen that emerges first cuts open the extra queen cells and after mating becomes the queen. More than one queens may emerge simultaneously and in that case there is a fight among the virgins and the dominant one kills the others and reigns the colony after mating.

It may be seen that the issuance of swarms is a method of further colonization. Unless some of the colonies perish new swarms from the daughter colonies in due course may lead to over population. Bees control their number through various devices as directed by their instinct. In nature, the number of colonies per unit area is determined by the nesting sites available. However, man by using wooden boxes has increased the sites artificially. So also he has improved the carrying capacity by intensive agriculture. Man has devised another method to increase the density of bee colonies. It is by practicing migratory beekeeping; the colonies are moved with the march of the bloom. In northern hemisphere the blooming of flowers is late as we move from south to the north and in southern hemisphere it is the other way round.

6.5 HIVING A SWARM

The practice of hiving a swarm is based on the basic bee behaviour discussed earlier, and by devising a few ingenious practices to reach a swarm it is put in a hive. No beekeeper can afford to or likes to lose his swarms. By catching and hiving a swarm he can increase the strength of his apiary. He can also sell them to prospective beekeepers but it is not fair to give to the new keepers old queens. Being inexperienced the new beekeeper with old queen may not succeed in maintaining the colony and may get discouraged.

It is important that the colonies sold to beekeepers contain young queens which can be ensured only by preparing nucleus colonies or package bees for selling. Therefore, the natural swarms that are hived and have old queens should normally be retained in the apiary and not sold to new beekeepers.

Catching a swarm from the wild appears to be an act of daring and agility but an experienced beekeeper knows that swarm can be handled with ease. The bees in a swarm are engorged and they are pre-occupied with the situation of sheltering their queen and establishing a new colony. There is no conflict between man and bees and the latter need to be assisted so that they can satisfy their instinct of colonizing. For this purpose a beekeeper is advised to acquire a few tools for catching, caging and transporting a swarm to the apiary, these are :

1. A strong 10 feet long pole with a heavy iron hook or a brush screwed into its end for dislodging swarm from a tree.
2. A funnel made of galvanized iron sheet 2 feet in diameter at the top, narrowing to 4 inches at the neck. A tea-leaf collector's basket open at one end and narrow at the other end can serve the same purpose. The funnel is placed on an open hive so that when a swarm is dislodged the bees will flow into the box below.
3. A box made of wood or wire gauze in which to pour bees before moving them from the swarm site to the apiary. A hive or a super with 2-3 frames can also be used.
4. Queen cages for confining the queens (if more than one in a swarm) and a piece of metal wire for hanging cages in the box.
5. Hedge scissors or pruners with long and short handles for cutting twigs and letting the swarm fall on to a box. Sometimes a small saw is needed to cut a thicker branch.
6. A light weight ladder, preferably of the type that can be extended.

Once a swarm is shaken into a box it will be very helpful if the queen is searched and caged. A queen cage hung in a box will bring the swarm under control and it can be taken to the apiary for transferring it to a hive. It is the pheromone odour of the queen that keeps the bees in near vicinity even when the queen is moved. Now artificial QMP is available to attract or catch queenless swarms or cluster of bees.

The swarms are generally hived without confining the queen. So long as the queen is there in the hive box the bees settle down within an hour or so. Some of the bees that might have been flying around at the time of dislodging the swarm come back to the same site and they are seen even the next day. Since the swarm is already moved they have no way of joining it, so they either go back to the mother colony or stick to the site tenaciously till they die of starvation.

6.6 HOW TO PREVENT EXCESSIVE SWARMING

Excessive swarming is contraproductive. The main purpose of beekeeping is to get a good crop of honey which is likely to be obtained from a strong colony, yet maximum hive strength is not an end in itself. An amateur beekeeper may gain professional pride when he has a hive boiling with bees. However, the purpose is to achieve maximum strength for gaining maximum yield per hive. At the same time swarm control is the most important element of hive management. How a swarm develops is now clear enough but what forces it to issue is not fully understood. Quite often it is observed that if a colony is "determined" to swarm, it is going to swarm. If queen cells are broken to prevent it, the swarm will take the old queen with it, leaving the colony behind to raise a new one. Lack of space and ventilation in hive is the foremost cause of swarming. It is corrected by providing more supers; so as to keep one super ahead of the requirement of bees. Too many extra supers are also not desirable because that induces the bees to move upwards in the central portion creating a "chimeny" pattern of brood raising on the combs and leaving the uncovered empty space on the periphery. Wax moth and other enemies appear in the empty space because it is not covered by bees.

The hive conditions that are popularly supposed to induce swarming are :

1. Lack of comb space for the spread of the brood.
2. Lack of storage room for honey leads to idleness of field bees.
3. Crowding of brood nest particularly with young bees.
4. Insufficient ventilation and unhygienic conditions in the hive.
5. High temperature.
6. Advanced age of the queen and its insufficiency in laying eggs.
7. Pedigree of the queen having blood of a race of bees prone to swarming (*A.m. adansonii*).

The outward signs in a hive indicating a swarming disposition are :

1. Young bees are in excess and they cover the brood rather than the nurse bees do it.
2. The nurse bees instead of taking care of the brood take up other duties like cleaning cells.
3. Pheromone activates the bees to feed the queen more abundantly so that queen may lay more eggs which produces more idle bees.
4. Many queen cells are constructed along the lower edge of the combs at the rate of one or two per day for a week or more. The cells are lighter in colour and appear to be off-white, and can be distinguished from the supersedure cells which are on the sides of the comb and are darker in appearance.

For the control of swarming well directed operations are required. Early in the spring when day-length is increasing the colonies should be examined once a fortnight. Any cause of discomfort to the colony should be removed. Ventilation should be improved by providing wire-gauze vents on the inner covers; more worker combs be given for the queen to lay eggs; combs containing pollen and honey be moved outwards; watering places be provided in the apiary; shade covers or wet cloth be provided above the hives if natural shade is scanty.

There are certain special operations which are to be undertaken for the control of swarming. There are a variety of operations devised by beekeeping experts and some are named after them. The Demaree plan of swarm control is the well known method. It is employed in an apiary when it is apparent that most of the colonies will develop queen cells.

6.6.1 Demaree Method

It was in 1892 that Charles Demaree described this technique. The hive is removed from the bottom board and in its place a new box is placed. In this box one comb containing unsealed brood, eggs, workers, and queen is placed in the middle and empty combs are placed on the sides. The chamber is topped with a queen excluder and, a super with empty combs, is placed above it. The original chamber containing bees, brood, food reserves of the original colony are placed on the very top, covered with inner board and the outer cover. Before the colony is closed the top most chamber is examined for any queen cells; if found those are to be destroyed or grafted elsewhere for raising nucleus colonies. After one week of closing the colony the top chamber is again examined for queen cells; if any is being raised it is to be destroyed. In three weeks all the brood shall have emerged out of the old combs at the top which will be available for honey storage.

Many variations of this technique have been devised. The idea is to confine the queen, some brood, and plenty of empty combs in a hive box at the bottom of the main brood stock and separating the two with a queen excluder in between. The result expected is that worker bees in the upper chambers will start collecting honey and those who pass through the grill of queen excluder and reach the bottom brood chamber will take care of the brood. Since there will be no queen cells, virgin queens will not emerge and swarming will be avoided. The colony will take care of the old queen.

This method is quite labour intensive and up to 20-50 colonies can be handled but if the number is large some other practices have to be devised. Moreover, condition of the colonies may be different from year to year.

6.6.2 New Queen Every Year

Snelgrove (1972) has described various other methods of swarm control. In commercial beekeeping the most common practice is to give new queens to colonies every year. It not only makes the colonies strong but also swarming is checked. Once the number of colonies increases beyond the home or hobby scale it is not practical to manipulate individual colonies so frequently.

Devices have to the improvised for natural swarm control in hives so that chances of swarming are reduced. Effort $hould be made to keep the colony in good strength (8-10 frames) throughout the winter months.

The various manipulations required to be followed in the apiary include :

1. Provide queens around one year in age.
2. In the spring move the colonies to a shady place with an open front yard.
3. Raise the front or the back side of the hive by one inch for air and water flow.
4. As soon as the late winter or early spring blossoms appear (fruits, oil seeds etc.) supers having combs should be given above the queen excluder.
5. When the super is well used up, place another one with partly drawn combs or ones having foundation sheets fixed; always give one super extra from their immediate needs to keep the bees occupied.
6. For a natural control of the swarm the queen cells close to being sealed should be covered with detaining screens, and later the queens be used either for preparing nucleus colonies or for making an artificial swarm having a new queen.

6.6.3 Destruction of Queen Cells

To avoid swarming the new queen cells are destroyed regularly once a week. Apart from the fact that it involves lot of work the method is considered undesirable and inhumane. In spite of the best of efforts some queen cells escape detection and consequently some swarms issue anyway. In modern beekeeping this method is not recommended.

6.6.4 Hawkin's Method

Without the queen excluder supers along with combs are supplied in plenty to provide lot of space for bees to raise brood and collect honey. In case there are three boxes, the frames containing young brood are moved to the lowest box, those containing honey to the top box, and sealed brood in the middle. Good honey yields can be obtained by this method but it is time consuming as the queen cells have to be destroyed frequently.

6.6.5 Removing the Queen and Requeening

In the swarming season queen is removed from a colony which is left queenless for three weeks. The queen cells are destroyed every 5th and 10th day. Alternatively, on removing the queen a newly raised queen is introduced in the colony. Apart from preventing swarming, the good pedigree queen will upgrade the colony.

6.6.6 Dequeening

Whether the bees have developed swarming instinct or not queen is removed and caged over the feed hole; any new queen cells are removed every 5th and 10th day. After 10 days the queen is removed altogether. Thus the bees have no means of raising their own queen. A piece of comb from a good stock is then grafted into a middle comb of the colony, for bees to raise a new queen. The queen excluder is placed on top and supers are given above it. After five weeks from the day of queen removal the new queen will take over the colony. During the five weeks the colony has a caged queen, or queen cells or eggs, therefore, laying workers, do not develop. A good crop of honey can be obtained and at the same time swarming is prevented.

6.6.7 Clipping Queen

Before the swarming time wings of the queen on one side are clipped with scissors. For doing this the queen is held from the thorax between thumb and a finger. Some beekeepers clip queens as a regular practice in the spring and feel secure. Even if a swarm is issued the queen cannot fly and the swarm returns to the hive. The clipped queen may fall to the ground and be seen with other bees for a day or so; it has to be searched and introduced back into the colony. Having satisfied with the swarming instinct the colony settles down.

6.6.8 "Shook Swarming"

It involves the shaking of all or most of the bees of a strong colony in front of a newly set-up hive containing frames. The type of frames to be given depends upon whether or not preparation for swarming had been made. In case there is no preparation for swarming the combs given are filled with recently gathered honey. One comb containing both sealed and unsealed brood is placed in the middle to encourage the shaken off queen to continue laying. A super is placed above the queen excluder and empty frames are then added in the chamber below.

In case the bees have prepared for swarming the hive is pounded upon so that bees are alarmed and they fill themselves with honey. Then the frames are shaken at the new site and no young brood is given in that hive and hence a new queen cannot be raised and the bees behave like a newly hived swarm.

A good number of bees will return to the original hive and take care of the colony there. If all the bees are shaken off and insufficient number go back to the mother colony the surplus frames of bees can be given to some other weak colony.

After studying the basic bee instinct and recognizing the signs and symptoms of swarming practices can be devised for their prevention. The presence of queen cells in a colony in the spring is a forewarning for possible swarming. In a new colony for the first year or two a queen reigns supreme and swarming does not take place if there is no crowding in a colony. So the best prevention of swarming is by keeping young queens in an apiary. Good management is necessary not only for obtaining high honey yield but also to off-set swarming.

Colony Life and Food Collection

7.1 STORAGE SPACE

Nectar and pollen of plants is a complete diet for the brood and adults in a honey bee colony. The former meets the requirements of carbohydrates and the latter provides proteins, fats, vitamins and minerals. The adult bees eat this food and convert some of it into a rich secretion called the royal jelly which is also used as food.

The bees store food in their nest which consists of vertical combs containing horizontal hexagonal cells of two sizes, the worker cells and drone cells, and sometimes irregular or imperfect honey cells. The cells are inclined upwards at the open end hence nectar does not flow out. Nectar is stored in all the three types of cells and pollen is generally stored in the worker cells.

Both old and new combs are used for food storage. When newly made from wax secreted by the bees the combs are white. When cells are used for rearing brood a number of times the cocoons and larval excrement are not cleaned by bees perfectly and some fragments remain deposited in the cells, giving it a darker colour. Some beekeepers use old frames for ten or twenty years, but it is not considered right. Ordinarily the combs should not be used for more than four years or so in order to reduce the chances of brood diseases appearing in the colonies.

A typical comb in the middle of a hive has brood, pollen and honey (**Plate 5a**), but away from the middle there may be a comb full of honey only, pollen only or mostly honey and some pollen. On the middle combs, brood is in the centre, spread like a half moon. On top of it there is crescent shaped band of pollen and above that is honey spreading like a roof and downwards to the sides. Honey cells and brood cells are sealed with wax, but the pollen cells are naked. Sometimes, a layer of honey is packed on top of the pollen and in that case the cells may be sealed and the good quality of pollen maintained for over a year. This pollen is used for early brood rearing after the winter. The middle three or four combs in a colony have bees and brood in the central area and together they form a spherical "brood nest". Above this nest forms a concave canopy of pollen and above it a roof of honey. This habit of the bees is exploited by beekeeper by collecting pure honey from the top. Above the main chamber of the hive are placed one or two supers where the bees store honey and if those are separated from the bottom chamber by a queen excluder, the queen cannot go to the supers for egg-laying, whereas the workers continue coming up for storing honey. Honey is harvested by the beekeeper through honey extractor.

7.2 FORAGING

As mentioned earlier the workers have to perform different duties as their age advances. After performing cleaning, wax secreting, guarding, and nursing duties, the bees become house bees or foragers at 3-week age. The last two duties are interchangeable and among the foragers the most experienced are the scouts who find new sources of food, bring samples in the hive for others to taste, and perform dances to convey the information on location. (Fig. 16). There is flexibility among the foragers as to what to collect; whether they collect nectar, pollen or both. Extensive literature is available on foraging which has been summarized by Eckert (1933); Eckert and Shaw (1960); Smith (1960); Crane (1990); Seeley (1978, 1985a, b); Free (1993); Gary (1992).

7.2.1 Floral Fidelity and Food Gathering

Honeybees specialize on one type of foraging at a time and after a number of trips they may switch over to another (Fig. 19). In general, workers tend to prefer nectar collection. In a study it was observed that 58% of foragers collected only nectar, 25% only pollen and 17% both nectar and pollen (Parker 1926, Free 1963, 1967). Both nectar and pollen collectors show a great variation in the number of flowers visited per trip, the number of trips made in a day, time spent on collecting a load, and the weight of the load. They need from 1 to 500 flowers to collect a load and make 10-15 trips in a day. During the full honey flow, however, the foragers may make as many as 150 trips in a day. Pollen collection generally takes about 10 minutes to prepare the load whereas a nectar collector takes 30-80 minutes. The weight of the pollen loads is 10-30 mg and that of the nectar is 25-40 mg. Since a full load of pollen is lighter in weight the pollen foragers travel further than the nectar foragers.

7.2.2 Influence of Environment

The environmental conditions, particularly weather and flora, influence nectar and pollen foraging. The field activity generally starts at 12-14°C but bees can work at much lower temperatures as influenced by light intensity. At temperatures higher than 43°C, the nectar and pollen foraging stops but water collecting continues. During rain and windy weather the foragers stop activity but as soon as the weather clears a bit the bees react quickly and start field work. Their flight activity more or less stops if the wind is blowing over 24 km/hr. Bees work more actively in the mornings and evenings coinciding with the secretion of nectar in flowers. There is much less activity in the early afternoon, perhaps because there is less flow of nectar at that time of the day.

7.2.3 Foraging Efficiency of a Colony

Foraging efficiency at the colony level is achieved by two sets of regulatory mechanism; those related to "estimates" of colony requirement and allocation of foraging tasks. The pollen requirements of a colony are 15-30 kg per year. Since the average load of pollen is 15 mg per trip and the average load of nectar is 32 mg. (= 16 mg of sugar), they must make one million trips in a year to collect pollen and four million trips to collect nectar. Under good weather conditions workers of a strong colony can make up to 163,000 trips in a day, and theoretically they should be able to collect their yearly provisions in about a month. In real situations, the period when flowers are blooming is quite short and the trips can be made within short duration during the hours and days when weather conditions are favourable.

7.2.4 Nectar and Pollen Requirements

Generally, in temperate climate the season is short, and in subtropical climate although the total season is long, yet the availability of single species of flowers or a crop is for a limited duration. If surplus honey is expected in a colony the availability of nectar has to be over and above the basic requirements. It has been estimated that a colony collects 5-6 times the amount of honey that is extracted as surplus honey. Although Kotze (1949) collected one tonne of surplus honey from a colony in a year the average range is 50-80 kg in USA. When there is no foraging an average colony needs one or two supers of 40-50 kg honey for the winter (Johnson & Johnson, 1978, 1979).

The basic requirement of a colony has to be translated in terms of field activity of the foragers (Dyer & Gould, 1983). A great deal of coordination is required in the colony to allocate foraging tasks to be carried out with maximum efficiency, for example, activity of recruits to search for resources, transmission of this information to foragers, and the decisions on shifting to another resource when one is exhausted. It appears bees have many sources of information based on which they determine the colony needs and they adjust their foraging patterns accordingly. It has been closely studied in regard to increased pollen collection when there is a greater brood rearing activity. Thus doubling the number of eggs and larvae in a colony increases the pollen collection three times and the bees react to the situation of requirement and collection very quickly. If dry pollen or pollen substitutes are provided, the foragers immediately cut down pollen collection from the field, as their needs have been met by alternative means.

7.2.5 Coordination

Seeley (1983a, b; 1985a, b) has called the "information centre" strategy of foraging as the basis of coordination of field activity with the colony requirement. There are daily changes in the number of workers visiting particular patches; it is guided by the scouts who search in a area 4-6 km radius around the hive. Foraging is so organized that bees concentrate on a few patches which are found to be the best and foraging is adjusted daily as the old sites are neglected and new ones having better resources are searched by the scouts. This information is passed on to the foragers by the vigour of the dances and by food samples (odour) provided by experienced scouts and, therefore, over all energy and time spent by the colony as a whole is utilised for maximum benefits.

7.3 FORAGING AREAS

Individual bees forage in a restricted area of a field, on a group of trees, or even on a single tree or a bush. On a given trip a bee forages on flowers of a single species and she continues to do so as long as food is available. Sometimes bees collect mixed loads of pollen also and those average 3% of the total; 12.6 per cent of bees visit several plants differing in odour and structure. The mixed loads are not necessarily made during the dearth period but those are also observed during the main pollen foraging.

In general the bees have a tendency to forage in limited areas of a field but in a single species crop where there are fewer land marks they may drift and forage in a much larger area. Moreover, when nectar and pollen production declines in a crop the bees spread out and forage with a wider coverage. Even when they have shifted to another field they keep visiting the old site, even though with much less frequency. The size of the foraging area may vary a great deal as determined by the number of flowers available, their nectar and pollen content and competition among bees. The intensity of foraging or the rate at which they visit flowers depends upon the richness of nectar and pollen crop, the type of flower and its stage of development, climatic conditions, that is, temperature and humidity etc. Singh (1950) observed that a given honeybee would constantly forage on apple blossom within an area of 4 m^2 day after day so long as nectar and pollen were available.

7.3.1 Distance and Speed of Flight

Although bees have been observed to fly as long as 13.7 km their tendency is to remain as close to the colony as possible provided forage is available (Eckert, 1933; Levin, 1959, 1961a, b). It was noticed that when a large area of honey plants was close to the colony most of the bees worked at a distance of 200 to 500 metres. In an experiment a colony was deprived of old workers and the younger workers had to take over foraging duties. Those who went out for the first time were recovered within 300 metres and at the end of the first day some of them ventured as far as 450 metres. Only when they grew older and became more experienced they spread out to wider areas of foraging. Some 5-day old bees were released 100 metres away from the hive, they could not find their way back. However, 80 per cent of the 10-day old bees returned to their hive and the same percentage of 35-day old bees returned from a distance of 500 metres.

Bees from hives placed in two separate apiaries tend to work in areas close to their own apiaries and they become less common as the distance increases. However, if a rich source is located at a considerable distance the bees will fly there, rather than work on a poorer source nearby. It has also been observed that bees may prefer to work on flora at a distance even though rich flowers may be available near the apiary. This diversity of foraging is contrary to the accepted habits and may be connected with nutritional requirement of the colony; in other words, food from some species of flowers meet the food requirement better. The bees going out for foraging fly at a speed of 2-4 km/h less than the loaded bees coming to the hive at an average speed of 24 km/h. It is possible that the bees going to the field look around here and there for some better prospects of foraging and that is the cause of their apparent slowness.

7.4 NECTAR COLLECTION AND STORAGE

Located within the flowers of many plants are nectaries that secrete liquid consisting of sugar and water, known as nectar. In addition there are extra floral nectaries on leaves and stems of plants which are visited by the bees. The honey-dew, an excretion of aphids and leaf hoppers, is also exploited by the bees as a source of sugar, particularly on mangoes, pines and palm trees.

7.4.1 Approach to Nectaries

The percentage of sugar in nectar varies a great deal (5-80%) and bees are attracted to flowers for the sake of nectar which is a kind of reward for their services in cross pollination. Once at the site, the bees are guided to the flowers by vision and smell and their activity will depend upon whether they are nectar collectors, pollen collectors or both. A nectar collector visiting a flower probably does not "know" nectar is there in a given flower. The nectaries have different odour than the rest of the flower and a bee is guided to them by the smell. The presence and quality of nectar is found out only when the bee dips her tongue. If the flower is large enough the bee may alight within but if the flower is too small or delicate she may sit on a convenient part of the plant and work from there. When nectar is found, a bee brings forth its tongue from the folded position below the head and, sucks it within her reach. If there is no nectar the tongue is withdrawn and the bee moves on to another flower. In case another bee has already alighted on the flower she may avoid it even though nectar may be present!

7.4.2 Time Spent on Flowers

Since bees keep moving in the field from one flower to another it is not easy to record the exact number of flowers from which a load is collected but it is estimated that many hundred flowers of a clover are visited for this purpose. Any plant species from which a bee can collect a load by making less than hundred visits is considered a very rich source. Moreover, large loads of nectar being brought to the hive indicate abundant and desirable flora. The time spent by a bee to collect a load varies according to the plant species; on sweet clover it ranged from 27-45 minutes and on *Limnanthes* 106-

150 minutes. In between the field trips bees spend approximately 4 minutes in the hive and on an average they make 7-13 trips depending upon honey flow. The weight of a large load may be 70 mg *i.e.*, 85% of the body weight of the honeybee. During honey flow the average load is 35-40 mg. In general, trips for pollen collection are shorter than those for nectar collection. The actual time varies according to factors like weather, distance of food source, and individuality of the forager (Park, 1949). When nectar flow is good a colony on two chambers can collect 5 kg of nectar per day and in one year collect 200 kg of surplus honey.

7.4.3 Unloading Nectar

A nectar forager on entering the hive moves to a place where other workers of the same kind are present (Fig. 18). When she meets a hive bee ready to take the load, she may give the entire load to her or distribute it to three or more workers. In case her nectar is from a rich source, being bountiful and having high sugar content, she may perform the communication dance, giving a taste of her sample and then handing over a bulk of the load to one worker bee. While giving the load the forager opens her mandibles apart and brings out a drop of nectar on the upper side of the base of the tongue, folding the distal portion downwards. (Fig. 18) The house bee stretches her tongue to full extent and sips the nectar. While this is going on the antennae of one are continually striking those of the other and the house bee is seen to stroke the "cheeks" of the forager. After nectar has been transferred, the field bee pauses to eat some food, wipes her tongue between the forefeet, rubs her eyes, cleans her antennae and takes off for the field.

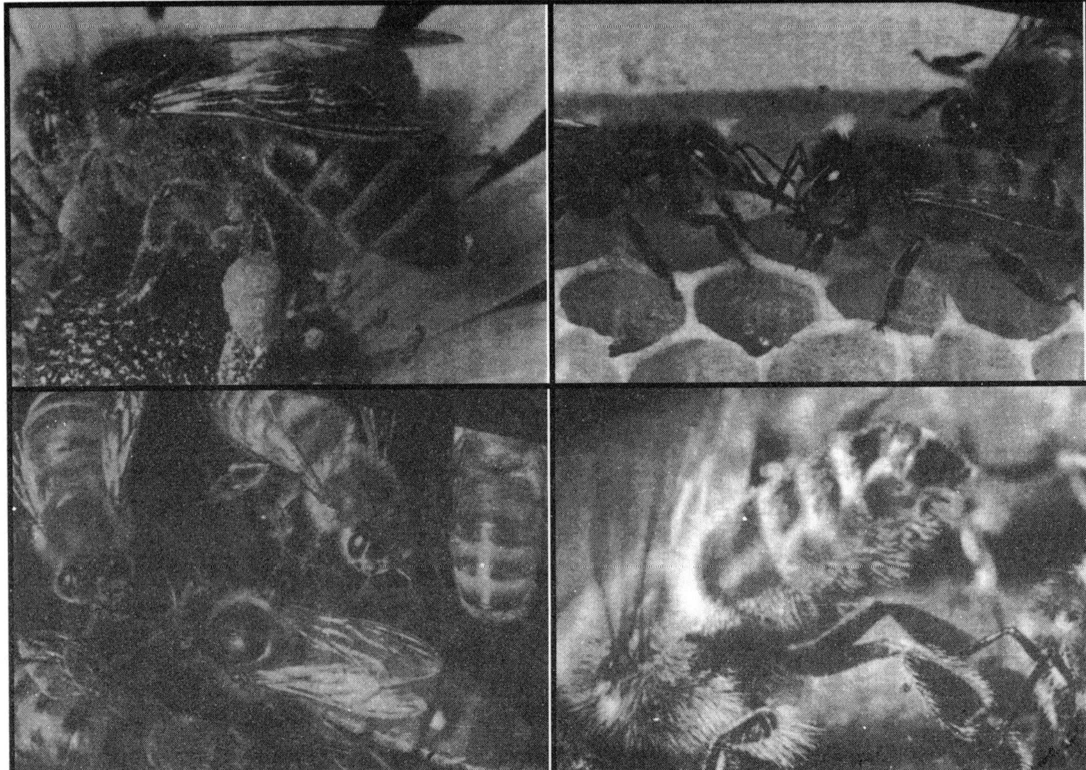

Fig. 18. Duties of honeybee workers : Clockwise from top, worker gathering pollen; worker transfering food on the tongue of another, worker exposing scent gland; workers attending on queen (adapted from McGraw Hill Study Prints).

7.4.4 Honey Making and Ripening

Depending upon the source, nectar brought from the field varies in sugar content ranging from 4 to 80 per cent, but more commonly from 35% to 45%. Honey is made from nectar by the hive bees by making two changes; one chemical and the other physical. The chemical change is brought about with the help of enzyme invertase (of salivary glands) and complex sugars (sucrose etc.) are converted into simple sugars, fructose and glucose, the former being always in a greater quantity. The physical change is brought about in ripe honey by evaporating excess water so that it is always less than 25% and close to 18%. When the honey is ripe it is sealed in cells with beeswax. The moisture is evaporated in many steps and until the unripe honey is made the liquid remains nectar, and later on it becomes ripe honey. At these early stages the liquid easily spills over when a full comb is pulled out and is tilted a bit.

During the honey flow when nectar is being brought at fast speed the house bees store it temporarily in the form of droplets hanging on the upper walls of cells. It already contains invertase and the spherical surface provides the maximum area for evaporation of water. Later on bees pick them up, fill their stomach and in a secluded area of the comb start the ripening process. A droplet of nectar is brought out of the pre-oral cavity and is held between the palpi and the distal proboscis of the tongue. It is increased in size and then reduced in size by sucking inwards and the process takes 5 to 10 seconds. It is repeated constantly by taking out and bringing in droplets for about 20 minutes. When the droplet is concentrated sufficiently the bee entres a cell with its head first and ventral side upwards, opens the mandibles, and brings the droplet on the dorsal surface of the tongue. With the help of her mouthparts and by moving her head side to side she "paints" the upper wall of a cell with liquid which may then run to the base of the cell. If there is honey already in the cell she may directly add the droplet without painting. Generally one fourth of the cell is filled at this stage and it takes many hours for the conversion of sugar (Park, 1949).

Evaporation of moisture continues from these cells and within 2-3 days the unripe honey become ripe honey ready to be sealed, the duration depends upon the initial concentration of the nectar. The availability of comb space is also important in the process of evaporation. The bees keep shifting unripe honey from one cell to another cell and eventually the ripe honey is stored on top portion of comb and is finally sealed. Temperature, humidity and colony strength play important roles in speeding up the process of honey ripening. Even after the honey (19% moisture) has been sealed it is possible to further reduce the moisture content to 17% or less by taking out frames and keeping them in a warm and ventilated room.

7.5 POLLEN COLLECTION AND STORAGE

A strong colony collects 35 kg of pollen in a year: pollen is the source of proteins, fats, minerals and vitamins required for the growth and development of brood and newly emerged bees. Older bees do not eat pollen but in times of great necessity they do secrete royal jelly, containing proteins and minerals etc., which is the outcome of their wasted body tissues.

Casteel (1912) described in detail how bees collect pollen while foraging on sweet corn. He states that a bee sits on tassel and clinging to the anthers bites them with mandibles, thus releasing the pollen grains. With the help of its tongue it licks the grains, which become moistened. The pollen grains also fall on hairy body and legs and this dry powder sticks to the branched hairs. After she has visited a few flowers her body is covered all over with pollen grains. She then brushes her head, body, and forelegs and transfers pollen to the hind legs and fills the pollen baskets located one on each leg. This task is accomplished either while the bee is sitting on a flower or while she hovers in her movement from one tassel to the other (see Michener *et al.*, 1978). While foraging workers choose pollen not by their nutritive value, age, moisture content or colour but on the basis of odour and physical configuration of pollen grains (Fig. 19). Invariably a colony collects mixture of pollens probably to meet nutritional requirements (Stanley and Linskens, 1974; Jay, 1986).

7.5.1 Pollen Basket

The wet pollen is removed from mouthparts with the help of front legs and dry pollen on the body is mixed with it. Pollen on the ventral side of thorax and other parts is removed with the second pair of legs and pollen on the first pair of legs is also moved there. Pollen is gathered on the inner surfaces of the broad tarsal segments of the middle legs. This pollen is patted upwards into the pollen basket. A greater quantity of pollen is transferred to the broad pollen combs located on the inner surface of the hind legs which is called basitarsus. The hind legs are moved up and down beneath the abdomen. The comb of one leg is rubbed on the pollen basket located on the outer side of the other leg on the opposite side. The pollen is pushed inside the basket and as the bulk increases the basket swells and pollen is held in position by a single hair that acts like a pin. Other hairs are pushed outwards and hold the pollen mass along the edges of the basket. Thus, compared to the body size of a bee a huge quantity of pollen is collected on a single trip. The weight of a pollen load is 10-30 mg which is 25% of the body weight of the bee (Fig. 18).

7.5.2 Unloading the Basket

The pollen gatherer when loaded fully flies back to the hive. She either walks quietly on a comb or is agitated and performs the communication dance. The pollen bearer searches for a suitable place to unload and puts her head in many cells as if to determine their suitability. When a cell is selected she holds on to the edge with her forlegs and puts her abdomen inside touching the base. The hind legs hang freely but the middle legs are used to push out the pollen load from the basket. After getting rid of the pollen pellets the bee leaves and a younger house bee comes along and packs the loose pollen by pushing it with its head towards the bottom of the cell. In this process the pellets break off and pollen is spread out evenly with mandibles and tongue. Honey or nectar and secretion of brood food glands are mixed with it and the stored pollen becomes darker than the original colour. It is commonly called "beebread". By the mixture of royal jelly the germination of the pollen is stopped and its keeping quality is improved.

7.5.3 Environment and Pollen Collection

The trips of pollen gatherers are shorter than those of the nectar gatherers. A bee makes a full load of pollen in 6-10 minutes and in a day visits the field 10-15 times. Temperature, humidity and wind velocity influence the number of flowers visited, time spent in making a load and the weight of the load. Pollen collection is reduced above 35°C and in wind velocity above 17.7 km/h. It has been noticed that bees from different colonies in an apiary may visit different species of flowers. In fact bees from the same colony may also visit various species of flowers (Park, 1926; Vansell, 1942; Ribbands, 1949).

Fig. 19. Various sources of nectar and pollen being exploited by four species of *Apis*.

Hirschfelder (1951) calculated that during good weather 50,000 to 54,000 loads of pollen were brought into the hive in one day, and a bee could make upto 150 trips. A colony requires 15-30 kg of pollen a year and may collect upto 55 kg (Eckert, 1942; Lauvedux, 1958).

7.6 WATER COLLECTION AND STORAGE

Honeybees need water to drink and to spray it in the hive for cooling. They also use water for humidifying the interior, otherwise, the young brood would die. They prepare the larval food by thinning honey with water. When fresh nectar is available they use it for the same purpose, but in the absence of nectar water is needed. Early in spring the water carriers are very active. As they bring water they start vigorous dancing on the comb and at quick intervals they give a sip of water to the bees who follow them. Sometimes a water carrier dances for a full minute and at other times the dance is very brief and then she distributes water to 3-6 bees (Park, 1923).

Water is stored on top of frames within cell areas made out of reused wax or propolis. The water droplets are also stored inside brood cells and on the cappings of brood cells. There are some bees in the hive that serve as water reservoirs; their stomachs are full of water and they remain inactive near the brood area. Whenever nurse bees have to feed larvae and they need water to process the food they get their sip from the reservoir bees. During the dry season when water is scaıce there may be many hundred or thousands of reservoir bees; they store diluted honey also (Lindauer, 1955).

The activity and efficiency of a water carrier depends upon how quickly her load is taken away from her in the hive by the house bees. It takes 2-3 minutes to transfer water and when the task is completed she may stay in the hive for a short while to feed herself by soliciting food from the house bees. Then she cleans her tongue between the front legs, rubs her eyes, cleans the antennae and leaves quickly. A foraging bee can be easily identified from these preparatory movements. It appears the water carriers are stimulated for foraging when unloading takes only 40 seconds or so. In case unloading takes 10 minutes it is an indication to the bee that water is not required anymore. During the short unloading times the water carriers perform dances also but if unloading takes more than 2 minutes they stop performing dances and continue bringing water. When water needs of a colony have been met the water carrier on arrival in the hive is ignored and in fact the house bees run away from her (Lindauer, 1954; Gary, 1992).

In the field a water carrier takes one minutes to sip water, one minute or so to fly 0.4 km from the source to the hive, and once she reaches the hive spends 2-3 minutes there. The weight of the load ranges from 25-50 mg. On an average around 50 trips are made by a water carrier in one day. The daily requirement of water for an average colony is about 1/2 kg but a strong colony may need 1.5 kg. It is understandable that when ambient temperature is 40°C or more, the water need will be much greater (Park, 1923, 1925, 1929; Gary *et al.*, 1973, 1979).

The least a beekeeper can do is to provide water near the apiary with some floats in it so that bees can sit on them and sip water from close quarters near the apiary. For this small omission so many bees in a colony might have to fly long distances and thus use up energy of the colony.

7.7 PROPOLIS

There are only a few propolis gatherers in a hive and they stick to this work for life, but during the dearth period they may join the nectar collectors temporarily and again become propolis gatherers. Propolis is made from gums and resins of plants mixed with a little wax. On finding the source of propolis the forager reaches there and bites into it with her mandibles and fore legs, tearing off a little piece. By using mandibles and one of the middle legs it is kneaded into a tiny lump and with the help of this leg it is transferred to the pollen basket on the same side of the body. Then more propolis is loaded on the other basket till those are full. It takes a bee 15-60 minutes to fill the baskets (Alfonsus, 1933; Meyer, 1956).

With the propolis load a bee returns to the hive, where she is met by a house bee. The receiver bee bites into the propolis and tears it off the pollen basket. It mixes some wax into it and then cements it at the right place (**Plate 5b**). Propolis is used for closing cracks and cervices for protection against weather and natural enemies. The propolis gatherers are active on warm days only when the natural gums are soft. On bringing propolis to the hive the foragers rarely perform dances, and if they do, no recruits follow them.

Propolis apart from filling cracks is also used for strengthening the combs. It may have antifungal and antibacterial qualities and is used for embalming the bodies of large predators like mice which cannot be thrown out of the hive through the narrow entrance.

The actual collection of nectar, pollen, water etc. is a complex process and is a kind of united effort of the colony. It is not a random activity of the individual bees who mechanically follow the dances of scouts and start collecting these food materials. The colony works in unison, in a well organized and orchestrated manner. As to how it is achieved is being understood by the most recent researchers. The messages on the needs of the colony are conveyed through pheromones or chemicals which are passed on from the queen to the workers, from one worker to another, and from workers, back to the queen. The seasons are the guiding factors and the bees synchronize their activity according to the availability of food and the suitability of weather conditions for rearing their brood, for strengthening the colony and storing food.

7.8 NATURAL SOURCES OF NECTAR AND POLLEN

In their evolution honeybees are dependent on natural forest flora. Through cultivation man has brought large stretches of land under crops or replaced the natural forests with single species forest trees. The man-made forests of such species as conifers, poplars, tea, teak, etc. are not suitable for bees but trees like Eucalyptus, rubber, Rubinea, coconut palm, oil palm, coffee etc. can be exploited by bees. The availability of field crops depends upon the cropping pattern which differs from area to area and these patterns change over time. The crops are grown for their economic value and not for the sake of bees. Intentional cultivation of bee flora for the sake of obtaining honey is a possibility. And yet, every keeper of bees, whether big or small, wishes to harvest surplus honey from his colonies, therefore, he must have a thorough knowledge of the major and minor sources of nectar and pollen in that area. Moreover, the colony strength should be at its peak well before the start of the major nectar flow. Thus he must know when the desirable flowers start blooming and whether they supply nectar or pollen or both.

With increase in human population and by the process of urbanization and monoculture of crops the natural flora of the honeybees has decreased. The use of pesticides by ground and aerial spraying have increased hazards to the bees many folds. These man-made changes though harmful to the bees need not turn entirely against beekeeping. With proper management of bees the available major and minor flora can be exploited for harvesting surplus honey. The week and the month in which a major flow starts should be known to the beekeeper. It can be best determined by keeping a colony on scales and recording the start of weight increase or decrease. By noting the data for a few years the exact time of nectar flow and the seasonal changes can be recorded. If the weight of the colony drops, a beekeeper will know that, something is wrong and he can investigate the cause.

The monocultures of wheat and rice in northern and peninsular India have greatly reduced the diversity of bee flora, but such alternate crops as Brassica, sunflower, oil seeds, pulses like chickpeas, *arhar* and soybean, and the summer fodders like maize, millets, ragi and winter fodders like clovers have widened vistas of nectar and pollen availability.

Since the cultivation of oilseed, pulses, and fodders is being encouraged throughout India, there is a good possibility that beekeeping will fit in quite well in future agriculture. There is a large cotton acreage in central, western and north-western India, which is heavily sprayed with insecticides and bees are killed in large numbers. With proper cultivation and management practices it might be possible to keep bees in those areas also.

Nectar secretion in flowers is a part of normal physiology of the plant. Its secretion is influenced by such environmental factors as temperature, humidity, sunlight, photoperiod, wind, soil type, species and even the variety of the plant. Nectar in some plants is secreted in the morning, in others in the evening, and it generally dries up in the hot noon hours. The bees regulate their activity accordingly.

Given a choice they go for nectar that has high sugar content even if they have to fly further away. They collect nectar of varying concentrations. The sugar content of some of the common plants is given in Table 5 just to show variation from species to species. A fuller account of bee flora found in India is given in Chapter 12.

TABLE 5. *Nectar-Sugar Concentration of some Common Flowers of India*

S. No.	*Name*	*Sugar Concentration %*	*Author(s)*
1.	Barberis sp.	48.0	Sharma (1958)
2.	Bombax ceila	15.0	Mishra (1995)
3.	Brassica campestris	38.5-53.5	Kapil and Brar (1971)
4.	Brassica juncea	52.0	Wakhle *et al.* (1981)
5.	Carvia callosa	35.0	Phadke (1962, 1964)
6.	Cidrella cerata	36.0	Sharma (1958)
7.	Citrus spp.	25.0-50.0	Many authors
8.	Citrus sp. (orange)	13.0-17.0	Vansell *et al.* (1942)
9.	Grevillea robusta	79.0	Deodikar *et al.* (1957)
10.	Impatiens balsamina	16.0-25.0	Singh and Sharma (1972)
11.	Litchi chinsis	61.0-78.0	Nair (1983)
12.	Moringa pterigosperma	12.7	Nair *et al.* (1962, 1974)
13.	Allium cepa, onion	59.0-75.5	Rao and Lazer (1980)
14.	Peach and Pear	70.0	Sharma (1958)
15.	Prunus persica (peach)	70.0	Sharma (1958)
16.	Plectranthus sp.	26.0-54.0	Gupta *et al.* (1984)
17.	Sapindus detergens	40.0	Sharma (1958)
18.	Tecoma grandiflora	14.0	Sharma (1958)
19.	Woodfordia floribunda	10.0-12.0	Mishra *et al.* (1987)

8 How to Start Beekeeping

After having studied the essentials of honeybee biology and behavior, the reader will naturally wonder how to make a start in beekeeping? Before taking the step a prospective beekeeper needs three things; some knowledge and training or exposure to bees, disease free hive and equipment, and a swarm or a nucleus of bees with a young queen. There are a number of excellent books on beekeeping and video tapes are also available. Many scientific and popular journals provide latest knowledge on the science and practice of beekeeping. Training can be obtained from Agricultural Colleges or Universities in various states, the Government Department of Agriculture, various centres of the Khadi and Village Industries Commission or even from the next door neighbour who might be keeping bees already. Some of the more popular books on the principles and practices of Mellifera beekeeping are by Smith (1960), Eckert and Shaw (1960), Wedmore (1979), USDA Bulletin (1980), Root *et al.*, (1990), Crane (1990), Graham ed. (1992), Mishra (1995) and others (See bibliography).

The hive and equipment can be purchased from the manufactures and suppliers; generally it is sold at subsidized rates by Khadi Commission or other Central or State Departments under various schemes meant for promoting beekeeping. Used and second hand equipment can also be purchased at a much less cost from retired beekeepers. In that case one should ensure that equipment is free from diseases, pests, or hibernating larvae of wax moths. The colonies of bees along with 2-chamber hives can be purchased from other beekeepers. If one has a hive he can catch a swarm settled on a tree or a building. The method of hiving a swarm has already been described. Some progressive beekeepers have started selling nucleus colonies. If the supplier has good reputation he will ensure that the queen is young and the colony strength is not less than 10,000 bees.

The best time to start beekeeping is early in the spring in the month of March. The reason is that major honey flow in Northern India starts in April-May and the colonies should be well established before the flow and the beekeeper should get familiarized with the practices of bee management and of handling the bees. Second hand colonies with boxes can be purchased at any time of the year but the months of June and July be avoided as there is dearth of flora and the temperatures are very high, making it difficult for the colonies to get established at a new place. A colony must have 5-7 kg. of honey store at anytime of the year, therefore, a new colony should be either given artificial feeding, or the autumn and winter flora should be available. Early Brassica (toria) and main crop (sarson) and some Eucalyptus are available at many places and the bees can tide over the winter months on these flowers.

8.1 HANDLING PACKAGE BEES

Package bees are sold by weight, generally 1.5 kg each, along with a queen. There are about nine thousand bees in one Kg, therefore, a nucleus should have 13-14 thousand bees. The wooden packages are covered with wire and contain a small can of sugar syrup. The queen along with 5-6 workers and sugar candy, is confined in a queen cage which is hung inside the package. The parcels of packages are sent by passenger trains or goods transport by road. If sent by air special instructions are to be given on the parcels to place the packages in a pressurized cabin which is not fumigated with an insecticide.

On receiving a package it should be kept in a dark place at 70°F and bees be released in a hive the same day. Before releasing, the bees should be fed sugar syrup (1:1); by brushing it on the wire gauze two or three times. The bees sip it readily and take in half kg of syrup in a few minutes. They become docile and easy to handle when well fed. They should be released in a hive late in the afternoon about one hour before dusk so that they do not fly around and get lost. They will settle down during the night, and the next morning they will make orientation flights, and get familiarized with the surroundings.

At first the bees are given 4-5 frames in the middle of the box and the rest of the frames have to be given later one. It is considered more hygienic to give foundation sheets to bees so that they can raise new combs but it would delay the process of settling and for that reason old frames are given generally. Before opening the cover of the package and releasing the bees the feeder can inside should be removed and the queen cage should be inspected and placed between two frames, with candy end upwards. At least half the candy should be removed so that bees can cut through the rest and release the queen within twelve hours. The bees are shaken out of the package box above the frames and then the box is placed in front of the hive so that the remaining bees can enter the hive later on. New feeding cans (one litre mobile-oil clean cans filled with syrup and the mouth tied with thick muslin cloth) or plastic jars are inverted over the frames and a super is placed on top, covered over by a piece of cloth, inner cover, and the outer cover. Bees need to be fed liberally in the next few days and they can consume 5-10 kg. of syrup within a week, in two or three feedings. If natural flora is plentiful the bees will start collecting nectar but if there is dearth of flora they may have to be given further feedings. Robbing by other colonies has to be avoided. If more than one packages are to be released in hives they should be placed at some distance from one another. It will also prevent bees from drifting into other hives. The package bees should not be left sitting in good weather; hiving them on the same day is no problem. If it is windy or raining lightly, effort should be made to hive them anyway.

The entrance of the hives should be three inches in length and the colonies should not be disturbed for a few days. Therefore, either the feeding cans should be of larger size or more of one-litre cans be given to the bees as food. The first opening of the colony should be within a week or so for the inspection of the queen and for giving more food. If the queen has not started laying eggs it probably is dead or is not a good one and is leaving the frames. In that case it is better to unite the colony with another package colony.

8.2 UNITING COLONIES

Colonies are united under special circumstances and there is a set procedure to do so. The queen of a colony may be lost and the workers are unable to raise a new queen. When a queen is lost the workers try to raise a new one from the eggs or young larvae (less than 3 days old) left by the old queen. If eggs and larvae are not present in the colony the workers will wait for a few days and some of them will start laying eggs. Thus the colony becomes "laying worker" in which sterile eggs are laid. Such a colony will produce drones only and no workers. With time the number of drones will go on increasing and the colony will be doomed to perish. Therefore, before the drones appear it is best to unite the workers with a normal colony.

In another situation there might be two weak colonies which do not have a chance to survive the heavy monsoons or the oncoming winter. Rather than let them struggle and die, it is better to unite them and give them one queen which may be better of the two.

Two colonies can be united only if their odour is mixed well. Any effort to suddenly mix them together will result in fighting, and, deaths will occur on a large scale. Before they are mixed together the colonies have to be brought close to each other gradually. One of the colonies is brought close to the other by moving it 2-3 feet every day; in case it is shifted abruptly a large number of workers, whom we are trying to save, will fly back to the original site and they will perish. By moving the colony slowly the field bees get accustomed to the changing position of the hive. When one or two feet apart the colony brought in is ready to be placed on top of the other. The top covers of the stationery colony are removed and the frames are covered with a sheet of newspaper making a few holes with a small nail. The brood chamber of the other colony is lifted from the bottom board and is placed on top of the other colony, a newspaper forming a partition between the two. The inner board and the outer cover are placed tightly so that bees cannot escape from above. They work downwards, cut more holes in the newspaper and in a day or two the odours of the colonies will mix. Thus the two colonies start functioning as one colony. After a few days all the frames can be placed in one hive, and the extra ones along with the upper chamber can be removed.

8.3 MOVING AND TRANSPORTING BEES

For moving a colony within an apiary the process of slow movement spread over a number of days is followed as described above. Sometimes there is robbing and the colony has to be moved a hundred meters or more, then the method of slow movement becomes too prolonged. In that case it is necessary to shift the colony more than three kilometers away, to keep it there for 3-4 days so that bees get orientation of the new place and forget the apiary site. The colony can then be moved closer to the apiary at a suitable and safe place. Before moving the colony the entrance is closed late in the evening when all bees have returned for the day, or early in the morning before they have started the day's work.

In migratory beekeeping and in following the bloom, colonies have to be shifted to longer distances. In that case the frames and chambers have to be properly secured so that they do not shift in position during transit and thus get damaged. The method of fixing the frames in position is to move all the frames to one side of the brood chamber and to press them tight. Against the last frame one nail is fixed on each side to keep all the frames secure in position. Then all the movable parts are nailed down making the hive completely bee proof. On top of the frames also a wire gauze covering is nailed. The entrance is closed at night with a piece of wire gauze. Many types of clamps are available which can hold tight the hive as one piece from top down to the bottom board. Alternatively a string can be tied around the boxes securely. The hive is now ready to be moved to any place. The colonies can be lifted manually or with the help of a hydraulic or mechanical lift fixed to the transport truck. In spring or early summer water should be sprinkled on the wire screens on top and colonies should be moved in early hours of the day when it is cool. It is not advisable to move the colonies when ambient temperature is above 37°C because the combs become soft and they break in transit. As an extra precaution the colonies should be so arranged on a truck that combs are parallel to the long axis of the vehicle; it will prevent damage from jarring caused by the frequent application of brakes.

8.4 EXAMINATION OF A COLONY

Before approaching a colony for examinations, a smoker should be lighted and be ready for use. The materials for burning in a smoker include rags, pieces of old jute sacs, dried leaves, pine needles or anything that will not burn fast with a flame. If those are lit and packed inside the smoker tightly with the help of the hive tool sustained smoke is generated. Some green grass on top helps in stopping the flame.

8.4.1 Opening a Hive

Having lit the smoker one should tie down sleeves of the shirt and legs of the trousers. A veil should be put on the head and overalls and gloves should be worn if felt necessary. The hive is then ready to be opened and the following procedure is adopted (**Plate 7**). Work the smoker and puff smoke into the entrance and wait for a minute or so. Stand on one side of the hive and with slow deliberate movements remove the outer cover and then the inner board without exciting the bees as the frames are sealed hard with the use of propolis (**Plate 5b**). The covers can be made to stand against the walls of the hive or placed close by. A few puffs of smoke on the frames directed inwards and a wait for a minute or two will calm down the bees. The first frame is loosened from its tight position with the help of hive tool and is lifted up in two hands in an upright position and bees are examined first on one side and then on the other side making sure that comb is not broken. It is made to stand against the hive, its length perpendicular to the ground. Extra puffs are given here and there along the edges. One or two more frames are removed so that the rest can be shifted one by one on the side. When queen is sighted (much bigger, more shiny than the worker bees) that frame is kept in the hive in the middle. All frames are examined for uniform brood, good food reserves, any wax moth or other enemies. The burr combs are removed without jarring the bees. It is most important to avoid abrupt movements and to keep the bees at a distance so that you do not directly breath on them. If you are not wearing a glove and a bee stings on your hand the hardest and most important task is to bear it with calm without violently shaking the frame and to replace it slowly in the hive. In case more bees get excited, give some smoke and let the bees settle down. Then place the frames in position and close the colony. The Italian bees are by and large docile and they do not sting even when the weather is not ideal.

8.4.2 Object of Examination

The examination of a colony be taken up very gently, by slow and steady movements. One should avoid crushing the bees while taking out frames and also avoid being stung because the odour of the poison irritates bees even more than the disturbance and makes them difficult to handle. Bees do not like too much of interference, as it upsets their normal working. During the spring or during the honey flow when strength is being built-up the colonies may be examined once a week but during the off season they should be disturbed as little as possible, may be once or twice in a month. A colony may be examined at a time when bees are busy collecting nectar and pollen. In summer it will be during morning and evening and in winter in the middle of a warm day. A hive should not be opened at night or on a windy and rainy day. In winter when the temperature is low the brood should not be exposed for too long as it is likely to be chilled. When there is robbing or any other disturbance in the apiary it is better to keep the hives closed.

The object of examination in general is to see the working of the colony and to determine whether their requirements are being met. The flow of nectar, pollen and water should be normal and brood rearing should be in a regular pattern; patchiness in the brood area indicates that egg-laying by the queen is not normal. Every time a colony is opened the following should be noted :

1. Whether the colony has sufficient food or it needs artificial feeding of sugar syrup and pollen supplement.

2. Whether the queen or eggs are present; eggs indicate that the queen is well and there is no need to search for her. If there are no eggs, a search should be made to locate her and see its physical condition. In case she is absent a new queen is to be provided.
3. Whether there are any queen cells; their presence in the swarming season particularly at the bottom of combs, indicates colony is making preparations for swarming. If new swarms are needed to increase the number of colonies the cells should be left intact till a queen emerges from one of them. The rest of the cells can be removed and grafted in some other colonies. In case there are too many queen cells in a number of colonies, some of them should be destroyed to avoid excessive swarming. During off-season if cells are found on the face of combs it would mean the queen is either missing or it is too old, and the colony intends to replace her. The colony should be watched till supersedure takes place, which may be imminent.
4. Whether there are sufficient number of combs in a colony for the queen to lay eggs and for bees to store honey and pollen; there should be one or two frames extra for the future expansion of the colony. If the main brood chamber is full, a super with some frames be placed on top of it.
5. Diseases or enemies found in the colony be dealt with according to recommendations made in Chapter 13 on Enemies and Diseases of Honeybees.

8.5 INTRODUCING A QUEEN

It is required to introduce a queen in a colony either when there is no queen or when the previous queen is old and not functioning and has to be replaced. In the latter case, the existing queen is to be removed first and then a new one is installed. Sometimes queenlessness is not detected for sometime and the colony becomes laying workers. Such a colony will not accept a queen and it should be united with another colony.

The first step towards the introduction of a queen is to destroy all the queen cells that many have been constructed in the colony. A queen is placed in a queen cage along with 5-6 worker bees. The entrance of the cage is plugged with candy made by mixing fine sugar and honey. With the help of a wire loop the cage is hung in between two frames. In two days or so the queen acquires the odour of the colony and at the same time bees from outside have eaten up the candy and the queen is released. Instead of candy a cork plug can also be used to close the entrance and after 24 hours when queen has gained the smell of the colony the entrance can be opened manually. In case the bees are hostile towards the queen and she is not accepted readily she should be recaged and suspended in the colony as before and released again after another 24 hours. Any queen cells built by bees in the presence of the caged queen should be broken. To ensure queen acceptance all the brood frames from the colony may be removed and given to another colony after shaking off the bees.

If there is difficulty in queen acceptance by the above method the queen can be introduced by "Push-in-cage" method. This cage consists of a wooden frame 20 × 13 cm., made of 1 cm. thick wooden frame on which is fixed wire cloth of 3 mm. mesh. This cage is fixed by rubber bands or string on an emerging brood comb after shaking off the bees. The queen with a few young attendants is released in the cage. The queen moves around on the comb and the young emerging worker bees are quite receptive to her. The older bees of the colony outside the cage cannot attack the queen. Wher she starts laying eggs the cage is removed and the frames are placed in position. Almost always this method works.

8.6 INCREASING THE NUMBER OF COLONIES

In an apiary, particularly in the third year of the life of colonies, swarms issue and settle on a tree, bush, hedge or a building. The best way of increasing the number of colonies is to hive these swarms by the method described already. There are other methods to make an increase in number without depending upon the swarms, which are unpredictable anyway. During the spring, 6-8 weeks before the start of honey flow, a colony on 8-10 frames is divided in two by transferring half of the farmes to a new box, along with the bees sitting on them. The queen is transferred to the new box and at night, after closing the entrance, the hive is moved to a place three miles away. After a week it is brought back to the apiary and is placed at a new site. It will start functioning from there as an independent colony.

The colony left behind at the old site is given a spare queen or, a queen cell from another colony is grafted on one of the middle frames. The queen on emergence will mate successfully as there are plenty of drones at that time of the year.

In another method the newly formed colony after division along with bees and queen is placed at the original site and the mother colony now without a queen is shifted to another site in the apiary. It is given a mated or a virgin queen or a ripe queen cell. Some of the bees of this colony will return to the old site and join the divided colony, but most of them will stay in this one and accept the queen given or will raise a new one. With timely division and sufficient food the divided colonies will build up strength to collect honey.

8.6.1 Colony Increase after Honey Harvest

In case of very strong colonies in two chambers the division can also be done after the honey flow provided good food reserves are left for the bees. The newly formed colony without the queen is shifted to another part of the apiary and is given mated or virgin queen or a queen cell. Since some of the workers are bound to return to the mother colony its productivity is not reduced. It may therefore be seen that queen cells, which are a sign of undesirable swarming can be removed and given to newly divided colonies. This way the queen cells need not be destroyed and the number of colonies can be increased. Since beekeeping in India is expanding fast the newly formed colonies along with hives can be sold to prospective beekeepers and established beekeepers can earn extra income.

8.6.2 Mating Problems

Early in the spring, particularly in the new beekeeping localities, sometimes there are insufficient number of drones, or they may not be of the ripe age to mate. In such situations many of the queens remain unmated. To overcome this difficulty the colonies should be given stimulation feeding quite early in the season and frames having drone cells be given to the colonies. On emergence the drones are still developing internally and their reproductive organs are not mature before they are 12 days old. By synchronizing their maturity with the appearance of virgin queens (3-5 day old) high mating success can be achieved.

Sometimes the virgin queens are lost and they never return to the hive. Invariably they are picked up by birds. The Drongo and the Green Bee-Eaters are the main culprits who should be controlled near the apiaries particularly in the breeding season.

8.7 ROBBING AMONG COLONIES

When it comes to food honeybees are greedy creatures. Give them a chance and in no time they will gather at a pot containing sugar syrup, an unattended honey comb, juicy fruits or sweet-meats dipped in sugar liquid. They have no regard for neighbouring colonies either; the strong ones rob the weaker colonies and consider their stocks as their own. As long as there is good nectar flow in the spring or early summer they are busy searching and collecting nectar. When there is dearth of nectar in the field their bad habit of robbing comes in the fore. At first there may be isolated robbing of one colony by another, then it may become wide spread in the apiary or between neighbouring apiaries.

Colonies are supposed to have 15-20 kg of honey stock all the time but 1-2 supers of honey of 40-50 kg is their need in winter time. Sometimes beekeepers over-harvest honey and leave lesser quantity for them to live on. When there is dearth of nectar flow the reserves are consumed fast and colonies are predisposed to robbing. Thus a greedy beekeeper will generally face the problem of robbing in his apiary because there will be shortage of food for the bees. In case it is observed that there is no sealed honey in the colony it would indicate that bees are approaching starvation. The weak colonies will have to be given sugar feeding which is done during the day time; it will further attract the robber bees. Thus the weak colonies not only need protection but have to be given food at night time to escape the notice of the robbers.

8.7.1 Recognizing Robber Bees

At first the robber bees look like ordinary bees, but once they start robbing and come in conflict with guard bees they loose body hair and become smooth, shiny, and almost black. They are nervously active and have an air of roguery about them. They do not alight boldly at the entrance of the robbed colony and try to glide by the guards without touching them. When caught by a guard the robber tries to pull away boldly and even fight. In a strong colony the robber is attacked by 3-4 defending bees and does not escape alive but in a weak colony the tables are turned and more of them escape with their bounty. There is commotion in front of the robbed colony and robbers hover around in increasing numbers. A new beekeeper may mistake the play flight of young bees for robbing but in this case there is no fighting or disturbance in the hive. Moreover, play flights are around noon on a nice sunny day whereas robbing may be at anytime.

8.7.2 Control of Robbing

In the control of robbing the first step is to locate the colony from where the robbers come. Talcum powder or wheat flour is sprinkled on the alighting board of the colony being robbed. The entrances of the strong suspected colonies are closed temporarily and the bees carrying the white powder are watched at the entrance. On locating the colony its entrance is kept closed and those of the other suspected colonies are then observed. In the evening both the robbed and the robber colonies are moved to localities three miles away widely separated from each other. In a few days colonies will be cured of the robbing menace. When settled down these can be brought back to the apiary after giving them food. A badly robbed colony itself can be shifted to a distance and in its place a hive with one frame of honey can be placed. The robbers can be tricked and they stop visiting the site after the honey is eaten up.

In case robbing has spread in the apiary the entrances should be made small, even one-bee size. By controlling traffic the guard bees can better protect the colony. A careful beekeeper, particularly in times of dearth, will manipulate the colonies to check robbing. The hives should be opened carefully, the examination conducted speedily and combs of honey should not be exposed. The tops of combs in a hive should be covered with a moistened piece of cloth and the frames taken out should also be covered. In addition to reducing the entrance size a broad wooden piece be placed across from one side of the bottom board to the other. It forms a sort of tunnel through which the robber bees have to pass and hive bees can obstruct them. It further helps to place some green grass in the tunnel and also at the entrance of the hive. The robbed colony should be opened as little as possible and if food is to be given it should be done when weather is good and the food jar should be placed inside the hive, not outside. Naturally all the holes and cracks in colonies should be plugged.

Sometimes only a few robbers enter the hive secretly and one at a time, and are not easily noticed by the beekeeper. There is no fighting or disturbance and yet the stranger bees enter and carry off honey. These bees sneak in and pass the guards one at a time and carry on their job stealthily. This kind of robbing is called "progressive" robbing and usually no attempt is made to control it.

In case of robbing from one apiary to the other both the beekeepers have to collaborate to stop the menace. The procedure of locating the colonies and shifting them should be undertaken with mutual understanding. If robbing persists there is whole-sale fighting, resulting in lot of deaths and loss of colonies. The badly robbed colonies are covered with grass in front and on top.

If honey is to be extracted at that time the super frames should be covered and taken to the screen house, which should be bee proof. Preferably extraction should be done at night. The extracted frames and supers should be replaced before the day break and all dripping and smeared honey on the outside should be cleaned up.

8.8 THE RULE OF THREE

For the sake of a beginner many of the activities, behaviour and biological characteristic of the honeybee have been simplified in order to remember them under the "Rule of Three". During the development of a bee there are three stages; egg, larva, pupa. The egg hatches in three days and a larva up to three days age can be raised to become a queen. There are three kinds of bees in a hive; queen, drone, worker. From egg to the adult stage a worker develops in three weeks. The worker bees start producing brood food (secretions of hypopharyngeal and mandibular glands) when they are three days old (also see Evans and Berrett, 1989).

There are three ways of obtaining bees; buy a nucleus, catch a swarm, buy a second hand colony. A nucleus or a swarm is allowed to settle down for a couple of days and then feeding is given on the third day. By that time the colony is well organized to defend itself from robbers.

Before swarming, the bees engorge themselves with honey which is just sufficient to tide over three days. Then again, if the position of a colony is to be shifted the hive cannot be moved suddenly because bees remember the old site and they keep going back there. The colonies are not to be moved more than three feet in a day or they should be moved three miles away.

A beekeeper becomes experienced and gains confidence in three years; the first year the bee colony become established, the second year it reaches the maximum strength and productivity and in the third year it swarms. A beekeeper may start from scratch or inherit equipment given free or on a nominal price.

Beekeeping is neither a full time occupation nor it is expensive, but certain amount of commitment and genuine interest is necessary. It will be helpful to visit Honey Shows and *melas*. Beekeepers are interesting people and they like to talk about their products and experience. There are many local Beekeeper's Associations or Beekeeping Societies which attract you to become their number and to share common interests of colony management, procurement of materials like foundation sheets and to sell honey. The suppliers of equipment when visited in their workshop readily give you information on other beekeepers in the area. Offices of beekeeping experts in Agricultural Universities, and Khadi Commission's field staff are always helpful. Training courses are run by them in the spring and that will be a good starting point for a new beekeeper.

8.8.1 First Year

While placing one colony the possible position of others should be foreseen. The colony should be placed on a metal stand and all the four legs should be dipped in earthen pots full of water as ant barriers. If bees are to be kept as a hobby three colonies make a good start but if beekeeping is intended as source of income and further expansion, five to ten colonies are recommended in the first year with some supervision of an experienced beekeeper or field staff of one of the Government or private agencies.

If colonies are installed in March and there is good honey flow, they will build good strength in the first year by the middle of April and given a super they will collect 10-12 kg surplus honey by the end of May. Sufficient stocks of honey reserves should be left with the colony to tide over the dearth period in June and July. Depending upon the crops in the area bees can collect more honey for themselves before the winter.

To successfully pass the winter colony strength should be strong, honey reserves should be plenty, and colony should be saved from damp cold and predators like birds and mice. During bad weather bees remain confined to the hive and faeces are retained in their body which becomes swelled up. They will come out on a bright sunny day for defecation which will discolour many hanging clothes.

8.8.2 Second Year

The second year will be the peak year when queen will be at its best and will start laying eggs in January or February depending on weather. The *Eucalyptus* and mustard flora will provide much stimulation and the colony will be ready to go on a super by the end of February. Therefore, on a warm sunny day in January—February the colony should be cleaned up thoroughly of bur combs, debris on the floor and excessive propolis. If needed, extra feeding is given for the bees to gain strength and, to give the maximum productivity that year on Brassica, Eucalyptus, *barseem* and subsequently on sunflower. Bees will not swarm and honey surplus should touch 50 to 100 kg, depending upon the locality.

8.8.3 Third Year

Normally, the colonies left to themselves will have a tendency to swarm in the third year which is not desirable from a beekeepers point of view. Some forward planning is required. The colony is divided and one or two nucleus colonies are formed out of it. Thus a new queen is given to the mother colony, hoping that it will pass over the third year without swarming. Even then the beekeeper should be prepared for swarming and take necessary precautions to handle the swarms as described earlier. The rearing of queens and raising new nucleus colonies will be described in a separate section. The aim in third year should be not to let the colonies swarm or swarming should be minimized. A good crop of honey should be obtained by taking these precautions and by managing the colonies well.

8.9 ESTABLISHMENT OF AN APIARY

Apiary should be set up in a level, well drained and clean piece of land, away from heavy traffic but approachable by road for moving hives and equipment.

The hives should be protected by natural or planted windbreaks of trees so that strong winds do not upturn them. Vegetation also protects bees from the hot summer winds coming from the desert areas. As far as possible the entrances of the hives should preferably be directed to the East or Southeast so that they take advantage of the morning sun and are protected from the afternoon sun and heat. Ideally, apiary should be within half a mile of the floral source, but for convenience of management it could be within one or two miles of abundant flora; bees can easily work up to that distance. If bees are kept in barren land near the cultivated area a make-shift approach road can be made which should be accessible to trucks, pick-up vans, or hive loaders. A scooter is a universal transport for the beekeeper and he can move to and from his residence at all hours of the day. It is important that the apiary is within a protected and confined areas so that wild animals like bear, wild boar and cattle grazing in forests cannot enter the apiary and damage hives. Planting of mulberry or willow trees is recommended. They provide shade in summer and permit sunlight in winter.

8.9.1 Size of an Apiary

The number of colonies that should be kept in an area will, depend upon the floral sources and the colony population. Rich sources like Brassica, clovers and sunflower can have a greater carrying capacity *i.e.*, the number of colonies and the distance between apiaries. Where there is rich pasturage and there are 75-100 colonies in apiaries kept 3-4 miles apart the bees have a tendency to spread in one or two directions and do not give a good coverage. If they are kept in apiaries 1-2 miles apart the bee coverage is thorough. In order to avoid robbing it is better that the distance is on the higher side. However, colonies are owned by different beekeepers and it is not always possible to maintain the safe distance.

The number of colonies in an apiary will also depend upon the choice of the beekeeper. Those who keep them as a hobby and want to harvest some honey should be satisfied with 10-20 colonies but those who keep bees to earn a livelihood may keep 50-100 colonies. As a commercial venture the number can be increased to many hundred and some of the more successful beekeepers in Punjab have 500-1000 colonies. It will all depend upon the crops in that area. Under conditions of irrigated mixed farming 0.5 to 2 colonies per acre can be kept. On fruit bloom one colony per acre does not produce much surplus honey. However one colony per acre in clover area (dairying) will give surplus honey. In cotton area (protected from spraying) not more than one colony be kept for every two acres. Crop of sunflower that blooms in April-May can support one or even two colonies per acre.

8.9.2 Placement of Colonies

The placement of colonies in an apiary depends upon the contour of the land. On level ground hives are arranged in rows, grouping 2-4 hives and leaving a space of 5-6 feet for working with the hives. In migratory beekeeping they are kept in groups of 10 or 12 with entrances on the outside in all directions. In this system it is easier to load the hives on trucks.

8.9.3 Bee Poisoning

There comes a day when a beekeeper is shocked to see a large number of larvae or adult bees dead or dying in front of the hive. There are certain plants like tea whose nectar is poisonous and when fed to larvae it causes mortality. It is necessary to make a survey of poisonous plants in an area and depending upon their abundance bees might have to be shifted during the blooming season. (also see Thakar *et al.*, 1962).

An excessive death of adult bees is primarily due to poisoning from insecticides. The neighbouring farmers spray their crops like cotton, soybean, Brassica, without giving warning to the beekeepers. To protect their bees the owners have to remain vigilant and before spraying is carried out by the neighbours the entrance of hives should be closed. There are some insecticides which are safer to the bees and those should be preferred. For a fuller understanding of the usage of insecticides the reader has to refer to more comprehensive works on the subject (Atkins, 1992; Kramer, 1985; Adey *et al.*, 1986; Verma, 1991; Mishra, 1995; Atwal and Dhaliwal, 2013).

Colony Management

9.0 COLONY MANAGEMENT

The honeybee colony is a city in itself. The colony works as a single organism though it comprises 50,000 or so individuals who work independently, but are linked together by behaviour pattern and by the smell of chemicals, secreted by workers, drones, and queen. They have a functional centralized "computer" system through which they remain in touch with one another and perform various jobs needed for the common good of the colony. It is achieved through colony organization, growth and behaviour in relation to environment.

The elements of environment consist of sun, rain, weather, seasonal changes and the occurrence of plant flora that provide the bees their food. Water is part of the food which they gather from the environment. No two seasons are alike and the bees adjust to the ever fluctuating environmental conditions to their best advantage. A beekeeper cannot direct the bees or help them directly. He can only help them indirectly provided he has a true understanding of the habits and activities of bees and of the functioning of a colony in the changing environment. Only that beekeeper will succeed who has an in-depth knowledge of the various phenomena associated with the life of the honey bee. In the management of a colony no strict sequence of practices can be followed but those are employed when the necessity arises. Therein lies the secret of success (Smith, 1960; Eckert & Shaw, 1960; Morse, 1973, 1975; Ambrose, 1992; Gojmerac, 1980).

The basic needs of a colony are suitable climate, good strength, proper behavioural activities good food resources, and protection from enemies. Some of these needs have already been discussed and an account of others will follow.

9.1 FLUCTUATION OF COLONY POPULATION

In animals and insects in general, the number of young-ones produced by females in a generation are taken into consideration for determining the size of the population after a given period of time. In case of honey bees, however, the calculations have to be modified because there is only one productive female in a colony and its life is long. Therefore the number of eggs laid per day and the number of adult bees emerged every day after three weeks, indicate the increase in population and the finite rate of increase will be determined by the number of bees that die after six weeks of adult life (Sakagami & Fukuda, 1968; Atwal & Bains, 1987; Andrewartha, 1961).

Suppose the queen lays one thousand eggs each day, three weeks later there will be one thousand adult bees emerging daily, who will live for six weeks and then die. As to how many bees will there be at the end of six weeks can be calculated by determining the total number of eggs laid during that period *i.e.* 1000 × 42 = 42000. By deduction of the number of bees that might have died, the actual population of the colony can be determined.

In nature conditions are not ideal and many factors limit colony population. Not all the eggs laid by a queen may hatch and many larvae may die without completing development. There may be mortality in the pupal stage and fewer adults may emerge. The length of life of the bee depends on how strenuously she works ; the harder she has to work the sooner she will die and may not complete the expected six weeks. High temperature and dearth of flora also wear them out. The mortality among immature stages and the adults may become high due to diseases. The brood diseases are noticed easily but adults suffer from parasitic mites and virus diseases which are not easily noticed but they reduce duration of adult life significantly.

The queen can lay eggs only at a certain rate; if she lays one egg per minute without stopping there will be 60 × 24 = 1440 eggs in one day. It limits the speed of population increase in the colony. A beekeeper can, with proper management, decrease the number of deaths by keeping good food stocks and by controlling diseases and bee enemies. Thus population of a colony fluctuates a great deal, depending upon time of the year, the weather and personal attention given to the bees.

9.2 TEMPERATURE AND HUMIDITY CONTROL

Honeybees are cold blooded animals and their body temperature changes with the change in ambient temperature. Bees are active in the field; they fly from flower to flower and come back to the hive using up energy. It keeps their body temperature higher than the ambient cold temperature. As compared to ordinary insects that live solitary life, the uniqueness of social honey bees is observed in the hive itself where they rear brood; eggs, larvae and pupae at temperatures maintained at 93-95°F. When the outside temperature is lower, bees generate heat by consuming honey and undertaking physical activity. Furthermore, the heat is conserved by insulation made possible by the formation of a nest. The nest in the shape of a ball is formed in the centre of the hive where bees on frames become denser in population; the colder the outside temperature the closer they cluster, and thicker the insulating layer. Thus in cold weather hive is kept warm, particularly in the centre where cluster is formed, and the queen is preserved. Studies on the colony life have been reviewed by Smith (1960), Winston (1987), Crane (1990) and Southwick, (1992).

In winter weather when there is no brood rearing and the temperature falls below 18°C (64°F) the bees begin to form a cluster and it takes a definite shape at 13°C. On the outside of the cluster bees remain still with their heads turned towards the centre. It is the bees inside the cluster that generate heat by their body metabolism. On the outside of the cluster temperature does not fall below 10 – 12°C (50–55°F) but on the underside it is lower and may even be 9°C (48°F). However, if the temperature falls below 8°C (46°F) the bees are chilled and they fall off the cluster. In the centre, temperature ranges between 20–30°C (68 - 86°F) and is never less than 17°C (63°F). Once the bees start brood rearing in cold weather they maintain warmth in the brood nest and temperature varies from 31 to 35°C (88 – 95°F). It has been estimated that in the consumption of one liter of honey about one liter of metabolic water is released which condenses on the colder parts of hive and may even flow to the bottom. In order to avoid chilling not only this water but the excess humidity has to be removed (Southwick, 1985, 1987; Seeley and Visscher, 1985; Seeley and Heinrich, 1981; Dyer and Seeley, 1987).

Keeping the hive warm is one aspect of air-conditioning. In summer it has to be cooled also and the humidity is to be raised. On the contrary, in cold humid and warm humid weather humidity is to be reduced. When the outside temperature rises above the brood-rearing temperature (31-35°C) the bees resort to ventilation. At the entrance of the hive on one side they stand and fan their wings vigorously, facing inwards. An air current is produced which passes through the hive and cools it. Furthermore, the bees bring in water and place the droplets in cells, on frames or on top of the frames; water evaporates with the help of air currents and it further reduces the temperature. Converting nectar into honey on their tongue serves the same purpose. Bees keep a watch over humidity in the hive and maintain it below the saturation point even during the rainy season.

In arid areas the requirements of water in the hive are great indeed and bees consume 1 pint of water per day. Sometimes bees abscond owing to the shortage of water in certain localities. The various activities of fanning, bringing in water, or working with nectar are so regulated that optimum temperature and humidity are maintained in the hive throughout the year. Depending upon the pressure of specific jobs to be carried out in the hive, the bee force is diverted accordingly, even cutting across their age related functions. The house bees can bring in nectar and water, the older bees start producing brood-food and the younger bees that normally do cleaning jobs can make honey from nectar and help air-conditioning the hive.

In the management of hives by the bees the beekeeper can play his part by providing food in plenty, and good quality water close to the apiary. Ventilation of the hives can be improved by widening the entrance and by making wire-gauze windows in the inner cover. The hives should be kept in shade in summer and they should be covered on top with wet gunnysacs. If hives are in the open where there are no trees or the shade of a building, an open thatched hut can be constructed and hives placed under it, on the ground or in a two-storey formation. When temperature is 40°C or above the entrance should be narrowed to stop the hot air going in. During the monsoon weather when it is warm and humid the entrance should be widened to improve circulation of the air so that humidity is lowered. The hives are placed with a one inch slope so that rain water in summer and water of condensation in winter may flow out of the hive. Shading of the hive is important in hot summer and these can be further cooled by various devices, the simplest being a wet cloth on top cover (Atwal and Goyal, 1974; Jay and Grankson, 1972).

9.3 CONDITION OF THE FRAMES

The condition of a colony can be judged by examining the frames, particularly those in the middle where there is the brood nest and the bees are more concentrated. As mentioned earlier the hive is opened after smoking with steady, deliberate, confident movements, taking care that no bee is crushed. A nervous person is easily spotted by the bees; sense of fear activates the adrenals of the person and perspiration oozes out which gives off body odour and the bees react to it adversely. A beekeeper should have so much self-confidence that he is convinced bees cannot hurt him. The first inspection after the winter is a thorough one and the subsequent ones can be brief.

After the winter dearth period the colony will start building-up and raising the brood. It should be ensured that their needs are met. Every frame is examined; those which are of standard size and fit nicely can be pried open easily with the help of hive tool. The frames which are not of standard size and fit loosely are sealed together by bees using propolis and are difficult to remove without jarring. When shaken the bees get excited and give a few stings. Every frame should be examined and those having too many drone cells should be removed and in their place new frames having foundation sheets should be given. If there is no brood on some combs those should be moved on the outside of the brood nest. Defective frames containing honey should also be moved to the sides and later on removed. There should be plenty of honey stored in the combs and if there are eggs and young larvae it means the queen is functioning normally. It is not necessary to observe the queen every time. It is sufficient to examine one or two frames from the middle of the cluster for the presence of eggs and young worker brood. If there is plenty of sealed honey on the side frames the food situation may be considered satisfactory. If not, honey frames from other colonies may be given to a weak colony. Similarly frames of sealed worker brood can also be given. If no surplus honey frames are available the weak colony has to be given artificial feeding.

During the honey flow it is to be ensured that there are plenty of empty frames to store honey either in the brood chamber or in the super. During the peak season, when honey is being stored in the super, it is filled approximately in one week. An extra super should always be ready to be given to the colony. In the honey-flow season some beekeepers extract honey more than once; others who have shortage of labour extract at the end of the season. After extraction the irregular or old frames are melted for wax and new frames are raised. For raising good uniform frames, full foundation sheets of the worker size should be fixed on wired frames and placed in the brood chamber, or in the super just above the brood nest. These combs when raised will be white, uniform, and a beautiful work of art. They should replace the old combs to be discarded.

The surplus frames after honey extraction should be stored in cabinets, trunks or almirahs which are insect-proof. The black ants will come for honey on frames and the wax moths for the combs. An easy and workable method is to stack them inside thick-plastic bags 4-5 feet high and one meter in diameter. The storage bags are placed in an attic or an isolated room where those can be fumigated when needed.

9.4 FEEDING THE BEES

9.4.1 Sugar Feeding

Normally it should not be necessary to artificially feed a good size colony in an area where there is abundance of flora round the year. Colonies should have 15-20 kg of honey reserves all the time. Feeding is necessary for package bees, nucleus colonies and newly divided colonies to make them get established. They can easily take 10-15 kg of syrup (1 : 1). Not all the seasons are the same and in certain years the food reserves may be depleted by the autumn. The winter will be ahead and then the colony will need to start early brood rearing. Such colonies have to be given food to tide over the difficult months.

As the colony strength increases in the spring there will be plenty of young bees to secrete brood food needed for brood rearing and they themselves will need to feed on honey and pollen. At that time if feeding is not given the bees can starve quickly because the queen will continue laying eggs and the young bees will feed brood food to the larvae for the first three days and pollen and honey subsequently. If there is no honey to follow up feeding the larvae there will be a sudden mortality.

It is considered that after the autumn honey flow a colony should be left with a full super of honey to pass over winter and to start spring brood rearing. Beekeepers do not like to leave that much honey for the bees and in that case they should be prepared to feed them ample quantities of sugar syrup, whenever the colony needs it. When there is starvation in the colony bees pull out larvae and pupae from the cells and throw them out of the entrance. The drones are also driven out and they die of exposure. Therefore, a beekeeper has to watch the needs of his colonies and not let them starve at any time. If there is no sealed honey in the colony and there is only nectar coming in it means the colony has reached the starvation level. A good beekeeper takes suitable steps long before that happens.

9.4.2 How to Feed Bees

After honey, sugar syrup is the best food for bees. It is prepared by dissolving equal quantities, by volume, of white sugar (not gur, cane guice or shakar) in hot water and is allowed to cool down. The syrup is stirred so that sugar does not stick at the bottom. For 50 kg of sugar used, one tablespoon of tartaric acid is added to the liquid so that sugar does not crystallize on cooling. There are various methods of feeding the bees.

There are division-board feeders available as part of the bee equipment. The feeder is a hollow container the size and shape of a frame and may be made of wood, metal or plastic and is water proof. It is placed at one end of the hive and can contain 4.5 kg of syrup. It is open at the top and liquid is poured into it. There are floats on the surface of the liquid so that bees may not drown but feed by sitting on them. The bees can use up one such feeder in a week during shortage but if some nectar is also coming in it may last two weeks. In case feeding is to be repeated the division-board feeders may be left in the colony and can be removed later on to give space to a frame. If the feeder is made of wood and it leaks slightly a coating of paraffin wax can be applied on the inside to make it leak proof.

Syrup can also be fed through plastic or metal jars of 4-5 kg capacity by inverting them over a feeding hole in the inner cover. The jars are filled and the screw caps containing a few fine holes are tightened, before inverting them and making them stand on their heads. The bees feed on the syrup slowly coming out of the holes. Around the jar a super box is placed tightly and is covered on top with a piece of course cloth, inner board and the outer cover. There should not be left any cracks or holes in the hive for the entry of robber bees. One jar of syrup is eaten up by the bees in one week. Smaller jars of 2-3 kg. capacity can also be used at a time if there is a history of robbing in the apiary. One or two jars can also be placed directly on top of the frames of the brood chamber and then sealed from the outside by super and top covers.

Perhaps the quickest method of feeding the bees is by pouring syrup over the combs. One or two combs at a time are placed flat in an open pan and syrup is poured over them to fill the cells. By inverting them upside down the other side of the combs is then filled. The frames are lifted and the excess liquid is allowed to drip; the frames can be dried with a cloth and then given to the colony, shutting the hive with speed before bees of the other colonies come to rob.

When a colony is not in a dire necessity but some reserved food is to be made available, granulated sugar can be placed on top of the inner board, opening the hole of the bee escape. The bees will be on their own to feed on this sugar by dissolving in their saliva. Using any of the methods there can be insufficient feeding or by misjudgment there may be over feeding. Bees will take up any amount of food and store it in their cells. It will not only mean wastage but also adulteration of honey when extracted subsequently.

9.4.3 Pollen Feeding

When colonies are developing in the spring and brood rearing is going on before the start of the honey flow there may appear a shortage of pollen in the hive. Unless pollen is provided there may be a setback to colony development. Pollen collected by pollen traps during the nectar and pollen flow can be dried at room temperature, in direct sun-light, under infrared light, by freeze-drying or stored as such in refrigerator. Its quality and nutritional value is fairly preserved if 2 parts of dry pollen is mixed with one part of granulated sugar and stored at room temperature. It can be fed to bees when necessary.

9.4.4 Pollen Substitutes

The pollen can also be supplemented with other materials and given to the bees as pollen supplement. Two parts of pollen mixed with one part of soybean flour is accepted by the bees for their brood and adult feeding.

Many pollen substitutes have also been tried. Haydak (1967), recommended a mixture of soybean flour 3 parts; medicinal, baker's or brewer's yeast one part; and dry skim milk one part. The addition of dry egg yolk improves the nutritional quality of the mixture. By using granulated sugar solution in water (2 : 1) the mixture is kneaded into a candy. One pound of this candy is wrapped in wax paper to keep it moist and the packets are punctured with a nail. These candy patties are placed directly on top of frames near where the cluster is. Haydak found that this pollen substitute was superior in brood production to the one year old dry pollen. Bees feeding on the supplement reared more brood than those fed substitutes, but the bees reared brood for a longer period on substitute.

Due to attractiveness of pollen the pollen supplement is preferred by the bees but pollen substitute has a higher nutritional value. It is indicated that attractiveness and nutritive value of dry pollen decreases on storage. The amino acids lysine, arginine and tryptophan are most affected and those can be added to the pollen supplements. Pyridoxine is the most limiting vitamin for brood rearing.

The pollen foods are fed in powder form placed outside the hive or in a candy form placed inside the hive. When placed outside in a dish the food is to be protected from rain, and direct sun. If a wind storm comes the dishes should be removed and taken inside.

9.4.5 Water Requirements

Bees need more than half kg of water every day for diluting the brood food and for cooling the hive by evaporation. During hot summer the need is most likely to be 1.5 kg and bees have to fly quite a distance to bring a tiny amount every time. During the period when there is dearth of nectar the requirement of water is further increased. Bees are seen collecting water at water taps, hand pumps, bird baths, lily pools or even village ponds. Static and stagnant water is not good for them and diseases like nosema are spread through foul water. Therefore, if colonies are not located near a natural source or taps of water some provision should be made near the apiary. Water can be made to drip from a tap or a container on to a sloping board or trough from where it may flow along the slope.

9.5 MANIPULATION OF BEES

The bees inside a hive although protected from daily fluctuations in weather, are restricted by the space available. At first the colony expands sideways in the brood chamber but when it is full they need more space. The only direction in which they can expand is upwards which is possible only if the top covers are removed and a super with frames is placed there, covered on top with the inner and the outer cover. For good management of brood rearing of bees it is essential to ensure that (1) queen is laying eggs ; (2) there is room for the expansion of the colony; (3) there are plenty of food reserves with the colony ; (4) and there is effective protection from disease and enemies of the bees.

9.5.1 Early Spring Manipulation

The manipulation of the bees will depend upon the type of bee hive being used. The two hives in common use are the two chamber Langstroth hive and the single chamber Modified Dadant hive. Brood in the MD. hive is reared in a single chamber that contains eleven standard frames. Its supers are shallow and are meant only for storing honey and not for brood rearing. This hive is not so popular. The langstroth hive has 8-10 standard frames (division board and division feeders can fit in). The brood chambers are also of standard size and one can fit above the other. Extra supers can be placed on top. Basically the bees and brood rearing is manipulated on two chambers in the spring and the summer and the colony retracts to one chamber before the winter.

Eight weeks before the start of the main honey flow in a locality the bees should start active brood rearing. In six weeks the new bees will start field work and an extra two weeks will give a full force of the experienced field bees for foraging required for building colony strength. If all the requirements of the bees are met it is expected that the colony will double in strength in six weeks and within eight weeks they shall have collected surplus honey. The tempo of the brood rearing will continue and surplus nectar will keep coming in. Thus there will be a good strength of hive bees as well as the field workers.

In the active season when the colony is on eight frames, a second box containing eight frames is placed on top. The colony should have plenty of food stores, at least two frames of sealed honey. If the reserve is insufficient sugar syrup should be fed. If drawn frames are not available the frames in the upper box should be given full sheets. Two brood frames from the lower chamber are brought up in the middle and in their place two sheet frames are taken down and placed on the outer side of the brood area. If pollen is not enough with the colony, pollen supplement or pollen substitute is given. By providing two chambers to the bees brood rearing will be stimulated and it should be done eight weeks before the beginning of the honey flow. The colonies which are not on 6-8 frames, even if given extra chamber and food will not reach the required strength in eight weeks and will remain non-honey producing. However, they will build up the strength and during the honey flow they will continue rearing brood. It is not advisable to extract honey from such colonies because it will be cutting through their own reserves.

If the stimulation of brood rearing is tried with insufficient food, and nectar from the field is also not forthcoming it will be seen that brood rearing activity is shifted to the upper chamber which is not good. In case the position of the chamber is reversed and the top one is taken below to serve as the brood chamber, it will have the effect of doubling the chambers and should be undertaken eight weeks before the start of the honey flow. The principle of feeding the colonies remains the same. It would be observed that colonies which had one super full of honey before the start of brood rearing would have reached the desired strength and collected abundant honey during a four week period of the major flow.

9.5.2 Manipulation during Honey Flow

One week before the start of the honey flow queen is confined to the brood chamber (whether one or two) so that she does not go to the super on top and does not lay eggs there. It is achieved by placing the queen excluder grill on top of the brood chamber. The queen excluder can also be placed on top of the lower brood chamber so that queen restricts her egg laying to the lower chamber. In that case the upper brood chamber is filled with honey and so also the supers above it. In most of the situations it should be the desired practice because after the honey flow when there is dearth of nectar and heat is intense the colony strength should not be too strong because that will mean more feeding requirements. In an area where there is an abundance of flora and there is a sequence of honey flows one after the other a stronger colony on two brood chambers will collect record honey. Such a situation in North India will arise in a locality where there is Brassica (Jan.-Feb.) ; followed by *Eucalyptus* (Feb.-Mar.) ; then Egyptian clover (Apr.-May); Sunflower (May-June) ; Acacias (Aug.-Sept.) ; Cjanus followed by *toria* (Oct.-Nov.).

Another advantage of giving the queen excluder is that pollen is not taken by the bees to the super but is stored in the brood chamber near the brood. Before placing the queen excluder the super should be in position so that bees get used to that area and start moving throughout the colony from lower chamber, through the grill to the super. The super containing frames may be shallow or a standard hive box may be used as a super. The shallow supers are used for convenience of weight. However, the standard boxes are easily interchangeable and beekeepers can move them from place to place while selling the colonies or the honey frames. The standard boxes can be taken to a central place of honey extraction without fear of mixing them with shallow supers.

9.5.3 Hive Space

Bees need space not only for storing honey but also for ripening it. Lots of empty comb cells are needed. As soon as the bees feel the need for more space a super should be already in position. A beekeeper should not wait till the honey chamber is full because that will mean not only loosing a part of the crop but the congested conditions may misdirect the bees and they may start preparing for swarming. As soon as the bees start working on the last two frames in the brood chamber a super should be given. When the honey flow starts extra super should be given ahead of time. As soon as one super is three-fourth full a new one should be given. The first super given should contain drawn combs so that the bees accept them readily and start storing honey in them. If drawn combs are not available full-sheet frames can be given, placing one or two drawn frames in the middle. Only when the bees have started working on the combs, the queen excluder should be placed between the super and the brood chamber.

9.5.4 Manipulation during Honey Extraction

When a super is full of ripe honey the cells are sealed with wax. It can be removed for honey extraction as soon as a greater part is sealed. Extracting honey during the honey flow makes the bees collect more. Some beekeepers leave the super on the hive and go on adding them till the honey flow is over. Before a super is removed it should be clear of the bees. An inner board with a bee escape is placed below the super. The bee escape is fitted on a hole meant for this purpose and within two days practically all the bees escape downwards towards the brood chamber. Whatever bees are left are smoked or removed by a blowing device and are released in front of the hive. After removing the super the hive is covered on top to avoid robbing. The supers containing honey frames should be quickly moved to screen houses otherwise bees will gather on them and start stealing honey. In general the manipulation of the bees is meant to help them increase in strength before the honey flow and for them to collect surplus honey. After the honey flow there is a decline of colony strength and the population of bees fluctuates in the pre-and post-monsoon seasons. The frames are so adjusted that the brood nest remains in the centre. When the population of bees is low there is no point in keeping the second brood chamber; the bees should remain in a compact mass in the main brood chamber where they can protect themselves from the enemies.

9.6 MANAGEMENT OF SWARMS

After the development and growth of a colony and its seasonal changes in bee population, swarming is the next reproductive and multiplication phase. Swarming is a basic instinct (see Chapter 6) which is activated under certain conditions of over population, congestion, idleness of nurse bees and age of the colony and of the queen. If a beekeeper has been vigilant he should be able to distinguish the mating swarm and the supersedure swarm, the latter invariably returning back to the mother hive. The reproductive swarms on the other hand either leave the apiary or settle on a tree or some other object nearby. These swarms are caught and brought back to the apiary and are hived there. It is quite a standard method of increasing the number of colonies. Even a spray called lemon balm has been developed to attract a swarm to a suitable site (Burgett, 1980).

Basically a beekeeper is interested in harvesting honey and swarming is contra- productive. Therefore his first effort should be to prevent swarming as described earlier. A beekeeper should remember the simple fact that a strong colony necessary for obtaining a large honey crop is also conducive to swarming. As compared to the South, in Northern India bees build up the colony strength earlier in the spring and hence more swarming should be expected there.

Swarm prevention and swarm control are two different things. To prevent construction of a queen cell leads to prevention of a swarm. In the early stage of construction of a queen cell a "cup" is formed and there may or may not be an egg in it. If the queen cup is destroyed at this stage, swarming is prevented. If a beekeeper notices that egg has hatched and the edges of the cup have been raised it is the stage of queen cell development, and he is to apply swarm control measures. The size, number and placement of queen cells on the combs are also important indicators to determine whether the colony is preparing to supersede or to swarm. If there are one to six small queen cells near the centre of the brood nest, supersedure is indicated. On the other hand if there are four to twenty queen cells, larger in size, rougher on the surface and lighter in shade, located towards the periphery of the brood nest or on the lower end of the frames, swarming is indicated. In that case swarm control is necessary which is rather difficult to achieve. Some beekeepers recommend clipping queen's wings as a method of colony control. It is neither a swarm prevention nor a swarm control method and, moreover, swarm may issue even if the wings have been clipped. Since the queen cannot fly with the swarm it may be lost or sometimes the swarm may return to the hive. Later on when the first virgin queen emerges she may depart with this swarm.

9.6.1 Swarm Prevention

Building the colony strength and not letting the colony swarm is the difficult task before an experienced beekeeper. It is believed that a good colony should have 50-80 thousand bees. A colony of 60 thousand is expected to produce twice the amount of honey produced by two colonies of 30 thousand each. The colony size beyond a certain optimum may not be desirable as one with 150 thousand bees may not produce twice as much honey as one with 60 thousand bees. Thus too large a colony or division of a colony into smaller units is not good for honey production and swarming should be avoided also for that reason. It has already been discussed in Chapter 6 on 'Swarming' but a summary is given again in relevance to colony management.

For the prevention of swarming it is required that congestion in the brood nest be reduced. There are three ways of doing it; reversing brood chambers, adding super with reversing and giving brood frame; to follow Demaree method of swarm prevention. When a colony has two brood chambers the queen has a tendency to move upwards while laying eggs. By reversing the upper and lower chambers the nest is split and queen gets more space for egg laying. The bees are put off making queen cups and queen cells. The reversing can be done 3-4 times before the start of the honey flow.

In the second method a super is placed above the two brood chambers which provides extra space. In another variation the brood chambers are reversed and after placing a super above, one brood frame is also given out of the brood chambers. This will give plenty of extra space by spreading the brood area but precaution has to be taken that chilling of the brood does not take place in case the weather is cold. The Demaree method has been described in Chapter 6. It involves confining the queen in the lowest chamber by placing a queen excluder on its top and the second brood chamber with one or even two supers above it. This drastic method of separating queen from the brood is very effective although it requires time and patience to locate the queen. It is the best method of swarm prevention in small apiaries but it is not practicable on commercial scale. Moreover the brood frame placed in the top super has to be examined within a week for any queen cells the bees might have built.

9.6.2 Swarm Control

When a beekeeper notices that queen cells have been constructed for swarming he will also observe some other changes taking place in the colony. Four to five days prior to swarming the queen bee loses one-third of its weight so that it can fly. Foraging in the colony slows down. A large number of bees engorge on honey. Thus the normal functioning of a strong colony is disrupted and efforts which should have been directed towards honey gathering get diverted. It is important at this stage to control swarming. As it has been recommended by some the cutting down of queen cells only delays swarming but does not control swarming and in the end bees may swarm anyway and take along with an unnoticed virgin queen.

There are four basic methods of swarm control, each with some variations. These methods are: removal of the queen ; removal of the brood; separation of queen from the brood; and dividing the colony into two. The separation of the brood and queen can be achieved by the Demaree method and it is effective both in swarm prevention and swarm control, depending upon the state in which a colony might be at the time.

The technique can also be modified. In a colony having two brood chambers, at first, the chambers are reversed so that the queen goes to the lower chamber. A queen excluder is placed on top of it and a new young queen is introduced in the upper chamber and supers, thus creating conditions of a two-queen system. The young queen being virile; builds up the colony strength without swarming. Later on the old queen is removed and the strong colony gathers plenty of surplus honey during the honeyflow.

For the control of swarming, a queen is sometimes removed temporarily or it is caged for 7-10 days and left in the colony. Owing to this gap in egg laying the swarming fever passes over but brood rearing is interrupted also and the colony may become weak in the long run. Before this operation the queen cells naturally have to be destroyed. Like removing the queen, brood frames can also be removed from a colony and empty frames given instead, after cutting or removing the queen cells. An extra super can also be provided for creating the same effect. The idea is to provide more space so that bees attend to brood rearing rather than queen rearing.

Dividing a colony into two also provides the bees with extra work and space. The method of dividing a colony has already been described. In India where *Apis mellifera* has been introduced in recent years there is lot of scope of multiplying and selling the colonies, therefore, beekeepers should develop this technique rather than destroying queen cells. The queen cells should be cut out and grafted on frames in nucleus colonies. The method of producing nucleus colonies and pedigree queens will be described separately, later on in this book. The important thing to remember is that an experienced beekeeper must learn to do the right operations at the right time. Colonies differ from one another in their performance and behaviour and the same colony may change from one season to the other. So, a beekeeper should be able to judge the state of the colony and outward signs and symptoms that signal swarming. Only then he can prevent or control swarming.

9.7 TWO QUEEN COLONY MANAGEMENT

With special management techniques two or more queens can be kept in a colony. The purposes is to build up colony strength in a shorter time than it is possible with the ordinary method of one-queen system. The colony strength is built up before the major honey flow so that bees can collect the maximum crop of honey. To make it a success special manipulations are required. A colony is divided into two, 7-8 weeks before the main honeyflow. The old queen with half the brood (mostly unsealed) and reserves of honey and pollen is placed in the lower chamber. A queen excluder is placed on it and two supers with combs are placed above. An inner board without any hole is placed on top. In place of the inner cover a double-screened (screens 1/4 inch apart) cover can also be placed. Above the inner board the second part of the brood chamber along with bees, honey and pollen is placed. One inch round hole is made in the front wall of the chamber below the hand-hold. A young laying queen is introduced. As soon as brood nest of the top colony is established the screened inner cover beneath can be replaced by a queen excluder. Above the second brood chamber two or more supers with combs or foundation sheets should be given and the colony covered on top. In the entire colony there will be plenty of space for brood rearing and food storage.

During the honey flow the supers of the lower and the upper brood chamber, are used by the bees for honey storage. If there is need for more supers those can be added below and above the second queen excluder. At the end of the brood rearing period in the spring the two queen colony should have 10-14 frames of brood in the lower unit and 8-10 frames in the upper unit. In the two-queen system, apart from the fact that ample space is provided for brood rearing and honey storage the instinct of the bees to produce queen cells is also suppressed so that swarming is averted. This system is very labour intensive and can be practiced only on hobby scale or on small farms. It is not practicable on a commercial scale.

9.8 QUEEN MANAGEMENT

Queen is the centre of the colony activity; its presence or absence can mean life or death of the colony. The behaviour of the queen determines the characteristics of the various races of *Apis mellifera*. Within a race the type of a queen will set pattern of the quality and performance of the colony. There is lot of variability from one colony to the other. During the breeding season quite a number of queen cells are made in a colony and the beekeeper can exercise his choice on the queens he wants to retain in his apiary and those he can sell with nucleus colonies. Although the basic virility and performance of a queen depends upon its genetic constitution yet some outward signs can help him judge her fairly well. Owing to individual variability various ways for the management of queens have been devised.

9.8.1 How to Find a Queen

Finding a queen in a large populous colony on two chambers and a super is not a small matter. Apart from handling the weight, controlling the bees with a smoker, the queen has to be searched, found, handled, and caged or marked as desired. The operation should be undertaken when the amount of brood is small and there is nectar flow so that robbing is not initiated. In the spring when brood rearing is increasing, the queen will be found in the upper storey of the two-chamber hive. During honeyflow on the other hand the queen is generally in the lower chamber. Owing to large population of the bees at that time it is more difficult to locate the queen. In autumn when the supers have been removed and the colony strength is not so great it is again easier to find the queen.

When trying to find queen in a colony the use of smoke should be the minimum and the frames should be lifted with slow and easy movements. Queen is not likely to be on the frames which have no brood or have very little brood. Moving such frames to the side, the central frames should be lifted and examined, one by one. While removing a frame the side of the next one should be seen quickly in case the queen is on that one. The lifted frame should be examined on both sides and if there is no queen it should be placed on the side and the next frame lifted, following the same procedure. Two persons watching on opposite sides of the frame can locate the queen more quickly. She will be found on a frame containing young brood and eggs and rarely on the sealed brood or honey combs. As the operator is working, sometimes, the queen runs away and hides among bees on side frames or even on the bottom board. The queen will generally be found on the first examination, if not, all the frames should be put back in the hive in the same order in which they were taken out. The hive should be closed for a few minutes and examined once again. In case the colony is on two brood chambers, a queen excluder can be placed in between the two and the chambers be examined one at a time.

In case queen is not located easily and it is necessary to find it and handle it, another method is used. The brood chamber is removed off the bottom board and is set aside. In its place an empty box is placed without frames. A queen excluder is placed on top and another empty box is placed above. The frames having bees are shaken above these empty boxes one by one. The bees will pass through the queen excluder and go to the bottom empty box and the queen will remain above the queen excluder as it cannot pass through the grill. If not required for a specific purpose it is not always necessary to find and handle the queen. The presence of eggs and young brood is sufficient evidence that queen is present in the colony and that it is in good health. The colonies can also be numbered and the records of queen introduction are kept in register from where age of the queen can be found out. The queen is generally changed when it is two years old, hence these records are a valuable management practice.

9.8.2 Queen Marking

For their easy recognition queens are marked with a dye and for determining their age their wings can also be clipped. A queen is held between thumb and forefinger of the left hand and its wings on one side are clipped with a pair of scissors ; wings gone on the right side can mean the even years and those clipped on the left side may indicate the odd years. Subsequently, if a colony has a queen with unclipped wings, it would mean that there has been supersedure. The best and most easily available marking fluid is the nail polish. Different colors are used denoting the year of emergence, the breeding stock, and the source of purchase etc. The liquid is applied with a fine hair brush on the centre of the thorax taking care that it is not applied on the base of the wings or the joints. The polish should become almost dry before the queen is released and it would matter little to her or the bees whether she is marked or not.

9.8.3 Queen Judging

A queen is judged not for her physical appearance alone; her performance as a mother of the colony is to be seen also. There should be a fast and uniform egg laying and brood rearing, starting 6-8 weeks earlier than the honey flow. A steady brood rearing during the honey flow for replacement of the old field bees is desirable and this quality should be maintained by her up to the autumn. She should lay eggs symmetrically, starting a little above the centre of the comb and spreading out from there in the form of a circle all round. The eggs should be slanted in the same direction and as the brood develops and bees emerge, concentric circles of empty cells in the centre, sealed brood around it and pollen in the outer circle will be seen on frames. The bees that emerge should be of a uniform colour although sometimes their colour varies depending upon the source of sperms from the drones with whom the queen might have mated. Thus a queen may start satisfactorily but produce mixed brood later on, and she might have to be changed for that reason. The bees that are produced should be gentle, good gatherers of honey and not prone to swarming. They should maintain abundant sources of honey and pollen and pass winter successfully.

The performance of a queen does not necessarily depend upon its age. A queen may be erratic in egg laying from the start or it may be good and continue to be so for many successive seasons. Some colonies may replace their queen more than once in a year and others may do it after several years depending upon their performance. There is no point in replacing queens every year although some commercial beekeepers who cannot follow the performance of individual colonies resort to this short cut method. The best way is to change only those queens who are not performing well and to change them all after two years.

A good queen is supposed to have a large and gently tapering abdomen and a large thorax. The colour of the queen depends upon its genetic constitution or race but it should have a uniform hue. A queen who is short and stubby and whose abdomen is tapered at the end as if it had a tail is not considered good. The one who is patchy in colour and erratic in movements is also considered undesirable. When a queen had laid eggs on 3-4 combs and the laying is not uniform it should be replaced. A queen that lays eggs slowly before the start of the honey flow will not build sufficient strength to collect honey even if the population increases subsequently. For replacing a queen there has to be a stock of good queens kept in a reserve. For breeding the desired quality of queens sophisticated genetic technique of closed population honeybee breeding have been developed (Page and Laidlaw, 1992) (5.1.1). It is possible to selectively breed honeybees that may guard or clean colonies (Robinson and Page, 1980; Rothenbuhler, 1964; Getz, 1983).

9.8.4 Reserve of Queens

For replacing queens in the spring and in the post-monsoon period, they should always be kept in the apiary as reserves of queens of good pedigree. Mass production of queens is done by the Dolittle method (10.6.3) (Fig. 13). Before introducing a queen into an established colony, it should be ensured that she is good at laying eggs and is in fact better than the one being replaced. The egg laying capacity of a new queen can be tested by keeping it in a nucleus colony for sometime. A caged queen can also be kept in a reservoir of a nucleus colony with combs of brood but no free queen. The caged queen is fed by the bees and is stimulated by them. It is then transferred to the regular colony whose queen is to be replaced. A nucleus colony having a queen can also be used as a reservoir. A queen excluder is placed on top and another chamber containing combs of brood, honey and pollen is arranged above. In this chamber queens in cages are placed temporarily or a queen may be made free as in the two- queen method.

9.8.5 Caging Queen

The cages are specially made for confining the queens on combs individually without the attendant workers. The candy holes should be protected from the outside with a plug. The queens are fed by bees and they increase in size, ready to lay eggs. The confinement of queens should not be for more than one or two weeks. If a queen is to be introduced in a colony lying in another yard the reservoir is taken there after closing the entrance. In the evening it is brought back to its original site. The method of introducing a new queen has already been described. Before introducing it the old queen is removed and the cork or plug outside the candy hole of the new queen is also removed. The bees will cut out the candy and release the queen. If the queen is ready to lay eggs she will be slow in its movements and will be anxious to be fed. The bees will accept her readily. On the other hand if queen is too active and agile, moving around fast on the combs, the bees will not accept her readily. They may even pull at her legs and wings or ball her to death. Therefore, it is desirable to release the new queen gradually and let the bees do it themselves rather than place the queen direct on the comb.

There is a method for releasing a new queen direct on the comb. It is done by placing it on the emerging brood. A 4 inch square push-in cage of wire gauze is embedded over the queen on the face of a comb. The young workers on emergence will accept her readily and feed her. When the queen is settled in a day or two the cage is removed and the older bees will take over the caring of the queen. A steady egg laying will be the indication of queen's acceptance.

9.9 ABSCONDING

Absconding is the complete desertion of the hive by the bees. The entire colony along with their queen fly away and leave the combs. Sometimes even brood, honey and pollen are left behind on the combs. The causes of absconding are lack of water, exhaustion of food stores, overheating of the hive, and persistent pest attack. Absconding is more commonly observed in *Apis cerana indica* than in *Apis mellifera*,

In desert areas or under drought condition when all sources of water dry up the bees have no choice but to leave the site even though reserves of food may be present in the hive. If a colony is robbed off its honey and no food reserves are left nor nectar is forthcoming it will desert the hive and fly away. Before leaving, the bees will eat whatever honey is available in the hive and sit at some distance on a tree. When the food in their stomach is used up the bees drop of starvation one by one. The swarm is generally too weak to collect nectar even if some is available in the new area. If such a swarm is collected in a box and is sprinkled with sugar syrup it may revive and start working again and commence foraging.

In the summer, colonies should be protected from intense heat, because in a weak colony combs in the upper chamber often become soft and honey drips on the brood. Bees often abscond under these conditions. The colonies should be kept under shade and wet gunny sac or wet cloth should be placed on the hives for cooling. The ground in front of hives should also be cleared of weeds and the under growth in order to improve the circulation of air.

Among enemies of bees, the Green Bee-Eater, wax moth and black ants are the worst. The birds should be killed by air gun or sling-shots. The frames should be kept cleared of the wax moths and the hives should be protected from the ants. Colonies in strong healthy condition, which can get plenty of water and food, are protected from pests and heat, do not abscond.

9.10 MIGRATORY BEEKEEPING

Bees kept in an apiary are very particular on the site of their hive and the specific odour of their colony. If the hive is moved 2-3 feet the bees gone foraging can locate it on return but hive moved further away is lost to them and the bees laden with food enter other hives. From then on they no more belong to the mother colony although they keep visiting the old site for sometime. When a colony is moved more than three miles away the bees are totally disoriented to the old apiary site and they take new moorings and immediately start working as soon as the entrance is opened. They orient themselves to the new site and its surroundings and start bringing pollen within 15 minutes. Thus the bees do not mind being moved and are happy at the new place so long as nectar and pollen are available. This habit of the bees is taken advantage of in migratory beekeeping.

Migratory beekeeping is practiced by experienced beekeepers and it is the most intensive form of honey production. The bees are moved round the country on various crops with the march of the bloom as the season changes. As soon as the bloom is over in a locality, honey is extracted and the beekeeper loads his colonies on a truck and moves to another place of honey flow on the same crop or a different crop (**Plate 8**). By surveying and mapping the crops, vegetation and climate and the resultant honeyflow in different regions a beekeeper can get one extraction after another throughout the year. Completing the circle he returns home for sometime for cleaning and repairing the hives, replacing the combs with new ones, and discarding or melting wax of the old ones.

9.10.1 Ancient Migrations

The art and practice of migratory beekeeping has been perfected in USA where large trailer trucks are used to move up to 20,000 colonies under one management. The movers also provide their services to other beekeepers who may have only a few thousand colonies. The wonderful adventure of migratory beekeepers across America has been described in Douglas Whynott's book "Following the Bloom" (1991).

In earlier times also migratory beekeeping was practiced by the nomadic or wandering beekeepers. In ancient Egypt pyramid shaped hives numbering up to 4000 were seen being moved on boats and rafts along the Nile, following the bloom of various flowers on the banks of the river. Lotus was introduced from India to the delta region of river Nile and the Lotus honey collected was the food of kings, nobles and priests. The Egyptians wandered on land also and went as far as Greece. The hives were moved on camels and donkeys. Solon in 600 BC mentioned bee caravans and bee floats and made the law that groups of bees be kept 300 feet apart. In the third century Romans reached the upper Rhine and floated down-stream their hives, harvesting honey as they went along. In the medieval times the gypsy like movements of bee-caravans were described in Greece, Italy, Germany, Austria and France (Crane, 1983,1985). In Switzerland bees were moved to harvest honey from buckwheat and in Scotland bees were moved from lowlands to highlands for gathering honey from heather. (Eckert and Shaw, 1960; Crane, 1976a, b, 1990, 1992).

With the introduction of modern hives in India in the early part of this century migratory beekeeping was practiced with the Eastern honey bee *Apis cerana* on a limited scale. In winter, bees from the higher altitudes in Himalayas were moved to the foot of the mountains where Brassica blooms were in abundance. The bees were also moved along the Western and Eastern Ghats in South India. These were the efforts of individual beekeepers who owned one hundred or more colonies. The hives were moved on mules, trains or trucks. These beekeepers hardly ventured deep down into the indo-Gangetic plains and after building the colony strength and extracting some honey in winter months they returned to the mountains early in the spring.

With the introduction of *Apis mellifera* in India migratory beekeeping has become more profitable and beekeepers have started making longer trips on trucks and pick-up vans reaching as far as the central districts of Punjab, Haryana, U.P. and Bihar.

9.10.2 Moving the Hives

Moving the colonies is an art and an adventure, and is a way of life with the migratory beekeepers. One or two colonies may be moved as such after closing the entrances and can be carried on head or tied on the back of a mule. Owing to propolis sealing the frames remain in position and no damage is caused to the combs.

A larger number of hives to be carried long distance on truck have to be properly prepared. The frames are held in position by nailing the last frame at the ends. The hive is covered on top with a screen of one eight-inch mesh made of metal or nylon. The screen frame should be one or two inch high and it should fit tightly over the brood chamber like a super. The hive is held together as one body with the bottom board by using two inch staples or tying it with metal straps or strings. Clamps can also be used. The entrances of the hives are closed with screen gates or pieces of rubber foam cut to fit the opening. While making preparations the size of the colony should be reduced as far as possible. The bottom boards should not be cumbersome and may be modified for easy transportation.

Before loading, the hive entrances are closed and, outer covers are removed. While loading the hives on a truck these can be carried by two men or lifted mechanically. In between the layers and rows of hives wooden planks two inch in depth should be placed tightly to facilitate ventilation in transit and the mass of hives should be tied together with a rope. The spaces in between will also facilitate water to reach the hives when it is sprayed on route in warm weather. When the hives are piled, care should be taken that frames are parallel to the length of the truck. If placed crosswise the combs are likely to be damaged by motion of the truck as the driver uses brakes. The truck should be wound gently at the sharp turnings particularly in the mountains. It should be covered on top preferably with cotton cloth or may even be with a plastic sheet. It is advisable to move the bees at night and a truck can cover 300-400 kilometers. On reaching destination the hives can be unloaded, and placing them properly the gates can be opened in the morning. With top covers in position the bees settle down in their new surroundings.

When floral conditions are favourable and the colonies have full strength honey collection is fast. In the right season it is possible to have extraction after a week or so. If more than one extraction is expected it is advisable to make prior contacts for the disposal of the honey.

9.10.3 Out Apiary

When migratory beekeeping involves moving bees 45-75 kilometers away it is convenient to establish an out apiary on purchased or rented land. The colonies are generally under the protection of an attendant or the owner of the land. Depending upon the number of colonies kept in an out apiary additional help can be hired. The site is generally on the edge of a forest amidst crops with plentiful flora. It is expected that there would be rich crop of honey at that place. Moreover, the site could be used for isolating diseased colonies or for selective queen mating, and breeding. An out apiary can be really useful for a beekeeper having a few hundred colonies.

9.11 BEE HOUSE

A bee house is generally a durable masonry structure inside which hives are kept. Its internal width is 8 feet (length of asbestos sheets) and length is 24 feet or more depending upon the number of hives to be kept. The ceiling height should be 7 feet or more (for insulation) so that there is sufficient working space for the beekeeper. The door should be 3 feet wide with a window at the opposite end. There should be at least four 4 feet windows in the side walls on each side. Along the length of the walls are constructed stands of cement slabs on which hives are placed. The hive entrances fit against galvanized iron pipes of 1.5 inches diameter which run through the wall and are about 20 inches above the floor. There are wire gauze screens on windows above the height of the hives and 3/8 inches bee escapes along the length of the ceiling. The plinth of the building should be above the ground level and the roof can be sloping on one side or two sides.

The hives are placed in pairs 5 inches apart, and there is a gap of 18-21 inches between one pair and the next. That space is sufficient for the beekeeper to open and work on the hives from inside the building. If he gives a little bit of smoke the bees can be handled easily; the guard bees fly outside the wall and there are comparatively fewer stings. The bees will identify their hives from outside from smell and colour. To assist the bees find their hives the wall area around each hive entrance can be painted white, black, yellow, blue etc. This way bees will find their own colonies which are otherwise placed quite close to one another. Colour marking is helpful in other situations also where colonies might be lying very close to one another.

The bee houses can be used permanently for security or those can be put to seasonal use. Bees can be given protection from cold in winter and from extreme heat in summer. The house can be warmed with electric heaters in winter or cooled in summer by desert coolers. During the honey flow the colonies can be moved out for working freely and for giving them supers. The supers full of honey can be stored in the bee house for extraction. Thus the building can also serve as a honey house. If the number of colonies is large a separate large screened room can be constructed for honey extraction making provision for bee escapes. In the snow bound areas the bee houses can be further modified The hives can be kept in two or three layers one above the other with entrances of their own. Such cozy bee houses can be extremely useful near the tree line in the Himalayas.

The bee houses are to be located near sources of water. Since in summer the temperature inside the house is lower than that outside, the bees have to do less fanning and they can conserve energy

The management of bees becomes easy and the beekeeper can work from inside even when the weather outside is not fine. A full demonstration hive with glass walls can be built inside and bees can be watched as a curiosity. The bee house can also be used for stocking nucleus colonies or those transferred from other places. Since 20-50 colonies can be accommodated in a bee house, a beekeeper can work with them efficiently. With proper management he can produce more honey and income, free from any fear of damage or theft. The problem to be faced is the concentration of bee-eaters who will fly around and congregate near the bee house. They can be controlled by shotguns or sling shots. Payment can be made to hunters by the numbers of birds killed by them, and thus the menace can be kept under check !

Management for Honey Production

For maximum honey production a colony should have the peak of brood and bee population just before the honeyflow, without showing any symptoms of swarming. If there is too long a gap between the peak and the honeyflow the colony will use up all the food reserves and reach a stage of starvation which will initiate a decline in population thus affecting the honey gathering capability. If there is no undue long gap, but the space is restricted causing crowding, the colony may prepare for swarming and this may happen even during the honeyflow. The period just before the honeyflow is as important as the honeyflow itself. A beekeeper must learn to know the state of a colony so that he can apply the appropriate management practices. For instance a colony that has acquired full strength sooner than the honeyflow and shows some symptoms of swarming may be divided rather than be allowed to swarm naturally and create commotion in the apiary. The two colonies can be kept as independent units or these can be joined together during the honeyflow. Normally steps are taken to prevent and control swarming during this period as described earlier.

10.1 STEADY WORKER FORCE

During the brood development stage, the bees have a natural tendency to move upwards, thus restricting the space for food storage above and, at the same time, vacating the chamber below. For that reason, in a colony of two-brood-chambers the reversing is practiced. That provides more space for brood rearing and for food storage. Now the brood rearing will be moved to the chamber brought below and honey will be stored in the space available above. It is to be remembered that continuous brood rearing is essential even during the honeyflow. The house bees rear the brood as the field workers bring in nectar and pollen. The old house bees are replaced by new ones that emerge and they themselves replace the field bees that die of hard work. By reversing, when the upper brood chamber is brought down, the house bees rearrange food near the brood that causes increased activity and excitement in the colony and the tempo of work increases. In a colony with young and prolific queen the reversing might have to be done every two or three weeks, or 3–4 times before the honeyflow. A colony having two brood chambers at the start of the spring will be on three chambers within 10–12 weeks. The bees will be on 12–16 combs, mostly on the second and third chamber, with an extra super on the top. The sources of pollen and nectar will be sequential, *e.g.*, Brassica crops and Eucalyptus. By coinciding the peak of the colony population and the start of the honeyflow (*barseem*), a good crop of honey can be gathered.

10.2 SUPERING THE COLONIES

For producing the maximum honey crop we need a prolific queen, good supply of food, favourable weather conditions and, proper management, providing protection to the colony from diseases and natural enemies. The most important part of management is to know when to give supers for honey collection, how many to give and how and when to remove them for honey extraction. Even before the honeyflow bees start secreting wax and deposit it on the top portion of the frames which gives them a whitish appearance. It is an indication that a super with drawn combs should be provided. Depending upon the timing, some combs would be completed and other frames may just have full foundation sheets. The bees will immediately start working on them and start moving stored honey from the brood chambers where it might have been stored for want of space. It will stimulate the whole colony. Delay in providing the required supers at this stage would mean poor collection of honey and unnecessary restricting the brood area.

The supers are of two kinds. Some beekeepers use shallow supers with shallow frames which are not too heavy to lift when filled; one man can easily lift a shallow super and place an empty one in its place. More commonly the standard boxes and full frames are used as supers. When full of honey they are heavy and can be removed with ease by two men without jarring the hive that may be on 3–4 chambers by that time. The advantage of using standard boxes is that they can be interchanged with brood chambers and no extra care is needed to keep the boxes and frames separate at the time of extraction. Moreover there is better circulation of air in the hot summer and during the warm weather prevalent in a greater part of the year in subtropical and tropical parts of India. As soon as a super is filled with nectar and honey another one is placed on top. The effect of supering a colony is not only that space is provided for honey storage, there is more space for the bees to congregate there while receiving nectar from field bees. Crowding in the colony is reduced and swarming is prevented.

When frames in a super are three quarters sealed those are removed for extraction and new ones placed instead. This process continues throughout the honeyflow period and the way beekeeper supplies frames or supers is important; if the speed is right the bees remain active and excited and produce a bigger crop of honey. It is a good practice to remain one super ahead of the need of a colony up to the peak period of honeyflow. After that stage, extra supers are not desirable as the bees will thinly distribute honey in all the supers and there will be extra work at the time of extraction.

10.3 USE OF QUEEN EXCLUDERS AND BEE ESCAPES

Without the use of a queen excluder a super can be simply placed on the top according to the need of the colony. It is hoped that the colony would use the lower two or three chambers for brood rearing and the top one for storage of honey. In case a super is delayed the bees would store honey in the brood area and, management would become difficult. The brood may be spread in all chambers and so also the honey. Extraction will then become difficult. The use of a queen excluder facilitates management of colonies, particularly during the honey flow.

When a queen excluder is used the queen can be confined to the bottom chamber with brood frames and some food. Above the queen excluder is placed an empty super with frames drawn and above it the rest of the brood chambers of the colony. For the next two weeks or so the queen will have space for egg laying in the bottom chamber and the empty super will be used by bees for storing honey. During that period the sealed brood in the third or the fourth storey on top shall have emerged providing more worker force and also space for honey storage. If the period of honeyflow is prolonged one of the chambers on top can be moved to the bottom and queen confined there. The honey stored in the supers or the chambers above the queen excluder can be extracted. Before it is possible to do so the honey super or the chamber has to be cleared of the bees. To clear the bees a bee escape is placed under the chamber to be removed from the top. The bee escape is fixed in the centre of the inner board which is placed under that chamber to be removed. The inner board so used is called the escape board. It will be noticed that the board will stop air circulation making it hot inside. Therefore, the escape boards should be used with care for one or two days till the super is cleared of the bees. Alternatively a screen can be placed below and the escape board above the honey super to be removed. If weather is hot it may not be feasible to use the escape board as the temperature rises inside and the combs start melting. In that case the honey super is cleared of the bees mechanically.

10.3.1 Removing Bees from Super

If the number of colonies is small the bees can be brushed off the frames with a soft brush or a twig of leaves. A few puffs of smoke will also make the bees move downwards to the lower chambers. The honey super is removed and bees are shaken or brushed off the frames one by one near the entrance of the hive. The cleaned frames are transferred to an empty box placed on one side and then taken to the honey house for extraction, avoiding the robber bees from other colonies.

The commercial beekeepers who have a large number of colonies and have to handle lots of honey supers use more effective and quicker methods of removing the bees. Certain chemicals used in moderation act as strong repellents for the bees without causing any serious harm. A cloth wetted with carbolic acid diluted with equal quantity of water when spread on top of the frames will make the bees run downwards to lower chambers and the honey super will be cleared of the bees within five minutes. This method has bean in use for many years but sometimes the acid smell taints the honey and if not used with care it may also burn the akin of workers and may even kill the bees.

Some of the newer repellents in use are propionic anhydride diluted with an equal volume of water, bezaldehyde, or formic acid. The repellents should not be allowed to drip onto the lower frames which can happen if an ordinary piece of cloth is used. There is a better control if the cloth is mounted on a wooden frames that can fit on top of the super. It is called the acid board. Sometimes cut out inner boards are used for fixing a piece of cloth in the gap. In all cases the cloth frames should be used for a few minutes only and after the removal of honey super the fumes should be expelled out of the hive by aeration. The chemical repellents should be used in clear weather when temperatures are moderate.

The use of air blowers for removing bees out of the supers is more recent. The blower fan giving 3,600 revolutions per minute is run by a gasoline engine or an electric motor. Around 1,500 cu ft of air moves through a plastic tube and a nozzle and this method can be used even when weather may not be ideal. The idea is not to push the bees down to the lower chambers but to blow them away. The front end of a honey super is lifted and tilted sufficiently to make the air blast effectively rush through the frame spaces and blow the bees away. The bees return to the colony from the front entrance. The speed of clearing the supers should be as fast as the process of extraction of honey from the supers (Jones, 1986).

The honey frames removed and placed in an empty super should be covered with a cloth and in their place new ones be placed on the hive for bees to store more honey. In order to avoid robbing the honey supers should be moved to the screened honey house as quickly as possible. Any stray bees that may come with the supers should be expelled from the house by providing honey escapes.

10.4 UNCAPPING THE COMBS

Ripe honey stored in comb cells should have 19% or less moisture. It is sealed with wax caps slightly raised above the cell walls. A comb having all the cells sealed is ripe and has a standard chemical analysis. Some of the combs in a honey super may not be sealed 100% but since they are in the vicinity of ripe honey it is presumed that honey in the unsealed cells is also ripe and only the bees were not given enough time to seal them. Thus a comb having 3/4 cells sealed is also used for honey extraction. Some of the beekeepers take extra precaution and dehydrate them further by blowing warm air through a stack of honey supers in the honey house. Thus moisture is evaporated not only from the unsealed cells but also from the sealed cells through pores in the wax cappings.

Before extracting honey with a centrifugal machine it is necessary to slice away the wax cappings with the help of an uncapping knife. While using it the knife is kept hot by dipping it in hot water or connecting it with a steam pipe or by having an electric heater inside. The knife should be sharp, clean and hot and only then it will cut off cappings cleanly without injuring the comb. While uncapping, a comb is placed on a support above a drum and is tilted slightly while working the knife up and down across the comb surface, taking care that the cappings (and some honey) may fall into a container. With practice the beekeeper will acquire the skill to work on a comb either from top to bottom or from one end to the other. A comb should be cut deeply enough so that its surface is in line with the top bar of the frame (**Plate 9**).

The uncapping process can be sped up by fixing a knife vertically or horizontally and running the combs across it, cutting off the cappings. Automatic uncapping machines have also been invented in which the cappings are cut with whirling knives. In some machines the frames are moved mechanically and knives attached to rollers cut the cappings.

Whether uncapping is done mechanically or by hand the combs should be drip dried. The frames can be placed in a deep tray or those can be held on a movable radial rack. In an automatic machine the combs can be placed on a moving chain on which those are carried while dripping and then moved over a tank upto the motorized honey extractor.

10.4.1 Handling the Cappings and Wax

There is always some honey with cappings which should be drained and recovered as liquid honey. The drum or the tank into which the cappings fall should have a false bottom of screen through which the honey will drain off to the bottom over night or even in a longer period. The combs may be uncapped directly over honey presses and when filled with cappings the honey can be pressed out mechanically. Whirl dry-machines have also been devised for extracting honey from cappings by centrifugal force. The cappings out of this machine or out of the press are practically dry. The honey goes to the bulk honey tank and the cakes of cappings can be melted in a solar wax melter or in a hot water bath or in a steam box (see Berthold, 1987). The cappings and the bad combs to be discarded are generally melted at the end of the season after the honey crop has been handled properly.

10.4.2 Wax Melting Tanks

Pure bees wax obtained form cappings, burr combs, and new combs is white, whereas that from old combs has contamination of dark materials such as old skins of pupae, resins, pollen, honey etc. which have to be separated from the wax. The wax from cappings forms 10–20% of the wax collected and is pure and white. Wax melts at 62–65°C and on a small scale the easiest method to handle it is by heating it in a water tank. Wax floats on top from where it can be drained off through an outlet near the top of the tank and cooled off in molds and converted to cakes of convenient size. At that time or when needed melted wax can be converted into comb foundation sheets by dipping a wooden board and scraping off the wax sheet in cold water.

Wax from cappings and combs can also be heated in steam chests and, impurities that along with water go to the bottom as "slumgum", can be drained off. In a greater part of India solar heaters can be used throughout the summer and even in autumn months. On a commercial scale large solar heaters and steam chests are fabricated for separating honey from wax cappings. Those can also be used for separating impurities from wax when old combs are melted. These tanks have a sloping bottom and a depression at the end where honey or water, as the case may be, accumulates close to an outlet. The wax can then be bleached to some extent by placing it in the sun in thin layers. Wax is a complex mixture of lipids and hydrocarbons and is used by industry for making, candles, ointments, polishes and cosmetics, beauty creams, lipsticks, etc. It should be sold at a remunerative price to regular traders who deal with industry. Small shopkeepers may also delve in the purchase of wax but not being conversant with the product may leave it unprotected. In no time it is attacked by wax moth and black ants and is reduced to a low quality product.

10.5 HONEY EXTRACTION

Honey production is the foremost purpose of beekeeping and it brings the greatest profit to this industry. During the earlier times the combs cut out of the hollow of the nesting places were squeezed through a thin cloth and the honey that oozed out was collected in a vessel or a jar for sale. In this process the combs were destroyed and the bees had to construct combs all over again. It was the practice both with *Apis cerana* in Asia and *Apis mellifera* in Europe and Africa. Up to this day, this is the only way practiced for gathering honey from the exposed single comb colonies of the two wild honeybees *Apis florea* Fb. and *Apis dorsata* F in the Asian countries. The latter honey bee is still the major source of honey production in India. In fact it is a great source of livelihood for the tribal people of Eastern and Central India who are exploited by the middle traders and are given only a fraction of the value of honey collected by them from the jungles at a great risk to their life under extremes of weather. The squeezed honey may or may not be ripe but it invariably contains impurities of wax, and body juices of the brood that is also squeezed along with the honey portion of the comb. In nature the *Apis dorsata* F. colonies are also smoked, burnt, or destroyed in this process. Not all the honey harvesters know how or from what portion of the comb the honey can be gathered without destroying the colony or risking their life.

10.5.1 Honey House

In modern hives there are comb frames which can be moved individually when they are full of ripe honey. Generally the entire super containing 8–9 combs is removed from the top of a colony and is taken to the honey house which is well-aerated room having screen windows and doors with a provision of bee escapes to get rid of the bees that may enter the room. It is a multipurpose room and its doors have springs with which they are shut automatically. There should be running tap water over a large sink and electric points for fans and heaters. The flooring should have platforms on one side and be well drained so that it can be washed frequently. The cement flooring in a room gets corroded in time as honey has an acidic reaction. Tile flooring is more durable and even wood flooring lasts longer than the cement or concrete flooring.

Inside the honey house cappings are removed as described above and the frames are then ready to be extracted in a machine. The liquid honey out of the uncapped cells is removed by means of a honey extractor that works on the principle of centrifugal force generated manually or with power from an engine or an electric motor. The principle was first discovered in 1865 by von Heuschkol, an Austrian, who lived in Venice, Italy.

10.5.2 Honey Extractors

All modern honey extractors of various sizes work on the principle of centrifugal force. The essential parts of a machine consist of a central spindle with fans or radii, which rotates on a fulcrum, as spun in one direction with the help of a handle and gear system. It is enclosed in a round tank or a drum having an outlet at the bottom from where honey can be drained out (**Plate 12**). A ball bearing at the bottom of the spindle makes the machine lighter to work as compared with those having bush bearings. Care has to be taken that lubrication oil does not fall into the drum and contaminate the honey. A radial type of extractor is most commonly used; the standard combs are placed in a radial position like the spokes of a wheel. As the machine is worked honey flows out of the cells and strikes against the wall and collects at the bottom. The cells of combs are not damaged. In some models three frames are placed in the form of a triangle and the machine also works on the principle of centrifugal force.

The hand operated machines are gear driven and speed is controlled by the operator. The capacity is 2, 3, 4 or 8 frames and warm honey (110 to 120°F) is extracted within 5–10 minutes from both sides of the combs. The cells become dry and combs can be reused or stored as such. In hot weather or when honey combs have been stored at too high a temperature the wax becomes soft and there is a danger of their breaking down as the machine is worked. Therefore, temperature at the time of extraction is important. The new combs break more easily than the old ones fortified with propolis. It is desirable to work the machine at slower speed at first so that most of the honey flows out and the combs become lighter. At that stage if the reel is turned fast the combs are safe and all the honey flows out of the cells, leaving them dry.

The large power operated radial extractors can accommodate 12, 30, 45 or 50 frames. The large machines having wide circumference have to be run at slow speed to start with and then it is increased to the top speed of 3,000 revolutions per minute for 15 minutes. Mechanical and electronic devices are available which maintain the slower speed in the beginning. In some of the newer machines the centrifugal force is applied at the horizontal plane and these units have built in devices for uncapping also. Such machines are used in large operational units. Electricity is generally the source of power for large honey extractors. At the village level manual honey extractors of 2, 4, 8 frames capacity are good enough in India.

10.5.3 Bottling of Honey

There is always a certain amount of wax floating on extracted honey, which can be removed by filtering it through a single or double layer of muslin cloth. Along with wax some of the larger pollen grains, air bubbles and other impurities are also removed. The operation becomes easier if honey is warmed but not heated up to the melting point of wax (62–65°C). On a small scale it is sufficient to pour filtered honey in a tank with a controlled outlet at the bottom. If allowed to stand for a few days wax and bubbles will float on the top as scum which can be removed. Glass jars with screw caps or the newer lug type metal caps can be filled, labeled and cleaned up manually and sold in the market. Squeeze bottles of plastic of various shapes and sizes are handy for users. For solid honey, artificially granulated in factory (cream honey) is packed in flat opaque plastic cartons having a well sealed caps on the top. The capacity of the jars is 250 or 500 gms and one or two kg.

The public has to be educated that all pure honeys left on a shelf during winter will granulate and that some honeys granulate sooner than others. Granulation can be avoided or much delayed by heating honey in water up to 150–160°F for thirty minutes or to 170°F for 5 minutes and then cooled quickly. It is to be remembered that by heating directly some of the original flavor will be lost. At a commercial scale processing and filtering unit is combined (Bryant, 1987).

10.5.4 Honey Processing

On a commercial scale, honey is processed and bottled in a honey plant with facilities for no–drip automatic filling. The various units of the plant consist of a receiving and storage space, warming room, a tank into which honey is poured by gravity or a honey pump, heating pipes through which honey is circulated, pressure filter, honey cooking unit, bottling and packing unit, marketing and dispatch (**Plate 10**).

The drums of honey reach the reception and storage area from beekeepers or wholesalers of honey. The steel drums of 50 gallon capacity can only be handled with proper mechanical equipment that consists of a small tractor or a workshop motor, with a lift in front, which can maneouver the drums up and down, foreword and sideways. After weighing the drums are taken to the store room.

For manual smaller 5 gallon tins of vegetable oil are used after proper cleaning and drying. Tins formerly were called kerosene oil tins. These tins if kept in storage during the winter contain granulated honey which is difficult to pour out of the narrow mouth even after warming. The traders in India have already devised 15 gallon round barrels with wide mouth; those are made of galvanized iron sheets and are fairly resistant to corrosion.

Yearly stock of honey is kept in storage at 40°F. When honey is to be taken out and processed the drums are moved to a warm room kept at 110–120°F and honey is allowed to flow out into a tank by gravity after passing it through a fine meshed cloth, metal or nylon strainer. Within a few hours the wax particles and air bubbles will rise to the surface. The scum is removed before honey is pumped into the heat transfer unit having jackets of hot water and steam. Honey is heated up to 150°F by controlling the temperature thermostatically thus maintaining its flavour and colour. Care is taken that no part of honey is heated directly.

The heated honey is then passed through a pressure filtering unit and is cooled very fast to a temperature of 120–135°F when it is bottled (**Plate 10**). There are a number of combinations of temperature and time at which honey can be properly pasteurized for killing the yeast. This honey does not ferment and is safe for persons for whom yeast is harmful. The most important factor in processing honey is that pasteurization should not overly affect natural flavour and colour of honey.

The honey so heated and bottled will remain liquid for several months. The bottles are stored at 52°F and granulation will not take place for a long time unless temperature of the storage room falls further in the winter. No fermentation will take place either. The size and uniformity of the granules in a bottle will depend upon the type of honey. Some of them granulate at the bottom only and the liquid portion will come on top; in unpasteurized honey fermentation may take place in the liquid portion where the moisture content is more. In pure honey, levulose content is always more than dextrose content, the more the quantity of the latter the sooner will the honey granulate. It is dextrose that crystallizes while levulose, sucrose and other sugars remain in liquid form. Honey from certain flowers will give very fine granules and it is used as a seed for producing cream honey.

10.5.5 Creamed Honey

In Europe and America solid honey of a creamy consistency sold in wide mouth plastic cartons is very popular. It can be easily taken out with a spoon or a knife and spread on a toast. It is a convenient form and remains so up to the temperatures of 80–85°F; at higher temperatures it becomes soft and may turn into liquid honey. This type of honey is basically a granulated honey and is prepared under controlled conditions. The granules are very small and are uniformly distributed in the body of the honey. It is generally prepared in commercial honey processing units, although the process of manufacture is quite simple and can be tried at a smaller scale by the beekeeper also.

In a processing plant after the honey has been heated, filtered and cooled, it can either go to the bottling unit or it can be diverted to a seeding vat where it is further cooled to 75°F (24°C). The vat or the tank is water jacketed and cold water that runs through it cools the honey rapidly. The honey used for creaming should have a moisture range 17.5–18.0%. It should be thoroughly strained and all extraneous matter including air bubbles be removed. It is not necessary to pressure filter honey. So creamed honey can also be made outside a factory at smaller scale.

The honey that has been heated to 150°F, filtered, and cooled to 75°F is ready for mixing in it the "starter" or the seed. The seed honey contains very fine granules of dextrose and is of the finest quality and flavour. To reduce the granules to the fine state, honey is passed through a meat grinder or a powerful mixing machine, and is filtered to remove the bubbles. To the warm honey (75°F) is added 5–10% of the seed honey and is mixed thoroughly but gently in order to avoid bubbling. It should be transferred to bottling tank and allowed to stand for a 1/2 hour and in case any scum appears on top it should be removed. The seeded honey is then transferred to the retail plastic or paper cartons of 1/2 to 1 kg capacity which are wide mouthed and opaque. The honey is allowed to granulate in these containers at 55–57°F. In order to obtain fine and uniform granules the temperature of the cold room should not fall below 50°F (10°C) or above 59°F. Within 4–6 days the creamed honey is set in the containers which can be capped and marketed. In a greater part of North India, the creamed honey can be prepared and sold during the winter months. It can also be prepared at home scale in batches of 5 kg and used during the winter. It is also possible to convert the ordinary liquid honey into a solid form by mechanically whipping it (Niiranen, 1985).

10.6 MANAGEMENT FOR COMB HONEY

Comb honey is sold as pieces of new combs full of ripe and sealed honey in the natural form in which it is stored by the bees. Before the extractor was invented honey was generally sold in this form in the European markets and was kept in houses for sweetening foods. In USA and Europe it is still marketed on the shelves of the liquid and creamed honey. It has many advantages over the liquid honey. The original aroma of the flowers is maintained and it does not ferment. Simple equipment is needed to produce comb honey and it fetches more price. The drawback of comb honey is that it has to be packed properly to avoid breakage. In case it is kept at low temperatures for too long honey inside the cells granulates and the product cannot be sold. Therefore, it is a seasonal crop and has to be sold soon after it is produced (Eckert & Shaw, 1960; Killion, 1992; Taylor, 1977).

Comb honey is produced in shallow supers and the frames have four sections, each being 4.25″ × 4.75″. Certain precautions have to be taken. Since the supers are compact, air circulation in the hive is not so good. There are also greater chances of crowding of bees which may lead to swarming. For these reasons special management practices have to be followed to produce comb honey.

10.6.1 Manipulation of the Combs

Foundation sheets are fixed on the wooden sections without wire supports. The foundations should be made of clean white wax and should be thin, all of worker cells. When the combs are drawn by the bees they should look straight and cells should show a beautiful pattern of uniform size. The sections and foundation sheets are prepared much before the honeyflow in warm room protected from dust. The top of the sections are painted with liquid paraffin so that bees do not stain the frames with propolis.

10.6.2 Manipulation of the Colonies

For producing comb honey a greater skill is needed to handle the colonies as compared with that for the extracted honey. Shallow supers are used which provide less space and bees are confined to the brood chamber for brood rearing and for storing the food. The colony strength is built up before the honeyflow. The conditions create crowing among the bees which may lead to swarming. In order to obtain a good crop of honey swarming has to be avoided at all costs. Therefore, colonies have to be manipulated with skill.

Colonies kept for the production of comb honey should be strong and the weaker colonies can be used for extracted honey. The queens should be superior so that they can produce brood both before and during the honeyflow. Just before the honeyflow a colony has to be properly prepared. If it has built up strength to occupy two or three chambers it is confined to one chamber only and the extra brood frames are given to weaker colonies. The one chamber colony should be full of frames with emerging brood. The bees on extra frames should be shaken in from of the hive so that a good strength of bees remains in the colony. The shaken brood frames can be given to other colonies.

The colonies should be examined every 7–10 days for observing any swarming signs. If queen cells are found those should be destroyed, the old queen removed and in its place a young queen be given ten days later. Every colony should be prevented from swarming and at the same time kept in good working condition.

10.6.3 Manipulation of Supers

Giving supers to a colony in right order and at the right time will not only prevent swarming but also stimulate the bees to store the maximum amount of honey in the combs. The supers are given one by one just before need is felt by the bees, and a beekeeper should not wait till one is filled and removed and only then to give another onc. If the honeyflow is good and there are good chances of a super being filled the new super should be placed below the existing super and above the brood chamber and the queen excluder. If on the other hand chances are that a new super will not be filled completely it should be placed above the existing super. The same principle will apply if more than two supers are to be given. It is a good practice to give a start to a new super above the brood chamber and then to move it to the top. There the wax secreting bees will continue building the combs and it will remain away from the main body of bees in the brood chamber. By splitting the bee force, chances of swarming are reduced.

As soon as the sections of comb honey are filled those should be removed individually or by the supers; if left too long bees travel over the surface and the pure white cappings are stained somewhat. Before removing, the supers are cleared of the bees by the same methods as described earlier. The moisture content of honey should be 17% and any excess moisture should be removed in storage. The honey room where comb honey is stored should be well ventilated, dry, and free from dust.

Comb honey can also be produced in shallow supers without the sections. The combs are then cut into convenient pieces with a hot knife or a caste of any shape. The cut cells are drained or extracted of the liquid honey and are wrapped in cellophane. Glass jars and clear plastic containers are used for packing.

10.7 QUEEN REARING AND NUCLEUS COLONIES

As mentioned earlier young and vigorous queens are required for the colonies to build good strength in the pre–honeyflow period. These queens are obtained from the reservoir of queens maintained in the apiary. They are produced from mother colonies having good history of honey production. Queen rearing and maintaining a reservoir is therefore an important part of colony management for honey production. Queens are reared from the multiple queen cells produced in the spring in many strong colonies. The queen cells are cut out without damaging them and are individually grafted on combs in queenless nucleus colonies. The nucleus colonies are either started in small 3–frames boxes or by dividing a one–chamber colony into three by placing two complete partition boards in between and giving three separate entrances in different directions.

The queens can also be mass reared by special techniques. The commercial beekeepers mass produce queens and sell them along with 2–3 packages of bees.The package boxes are so made that bees are safe even when sent by parcel over long distances.

10.7.1 Natural Queen Cells

Out of the three types of queen cells supersedure, requeening and swarming, the last mentioned is believed to be more suitable for producing the queen stock. At the time swarming cells are produced the food, flight, and mating conditions are favourable and best queens are produced under those conditions. It is through the inherited characteristics of the queen and those acquired through mating that the future quality and performance of the colony will depend. The important colony characteristics are industry, temper, comb building, foraging, honey collection, disease resistance, colour of workers and robbing habits. Colonies headed by young queens swarm less which is desirable. New queens are given to colonies every second year and in localities where brood rearing period is prolonged and colony activity is more intense it would be desirable to change queens every year. In addition, new queens are needed to replace injured, lost, or non–performing queens. Even with best precautions the queens are sometimes lost during the routine inspection of the colonies. Spare queens are needed to meet all such needs in an apiary.

Natural queen cells are obtained from colonies during the swarming season. Those colonies are selected which in the year before have given good performance. The queen cells in the swarming season are generally hanging downwards at the bottom of combs. Those sealed cells are selected in which the queens are ready to emerge in a day or two. A portion of comb around a cell is cut alongwith, taking care it is not damaged. It is then taken to the sealed brood portion of a comb in a nucleus. The cell is attached to the comb by pushing in the base of the cut out cell. If the weather is warm it can be simply pushed in between the end bar of a frame and the comb, maintaining its natural hanging position. At times it many be necessary to cut a hole in the surface of a comb and insert the cut out cell. Coinciding with the emergence of young queens mature drones should be ready in other colonies. The nucleus colonies may even be kept at a distance in another apiary having drones with desirable characteristics and uniform colour. Such a mating is expected to result in good progeny in the new colony.

10.7.2 Controlled Cell Building

In the process of natural building of queen cells a beekeeper is dependent on the circumstances under which colonies are allowed to raise their own queen cells and also the number that is raised naturally. By controlled rearing of queen cells the beekeeper can exercise judgement in selecting a "cell building colony" and can regulate and improve the future stock. Moreover, he can control the number of queen-cells to be raised as and when necessary. The colony selected is called "queenless cell builder" (Fig. 14).

There are certain desirable conditions under which good queen-cells are built. Generally the conditions of queenless or swarming colonies are stimulated; there should be a strong force of young bees, plenty of stored or fresh pollen available and a constant supply of honey for bees to utilize and convert it into royal jelly and bees wax. The queenless cell builder can be made from one, two or three storey colonies. In an ordinary apiary one chamber colony is generally used but commercial beekeepers prefer to use three chamber colonies. In two or three chamber colonies queen excluders are placed below the top chamber and, in a one-chamber colony above the bottom board, to stop the entry of a stray queen that might enter the colony and cut the queen cells. In these colonies two or more batches of queen cells can be raised in a season.

The basic technique used in this process is that queen is confined to a small section and major portion of the colony is rendered queenless. In the queenless portion frames contain sealed brood, plenty of honey and pollen and there are combs of young brood in the centre. In a strong colony on two chambers the queen is confined in the upper portion on two or three frames of sealed brood and bees by using a partition board. A queen excluder is used below and nine frames of bees, sealed brood, young brood and food, are placed in the lower chamber. Bees from extra frames are shaken in front of the entrance or extra bee force is provided by uniting a nucleus. Additional feeding is also desirable. The colony will start building queen cells in a few hours and sealed queen cells from the other colonies can also be given. In a two storey colony if most of the sealed brood is given below and young brood, sealed brood, and food in the upper chamber, new queens will be raised there. The emergence of the queens is to be watched most carefully as the first virgin queen is likely to cut down other cells. For safety queen–cell guards should be used so that virgin queens can be handled one by one and given to nucleus colonies. A more practical way is to keep track of the age of sealed cells and remove them before the emergence of queens and to introduce them in nucleus colonies.

10.7.3 Mass Rearing of Queens/Manipulated Queen Rearing

Bees can be induced to build queen cells around natural worker cells containing eggs or young larvae and many methods have been described in books on queen rearing listed in bibliography at the end. More practical method of mass rearing of queen cells is by the use of artificial queen cells cups made of wax, wood or plastic of the size and shape of the base of a natural queen cell. These cups are embedded into the surface of a comb and, eggs or young larvae are transferred (grafing) on to them along with royal jelly. The bees then take over, construct the queen cells, and feed the larvae inside, eventually raising new queens from the desired stock in the numbers required.

Attempts at starting such manipulated queen rearing were made as early as the beginning of the last century. Making use of early experiences, G.M. Doolittle in 1870 standardized the technique, which with minor changes here and there, is still being used till today. The Doolittle method was described in his book "Scientific Queen Rearing" published in 1888. He used queen-cell cups made of wax and instead of embedding them in comb surface he fixed them on a wooden bar held across the middle of a frames. The cups were made with the rounded end of a stick of the same thickness as the base of a queen cell. The end of the forming stick is dipped in cold water and then into melted bees wax, up to the desired depth. Two or more shallower dippings are made in the wax so that the cup becomes thicker at the base and thinner above. The wooden bar on the bottom end of which the cups are to be fixed 3/4 inch apart, is also dipped in melted wax. In these cups are placed small droplets of royal jelly taken out of a newly sealed queen cell. With the help of a wire spoon young worker larvae 36 hours old or less are grafted on to the droplets of royal jelly. The frame is then placed in the centre of a cell building colony. Ten days later, the queen cells now formed and sealed, are removed and grafted one in each of the nucleus colonies. The virgin queens emerge in 11.5 to 12.0 days after the larvae are grafted in the cups. After mating they start laying eggs and establish the nucleus colonies.

If adequate precautions are not taken the larvae in the process of grafting are likely to be chilled, overheated or desiccated. Therefore grafting should be done in small rooms which are kept warm and humid. The floor of a room can be sprinkled with water or a pan of water can be heated to increase humidity. Strong neon light is needed to observe the larvae in their natural cells and also the inside of the cups. The age of the larvae should be uniform so that queens develop within the same time. Fifteen queen cells can be raised on one bar of cell cups and those will be sealed within 4–5 days. If two or three bars of cell cups containing larvae are given at 4 day intervals, 30–45 queens can be raised.

The number of queens that can be handled at a time will depend upon the number of nuclei prepared for the purpose. The method of handling will also depend upon whether the number of colonies is to be increased in the apiary or those are to be sold to other beekeepers nearby or shipped over long distances.

10.7.4 Queen–Mating Nuclei

The queen mating nuclei are made by dividing a colony into two using a solid division board and making separate entrances, one in the front and one at the back. Each nucleus should contain one comb of honey, one frame of brood, an empty frame and, at least two frames of bees. A good strength of bees is required to cover the brood frames and to take care of eggs and young brood when the queen to be introduced lays eggs. The nucleus colonies can be given extra bees by shaking frames over them from other colonies. The nucleus colonies may be kept in the same apiary or another apiary 2–3 km away. Within a few hours of forming a queenless nucleus, the colony will accept a ripe queen cell from which a virgin queen is ready to emerge. Generally nuclei are made one day and queen cells are given the next day.

After giving a queen cell the nucleus is not to be opened or disturbed for a week or ten days. A queen should be laying eggs on the twelfth day after emerging and the nucleus can be opened for examination. Smoke should be used sparingly. If the queen is laying eggs regularly, placing single eggs at the bottom of the cells it is considered to have properly mated. It is ready to be caged for introducing in another colony in the apiary, or to be transported and shipped with 8–10 young bee attendants with the entrance hole plugged with sugar candy. If so required the two nucleus colonies in a box can be separated and established as new colonies. Apart from selling and posting the queens in queen cages they can also form a part of 2–3 lb package of bees sold and shipped long distances.

10.7.5 Classification of Queens

A virgin queen on emergence mates within 5–10 days and then starts laying eggs. Its progeny has not yet developed to maturity and at this stage the queen is known as "untested". After three weeks its progeny emerges and the colour of workers can be noticed. If the colour is uniform and of the same race as the queen it means mating has been with the right drone and the queen is called 'tested". Further, the performance of the queen is observed from the colony strength and honey production; if performance is good the queen is called "select tested". With persistently good results with a queen it is marked for further breeding and is retained as a "breeding queen". Its characters are noted and records are kept for further improvements of the stock.

Prolificity, gentle behaviour, less swarming, are some of the desirable characters. Those can be incorporated in daughter queens by keeping drones of good stock in separate mating yards. To ensure that drones are mature when queens emerge they are produced in selected colonies early in the season by providing drone combs. Techniques for artificial insemination have also been developed and desirable characters from male and female lines can be combined by controlled fertilization. Therefore, it is desirable to keep records of both the queen and the drone lines so that progeny of pure colours can be produced. In order to offset the undesirable effects of inbreeding within an apiary, three or four desirable lines of queens and drones should be maintained and males and females should be used alternately. It can be successfully achieved by artificial insemination (see; Laidlaw, 1987; 1992; Morse, 1979; Page & Laidlaw, 1992; Ruttner, 1987).

10.7.6 Sale of Colonies, Nuclei, and Package Bees

Sale of bee colonies is a source of income for the beekeepers. In the absence of big commercial beekeepers in India it is also a practical way of propagating beekeeping to various regions. Bees are generally sold during the pre-honeyflow period. No one sells a colony in two chambers. The bees are sold as a full colony on 8–10 frames or a half colony prepared as a nucleus on 5–6 frames. In both cases the colonies should be queen right with new queens that are laying eggs. The buyers bring their own new standard Langstroth hives and transfer the frames to them, leaving with the beekeeper an equal number of frames out of their new hives. If the weather is not hot and colonies are to be taken short distances about 100 km connected with good roads the hives are taken on a pick-up truck after nailing the frames so that they remain in position. If the weather is warm and bees are to be transported on a truck many hundred kilometres, screen covers are placed on hives for proper aeration and water is sprinkled over them for drinking and cooling. If the colonies are to be carried 1000 km or more by road it is advisable to cage the queen and hang the cages in between the frames. Queens are always caged in packages to be sent by parcel mail, goods road transport or air freight, giving special instructions on the consignment. While caging the queens with attendants, care should be taken that there is sufficient candy blocking the entrance so that queen is not released on the way. Sometimes queen excluding bars are fixed in the cage so that workers can go in but queen cannot come out unless released manually. During this process the bees in the colony get familiarized with queen and its acceptance is easy.

10.7.7 Sale of Mated Queens

The queen cages are made of wood or plastic and there are three cells in it which open to the outside through the candy hole. A wire screen cover fits tightly on one face of the cage. A queen can be pushed in by a thumb and finger by placing the opening in front of her. In a mating nucleus or a colony, queen has to be searched by using the minimum smoke. Using thumb and index finger the queen is caught from behind at the four wings and is made to run into the hole of the queen cage. Young bees that are found feeding on honey are caught and pushed into the cage as attendants. The candy hole is then filled and passage blocked. Before caging, the queen can also be marked with paint or can be clipped by cutting two wings on one side.While doing this operation queen can be held from legs on one side in one hand and the operation carried out with the other hand. The queen cages can be packed one above the other and made into a parcel for sending by post.

10.8 PREPARING PACKAGES OF BEES

In USA particularly in California and Florida it is a multi-million dollar industry to prepare packages of 2–3 lbs of bees with a caged queen and to sell them to the beekeepers in Northern States of USA and Canada. The first time package bees were put up on sale was in 1913 and since then the industry has seen many ups and downs. With the appearance of Africanized "Killer Bee" in New Mexico and southern California and the consequent fear that through the genes of the queen in a package the undesirable characters might be spread throughout the United States, the import of Package Bees into Canada has been banned by law. Within the United States the sale of Package Bees is still common and the beekeepers in snow bound regions kill their bees after the honeyflow and start afresh with the Package Bees next spring, to build the colony strength, extract honey, and to kill the colonies again. However, in India such a sophisticated level of life-management is not likely to be achieved for quite some time.

The packages contain 2–3 lbs of bees collected from various colonies and sold in wood and wire cages. Those may be queenless or queen–right nuclei which are used for starting new colonies. Those are also purchased to make-up for winter losses and to increase the colony strength or to add to the bee force for pollination. In the packages the queen cages are hung with a wire and a can of sugar syrup is well secured so that it does not move about in transit. In some packages there may be 2–3 combs partly filled with honey. The bees in the package boxes are poured en masse through a funnel shaped shaker box and weighed on a scale. The bees before shipment should be given plenty of food so that they can pass the journey without undue mortality. The bees should also be of young age collected from the upper chambers. However, when sold for pollination, bees of all ages are to be included.

10.9 HANDLING OF COLONIES AFTER HONEY FLOW

At the end of honeyflow a few days are required to ripen the honey. The combs with sealed cells can be extracted straight away as the moisture content of honey is expected to be 17.5 to 18.0%. The combs with unsealed cells should be left in the hives for 7–10 days so that moisture is reduced and cells of the ripe honey are sealed. Then honey is extracted.

After the honeyflow there may be robbing among the bees. The instinct of the bees to collect honey from the field is not suppressed immediately although nectar flow may end suddenly. To satisfy this instinct bee colonies may resort to robbing one another. Therefore hives should be closed properly and entrances should be regulated. In areas where, barseem or Egyptian clover, is grown the flowers suddenly dry up in the first week of June. New strains have now been developed which can withstand the heat better and the crop can be kept green by more frequent irrigations. Thus flowering period can be extended up to the third week of June. By this practice not only robbing is avoided, the bees also collect extra stores of honey for themselves, even after the extraction is over. If the chambers are full one super on top is sufficient for this purpose. With the extra space available controlled aeration of the hive is possible. The maximum ambient temperature in Indo-Gangetic plains at that time is above 45°C and hot winds come from the South and the West. To avoid the effect of intense heat the entrance should be controlled and its direction be changed towards the East or North–East. A double row of trees and shrubs planted on the South side give effective protection to bees in the hot summer and vegetation serves as a good wind break. Temperatures on the northern side of solid plantations are at least 5°C lower than on the southern sides.

When the combs are heavy with honey and, temperature outside is between 40–45°C wax becomes soft and honey may start dripping within the hive. Wet cloth or gunny sac on the hive and a sprinkling of water in the apiary grounds reduces temperature by 10°C. The softness of the combs is rectified and honey extraction is facilitated.

10.9.1 Security

Honey crop in the apiary is a great temptation for vandalism. If the apiary does not have a secure enclosure a beekeeper has to guard his hives. Vigilance, particularly at night is very important. The yard should be kept well lit and one or two persons should sleep in the apiary at night. Since an experienced beekeeper can steal honey at night the beekeepers should from local cooperatives and societies for their mutual good and understanding. The outsiders who attempt vandalism will then be a common enemy of the beekeepers in that locality.

Seasonal Management

11.0 SEASONAL MANAGEMENT FOR INDIA

The climate and vegetation in the Indian subcontinent is extremely variable and ranges from the Arctic in Himalayan region, to that of Sahara in Rajasthan desert; from the highest monsoon rainfall in East India, to the Tropical rain forests of South India (Gopal Singh, 1995). The two hive honeybees, *Apis mellifera* and *Apis cerana* can survive throughout the country and also in the Himalayan region beyond the tree line up to 11000 ft altitude. Under such diverse conditions the management of honeybees would naturally be very variable and it would not be possible to list the month-to-month management practices applicable throughout the country.

Depending upon the type of agriculture, vegetation and floral conditions in tropical, subtropical and temperate India, management practices have to be devised separately for those regions. The practices have to be further devised for sub-regions, zones, country areas and even localities. Really meaningful practices can be developed by beekeepers themselves after observing the weight gain and loss of a colony on a balance and by recoιding the seasonal availability of nectar and pollen of flowers. There are certain broad principles common to the practices for all situations. Then again, certain practices may be common for various places but the calender timings may be different owing to the seasons and the local conditions.

Seasonal climate plays a dominant role in bee management. In the temperate and subtropical areas seasons are distinct and so is the appearance of flowers and therefore the colonies develop certain patterns of activity which determine the seasonal management practices. In the tropical areas, seasons are not distinct but there are patterns of rainfall and drought and also the appearance and disappearance of flowers. In the tropics, management become more meaningful if developed for the dearth period, built-up, honey flow, honey extraction periods etc.

Wedmore (1979) defines management as the planning of operations to secure maximum return with minimum effort under given circumstances. Without proper management bee colonies would stray, swarm, or resort to robbing, causing problems for the other beekeepers. Scientific management of bees in various countries has been dealt with suiting the conditions there, in particular the climate and bee pasturage (Smith, 1960, Akratanakul, 1986; Crane, 1990; Rawat, 1982; Mishra, 1995).

11.1 TROPICAL INDIA

The climate, vegetation and floral cycles for the tropical and other parts of India are described further in Chapter 12. The tropical forests in South India are not in large stretches, but are interspersed with crops and plantations. There are no single-species flora that provide both nectar and pollen to the bees for prolonged periods. Therefore bees have to build strength on mixed tropical flora and are then taken to plantations for the exploitation of nectar or honey-dew (Table 6).

TABLE 6. *Seasonal Brood Rearing and Colony Management of Apis cerana in the hills of Kodai Kanal bordering Tamil Nadu and Kerala.*

S. No. Season	Brood and Colony Activity
1. Dry Winter ; (January to March)	Ground becomes frosty and temperatures do not go below 7°C; weeds and grasses dry up and there is no bee flora till March, when *Celtis* and *Michelia* etc. bloom after rains; brood activity declines and absconding is common. After the rains colonies start building up strength.
2. Hot Season; (April to June)	Frequent showers with intermitant hot sun, high temperatures 22-23°C at high altitude (2000 m) and 35°C in the lower plains; strong colonies collect surplus honey at first from *Gmelina,* followed by *Syzygium,* and *Ligustrum* in June. New queens are raised from swarm queen cells; weak colonies abscond and the feral colonies are hived.
3. South-West Monsoons; (July to September)	It is a dearth period and rainfall is variable; colonies sustain themselves, on reserves in brood chamber; brood rearing is the minimum till *Erythrina* blooms in September. Generally colonies are migrated to plains to exploit cultivated crops like *Sesamum, Pennisetum,* coconut, cucumber. Colonies are brought back in February.
4. North-East Monsoons; (October to December)	Heavy rains; colonies retained in hills suffer heavy attacks of wax moth and wasps; there is no egg laying or brood rearing; sugar feeding becomes necessary. Weather warms up in December and bees start activity.

*From Chandran and Shah (1974).

The management can become precise based on the value of specific flowers and the exact condition of the colony. Any delay in moving the colonies means exhaustion of the food reserves, set-back to colony build-up and consequently to honey extraction. The management in this section is discussed under the Dearth Period, Build-up Period, Honey Flow Period and Harvest Period.

11.1.1 Dearth Period

During the dearth period there are hardly any flowers and there is no nectar and pollen for bees to collect. Bees depend upon food reserves for their survival. Dearth is caused by prolonged dry season, or heavy rains or a combination of the two in monsoon areas where drought is followed by heavy rains. During the hot and dry season, apart from the shortage of food, shade is required for the colonies and water is required for drinking and cooling down the hives.

During the prolonged heavy rains bees remain confined inside the hive where humidity is high and the growth of fungus on the combs is promoted. Wax moths appear in hives, they lay eggs on debris at the bottom and the growing larvae start boring into the wax of combs. Their appearance is to be watched from the beginning and the larvae are to be killed mechanically with the help of needles or the hive tool. The combs so damaged are repaired by the bees themselves.

During the dearth period, in fact all the year round, colonies have to be protected from the attack of predators like ants, wasps, and birds. The legs of the hives are dipped in cups full of water which serve as ant barriers. During this period bees remain confined inside the hive and under those contagious conditions the endoparasitic mite (Acarine disease) and the ectoparasitic mites flourish in the colonies. Swabs of cotton at the tip of iron wires or on absorbent pads of cloth, paper or plastic dipped in menthol, formic acid or pyrithroids are placed in the hives. The fumes of these volatile chemicals selectively kill the mites without harming the bees. In case attack is detected at an early stage one treatment gives the relief; if need be the operation can be repeated after a week or so. It is now a general practice to give fumigation as a routine at the start of the dearth period and once again at the end of the period.

In case food reserves seem to be depleting, sugar feeding is given, taking care that robbing does not start in the apiary. In dry season the water-sugar ratio of the syrup is 2 : 1, but during the rains it can be increased to 1 : 1. Even if some reserves of food are left in the hive heavy feeding is desirable before the end of the dearth period so that bees are stimulated to take up field duties and to start building up the colony strength. As a rule-of-thumb one super full of honey at the start of dearth period is considered sufficient to tide over the period (Smith, 1960; Crane, 1990).

11.1.2 Build-up Period

With the flowering of trees, herbs, and crops, bees start bringing in nectar and pollen and rearing brood. The queen in a hive starts laying eggs, bees consume honey reserves and loss is registered in the weight of balance colony. As more nectar and pollen are brought in the colony weight becomes constant or it fluctuates within narrow limits. It is a sign that the incoming food is being used for rearing brood. All facilities should be provided to the bees to build up good strength. Good food reserves and protection from enemies and ample space are the key factors in management practices (Thakar, 1976).

The build-up period should start eight weeks before the expected honey flow. There should be sufficient space for the bees to expand sideways and not upwards. If a colony is on one brood chamber a second be given on top. If on the other hand colony is already on two chambers it would be noticed that the queen would have a tendency to move upwards for laying eggs, neglecting the lower chamber. The position of the chambers should be reversed so that brood rearing starts in the chamber that is now lying below (Wedmore, 1979).

The build up period starts eight weeks before the honey flow. The colony should be examined thoroughly for the health of its queen. Its egg-laying pattern should be steady in solid areas without patchiness. If the queen is defective or is missing altogether the colony should be mixed with another medium size colony having good queen. Requeening is done if there is no other choice and in so doing the colony will naturally lag behind in building its strength.

During the build-up, food frames are moved outwards, empty frames are placed next to them and the brood frames are placed in the centre. When only two empty frames are left to be filled with brood, another chamber of empty frames is placed on top for the expansion of the colony. The queen will soon move there and start lying eggs. The sealed brood in the lower chamber will emerge soon and when that happens the position of the chambers should be reversed so that brood rearing continues in the lower chamber. The reversing of the chambers should be done at 2-3 week intervals and during the build up period there would be two or three occasions to do so.

Towards the end of build-up period there will be lot of bee force which should be provided with work otherwise they would have the tendency to build queen cells in preparation for swarming which should be avoided. One method of giving the bees work is to place a chamber on top with frames of foundation sheets and one or two partially drawn combs. The young bees will secrete wax and raise new beautiful white combs.

If the bees continue raising queen cells those should be cut with strong base and given to nucleus colonies. The idea is to avoid swarming by whatever means and not let the bee force of the colony divide into swarms. It is a test of good management. Every effort should be made to maintain strong colonies which have a great worker force for honey collection during the honey flow.

11.1.3 Honey Flow Period

When the honey flow starts queen cell building may continue and thus the swarming tendencies are to be watched and suppressed by the methods discussed in detail earlier. Another undesirable tendency to be watched is excessive brood rearing during the flow of nectar and pollen. Moderate brood rearing is required for the replacement of bees that die of old age but heavy brood rearing is to be discouraged. If this tendency is observed the queen should be confined to the lower chamber, placing a queen excluder in between. The additional advantage will be that combs will be available for honey storage in the upper chamber as brood emerges from the sealed cells. When the upper chamber is three-fourth full of nectar another super having drawn frame should be placed for honey storage. Thus the tempo of honey collection will be maintained among the bees.

Timing of placement of supers is very important. A delay in giving a super would mean that the lower super when full of honey will create congestion. Apart from losing one-third of the honey crop the bees may start making preparations for swarming, particularly if the queen is not confined to the bottom chamber below a queen excluder. Therefore, as one super is being used by the bees and approximately two frames are left in it a new super should be placed on top.

In the early stages of honey flow, quite often queen cells are observed, particularly if the queen is three years old. It is the time for removing queen cells and giving them to nucleus colonies which should be available in the apiary. The nucleus colonies are raised by taking out of other colonies honey frames and sealed brood; 2-3 frames of brood and bees per nucleus are sufficient which can be taken out of large colonies. Some extra frames of young bees will give a strength of 5-6 frames. The nucleus is now ready to receive a sealed queen cell from a colony with good history of brood rearing and honey collection. The nucleus colonies if not required for the apiary can be sold to prospective beekeepers. At the same time swarming is prevented and commotion in the apiary is avoided.

In tropical India the forest areas are interspersed with crops and plantations. The plantations of cardamom, coffee, banana, palms etc. are in shorter stretches and they alongwith field crops provide some nectar but good pollen. Colonies build up strength in these areas and are then shifted to rubber plantations for the main honey flow. The nectar from rubber is from terminal leaf buds and forms a rich source of honey. The movement of colonics to these plantations and other sources such as oil palm is an important part of bee management. (Thakar *et al.*, 1962; Chandran and Shah, 1974; Chubal and Kotmire, 1980).

11.1.4 Harvesting Period

When the supers are full of honey those are left in the colony for 7-10 days for further reduction of moisture. Harvesting is done when moisture content of honey is 18-19 per cent. Some bee keepers have a tendency to extract honey as soon as combs are 2/3 or 3/4 sealed which is not a desirable practice in tropical area particularly during the rainy season. Honey absorbs moisture very quickly and moisture content of unsealed honey becomes high.

Some beekeepers feel that extracting ripe honey during the progress of honey flow stimulates the bees to collect nectar with a greater tempo. In any case there should be no objection to extracting honey more frequently if it is ripe and extra labour is available.

After the rubber honey flow is over colonies have to be moved out. From the east plantations colonies are moved out to forests where minor flow is available from the multiflora. Later on they are taken further east to cultivated areas to forage on crops like sesame, niger, coriander etc. From the west plantations colonies are taken towards the coastal areas where palms are available and cardamom plantations are spread all over. Nectar and pollen from cardamom flowers is exploited by the bees from March to September with the peak period of mid-May to mid-June (Chandran *et al.* and Shah, 1974, 1980; Siddappaji *et al.*, 1980). On cardamom flowers bee activity continues practically all day with predominance of pollen collectors in the forenoon and that of nectar collectors in the afternoon. It was recorded that concentrations of sugar in nectar was 37% at 7 am and 43% at 2 pm which perhaps explains the collection of nectar in the afternoon hours.

In south-eastern part of India, Kanyakumari area of Tamil Nadu is most suitable for beekeeping and, one half of honey production of the state is from that area. The major honey flow is from October to March from such sources as coconut, coffee, Terminalia and other flowering trees and oilseed crops (Mishra and Kaushik, 1992).

In coastal areas of tropical India there are a great many small plantations of coconut and oil palms which give only a small income to individual farmers. Where these plantations are interspersed with crops like niger, sesame, banana, pineapple and other fruits there is a better scope of beekeeping than in the isolated coconut plantations. In the presence of these minor sources, the coconut palms which provide pollen and some nectar throughout the year can be a source of major honey flow and can yield surplus honey. Makhdzir and Osman (1980) made similar observations in South Malaysia where *Apis cerana* bees were kept under these situations and surplus honey was collected.

11.1.5 General

In tropical India the key points for *A.m.* bee management are: proper comb space for brood rearing and food storage; artificial feeding of sugar, pollen supplement or pollen substitute whenever there is dearth; migrating bees to areas of availability or abundance of pollen, nectar or both; shifting bees to plantations or crops for harvesting the major honey flow. Throughout the tropical India there is a mosaic of all types of vegetation which can provide to the bees food for subsistence, colony build up, minor honey flow, or a major honey flow. The colonies have to be moved from one locality or area to the other, taking advantage of the floral conditions there. In order to maintain colony strength during the dearth period in a given locality bees have to be given feeding. Even during the major nectar flow in rubber plantations colonies have to be given pollen substitute otherwise the colony strength would dwindle, partly owing to overwork and exhaustion of the foraging bees and partly due to the lack of pollen required for brood rearing.

There is an area of 2,30,000 ha under rubber plantations in south-western part of South India spread over in the states of Kerala, Karnataka and Tamil Nadu. Only a fraction of this source of honey is being exploited at present. On the rubber tree after the trifoliate leaves are shed from December to February, flowers appear which continue upto April. There are no nectaries in flowers but small quantities of pollen are available. The hive honeybees do not visit flowers. When the light green leaves appear nectar is secreted in the three extra floral nectaries situated at joints of petioles. Honeybees exploit that nectar for over two months from March to May. The beekeepers move an estimated number of 1,00,000 colonies of *Apis cerana* to the plantations and collect most of the surplus honey crop in Kerala (Suryanarayana, 1966, 1980). For want of pollen on rubber trees only those sections of the plantations can be exploited successfully near where there are coconut and oil palms available for the supply of pollen (Devadason, 1971).

Along the Western Ghats extend tropical rain forests interspersed with crops, fruits and plantations. The forests are spread roughly north to south on slopes both on the western and eastern aspects. Growing in almost parallel strips on the eastern side is a sequence of various types of vegetation including tropical wet evergreen forests, semi evergreen forests, tropical moist deciduous forests, tropical dry deciduous forests and tropical thorn forests, the last mentioned extending upto the eastern districts of Karnataka (Rajan, 1980).

Vegetation of the tropical wet evergreen forest on the western and eastern slopes of Mahabaleshwar hills was studied for many years. Thakar *et al.* (1962) surveyed over 700 flowering trees and plants and found that only a few were of importance to honey bees. Based on the flowering period and pollen and nectar values annual floral calendars were prepared in order to help the beekeepers. It was observed that most of the forest trees, shrubs and annuals flowered during the months of March-May and again during October-December. There were floral gaps during January-February and the monsoon months of June-July. These were the periods of acute scarcity of bee forage. With a view to filling these gaps they made valuable suggestions for planting useful species.

It was rather ironical to observe that during the gap periods a number of wild plants that flowered scantily were poisonous to bees, and colonies had to be shifted to safer places. The poisonous plants included *Lasiosiphon, Colebrooked, Lobelia, Clematis, Argemone* and others (Thakar *et al*., 1962).

Towards the East of the Ghats, a greater part of Karnataka is under dry deciduous forests and tropical thorn forests. Rajan (1980) has suggested the plantation of flowering trees in the eastern districts which are highly degraded at present due to over-exploitation of timber and overgrazing of the under growth. Apart from planting trees on Government lands the solution lies in encouraging farmers to plant multipurpose trees on their farms so as to augment their income. The trees planted should supply timber, small timber, firewood, fodder, oil, fruits and minor forest products, including bee flora. The choice of trees for this arid region falls on species like *Eucalyptus, Acacia, Moringa, Casuarina, Syzygium, Leucaena, Tamarendus,* etc., all of which provide rich bee flora. With the coming up of mixed plantations and of the existing field crops there is a tremendous scope of beekeeping and harvesting surplus honey. Thus agriculture, farm forestry, and beekeeping could be combined to augment farm income.

Further east in Tamil Nadu cotton is the prospective crop for beekeeping but the opportunity is thwarted due to heavy spraying of insecticides on this valuable commercial crop.

11.2 SUBTROPICAL INDIA

The tropical climate and vegetation gradually changes to subtropical crops, fruits, forest trees and undergrowth. The subtropic regions have their own gradient of temperature, humidity, plant life and the time and season of flowering. The tropic of cancer passes south of Rajasthan, through Madhya Pradesh, South Bihar, West Bengal and Bangladesh. The climate and vegetation south of it gradually transforms into that of the tropics and that to the north eventually join the temperate, showing distinct seasons of spring, dry summer, the monsoon, autumn and winter.

The subtropical India is a large stretch. There is dominant hot and dry climate in the western region that extends from Punjab, through Rajasthan, Gujarat, Northern Maharastra. In a greater part of Madhya Pradesh and Andhra where climate is hot and dry but towards the East humidity is high and vegetation characteristics change a great deal particularly in West Bengal and Eastern India (Gopal Singh, 1995). The valleys of eastern Himalayas and the slopes and valleys of the north-western Himalayas are subtropical in their characteristic from the point of view of beekeeping.

The Indo-Gangetic plains have distinct seasons, including cold winter when the minimum temperatures observed range from 0°C to 7°C. There is underground water and river-canal systems for irrigation that changes the whole picture of vegetation and field crops. The Himalayas in the North stop the cold winds from central Asia and therefore the climate on an average is 3-8°C warmer than it would have been otherwise. Even so there is distinct winter but no sub-zero temperatures in the plains, although there are over 20 frost nights in a year.

The honeybee colonies kept in northern India have a chance to build up their strength in the postmonsoon period on crops like sesame, desi cotton, cucurbits, solanacious and other vegetables, forest plants and fruits like guava. In many areas *toria* is sown in September and it flowers in October and November and provides a minor source of nectar. The colonies build up their strength, some of them supersede their queens and are ready for the winter in the months of December and January, when the temperatures fall low enough that honeybees form nests in their colonies. The formation of a nest does not mean the colony has become dormant with practically no activity. The nest is a living, eating, moving, brood rearing, heat generating cluster of bees. It expands and contracts according to the degree of cold. In north Indian plains where winter is mild the cluster may remain very loose. If the colony strength is good and locality is rich in *Eucalyptus* and *Brassica* etc. the cluster may remain active, rearning brood and even storing food in the winter. A weak colony on the other hand will need good feeding before the winter and sufficient packing both inside and outside to protect it from the cold. Within the subtropics there are a series of management practices suiting the area or a locality that are mentioned under seasons of the year. What happens to a colony in spring is determined by the preparations made in autumn and what it went through in winter (Atwal and Sharma, 1970a, b; Atwal and Goyal, 1973; Chahal *et al*., 1995; Atwal, 1987; Mishra, 1995).

11.2.1 Spring Management

After winter, honeybees start activity in the spring, and colony inspection can be performed before foraging starts and before the winter packings are removed. Bees themselves show signs of the approaching spring when some bees as a prelude to regular foraging start bringing in pollen. It indicates that brood rearing has begun in the hive.

In subtropical India there is always some activity as winter flora are available to a lesser or a greater extent. At places there is even a minor flow of *Eucalyptus* and *Brassica* flowers. Thus even in the presence of a nest foragers remain active and brood rearing continues. Even extra frames are required for honey storage which should be given with a minimum of disturbance to the colony but no thorough inspection be made.

When the season opens up and spring comes it is necessary to make an intensive inspection of the colony. The weather should be sunny and windless and temperature above 18-20°C. In North India such days are common in the second half of January or February depending upon the region. Before approaching a colony the outer and winter packings if used are removed, the top cover is lifted and inner coushin if given is taken off. The colony is approached as described earlier, puffing smoke on to the frame tops and waiting for two minutes till the bees are subdued and they start engorging honey. It is now known that smoke masks the effects of alarm pheromone (Nasonov gland) generated by the guard bees. Hence the, bees are confused and can be handled with minimum of stings. In many localities in the Indo-Gangetic plains and southern slopes of Himalayas it is advisable to keep the outer covering intact in case there is unexpected bad weather. The inspection includes determining the queen rightness of the colony and some spring operations, medication and shifting of the frames and the colony site. Depending upon the condition of the colonies sometimes it is possible to handle them at once but quite often it is convenient to take it up in two or three steps. It is important that hives are not given too much jarring and therefore it is better that those are lifted by two men, with smooth motions. The various steps and operations to be undertaken are discussed here not necessarily in the order of preference.

11.2.2 Inspection and Operations

1. Colony inspection is a slow, cautioned and determined process. It is desirable to delay it somewhat rather than hurry in the spring. Good weather is a must. The first observation to be made is on the food stores because on it depends the future development of the colony. There should always be a store of 7-15 kg of honey and pollen in a colony. If the reserves are low sugar and pollen substitutes have to be fed even though bees might be bringing in some nectar and pollen. The method of spring feeding and handling of colonies has already been discussed in some detail. The important point to remember is that food stores are required for building colony strength, avoiding congestion that may lead to swarming.

2. The undesirable frames which were moved to the sides in autumn should be removed as those would be empty by now. These frames may include broken combs imperfectly repaired by bees or those having irregular honey cells; all should be taken away. These frames should be replaced by new white well drawn combs, having the cells size of worker brood.

3. Cleaning of propolis is an important operation. First scrapping with a hive tool should be done on the top of frames and some of it is bound to fall in between frames on the bottom board. Frames when separated one by one with caution should also be cleaned and placed in a pre-cleaned box on another bottom board which would replace the overwintered sections. The old box and bottom board can then be cleaned thoroughly at leisure. If in autumn a thin paraffin layer is made on top of frames there is less propolis deposit and it comes off easily.

4. It is essential to determine the presence of queen not necessarily to observe it. If young brood and some eggs are present it would mean the queen is in good shape. In the beginning of the season bees do not like to be disturbed and if there is improper handling the queen is likely to be balled.

A colony that has drones among the overwintered stock is likely to have its queen missing. Therefore, a thorough search is required. In case the colony is queenless it should be united with a weak colony by the newspaper method. Likewise two weak colonies in which brood rearing has not started or is nominal, should also be united, caging and suspending the poorer queen as a reserve in case the other queen is not accepted by the newly united colony.

Apart from ensuring the presence of queen it is the right occasion to determine the prosperity of the colony which is indicated by foraging bees, in-coming stocks of food, presence of eggs and brood, particularly the sealed brood. If records show that queen is more than two years old it is advisable to change it and give the colony a young queen out of the reserve stock or by raising a new one as soon as the opportunity arises. The new queen should be marked with a colour in order to determine its age when required. The process of giving a young queen ensures that it is viable, can lay a large number of eggs, can build colony strength before the major honey flow. Hence there would be less tendency of the colony to swarm.

5. During the winter if ventilation in the hive was inadequate lot of condensation of moisture would be observed. Sometimes there is water accumulated on the floor, which should be drained off by tilting the hive towards the entrance. It also indicates that the solid inner board and the outer packing had no holes for ventilation. In spring new inner boards having wire-gauze ventilators be placed instead and ventilation be improved. In fact, for the summer, a collection of complete two-layer wire-gauze screen (boards) should be ready. There should be a gap of 1/2″-3/4″ between the two layers so that bees do not try to close the holes with propolis. These screens are invaluable in hot weather experienced late in spring and in dry hot summer.

6. Spring is the time to check health of the colony. A thorough examination will reveal the presence or otherwise of diseases and parasitic ailments. Appropriate or even prophylactic medication

is given at this time. A full account of the diseases and parasites of the honeybees is given in chapter 13. Here are mentioned the operations to be undertaken.

(*a*) As the colony is opened it should be examined for any sign of dysentry. Dirty yellow marks on bottom board and the inside walls of hive or on the alighting board indicates this disease. It is of two types. The Nosema dysentry is caused by protozoa which contamination comes from stagnant water source and it can be confirmed by staining the smear on a slide and examining under a microscope. The other type is from unsuitable or fermented food. Gur or shakar syrup or fermented sugarcane juice when fed to bees generally cause dysentry. The former type of dysentry is controlled by feeding to bees Fumidil-B alongwith food and the latter by feeding Terramycin (1/4 tea spoon in a gallon of sugar syrup).

(*b*) Dead larval brood might indicate merely chilling or the result of bacterial infections (see chapter 13). When a queen is viable and is laying eggs in excess of the bee coverage, the larvae at the periphery are neglected by bees and they die. In such a situation colony strength is improved by giving frames of sealed brood. As explained in chapter 13 there are no reports of bacterial brood diseases in India so far but in USA and Europe they are a serious problem and are kept under control by Terramycin given with food.

(*c*) Acarine disease is caused by a microscopic mite that lives in the thoracic tracheae of the adult bees. The outward symptoms are that bees crawl around on the bottom board, alighting board or on the ground in front of the hive. The side wings are disengaged and when separated from coupling look distended and are held like the letter K. When insufficient control measures are taken in the autumn the bees during winter confinement pass on the disease from one to the other particularly to the young bees. The disease is so prevalent in India both among the *A.c.* and *A.m.* bees that routine prophylecic measures are recommended. The earlier recommendation was 3-4 weekly fumigations by burning chlorobenzilate (Folbex) dip-dried on paper strips. The newer method includes the release of fumes of menthol or formic acid on a swab of cotton wool held inside the hive on an iron wire pushed through the entrance hole. Two fortnightly treatments are generally considered sufficient but the treatment can be repeated if the symptoms continue.

(*d*) There are two ectoparasitic mites that are common in the hives and they cause serious damage to brood and the bees. *Varroa* is a flabby flesh coloured mite which is seen attached to adult bees on the inter-segmental membranes in the neck and abdomen regions. *Tropilaelaps* is an active rust red mite that is seen moving fast on combs and adult bees. Both are visible to the nacked eye, and they breed on young brood killing them. The menthol and formic acid treatments are also effective against these mites as recommended for Acarine disease. Earlier, dusting of fine sulphur (300 mg or less) on combs or top of frames was practiced widely but sulphur contamination of honey was not accepted in international trade.

(*e*) There is a complex of virus diseases of the adult bees which cause paralysis. The afflicted bees are seen crawling in front of hives and it appears as if they cannot coordinate their limbs. Much before this final stage, the bees harbour viruses and their life is shortened owing to the infection. There is no specific cure for this ailment and the diseased colonies should be isolated so that there is no further spread in the apiary.

(*f*) On opening a hive, if the colony is found to have dead bees and is in poor condition, it should be closed immediately and its entrance reduced to avoid robbing. The colony should be isolated and shifted to a distant place. The cause of ailment should then be determined with the help of experienced beekeepers, or by an expert in a scientific laboratory.

7. In winter, colonies are generally in a single chamber as there is dearth of forage. In areas where there is winter forage colonies need more space and a super is added on top. In the spring inspection

availability of space should be checked. Once a bee cluster has moved upwards it is not possible to move it downwards. A shortage of space is created and bees cannot expand brood rearing area any further and hence crowding occurs. Crowded conditions lead to swarming which is not considered desirable if a good honey crop is to be obtained subsequently. The natural solution is to give a super on top. If the colony is already in two chambers and the lower portion is empty by this time their position should be reversed so that more space is made available for the colony to expand. How to prevent or control swarming which takes place in February-March, has already been discussed in detail. It is all important to always provide plenty of space which is manipulated by reversing the position of chambers every 2-3 weeks. It is a very effective method of swarm prevention and swarm control and at the same time giving more space for brood rearing and food storage.

8. The primary objective of spring management is to build colony strength for which bees require food and comb space. In the first inspection colony strength in the apiary is equalized and weak colonies are given frames of sealed brood taken out of strong colonies. The food stores are checked and, if found less than 7 kg in a colony, artificial feeding of syrup and pollen substitutes are given before regular foraging by the bees starts. One frame of sealed honey with some pollen is required for producing one comb of brood, therefore, a plenty of reserves are required.

In subsequent inspections progress of the colony is observed. The progress can be judged by noting the relative amount of eggs, unsealed brood and sealed brood. The ratio of the area of eggs and the unsealed brood to the sealed brood if 1 : 1.5 indicates that no expansion is going on. A ratio of 1.5 : 1 indicates rapid expansion and 1 : 3 is a sign of rapid reduction of brood rearing (Wedmore, 1979).

After brood rearing starts, adequate space is required in the hive. At first space is provided by reversing the chambers 2-3 times but as the season progresses and colony strength builds up extra super are to be given for the storage of honey. Researches with Department of Agriculture of the United States showed that "honey hoarding" instinct of the bees was increased if surplus of drawn combs was provided to a colony. It was observed that a colony with two or three empty supers stored more honey than a colony with one extra super. These results were obtained when drawn combs were given and not frames with only foundation sheets.

In general, it is recommended to over-super in the spring and, to under-super at the end of honey flow. It will ensure that bees are stimulated to collect more surplus honey at the beginning of the honey flow and to consolidate honey stores at the end of the season.

9. The extra supers which are meant to be used for honey storage should not have the complete 10 frames because it causes congestion. It is advisable to place 9 or even eight frames so that they can be moved freely and when honey cells are sealed the cappings are somewhat raised. The raised cappings are easier to cut with a knife at the time of honey extraction and also more honey is stored per frame. Moreover it is easier to remove frames from the box as compared with the 10 frames from a box which get packed tightly.

10. In winter colonies are moved close together and some beekeepers prefer to give outer covering in pairs. In the spring the colonies are to be opened out in the apiary. Their arrangement in rows looks good to the eye but at the time of examinations the alarm pheromone from one colony to the next in a row also alerts the bees and their handling becomes more difficult. Colonies placed at random in the apiary can be handled with comparative ease as there is no bee commotion. While placing the boxes, spots should be selected near shade, keeping in view the coming hot months.

11. During spring it is quite a pre-occupation for the beekeeper to watch for swarming and to prevent it. It is a natural instinct of the bees to raise new queens and issue swarms for the multiplication of their race. As soon as colony build up starts in the spring there appear congestion for

space. If the mother queen is over two years of age in its third year it is almost certain swarming will take place. In the apiary if most of the queens happen to be of that age the beekeeper should be well prepared for it as his hands will be more than full. Knowing the position before hand he should speed up the rearing of new queens well in time, have them mated and then to change all the old queens in his apiary. For the successful mating of queens mature drones should be present, which is managed by giving drone frames early in the season for the queen to lay eggs and to raise drone brood. If he cannot manage it the next best strategy is to take up operations that would prevent and control swarming. The methods that can be adopted have already been discussed.

In subtropical India the months of February and March are the swarming season. Therefore whatever steps are to be taken for raising new queens or for the prevention and control of swarming, should be timely. If successful, the colonies will build up good strength and will have the foraging force to collect a good harvest of honey.

12. There are many areas where winter flora and early spring flora may provide minor honey flow. As the bees are rearing brood they are also collecting nectar and pollen from chick peas, *Brassica,* fruit trees, early flowering *Eucalyptus* and other trees like shisham, soapnut, tun etc. Honeybees may store good quantities of honey and, if it is found surplus, it can be extracted. It would be the first extraction of the year and moreover the empty frames would provide space for brood rearing and further expansion. It would also reduce the chances of swarming. While determining the surplus, good judgement is required so that extraction does not cut into the reserve food stocks required for brood rearing, prior to and during the main honey flow in the months to follow.

11.2.3 Summer Management

From the beekeepers point of view summer has three distinct parts, honey flow period, hot and dry dearth period, and humid monsoon with limited forage. By the close of spring the objective is to have in the hive the largest number of bees of harvesting age (Table 7).

TABLE 7. *Brood Rearing by Apis mellifera (yellow, California) at Ludhiana (Punjab)*

(Average of 14-20 Colonies during 1967 & 1968 combined)

Month	*Ave. frames of bees*	*Ave. brood area sq. cm.*	*Month*	*Ave. frames of bees*	*Ave. brood area sq. cm.*
January	3.3	693.7	July	8.5	2253.1
February	3.9	1146.9	August	6.8	1600.0
March	5.5	2058.7	September	5.3	1385.6
April	6.2	2789.4	October	4.8	1408.7
May	9.7	3456.2	November	3.9	827.5
June	10.4	2727.5	December	3.2	517.5

Note : 1. From Atwal and Sharma (1970a).

2. Observations were made twice a month, except in Nov. and December.

When the major flow comes from crops like clovers and sunflower, the attention of the bees should be diverted to honey gathering. The general management should suit the local conditions and the weather has to be watched. In forest areas where there are no standard crops, there are specific trees providing main honey flow, which might be richer in certain localities than in others. Taking advantage of such localities may mean all the difference between no honey and a good crop of honey. Therefore a beekeepers has to be ever watchful and vigilant not only of his apiary but also of the flowers and honey flow in his locality, and be ever prepared to move colonies wherever there is honey. He can achieve this by visual observations alone or with the aid of records of gain or loss on the balance colony.

11.2.4 Honey Flow

The months of April and May indicate the period of major honey flow in a greater part of North Indian plains. Temperatures rise quite high during that period. The bees are busy foraging and the hive temperature has to be reduced to 95°F (34°C). In order to save the bees extra work hives should be kept in the shade of trees of walls or buildings. In desert areas where there are no trees colonies are kept under thached huts open from all sides. In May, temperatures often touch 38-40°C, even where there is vegetation. Wet gunny sacs or canvas cloth on top is very helpful in keeping the boxes cool. The hot wind that blows at that time, in fact, hastens the cooling process. Fresh running water from a tubewell or a stream is necessary not only for wetting the cloth but also as food for the bees. Water is needed for diluting honey before it is fed to the brood. Bees also use water in ventilating and cooling inside the hive.

Storage Space. Supers with fully drawn frames, should be in readiness before the honey flow and one unfilled super should always be available to the bees. If there is no place to store no honey will be stored. It is also not the time to start raising foundation sheets and spending bee energy on this work which should rather be spent on foraging. During honey flow bees will raise combs if they have to. If a beekeeper is short of raised combs he should give only 2-3 foundation sheets at a time, placed in between the raised frames.

How many supers to give and where to place them has already been discussed. The main consideration is whether to place a super on top or in the middle ? In the latter case there is more work and a queen may start laying eggs in this section if queen excluder is not used. If placed on top there is no harm done in over supering. Lot of space is needed for storing ripe honey and much more for nectar that contains 80% water. So surplus of storing space is always welcomed by the bees.

Queen Excluder. For a beginner the use of queen excluder is very helpful and he can confine the queen in brood chamber free from any worry that brood rearing may be spread to the honey supers. Some experienced beekeepers think a queen excluder hinders the traffic of bees. They consider 'queen excluder' is a 'honey excluder'. Therefore the use of this equipment is a matter of choice. On the whole it is a very useful device particularly with the new beekeepers and in new areas. The brood rearing remains confined to the lower chambers (one or two, as planned) and above the queen excluder the honey supers remain free of brood. In the honey flow season good ventilation is important which can be created by double wire screens. The super can be cleared of the bees by bee escapes or other methods already described. A screened inner board having a bee escape fixed in the centre is placed below the honey super so that bees move downwards.

Screen and bee-escape. Heat during the months of April, May and June is very intense and inside the hive bees have to work hard to lower the temperature. By establishing a natural ventilation bees can be helped in this activity. The entrance should not be towards the South or the West from which directions the hot winds come. Natural ventilation is created by adjusting the entrance size and using double walled screen in place of the inner board.

The screen in conjunction with bee escape can also be used for clearing honey supers of the bees. In that case the screen has to be placed below the super. The standard method of using bee escape on the inner board is not suitable in hot climate as it cuts off air circulation and it becomes too hot inside the hive. It also means extra work for the bees to cool it down.

Super clearing. Apart from the above method of clearing the super there are other devices. The basic technique is to pick up the super, place it beside the colony and brush off the bees from frames near the entrance with the help of a soft hair brush or soft leaves of a twig. After clearing the frames those can be placed in a box and taken to the honey house.

Fume boards or clearing boards have been used for this purpose for the last many decades. A guaze cloth or padding is fixed on a wooden frame and is placed on top of a super. Diluted carbolic acid is smeared on this cloth with the help of a cotton swab. Within five minutes or less the bees affected by fumes run downwards to the lower chambers and the super is clear of the bees.

The use of carbolic acid is being discouraged as it taints honey with its smell. Instead two other chemicals have come into use in USA recently and they are applied on fume boards in the same manner. Benzaldehyde should be applied when temperature is 60-70°F and butyric (proprionic) enhydride when temperature is 80-95°F. In India temperature are much higher in the honey flow season. Formic acid is the answer and it should be applied in very small amounts either on a fume board or on a piece of moisoned musline cloth on top of a super. Bees run downwards in no time.

The new device for large apiaries is an electrical blower machine. The super to be cleaned is tilted on its front end by one man and a blast of air is forced at the bottom end of the frames by the other. Bees are pushed out towards the entrance and they rejoin the colony. The supers can be cleared at the same speed honey can be extracted from the frames.

Protection of bees. The mite diseases of bees if not properly controlled in the spring can again raise their ugly head during honey flow. Colonies should be protected by adopting proper control measures.

The other bee enemies include garden lizard and birds. The Green Bee-Eater glides around over apiaries picking up the foraging bees. The Black Drongo comfortably sits on objects in apiary and quite often on the hive itself and catches bees approaching the colony. These birds can be a great nuisance and should be controlled by air guns. The lizards can be killed by beating and their progeny should be controlled systematically.

Control of drones. The presence of some drones all the year round in an apiary is a biological necessity. In case of a chance death of a queen or as a result of severe robbing a colony might be left queenless. If queen is old sperms in spermatheca may be getting exhausted and she might start laying fewer and fewer eggs. Fearing danger to the colony bees themselves kill the queen to raise a new one. It is not a seasonal phenomenon and can happen at any time of the year, in autumn in particular. The young queens can mate only if there are drones in the apiary or somewhere else closeby.

It has been observed that after the breeding season in spring too many drones may be present in hives. The bees and the queen generally can sense how many drones can be accommodated and they regulate their numbers. When surplus in number the bees do not feed them and bump them off the hive and they starve to death.

Sometimes a beekeeper himself feels the number of drones should be kept in check. In spring he should remove the frames having too many drone cells or move the drone brood on the outside where bees might not cover it. In extreme cases the drone brood can be pulled out of cells and killed. Having missed those opportunities there might be too many drones in the summer season. They can be caught in drone traps placed at the entrance. To hasten the process frames of worker bees and drones can be shaken at some distance from the colony. The worker bees fly back to the hive and can pass through the screens of drone trap whereas the drones are left out and they can not rejoin the colony. Whether or not these discarded drones are accepted by other colonies is another matter.

11.2.5 Migratory Beekeeping

Migratory beekeeping is the secret of honey collection. With the coming of winter, temperatures at the higher altitudes fall suddenly and there is no bee flora left. Honeybees subsist on stored food. Down the Himalayan valleys into the plains winters are mild and many crops of the *Brassica* group are grown. Eucalyptus and other trees also flower in autumn and winter. Beekeepers throughout the Himalayan region from Jammu & Kashmir, Himachal Pradesh, Uttar Pradesh, Bihar and Eastern India move their beehives down the valleys and to the plains of North India all along the mountain ranges. Bees forage on winter flora, build their colony strength and also collect surplus honey. Before the summer comes those are taken back to higher altitudes where the crops and pome fruits, soapnut, tun and wild flowers appear in summer.

The march of summer from western and central India towards the northern plains is very interesting. Taking maturity of wheat crop as the indicator there is a difference of 2-3 weeks from south districts to the north, reaching the base of mountains. Bee flora also appear earlier in the south than in the north and the crops of importance to the bees are barseem, sunflower, fruits like citrus, guava, litchi, mango and trees like *Eucalyptus,* shisham etc. Beekeepers can move their colonies northward advancing ahead of the scorching heat that dries up nectar in flowers. Honey can be gathered from place to place on northward journey.

Long distance and elaborate system of moving boxes as handled in the Western countries is not the only migratory beekeeping recommended. It is possible to migrate colonies within a state or a smaller area where there is variation in the determination of nectar. Chahal *et al*. (1995) have suggested one such scheme for barseem crop within the state of Punjab. It is a rich source of nectar from mid-April to mid-June at various places. In the southern districts honey flow ends in the last week of May whereas in the northern districts of Gurdaspur, Jullundur and Hoshiarpur, nectar flow continues upto the middle of June. By migrating colonies to the north on barseem and sunflower crops one or two additional extractions are possible (Goyal and Atwal, 1973a, b). Shain (*Plectranthus*) is another nectar source in Kashmir and Himachal Pradesh that is a nectar source in September-October and colonies are migrated to collect honey from this wild bush (Singh & Singh, 1971; Shah & Shah, 1976; Hameed and Adlakha 1973; Atwal and Goyal, 1974a).

Honey Harvesting. During the honey flow of a rich crop like barseem bees are very active and they fill a super with sealed honey in 7-10 days. It is considered a good practice to extract honey from these supers and to give new ones for bees to work on. It maintains the tempo of bees to collect more honey and in the end a bigger crop is harvested. In large apiaries with limited labour available it is not easy to handle these supers as soon as filled, therefore, beekeepers have no choice but to let 2 or 3 supers accumulate. If the last super is placed on top, bees have to pass through the lower supers and the bee traffic is slowed down. Some beekeepers suspect less honey is collected by bees this way. The extra super placed in the middle does not affect the bee traffic but it is more work to handle. To overcome these situations, regular extraction of honey as the season progresses, two or even three times is the right approach.

Some people like to have mono-floral honey with distinct flavour. A small March extraction gives sarson honey and the one in April will give *Eucalyptus* honey. The honey extracted in May and June will be clover honey from barseem. Colonies lying in litchi rich areas of Dehradun, Saharanpur, Ranchi, in February-March will give extraction of Litchi honey. Likewise special honey can be extracted from citrus, apple and shain areas.

In North India the last extraction from barseem should not be delayed beyond the end of May or the very beginning of the June, when the honey flow has declined but not completely stopped. So long as the nectar is coming, bees would not be demoralized and they will collect honey in the last super which has a great importance as their food reserve. A beekeeper should never extract honey from this last super.

At some locations sunflower crop may still be providing nectar and in that case another extraction might be possible, but one super of honey fully or partly filled should always be left with bees. The coming monsoon time is a dearth period and if good reserves are not left the colony would decline suddenly. The monsoon and the post-monsoon periods are long and to tide over that time honey reserves left over after the harvest would be the only food for bees.

11.2.6 Management During Monsoons

Rains start on the West and East coasts of India in the end of May or early June and the winds move inwards reaching the great plains in the end of June or early July. The scorched lands and withered vegetation gradually turn green. There is hardly any bee flora. Moreover during the rain bees are confined inside the hive for hours and sometimes for days. When there is no rain bees venture out in search of food and are often lost in sudden shower of rains. The food reserves left over from the tail end of honey flow is the only source on which to live. The bees cut down brood rearing and as the old bees die the colony strength dwindles. It can be visualized how important it is to leave good food reserves in the colony and not to resort to over extraction at the end of honey flow.

Depending upon its strength a colony may be in a single chamber or in double chambers. In the latter case there is more space for ventilation. The atmosphere inside the hive becomes very hot and muggy, particularly when there is no rain, and the sun shines brightly. The bees try to reduce temperature by beating their wings and creating air currents. They can be helped by making the entrance wide open and giving the inner boards with screen ventilators or complete screen covers. To further improve air circulation the supers can be slightly moved ahead to let air pass through. In order to give more air space to the bees the number of frames may be reduced to nine.

The appearance of bee flora is a long way away yet. It is in August that trees like *Acacias, Eucalyptus* and crops like cotton, niger, tobacco, banana, maize, ragi, produce flowers on which bees can forage. In the meantime food shortage may appear in hives and in that case sugar feeding should be given taking precautions that robbing does not occur. Owing to over all food shortages the colonies are in a mood to look for a bounty to be gathered by robbing. The weak colonies that are to be given food are already targets of robbing by the stronger colonies. Feeding should be given to all the needy colonies the same evening, keeping vigilance against robbing.

This is the time when there is shortage of food for bee enemies also. Black ants are crawling all over, ready to reach inside the hives, which can be prevented by placing the legs of hive stands dipped in cups full of water. Predator wasps try to catch bees at the entrance and they should be killed by swatting. The predator birds described above suddenly appear in large numbers and should be controlled. The garden lizards should be scared away and not allowed to get into the colonies.

Debris lying on the bottom board provides a tempting material for caterpillars of wax moths to make a start and later on to bore into the wax of combs. The debris should be collected and burnt or burried deep in the soil. The extra frames vacated by the contracting colony should be removed and stored safely inside a sealed bag of heavy plastic. The wax moths if driven from the hives are sure to reach the combs in storage. Their attack is not noticed at first but when grown up caterpillars are seen boring through the wax it is already too late. For safety, the frames in bags should be fumigated with phostoxin in an isolated room or a store away from the habitation of farm or domestic animals.

Sometimes the rain storms can be quite violent and it may rain very forcefully. The top cover of the hive should have proper metal sheet and there should be no cracks from where rain may enter.

Even so, the boxes should be slightly sloping to the front for water to drain out. The boxes should be placed at raised ground away from ground pools of water. Layout of the apiary should be such that the ground can be drained easily. It is quite a common practice in India to place heavy stones or bricks on top of boxes so that top covers may not fly off in strong winds. Also individual boxes can be recognized from a distance from the type of stone on top. It is a useful practice.

During the monsoons there is hardly any flora out in the fields. When it stops raining and weather opens up the field bees come around in search of food. In this season they are often seen on over ripe fruits in markets and sweetmeat shops, or on open sacs of wet gur and sugar. From their point of view bees are just foraging as they cannot remain idle. In this persuit they also reach the food stores of weaker colonies and start helping themselves. The initial stealing becomes a bad habit of robbing. Once it starts in an apiary it spreads fast and becomes uncontrollable and persists for days and weeks till some colonies are killed. Then robbing may spread to neighbouring apiaries and all beekeepers in that locality have to join hands in controlling the outbreak. At the very first sign of robbing hive entrances should be narrowed and all the steps described earlier should be taken to curb robbing in the very beginning.

11.2.7 Control of Dwindling

Odds of the monsoon weather will lead to dwindling of the colonies. The second half of June and whole of July are very difficult for the bees. The kharif crops are sown with the moisture from rains and they do not flower till August or September. The major crops in central and north India are maize, ragi, cotton, niger, soybean, tobacco, jute, arhar, summer vegetables, sesame, guava, *Eucalyptus*, *Acacia* etc. Before bees start foraging their colony strength has to be built up by feeding and brood rearing. With good colony strength bees can even collect surplus honey in certain areas after the rains on crops like cotton (unsprayed), niger, soybean, tobacco. In East India multifloral sources include wild flowers, banana, pineapple. In Plains of North India colony strength is built on *Eucalyptus,* arhar, sesame, maize. Subsequently, in the postmonsoon period there is a minor crop of honey from toria which is sown in September and flowers in October and November. It is a *zaid kharif* crop and is most useful for the bees in postmonsoon period for building up colony strength and in good localities even providing surplus honey. Cotton, sesame, and soybean may also yield some honey. In this season obtaining honey surplus is not as important as observing the build up of colony strength for the cold winter months ahead.

11.2.8 Autumn and Winter

In a good beekeeping area the postmonsoon flora should be sufficient to build the reserves so necessary for passing winter. Autumn brood rearing is necessary to build up the colony strength. Areas where toria is sown and autumn bloom of trees is available or areas where cotton (unsprayed) niger, tobacco are sown and plantations of banana and sugarcane (honeydew) are available surplus honey should be expected. As a result of strenuous foraging the old bees and those suffering from diseases like acarine and paralysis will die, leaving the colony comparatively clean. The new bee force apart from being young will have a different physiological constitution with expectancy of long life of 3-4 months to pass over the winter. In areas of inadequate flora the dwindled colony strength will not build up fast. Moreover colonies will be full of diseases, particularly the parasitic mites. These colonies need attention and sugar feeding.

The basic principles of management for honey production and extraction are the same as described under spring management. Sufficient honey storage space is provided in the super and brood rearing is confined to the lower chamber. The extra side space in the brood chamber is also filled with honey as the winter reserve. The procedure of extracting honey is the same but it is combined with examination of the colony for assessing its winter requirements.

The broken and misshaped frames and those having too many drone or honey storage cells are removed if empty, otherwise, those are placed on the sides to be removed in the spring. All the frames, boxes, covers, bottom board etc. that might have been damaged during the year should be repaired or replaced. The piece of jute cloth covering the frames should not be ripped and should fit well covering the entire area. It should not be gauzy but thick to provide insulation on the colony in winter. The queen excluders in two chamber colonies if not removed during the postmonsoon period should be taken out now. The screen inner boards should also be removed in the postmonsoon period and, for the prewinter, solid inner covers having screen ventilators in the centre should be placed in the hives.

The examination should reveal most importantly the strength and health of the colony, the condition of the queen, and the food reserves. It will be observed that all the colonies in the apiary exposed to similar external conditions will have a great variation in their strength. If there is no disease in the colony the queen might be the cause of poor strength. Her inherent egg laying ability might be poor or she might be aged and in either case she has to be replaced. In this season natural supersedure is common and if the old queen is poor her daughter raised by bees themselves may also not be much superior. Therefore a good beekeeper raises his own queens of the right pedigree using eggs of a superior queen. When colony is given such a queen we would be sure of her performance. With a young virile queen the colony would have a much better chance of passing the winter by keeping the bees busy rearing brood. Strengthening the colony with brood frames from strong colonies would further improve their ability to winter over.

The colonies of moderate size which have good queens but have dwindled bee population should also be given sealed frames of brood taken out of colonies on two chambers. For the winter, colonies should invariably be on one chamber of full strength. Such a consolidated colony with good food reserves near the nest in winter passes the cold period more successfully. An overall effort should be made in the apiary to equalize the size and strength of colonies which will facilitate the beekeeper in general supervision and in keeping watch over the condition of his stock. Weak colonies that cannot be suitably strengthened by brood frames should be united and one of the queens be kept as reserve. As an alternative to uniting two weak colonies they can be kept in one hive with partition board in between and two separate entrances. In the spring they can be united and the spare queen given to a nucleus.

As regards food resources for bees in autumn and winter in the subtropics there are many situations ranging from insignificant forage to a good bee flora. At places where there is insignificant to poor forage, colonies have to be well prepared for the winter. Assessing the existing food reserves additional sugar feeding has to be given. Depending upon the local floral conditions colonies must have 7-15 kg of honey reserves and sufficient pollen as brood rearing continues during a greater part of winter in most areas. In areas where there is winter forage there should be at least 7 kg of honey stock in the colonies. Bees will no doubt keep bringing in nectar and pollen but in bad weather they always depend upon the food reserves in the colony.

The method of sugar feeding has already been described. The syrup should be more concentrated (1 : 1) as compared to that given in spring or summer. As an additional source, solid sugar of fine grain, can also be placed on top of the inner board. During winter bees help themselves to it. In case of sudden change of weather at the time of syrup feeding the comb cells are quickly filled manually and when dripping stops they are placed in the hive. The frame feeders can also be used.

The placement of colonies is important. In India there is plenty of sun shine in winter and colonies which were kept in the shade of houses or trees in summer should now be brought in the open. At the same time they should be protected from northerly cold winds. The colonies should be placed on the south or west side of a protective wall, a house, a temporary barrier or a regular planting of hedge or trees as wind breaks. The boxes should be made wind proof by repairing and closing cracks or crevices which will also make them secure against lizards, cockroaches and mice. The entrances should not be facing the North.

During the examination of the colonies those suspected of having mite parasites were supposedly marked. Formic acid or menthol treatment is given to the diseased colonies, 2-3 times till the mites are controlled. The rest of the colonies are given one treatment to ensure that they remain free of the mites. The colonies suspected of paralysis disease should be isolated and kept at a considerable distance. The menace of wasps does not exist in winter as they go in hibernation. There are no ants in the apiary either. The greatest enemy of bees is cold, particularly the wet cold in winter. Bees should be protected by packing the hives properly.

Lighter packing is given to hives in the subtropical parts as compared with that in temperate Himalayas. Packing is given inside as well as outside the hive. A strong colony that is on 8-10 frames in one chamber will have no space left. Packing is given to provide insulation and protection from cold coming from the sides. The packing materials used are rice straw, wheat straw, saw dust, newspapers, pallets or even sheets of the thermocole. Cushions of these materials are made by packing them in a newspaper or jackets are made of plastic covers sealed with a candle. Before placing these cushions in the box, frames are moved in the centre and empty spaces are created outside the dummy board. These spaces are filled with the insulation materials. If there are 7-8 frames in the colony all the frames should be moved on one side of the box and cushions placed on the other side. If there is no space on the sides of frames a quilt of proper size be placed on top between the inner board and top cover.

The outer packing consists of some kind of padding under the hive and a plastic sheet for wrapping round the box. The padding of rice straw layers one inch thick are placed between the bottom board and the stand as protection from wind coming from below. A plastic sheet is neatly wrapped round the box tucking the open ends under the bottom board. If the sheet is large enough to go across the board it will give full protection and there is no need of straw padding. In windy places the sheets can be tied with a string. Care should be taken that a little ventilation is left at one corner. Without this precaution the metabolic water and other moisture may accumulate inside the hive causing dampness. The damp cold is very damaging to the bees.

Ordinarily boxes are not opened in winter and bee activity is watched at the entrance. From the quantity of nectar and pollen coming in the amount of brood being reared inside can be judged. In areas where there is good winter flow from sarson the activity of bees will be very noticeable. Such a hive may need more space for storing and expansion. It would probably have no inner packing and outer packing can be removed on a nice sunny windless day. Judging from the colony size a super or a half super with fully drawn frames can be given, including some of the honey and pollen from below. The outer packing can again be given to provide favourable environment inside. In areas of abundant sarson most of the colonies would need frames for honey and pollen storage. At the end of winter with the earliest signs of spring these colonies should be examined again. It may be possible to even extract honey but quite a good quantity should be left with the colony for brood rearing in the spring. The monofloral honey from *Brassica* which is white has a characteristic mustard flavour liked by many people. The honey granulates easily and if left in the hive bees may not be able to utilize it fully hence it is a good practice to extract this honey for sale.

In non-sarson areas also winter activity may continue on *Eucalyptus* and other flowers. There may not be any surplus honey and there may not be any rest for the bees either. The life of the foraging bees is shortened, therefore, it is important to stimulate brood rearing early in spring by providing pollen supplement. Young bees will then emerge to build up the colony further.

Right Strain of Bees. Among *Apis mellifera* there is a great variety of brood rearing patterns. Some strains suddenly stop brood rearing in autumn with the decline of flowers. Others continue brood rearing though at a reduced rate. It is regulated by egg laying by the queen. Taking advantage of this variability beekeepers can develop their own strains suiting the floral conditions of that area. At places where *Brassica* and *Eucalyptus* flowers bloom in winter and bees are expected to forage on them the colonies should continue brood rearing. The selection of queens for such places should be made accordingly. Young worker force would be added in the colony and thus winter flow would not adversely affect the ability of the colony to build up in the spring. At places where winter is severe and there are no flowers, arrest of brood rearing would be a desirable attribute. For those areas queens of the desired type should be selected.

11.2.9 Potential of Beekeeping

In the irrigated areas of subtropics it is possible to grow two or even three crops in a year. Farmers grow crops for their own benefit, not keeping in view the needs of the honeybees. In certain areas the sequence of cropping and the availability of tree flora are such that honeybees can build their colony strength practically throughout the year and make use of two major honey flows, one in spring or early summer, and the other in autumn. With good management, beekeepers can have more than one extraction, during a given honey flow. Thus annual production per colony can easily range between 50-100 kg. Considering this potential for the subtropical parts, particularly the northern plains of India, may be taken as the most suitable for Mellifera beekeeping. The secret of success is the right sequence of bee forage during the year. The dearth periods should be brief and the bees should always have sufficient food to tide over that time (Atwal, 1987; Goyal, 1978; Goyal and Atwal, 1979; Chahal *et al.*, 1995).

Much relief can be given to bees even during the dearth period. There are a number of forest tree species for planting in blocks or as avenue trees on private farms. These trees include *Eucalyptus*, soapnut, tun, amla, moulsari, jamun, tamarind, and a number of fruit trees. All the species are themselves a source of income to the farmer and in addition they can provide forage to the bees.

The major honey flow is generally from field crops and not from trees, with the exception of solid plantations of *Eucalyptus*, and a few other species. By noting the sequence of crops in an area the potential of honey production per colony can be worked out. The acreage of a crop in a locality would determine the number of colonies that can be kept there. A successful beekeeper is ever observant of the potential of honey harvest of his locality and other surrounding areas where he can shift his colonies for obtaining additional extraction in that season.

Another profitable method of getting additional extractions in a season is by practicing migratory beekeeping. In winter, colonies are brought from the mountains to the plains. It gives not only an escape from cold and winter feeding but also the colonies gather honey from *Brassica*. With the coming of blooms of spring fruit trees in the mountains the colonies are taken back there for pollination and honey extraction.

Early in summer the picture is reversed in the plains. Hot weather moves from south to the north and within a distance of 300 km there is a difference of 15-20 days in the disappearance of honey flow. Flowers of honey plants like *Eucalyptus*, fruit trees, barseem, wither in the south first and then in the north near the Shivaliks. By migrating honeybee colonies northwards one or two additional extractions can be obtained.

At present only a small acreage of *Eucalyptus*, *Brassica* and barseem are being exploited for honey production. There is a tremendous scope of developing beekeeping in the Indo-Gangetic plains.

11.3 TEMPERATE INDIA

In India, temperate climate exists in the Nilgiri Hills, and the Himalayas. On the southern and south-western valley slopes of Himalayas there is moderate climate and savana type vegetation occurs upto an altitude of 1500 meters. Above that upto the height of 3650 meters, grow temperate forests and, further up is the alpine vegetation (Gopal Singh, 1995). Human habitation is generally upto the tree line and the villagers generally keep honey bees. At the higher altitudes a dark strain of *Apis cerana* occurs naturally and people harvest honey collected by these bees.

At lower altitude in the temperate zone there is resident population who grow field crops and plant orchards of apple and other pome fruits. Traditionally they have kept *Apis cerana* in skeps and wall hives for the pollination of fruits and crops like buckwheat. Being familiar with bees they are now adopting Mellifera beekeeping. Many of them practice migratory beekeeping and are now anxious to learn the scientific practices.

Cold presents the most difficult situation if colonies are to be kept in the winter after the autumn foraging. The queens should stop laying eggs and the bees should prepare for winter. Sufficient food reserves should be available for bees to form a cluster near the food and to pass the winter with minimum mortality. The beekeepers who migrate their colonies down to the plains should have a selection of queens that do not stop egg laying in autumn but continue brood rearing according to the forage available. In the plains *Eucalyptus* and *Brassica* flowers provide sufficient forage to continue brood rearing. Therefore, the colonies to be taken for winter foraging should have the right strain of queens and bees that continue brood rearing. The colonies should be moved out of the mountains three weeks before the first frostly winds start in the autumn.

The climate in temperate India is comparable to that of southern Europe and many of the management practices that have been developed there can be adopted in Mellifera beekeeping. There is an old European saying :

A swarm of bees in May
Is worth a load of hay;
A swarm of bees in June
Is worth a silver spoon;
A swarm of bees in July
Is not worth a fly.

In Europe natural vegetation and forest flora is better preserved than in the Himalayas where the influence of man through overgrazing by his goats and sheep has degraded soil. The undergrowth is over exploited for fodder and fuel and the forests that could have supplied abundant bee flora have been degraded. On the southern slopes of Himalayas where there is plentiful summer rainfall there should have been lush forests but there are now naked hills. The potentials of beekeeping have been much reduced. There has been further cutting of such valuable trees like tun and soapnut, wild pear and wild cherry. Much can be done to improve conditions by replanting honey producing trees on the roadsides and at the farm boundaries. In the lower Himalayas and the Shivaliks, forests of Eucalyptus, tun, amla and other trees can be planted.

In temperate India seasons are very distinct and management practices for spring, summer, rainy season and winter are also well defined. The performance of a colony in the spring depends upon the care and treatment given to it in the autumn and the way it might have passed winter. As a protection against cold the inner and outer packings are necessary. Temperatures in the spring can also be quite low, therefore, due precautions are taken while opening a colony.

11.3.1 Winter Packing

How to prepare colonies for winter packing has been explained earlier and the same procedures apply in the temperate zone. The climate demands heavier insulation and protection from cold winds. Colonies can be packed in double walled hives or single walled hives. They can be kept for the winter above ground or below ground level in cellars. In Europe it is quite common to keep colonies in cellars built below houses and barns, 6-7 feet in height in their ceiling, and 30 inches below the frost line. In a covered area of 10 × 20 feet nearly 100 colonies can be kept for the winter. The winter and early spring management of colonies in cellars is quite different (Wedmore, 1979).

In the Himalayas upto the tree line colonies are packed and kept besides a wall, a house or a mountain side protected from the northern cold winds. Heavy packing is given on the sides of the farmes and a quilt slightly turned on the sides is placed between the inner board and the outer cover. The boxes are also covered on the outside with paddy straw tied with strings. Water proof paper or plastic sheet is wrapped securely leaving a small hole for ventilation through which moisture from inside can escape but rain water from outside cannot get in. Two wooden flats one inch wide are placed above the frames below the inner board for air circulation. A hive that has a shallow super full of honey on top has in it ventilation space and the wooden flats are not required. Wet cold is most harmful to the bees and every care should be taken that neither moisture accumulates within the hive nor it can get in from the outside. If the straw packing outside the box cannot be kept dry it is better not to give it and instead to secure plastic sheets tightly.

Inside the hive bees consume honey, and with the metabolic combustion, water vapour and carbondioxide are given out. The relatively warm air charged with moisture moves outwards and, if the temperature of the hive walls is below that of the dew point of the air, condensation occurs on the walls. The moisture on accumulation may gather at the bottom. In order to avoid it air circulation and ventilation inside the hive is necessary in the winter time.

Metabolic activity is generated by the bees inside the winter nest. The cluster of bees shrinks in size and takes up a spherical shape with combs passing through it. A sphere on the outside has the least surface area allowing the minimum cooling down in winter. This is what happens in a nest. When the weather warms up in spring the cluster becomes loose, covering a greater surface area of the combs, where brood rearing is started. Literature on the management of *Apis mellifera* in Europe, USA and other temperate countries has been reviewed by many authors and our experience in the Himalayas has also been mentioned in the pages that follow (Eckert and Shaw, 1960; Wedmore, 1979; Winston, 1987; Crane, 1990; Verma, 1991; Graham ed., 1992; Sakagami, 1959; Sakai and Matsuka, 1982).

11.3.2 Spring Inspection

The spring inspection of colonies and the operations to be carried out come in phases and not all at once. The first inspection is not thorough and should not involve intensive manipulations of the hive. The main consideration is to avoid chilling of brood and the colony can be opened when the minimum temperature is not below 40°F. By lifting one corner of the hive food reserves can be estimated which should not be less than 15 kg. Using these reserves honey and some pollen the colonies will start brood rearing in March without foraging.

In the second half of March colonies are opened to assess the condition of queen and to observe whether brood rearing has started. The inner packing is removed and more space is provided for the brood to expand. Bees start foraging on wild flowers like shisham, wild pear, vitex, soapnut etc. to raise their brood. If the food reserves are low artificial feeding is to be given. The sugar syrup should have low concentration for the stimulation feeding and should contain Fumidol and Terramycin at the rate of 1/4 teaspoon per gallon of the liquid. Depending upon the prevailing weather conditions the hives should be cleaned of debris at the bottom board and propolis from the frames. The broken combs and those having irregular cells should be removed. The colonies should be prepared and manipulated as described under spring management. At this time treatment is also given against parasitic mites.

As the season opens up foraging appears in the field in March-April (Table 8). At the lower altitudes Brassica, soapnut, tun, shisham, amla etc. bloom at different times and the bees exploit them for brood rearing and for surplus honey gathering. At the higher altitudes in Jammu & Kashmir, Kulu valley (Table 9) Uttar Pradesh, Bhutan, Sikkim, Arunachal, apple and other pome fruits blossom in

TABLE 8. *Brood rearing by Apis mellifera (California yellow) as compared with that of Apis cerana in the lower Himalayas (at Nagrota, height 3000 ft.)*

(Average of 9 colonies during 1967 and 1968)

Month	*A.m.*		*A.c.*	
	Ave. No. frames of Bees	*Ave. brood area (sq. cm.)*	*Ave. No. frames of Bees*	*Ave. brood area (sq. cm.)*
February	3.9	2288.1	3.7	1517.5
March	6.9	4731.3	5.3	1765.6
April	12.0	5380.0	6.6	2451.9
May	15.3	4431.3	8.4	1946.8
June	12.3	3673.7	6.9	1333.1
July	11.1	2783.8	5.9	956.9
August	10.8	4345.0	5.7	1405.0
September	9.8	3994.4	5.5	1413.1
October	7.2	1815.0	5.1	1099.4

Note : 1. From Atwal and Sharma (1970b).
2. November to January colonies were given packing.

TABLE 9. *Brood rearing by Apis mellifera (California yellow) as compared with that of Apis cerana in the lower Himalayas at Kulu-Katrain (height 3300-4900 ft.)*

(Average of 8 colonies during 1971)

Month	*A.m.*		*A.c.*	
	Ave. No. frames of Bees	*Ave. brood area (sq. cm.)*	*Ave. No. frames of Bees*	*Ave. brood area (sq. cm.)*
March	4.9	2264.7	3.5	478.6
April	7.1	4053.7	5.8	2433.1
May	9.8	3640.0	6.0	2192.5
June	10.1	3383.7	5.3	1765.0
July	10.3	3009.6	4.5	1038.8
August	10.0	4862.1	4.6	2271.3
September	9.3	4742.5	5.4	2414.8
October	6.1	4101.3	5.5	1025.0
November	5.8	928.5	4.0	75.0

Note : 1. From Hameed and Adlakha (1973).
2. December to February colonies were given packing.

April-May. Of all the fruit-area in the region two-third is under apples. According to one estimate this fruit covers 250 thousand hectares. In Himachal Pradesh alone 75000 hectares are under temperate fruits which are dependant on bees for pollination required for getting higher yields and better quality fruit. To provide adequate bee services over 2 lac colonies are required as against the present number of 10,000 available (Verma, 1991). The honeybees apart from providing pollination services, also collect honey from apple which is amber in colour and has a special flavour. Other supplementary flora in Kulu area include *Robinia, Berberis, dral*, etc. and bees collect dark honey that has a strong taste. Some beekeepers extract this honey too, but it should be left to the bees for the rainy season that starts in July.

11.3.3 Rainy Season and Autumn

In the Himalayas rainy season is humid but not hot. The rains are heavy and sometimes go on for days from July to September in a greater parts of the Himalayas and, sometimes in August and September in Jammu and Kashmir area. The bees are confined to their hives and cannot go out for foraging. It is also a dearth period and bees have to depend on the food reserves with them. Wax moth, tracheal and ecto-parasitic mites also appear in serious form and they should be controlled. By the time rains stop colonies might have dwindled. Pollen from maize is available but there are hardly any nectar sources. The wild and forest flora are the main source in addition to cucurbits and other vegetables. It is important at this stage to give sugar feeding so that colonies can build their strength before the forthcoming autumn. It is necessary for two reasons. Colonies might have to be shifted to shain (*Plectranthus*) which provides major honey flow in September-October in certain areas, and they would have to pass the long winter months later on.

The *Plectranthus* honey flow is available in certain areas of Gharwal, Himachal Pradesh and Jammu and Kashmir. The stand is rich in honey source and *A.c*. bees work on it strenuously (Shah and Shah, 1976; Sharma, 1958). During that time unexpected cold weather is also experienced. There can be a heavy mortality among the bees particularly in Mellifera colonies (Atwal and Goyal, 1974; Sharma et al. 1980). The Cerana bees comparatively adjust better to the sudden changes in weather. Another reason for bee losses is that tempo for brood rearing is slowed down mainly due to the non availability of pollen. In such areas late sown maize crop can provide pollen. Provision for the supply of pollen supplement and pollen substitutes should be made to overcome the possibility of colony dwindling.

Plectranthus and Rabdosia honey is white and clear and is prized for that reason. The beekeepers have a tendency to over extract it from the colonies. If this temptation cannot be avoided, the colonies must be given sugar feeding after honey extraction so that bees are not demoralized. The beekeepers must know the food requirements of a colony for successful over-wintering. For a strong colony at the higher altitudes it is 20-25 kg, equivalent to a full shallow super. For the lower mountains a colony should have 10-15 kg of sugar and honey as the reserve sources.

The placement of reserves is also important. Honey should be so placed in the hive that bees can reach it after the formation of a nest. The cluster cannot move freely owing to low temperatures. The honey frames should be placed close to the area of cluster formation. Once the reserve food is given the colony would be packed for the winter and it would not be opened till the next spring. If cluster is forming near the top of frames, honey frames should be placed on the sides near the cluster where bees can reach them.

For the successful over-wintering of a colony, it should have good bee strength, preferably of young age. The queen should be productive, healthy and, less than two years old. The older queens should be replaced by young virile queens and the beekeeper should be prepared before hand by raising new queens in nucleus colonies just for this purpose. He should not wait till the bees in a colony prepare themselves for supersedure. From his own records he should be able to determine the age of the queen and its performance during the past two years. Even if it has given a good performance and in age it is nearing three years it should be replaced. Next spring the new queen will be expected to give a good lead to the colony and start early brood rearing.

In the mountains brood rearing and increase of the colony strength is not as fast as in the subtropical areas where there is abundance of spring flora. In temperate climate the brood rearing area in the nest does not move upwards so fast and for that reason it does not become necessary to use a queen excluder. Near the autumn honey flow, if the colony is on two brood chambers and most of the brood is on upper one, the chambers can be reversed. If required the third chamber is placed for honey gathering on top. With this manipulation it does not become necessary to use the queen excluder. Even if it is used it should be removed after the autumn honey extraction, definitely before the winter packing of the hive.

After the winter packing is given the bees become quiet within the hive. Once they are confined inside in the cold weather they do not come out for days. After two or three weeks on a clear day a few bees may come out occasionally to void faeces. Bees do not like to be disturbed, shaken, or jarred in winter. This is the time also when the apiary yard can be cleaned, levelled and modified according to the needs. The hives can be moved to clear the place. They should be lifted gently by two persons. When the bees come out at the new place they make orientation of the new site. There are some losses and quite a few of them drift to other colonies but it may be considered normal in order to complete the required jobs at the apiary.

If adequate measures are not taken, up to 40 per cent of the colonies may be lost, but with proper management the losses can be reduced to 20% or less. As mentioned earlier, a colony needs a young productive queen, good food reserves, freedom from diseases and enemies and a good packing. By meeting other needs of the bees in winter even a light packing may be good enough. If the other needs are not met properly even a heavy packing may not save the bees. If a beekeeper finds lots of dead bees, he must find the cause and take due steps to deal with that colony and to save others in the apiary in case it turns out to be suffering from an infectious disease.

11.3.4 Comparative Performance of *A.m.* and *A.c.*

When *Apis mellifera* was introduced in the Himalayas in mid sixties (1962-64) at Nagrota the Indian honeybee, *Apis cerana indica (= Apis indica)* was already established and scientific researches on this species had been going on since the thirties. It was natural to think as to how these two species would live together. Will they compete to the stage of elimination of one or the other or they would become tolerant? Apart from the long range ecological considerations it was important to know their immediate competition at the apiary level in terms of robbing or co-existence. At the very early stages their comparative behaviour and, more importantly, their comparative performance in terms of honey collection was studied in the mountains as well as in the Punjab plains.

In one of the earlier studies in the lower Himalayas the comparative brood rearing capacity and the build up of colony strength of *A.c.* and various strains of *A.m* were observed (Atwal and Sharma, 1970a). Both species produced the maximum brood area in the month of April (*A.c.* 2451.9 sq cm ; *A.m.*, 5380.0 sq. cm; Table 8) and gained the maximum strength in May (*A.c.*, 8.4 frames and *A.m.* 15.3 frames). Among the various strains of *A.m.*, the yellow race gave a significantly better performance as compared with the English strain and the hybrids from Florida. Commensurate with the brood area and colony strength the honey yield per colony of *A.c.* ranged between 3.7–4.7 in various years and that of yellow *A.m.* between 10.6–19.3 kg. The latter honeybee worked strenuously gathering honey from tun and soapnut and gave three extractions whereas the former divided its energy between brood rearing and honey gathering and barely gave one or two extractions.

At the higher altitudes in Kulu area comparative studies established the superiority of *A.m.* over *A.c.* in brood rearing and building up of colony strength (Hameed and Adlakha, 1973; Table 9). *Apis mellifera* colonies were taken from Nagrota for experimental purposes, their performance in winter in Kulu was not reported, nor their survival rate by next spring was mentioned. The performance of *A.m.* was also studied in Punjab plains at Ludhiana from 1966 to 1973. The maximum brood area of 4087 sq. cm. in a colony was observed in April and the maximum gain in weight in the balance colony observed was 37 kg. (Atwal and Sharma, 1970b ; Table 7; Atwal and Goyal, 1973).

Comparative aggressive behaviour of the two species was studied by confining their workers in a small space for confrontation, by giving brood frames of one species to the colony of the other, and by observing robbing among them in mixed apiaries (Atwal and Dhaliwal 1969a; Dhaliwal and Atwal, 1970; Sakagami, 1959; Kallogg, 1941, 1959). In these experiments it was observed that face to face *A.m.* workers intimidated *A.c.*, were aggressive against their brood and in active season noticing their weakness of strength resorted to robbing of the exterminating type, killing queen and eliminating the colony. It was interesting to observe that *A.c.* guards did not scrutinize the robbers thoroughly and even if they tried to do so they were over powered and were bodily lifted and taken some distance away. *A.m.* guards on the other hand thoroughly scrutinized the *A.c.* robbers and resisted them but in cold weather the latter not only bypassed them but also resorted to regular robbing and even finishing the colony. The most intriguing part was that demoralised or drifted *A.c.* workers were accommodated by *A.m.* colonies in hundreds over a long period of time, whereas the reverse tolerance was not observed. It was most significant to note that only weak colonies were attacked between the two species, as well as within the species.

Temperature played an important role in the comparative field performance of the two species in autumn season when *Plectranthus* offered a rich honey flow. *A.c.* gave a better performance and maintained good colony strength in areas where temperatures were generally low. They remained active for a longer period of time during the day and incidentally gathered pollen also while collecting nectar. They continued brood rearing and maintained colony strength. The activity of *A.m.* slowed down when temperature fell below 18°C and the bees practically stopped working at 13°C. In certain localities where temperatures were suitable it successfully collected nectar from this source but the non-availability of pollen restricted brood rearing and the field losses were not compensated by the birth rate, hence the colonies dwindled (Sharma *et al.*, 1980).

In many parts of western Himalayas wild cherry (*Prunus* sp.) blooms in October-November. The comparative performance of the two honeybees was the same as on *Plectranthus*. However, the field performance of *A.m.* in suitable weather and floral conditions was very impressive in terms of honey yield owing to its long flight range, large stomach and its ability to build up colony population fast (Atwal and Sharma 1968, 1971, 1976; Adlakha and Sharma, 1975; Goyal, 1974, 1978). This species was most suitable for migratory beekeeping when colonies were moved before winter from the mountains to favourable floral conditions in the plains of North India.

Apis mellifera has been with us now for over thirty years and no bacterial brood diseases have been observed so far. Since late seventies *A.c.* has been devastated by the Thai Sac-Brood Virus (TSBV) and for that reason also *A.m.* has been widely accepted by the beekeepers in India (Atwal, 1987; Mishra and Kumar, 1995). Both the species suffer from Acarine disease and prophylactic measures are required to control it. The ectoparasitic mites can be controlled by releasing the fumes of menthol, formic acid or pyrithroids in the hives. *A.c.* is also highly susceptible to the attack of wax moths whereas *A.m.* can resist it more successfully. The moths should be kept under control by methods described in Chapter 13.

Both the honeybees are attacked by the predatory wasps namely, *Vespa magnifica, V. orientalis, V. auraria, V. cinta, V. valutina*. The characteristic behaviour of *A.c.* in combatting *V. madarina* has been described. Apart from the shimmering behaviour this honeybee presents a well organized defence and can even over power the giant wasp (Ruttner, 1988; Matsaura and Sakagami, 1973). *Apis mellifera* bees on the other hand are less adapted to defend themselves against this wasp. To protect the colonies it is recommended to use wasp traps, poison baits, and to locate nests of the predators and to destroy them (Mishra, 1979, 1989).

There is difference of opinion as to the future of *A.c.* in India. One extreme view is that *A.m.* should not be allowed in *A.c.* areas and the other is that they should be kept in separate apiaries, and the third is that *A.m.* should be kept as a commercial honeybee and *A.c.* be encouraged to live in its natural habitat in the wild or in hives at higher altitudes. Taking a more realistic view some authors could foresee the complementary nature of the two honeybees for beekeeping in India (Goyal, 1974, 1978). The fact cannot be ignored that in a greater part of Asia including Japan and China, *A.m.* has been accepted for commercial beekeeping and *A.c.* in those countries has receded to the higher altitudes towards the mountains, finding a natural habitat and nitches where they can survive. They have subtle differences in their temperature and floral requirements; they can live and forage in the same locality (Goyal, 1974; Mishra *et al.*, 1984).

Unfortunately *A.c.* has not recovered from the onslaught of TSBV so far. In due course resistant races might be selected which would be good gatherers of honey also (Mishra and Singh, 1986; Mishra and Dubey, 1996). When that happens it would be appropriate to make comparisons of the performance of the two species of honeybees at various altitudes. The studies would be particularly interesting when made on Kashmir race of *A.c.* which is larger in size and is somewhat comparable to *A.m.*

Bee Pasturage

12.1 CLIMATE AND VEGETATION OF INDIA

Honeybees live on flowers of plants, shrubs and trees. The type of vegetation of a place depends on its temperature, rainfall, influence of man and soil. In India the tropic of cancer passes through the middle part and therefore the southern part of the country has a tropical climate. The northern part of India is also warmer than areas of the world located at similar latitudes; the temperatures are higher by 3° to 8°C. The cold polar air is checked by the Himalayas and the great plains are thus mildly cold during winter and hot for a greater part of the year. Throughout India, below the height of 900 meters, the vegetation is tropical monsoonic in character. The variety and luxuriance of vegetation from place to place depends on the distribution of rainfall. In various regions rainfall is unequal, and owing to varied edaphic and biotic conditions, India is rich in very large number of plant species and, the types range from drought resisting thorny scrub to tropical wet ever green forests (Gopal Singh, 1995).

The Indian subcontinent has a large expanse and its interior parts become very hot in summer. The plains of North-West India record very high temperatures in this season, the maximum ranging from 43° to 50°C. From mid-March to mid-June it is the hot and dry season in a crescent shaped part of country from Punjab, Rajasthan, dry Karnataka, S.W. Andhra Pradesh. Vegetation type is dry savana with open spaced short trees of *Acacia,* shisham, neem & ber all useful for the bees. The rainfall is 60 cm. The rabi crops mature in this season and under irrigated conditions the period of flowering is prolonged somewhat (Jaiswal ed. 1980). The flowers of some trees, shrubs and crops are rich in nectar and pollen and this is the most important season and area for the production and harvest of honey.

12.2 VEGETATION IN VARIOUS REGIONS OF INDIA

In May the mean monthly temperature throughout India is 32°C and the maximum ranges from 40°C to 50°C, except in costal areas where it is below 38°C. In general the rabi crops in southern and central India mature 3-4 weeks earlier than they mature in the North. The atmosphere is windy laden with dust and four such storms appear in April and two in May and generally they bring rain from the Arabian sea. The bee management practices have to be adjusted accordingly. The temperature is high, drinking water is scarce and bees need special care. Wet pieces of cloth on top of the hives are very helpful.

After hot and dry season starts the rainy season in the end of June but in South India the south-easterly winds from the Arabian sea bring rains in May. Similarly monsoon winds rise from the Bay of Bengal and move towards North and West bringing rains to the great plains and Himalayan region in

June-July. In South India above 900 meters where annual rainfall is 250 cm conditions are created for the tropical wet evergreen forests. There is a profuse growth of vegetation and a large number of species grow in 3-4 tier of trees covered with climbers and a profuse undergrowth cover of palms, bamboos etc. The closed forest conditions are not suitable for hive bees. However plantations of rubber, coffee and spices provide good flora for the bees.

The tropical evergreen forests in East India are also dense and the rainfall is very heavy. The honeybees are adversely affected under such situations, but in the open valleys of Assam bees do quite well, and gather good honey. Along the coast line of tropical India plantations of coconut and oil palms provide good pollen to the bees.

The subtropical and temperate wet hill forests of South India occur between 900 and 1,850 meters above sea level. These are highland forests of Orissa, Karnataka and Western Ghats. There is a great variety of herbaceous ground flora in open spaces in Bastar District, Mahabaleshwar, Nilgiri hills etc. There is rich bee flora and scientific beekeeping with *Apis cerana indica* has flourished there for the last many decades.

The Himalayas rise abruptly from the tropical heat to the Arctic cold. Going from the lower to the higher altitudes one notices varied vegetation; at the foot hills there is luxuriant evergreen type, and the temperate forests grow from 1,500 to 3,650 meters, and alpine vegetation is found at the higher altitudes. In addition to the pome fruits and crops grown in valleys the natural vegetation consists of useful bee flora of many wild shrubs and trees including, shisham, *Acacia,* tun, simal, mahua, *Barberis,* in the spring and summer seasons, and *Plectranthus* in the autumn in Himachal Pradesh and Jammu and Kashmir.

In eastern Himalayas tropical wet evergreen and semi-evergreen vegetation is found upto a height of 900 meters. There is a rich undergrowth of shrubs, ferns, bamboos, grasses etc. At higher altitudes there are wet hill forests. The pines and oak trees provide honeydew to the bees, and chestnut provides nectar and pollen. In the western Himalayas vegetation is of different type. Dry savanna is dominant over vast areas in the forest hills and above 900 meters there is subtropical vegetation upto 1,500 meters. The temperate belt is further up. Above the tree line in the alpine valleys rich flora of wild flowers appear after the snow melts. Many such areas are a haven for migratory beekeepers although the terrain is difficult. Throughout the Himalayas indigenous beekeeping has been practiced from time immemorial. With the advent of modern beekeeping the hives are moved in winter to the plains of North India where there is rich crop flora. After extracting honey the bees are taken back to the mountains to avail of the spring and summer flora there.

12.3 CROPS AND PLANTATIONS AS BEE FLORA

Many of the cultivated crops and commercial plantations provide nectar and pollen which are food for the bees. Those are generally grown on large stretches of land and their contiguity is desirable for successful beekeeping. The crops that need cross pollination benefit from the foraging activity of honeybees. The evolution of the symbiotic relationship is over 60 million year old; the crops and honeybees mutually benefit from each other. Honeybees may also forage on certain crops which are not dependant on their services for pollination, for example, maize is wind, rain, or gravity pollinated and yet honey bees exploit this rich source of pollen.

In Chapter 15 are discussed some crops and plantations which provide forage to bees, mostly those that provide nectar and pollen and at the same time benefit from cross pollination. Invariably cross pollination improves crop yield and thus man is doubly rewarded for keeping honeybees by way of obtaining higher crop yield and extracting honey from bee colonies.

12.4 THE WILD BEE FLORA

Honeybee flora, primarily for *Apis cerana indica* F. have been surveyed in various parts of India. Plants have been evaluated as sources of nectar and pollen and the authors are listed in Table 10. After the introduction of *Apis mellifera* L. in mid-sixties, the first survey of the flowers visited by the species was conducted at Ludhiana in Punjab (Atwal *et al*., 1970). Subsequently, its activity on various field crops, fruits and vegetables was observed by many authors as mentioned under pollination of crops (Chapter 15). Temperature and photoperiod play important role in the availability of nectar and pollen from a particular source.

TABLE 10. ***Floral Surveys in various parts of India for the Honeybee (Apis cerana indica F.)***

S. No.	*Regions of India*	*Author(s)*
1.	Kerala	Devadason (1971, 1972)
2.	Western Ghats	Deodikar and Thakar (1953, 1966)
3.	Western Ghats	Subramaniam (1979)
4.	Western Ghats; pollen grains	Chaubal & Deodikar (1965)
5.	Karnataka and Kerala	Khadi V. Ind. Commission (1959)
6.	Karnataka	Rajan (1980)
7.	Karnataka; Coorg	Suryanarayana (1966)
8.	Karnataka; Coorg	Divan and Rao (1972)
9.	Karnataka; Chikmaglur	Krishnaswamy (1980)
10.	Tamil Nadu; Kodai Hills	Chandran and Shah (1974)
11.	Sagarmal	Chaubal and Kotmire (1980)
12.	West Bengal	Tripathi (1971)
13.	West Bengal; Sunderbans	Chakarbarti and Chaudhri (1972)
14.	Bihar; Pusa	Naim & Phadke (1976)
15.	U.P; Eastern and Himalayas	Chaturvedi (1965); Singh and Singh (1987, 1988)
16.	Northern India	Kohli (1958/59)
17.	Bhopal	Khan (1948)
18.	Delhi	Vasu (1967)
19.	Pre-partition Punjab	Rehman & Singh (1941)
20.	H.P and Punjab	Singh (1948)
21.	H.P. and Punjab	Sharma (1958)
22.	Punjab Plains (and Ludhiana)	Chaudhari (1977); Atwal *et al*. (1970)
23.	Western Himalayas; pollen	Nair (1965)
24.	Shimla Hills	Mattu & Verma (1985)
25.	Kashmir	Deodikar (1962, 1970)
26.	Kashmir	Saraf (1972)
27.	Kashmir	Shah & Shah (1976)
28.	Pakistan	Ahmad (1984)
29.	World Honey Sources	Crane *et al*. (1984a, b)
30.	India; honeybee maintenance	Indian Standard Institution (1973)
31.	India floral maps	Mishra & Kaushik (1992); Ganguly and Chanda (1980)

Among the weeds, shrubs and trees growing wild some are selected for their commercial or aesthetic value and are planted with care in groves or on the roadside. They gain importance for honeybees and come closer to the cultivated crops and plantations. Examples of the trees are *Acacia*, neem, bottle brush; chestnut, shisham, *Eucalyptus,* silver oak, Parkinsonia, soapnut, tun, jamun, arjun, tamarind etc. Taking into consideration such potentialities a list of plants has been complied (Table 11). Naturally, all the weeds and other plants exploited by honeybees could not be included for fear of making the list too long. For the some reason all the annual or perrenial flowering plants grown for landscaping have not been listed. At places where both *Apis mellifera* and *Apis cerana* are available they forage together. In fact *A. florea* and *A. dorsata* also forage on the same plants, particularly where nectar flow is abundant.

The list is arranged alphabetically according to the scientific names, and the common names are also given. The numerals given as i, ii, iii etc. indicate the month of the year when they flower and those given as 1, 2, 3 etc. indicate in descending order the abundance of nectar (N) or pollen (P) in terms of exploitation by the bees. The climatic requirements of plants are shown as temperate (te.), subtropical (str.) and tropical (tr.).

12.5 AVAILABILITY AND EXPLOITATION OF FOOD RESOURCES

Nectar and pollen are collected by bees as food for the colony. In a given area or a locality, whether there is a dense crop or wild flowers sparsely distributed, the honey bees exploit them to their advantage. Their search for the sources of nectar and pollen is thorough and even though they seem to be moving at random from flower to flower, they are always oriented to return back to their colony at the end of every trip. As to how they work alone and yet communicate with the colony has been described earlier. It will be discussed here what are the factors that determine the availability and exploitation of resources and how bees economize on their energy while collecting food for the colony. They try to select the best flowers and the definition changes according to the situation at a given time. Unless there is some over-riding factor they exploit the flowers nearest to their hive first and then move outwards. Their flight range can extend upto 2 or 3 miles during times of dearth but generally they do not fly out that far. The site of apiary is so determined that bees can exploit food sources close by. It was observed on two crops, cauliflower and Berberis, that 60-80 bees of *Apis cerana* per count visited flowers within a distance of 200-400 meters and the number came down to less them 10 at a distance of 800-1000 meters (Dhaliwal & Sharma, 1972).

Bees are attracted to flowers by their colour, shape, and odour; they can perceive yellow, blue-green, blue, and ultraviolet colours reflected by the petals of flowers. Bees get conditioned to a particular species of flower at a time ignoring the others. Conditioning is through odour as bees have a very acute sense of smell much greater than that of man. When a flower is fertilized, petals fall off, nectar and scent secretion ceases and bees move on to some other flower.

12.5.1 Nectar Secretion

The nectar secretion glands are located on various parts of a flower and many plants have extra floral nectaries on stems, leaves, leaf-buds etc. When a flower opens secretion is profuse on the first day or the first few days and then it declines. The duration of secretion varies with various species of plants (Shuel, 1961).

Nectar is a solution of sugars secreted in floral or extra floral nectaries of plants by the action of enzymes. The common sugars present in nectar are glucose, fructose, maltose, sucrose, monosaccharides and disaccharides, acids and oils. Some other compounds found in minute quantities

TABLE 11. *Selected Bee Flora of India* (Fig. 20)

S. No.	Name/Local name/Type	Family	Bloom Months	Status of Nectar N	Status of Pollen P	Region	Remarks—Colour of Flowers; of Honey
1.	*Acacia modesta* Wall phulai, tree	Leguminoseae	iv—v	1	4	subtropical (str.)	Flowers white; low nectar flow in heat
2.	*Acacia senegal* (L) gum acacia, tree	Leguminoseae	vii—ix	1	3	subtropical (str.)	Whitish spikes; granulates, amber
3.	*Acacia nilotica* (Linn.) kikar, tree	Leguminoeae	viii—ix	—	3	subtropical (str.)	Yellow bunches; amber
4.	*Actinodaphne angastifolia* Nees pisa, tree	Lauracae	xii—iii	1	—	subtropical (str.)	Yellow; black
5.	*Actinodaphne hookerii* Meissn pisa, tree	Lauracae	xii—i	3	1	tropical (tr.)	Pale; dark amber
6.	*Anacardium occidentale*(L.) cashew nut, tree	Anacardiaceae	ii—iii	1	3	tropical (tr.)	Pink or striped; yellow
7.	*Azadirachta indica* A. juss neem, tree	Meliaceae	iii—iv	1	3	str, tr.	White, fragrant; light golden
8.	*Berberis* spp. rasaunt, bush	Berberidaceae	iv—v	1	—	tropical (tr.) widespread	Yellow; dark
9.	*Bombax ceiba* L. simal, tree	Bombacaceae	i—iii	1	2	tropical (tr.) widespread	White and bright crimson; pale
10.	*Brassica campestris* L. var; toria, annual crop	Cruciferae	x—xi	1	1	tr., str.	Yellow; clear, yellowish white
	var; sarson, annual crop.	Cruciferae	xii—ii	1	1	tr., str.	Yellow; clear, yellowish white granulates easily.

11.	*Brassica juncea*(L.) raya, annual crop	Cruciferae	xii—ii	1	1	temperate (tr.) subtropical (str.)	Yellow; whitish, granulates easily
12.	*Brassica nigra*(L.) lahi, brown mastard	Cruciferae	xii-iii	1	1	te., str.	Yellow; light
13.	*Butea monosperma*(Lamk) dhak, tree (flame of forest)	Papilionaceae	ii—iii	1	1	str.	crimson red; amber
14.	*Cajanus cajan*(L.) red gram, annual or biannual shrub	Leguminosae	x—xii	1	3	tr., str.	Yellow, spotted red; greenish hue
15.	*Calliandra calothyrsus* red calliandra, shrub	Leguminosae	all year	1	1	tr.	Red; bitter sweet
16.	*Callistemon citrinus* Curt. bottle brush, small tree	Myrtaceae	iii—ix	1	4	tr., str.	Bright red; deep amber
17.	*Carica papaya* L. papaya, small tree	Caricacae	all year	3	3	tr., str.	Yellow, small; nectar inaccessible
18.	*Carvia callosa* (Nees) karvi, shrub	Acanthaceae	vii—ix	1	3	tr.	Bluish purple; dark red
19.	*Castanea sativa* Mill chestnut, tree	Fagaceae	vi—vii	1	1	te.	Greenish yellow; amber to brown
20.	*Centaurea cyanus* L. blue cap, corn flower	Compositae	iv—vi	1	1	te., str.	Blue; greenish yellow
21.	*Cicer arietinum* L. chick pea, gram crop	Leguminosae	ii—iii	1	3	te., str.	Bluish purple; amber
22.	*Citrus aurantifolia* Swingle Lime, small tree	Rutaceae	ii—iii	1	2	str., tr.	Whitish; amber, yellow
23.	*Citrus deliciosa* Ten. tangerine, tree,	Rutaceae	ii—iii	1	4	str., tr.	White purplish; pale yellow
24.	*Citus reticulata* Blanco santara, tree	Rutaceae	ii—v	1	3	str., tr.	White, fragrant; pale yellow

S. No.	*Name/Local name/Type*	*Family*	*Bloom Months*	*Status of* Nectar N	*Status of* Pollen P	*Region*	*Remarks—Colour of Flowers; of Honey*
25.	*Citrus cinensis* (L.) orange, tree	Rutaceae	ii—iii	1	1	str.	White, fragrant; pale yellow
26.	*Cocos nucifera* L. coconut, palm	Palmae	i—iii	1	3	tr.	Cream, small
27.	*Coffea arabica* L. Arabian coffee, evergreen	Rubiaceae	ii—iii	1	3	tr.	White, fragrant; amber
28.	*Coriandrum sativum* L. dhania, herb	Umballiferae	iv—vi	2	2	str.	Whitish pink aromatic; dark
29.	*Crocus sativus* L. saffron crocus, bulbous	Iridaceae	x—xi	3	3	te.	Orange; amber
30.	*Cucumis melo* L. melon, trailing	Cucurbitaceae	iii—vi	2	1	tr., str.	Yellow; light amber
31.	*Cucurbita pepo* L. pumpkin, trailling	Cucurbitaceae	iv—vii	2	3	tr., str.	Yellow, large; light
32.	*Dalbergia sissoo* DC shisham, tree	Leguminosae	iii—iv	1	4	tr., str.	Yellowish, small; amber
33.	*Daucus carota* L. carrot, cultivated	Umbelliferae	vi—ix	1	3	te., str.	White, small; dark yellow
34.	*Dimocarpus longan* Lour Longan, crop	Sapindaceae	ii—v	1	—	tr., str.	Yellowish white
35.	*Ehretia acuminata* R. puna, small tree	Boraginaceae	iv	1	3	tr., str.	White, fragrant; amber
36.	*Elaeagnus angustifolia* wild olive shrub	Elaeagnaceae	v—vi	1	3	te.	Yellowish, fragrant;
37.	*Elaeis guineesis* Jacq. cultivated oil palm	Palmae	all year	1	1	tr.	Small, cream; dark amber

38. *Erobotrya japonica* evergreen, fruit tree	Rosaceae	viii—i	4	1	str.	White, woolly; amber
39. *Erythrina indica* (Burrm) coral tree	Papilionaceae	ii—iii	2	2	str.	Crimson red; amber
40. *Eruca sativa*, taramira crop	Cruciferеae	xii—ii	2	2	te., str.	Greenish yellow; clear
41. *Eucalyptus* spp. Mysore hybrid, tree	Myrtaceae	xi—iv	1	1	str., tr.	Cream white, fragrant; light amber
42. *Fagopyrum esculentum* Moench buckwheat crop	Polygonaceae	vii—ix	1	2	te.	Sepals white, fragrant; brown
43. *Glycine max* (L.) soyabean, crop	Leguminosae	vi—viii	2	—	str.	White to purple; light amber
44. *Gossypium hirsutum* L. American cotton, crop	Malvaceae	vii—ix	2	4	str.	Pink to yellow; light
45. *Gossypium arboreum* L. desi cotton, crop	Malvaceae	vii—x	2	4	str.	Pink to yellow; light amber
46. *Grevillea robusta* A. silver Oak, avenue tree	Proteaceae	iv—vi x—xii	1	4	str., tr.	Orange yellow; dark amber
47. *Guizotia abyssinia* Cass. niger, oil crop	Compositae	ix—x	3	3	tr.	Yellow; light amber
48. *Helianthus annus* L. sunflower, year round crop	Compositae	all year	1	2	te, str., tr.	Yellow large; Yellow or golden
49. *Hevea brasilienis* Muell. rubber, tree	Euphorbiaceae	all year terminals	1	4	tr., ste.	Yellow; translucent
50. *Ipomoea acuminata* (Vahl) blue flower, shrub and climber	Convolvulaceae	all year	1	2	str., tr.	Blue dawn; amber
51. *Lagerstroemia* spp. harshingar, queeen flower, shrub to tree	Lythraceae	v—ix	2	—	te., str., tr.	White to pale violet; amber

S. No.	Name/Local name/Type	Family	Bloom Months	Status of Nectar N	Status of Pollen P	Region	Remarks—Colour of Flowers; of Honey
52.	*Lucas aspera* Link tumba, forest weed	Labiatae	x—xi	1	4	tr.	—
53.	*Ligustrum walkeri* Dcne pungalum, shrub	Oleaceae	v—vi	1	1	tr.	—
54.	*Ligustrum perrolletii* privet, shrub	Oleaceae	x	2	2	tr.	—
55.	*Linum usitatissimum* L. linseed, crop.	Linaceae	xii—ii	2	2	te., str.	Blue/ white, amber
56.	*Litchi chinensis* Sonner litchi, orchard fruit	Sapindaceae	ii—iii	1	4	str., tr.	Whitish, insignificant; light golden
57.	*Litsea stocksii* Hook pisa, betal nut, laurel tree	Lauraceae	ix—x xii	2	2	tr.	Cupshaped; dark amber
58.	*Makenziea integrifolia* wahiti, shrub	Acanthaceae	x—i	1	4	tr.	Honey colourless
59.	*Madhuka longifolia* (Koenig) mahua tree	Sapotaceae	ii—iii	1	4	tr., ste.	Yellowish brown; amber
60.	*Malus domestica* Borkh. apple, tree	Rosaceae	iii—iv	1	1	te.	Pinkish white; amber
61.	*Mangifera indica* L. mango, tree	Ancardiaceae	i—iii	3	3	str., tr.	Light cream; dark amber from honey dew
62.	*Medicago sativa* L. lucerne, crop	Leguminosae	iii—v vii—ix	1	3	te., str.	Purple; light amber
63.	*Melilotus alba* Desr. sweet clover, crop.	Leguminosae	mid-summer	2	3	str., te.	White; slightly green
64.	*Mimusops elengi* L. moulsari, tree	Sapotaceae	v—vi	1	4	str., tr.	White; amber

65. *Moringa oleifera* Lamk. sohanjna, tree	Moringaceae	ii—iv	1	2	str., tr.	White; light golden
66. *Musa* spp. banana, plantain	Musaceae	all year	1	1	str., tr.	yellowish or pinkish; dark
67. *Nicotiana tabacum* L. tobacco, crop	Solanaceae	xi—iv	2	1	str., tr.	Pink; dark
68. *Parkinsonia aculeata* L. parkinsonia, tree	Leguminosae	all summer	2	4	str., tr	Bright yellow; amber
69. *Phaseolus multiflorus* Lam. white kidney bean, crop	Leguminosae	i—iii	2	3	str., tr.	Red or white; light amber
70. *Phlebophyllum kunthianum* Nees kurinji, wild plant	Acanthaceae	x—xi	1	1	tr.	—
71. *Pithecellobium dulce* Madras thorn, tree	Leguminosae	xii—iv	2	2	tr.	White or yellowish; amber
72. *Plectranthus* spp. shrub, not shain	Labiatae	ix—xii	1	4	te.	White; clear
73. *Pongamia pinnata* Pierre karanji, tree	Leguminosae	iii—vi	2	1	str., tr.	Pale pink, dark
74. *Prosopis* spp. mesquite, tree, prickly	Laguminosae	xii—iii iii—ix	1	3	str., tr.	Creamy or yellow; light amber
75. *Psidium guajava* L. guava, tree	Myrtaceae	iv—v ix—xii	2	2	str., tr.	White, fragrant; amber
76. *Rabdosia coetsa* Hara *Robdosia rugosa* Hara plectranthus, shain, shrub	Labiatae	ix—xi	1	4	te.	Lavender-blue; clear watery
77. *Robinia pseudacacia* L. honey locust, avenue tree	Leguminosae	v—vi	3	4	te., str.	White, fragrant; pale yellow

S. No.	Name/Local name/Type	Family	Bloom Months	Status of Nectar N	Status of Pollen P	Region	Remarks—Colour of Flowers; of Honey
78.	*Rubus* spp. L. blackberry, bramble	Rosaceae	vi—viii	2	2	te.	Pink or white lavender; dark
79.	*Saccharum officinarum* L. hybrid sugarcane	Gramineae	all year sap	2	—	str., tr.	Sap and honey dew; dark
80.	*Salix alba* L. white willow, tree	Sallicaceae	iv—v	3	4	te.	Cream; minor honey, honey dew
81.	*Sesamum indicum* L. sesame, crop	Podalliaceae	vi—vii	1	1	str., tr.	White; light amber
82.	*Sapindus mukorossi* Gaestn soapnut, retha; tree	Sapindaceae	v	1	3	tr.	Greenish white; light golden
83.	*Schefflera* spp. lodabettu, tree	Araliaceae	i—ii iv—v	1	1	tr.	—
84.	*Schefflera wallichiana* pongabettu, tree	Araliaceae	v	1	3	tr.	—
85.	*Solidago canadensis* L. golden rod, herb.	Asteraceae	ix—x	2	2	te., str.	Golden; amber
86.	*Syzygium aromaticum* Perry clove tree	Myrtaceae	iv—v	1	4	tr.	White or reddish
87.	*Syzygium cuminii* Skeels jamun tree	Myrtaceae	ii—iv on location	1	1	str., tr.	White; light amber
88.	*Tamarindus indica* L. tamarind, tree	Leguminosae	vi—vii	2	3	tr.	Yellow spoted red; dark
89.	*Taraxacum officinale* Weber dandelion weed, short	Compositae	iv—v	3	1	te.	Yellow; rich golden

90.	*Terminalia arjuna* Arn. arjun, tall tree	Combretaceae	iv—vi	1	4	str., tr.	Greenish white; amber
91.	*Terminalia bellerica* behda, tree	Combretaceae	iv—v	2	4	tr.	Light yellow; nausaeating
92.	*Terminalia chebula* Retz. inknut or hirda, tree	Combretaceae	iv—vi	1	—	tr.	Pale yellow, light yellow
93.	*Thelepaepale Ixiocephala* Bremk darmori, shrub	Acanthaceae	xi—i	1	1	tr.	White;
94.	*Toona ciliata* M. tun, tree	Meliaceae	iii—iv	1	3	str.	Cream fragrant; whtite amber
95.	*Trifolum alexadrinum* L. barseem, crop	Leguminosae	iv—vi	1	1	te., str.	Cream; light yellow
96.	*Trifolum pratense* L. red clover crop, pasture	Leguminosae	vii—viii	1	1	te., str.	Reddish purple; reddish tinge
97.	Vitax spp. shrub	Verbenaceae	xi—vi	2	2	te., str.	White and pale blue; amber
98.	*Wendlandia notoniana* Wall renda, tree	Rubiaceae	ii	1	1	tr.	White
99.	*Woodfordia floribunda* Salisb dhain, shrub	Lythraceae	v—vi	2	3	tr.	Orange red
100.	*Ziziphus mauritania* indian ber, tree	Rhamnaceae	vii—x	3	3	str., tr.	Greenish, small. yellow brown

Notes : (*a*) Months of the year 1 – xii;
(*b*) Nectar and pollen 1 = Major source; 2 = medium source ; 3 = minor source ; 4 not significant
(*c*) te = temperate ; str. subtropical; tr. = trcpical.

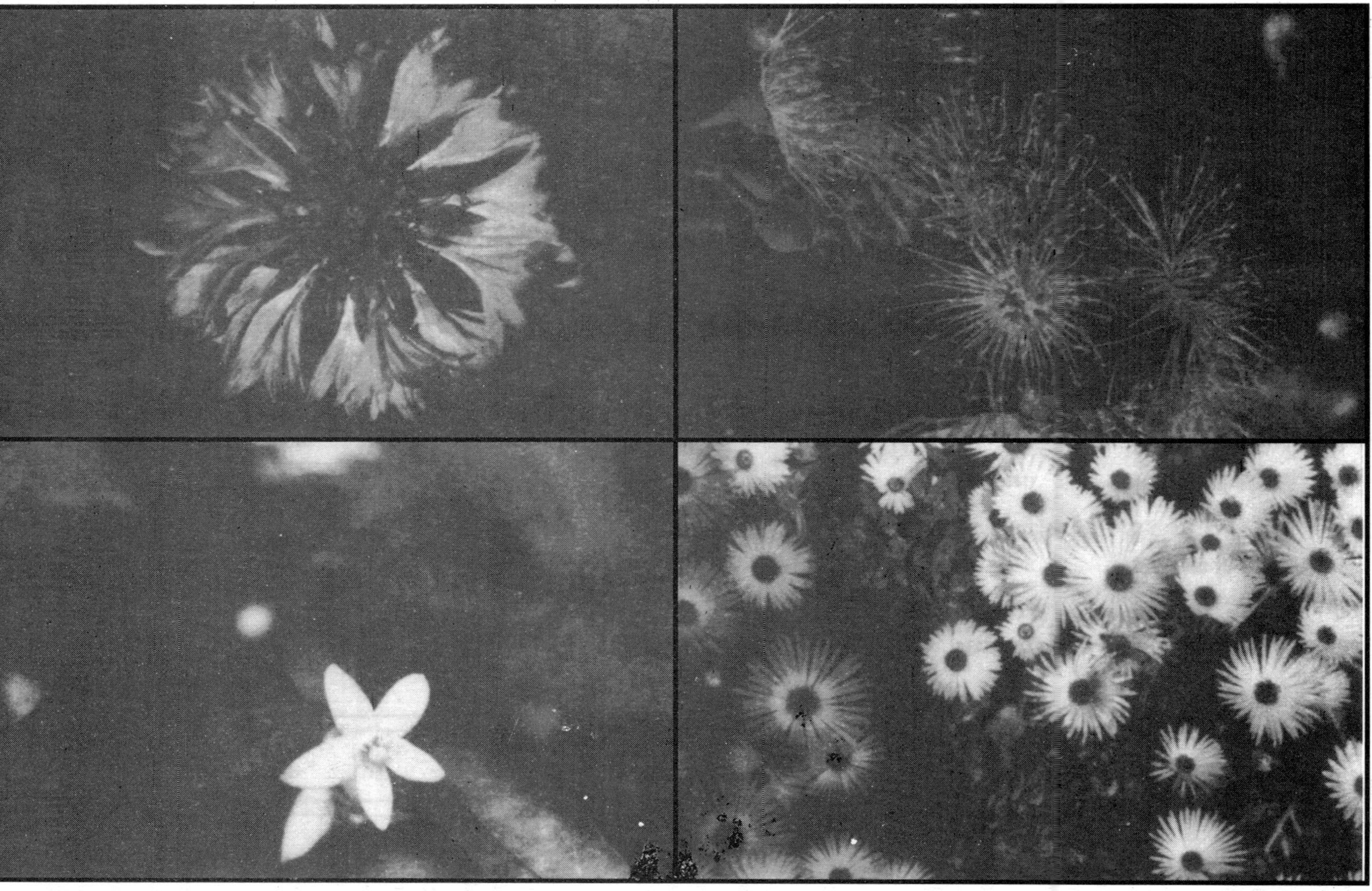

Fig. 20. Some flowers visited by honeybees; clockwise from the top left, corn flower, *Eucalyptus*, ice flower, citrus.

are amino acids, minerals, essential oils (aroma), organic acids etc. Minute quantities of pollen grains, yeast cells, fungal spores, and bacteria are also found in nectar. Moreover bees prefer flowers which secrete nectar having equal amounts of glucose, fructose, sucrose and all in high concentration; the higher the concentration of sugar in nectar the greater the preference; but abundance of flowers is all important for effective foraging by honey bees and for bringing to the colony ample stores to be used as food.

The value of flowers for bee foraging depends upon a number of factors. The duration for which nectar is secreted and the amount of nectar produced in a flower are important. Some flowers produce nectar for a very short time, others for a day or two and still others like *Brassica* spp. have a sequence of flowering which lasts for about one month. In silver oak and bottle brush also there is sequence of flowering lasting many weeks and cardamom panicles keep bearing flowers from July to December (Siddapaji and Channa-Basavanna, 1980). In addition, the composition of sugars and nectar-sugar concentration are also important factors for the evaluation of flowers.

12.5.2 Nectar Preferences

The value of nectar for the bees is determined both by its composition and sugar concentration. By analytical and chromatographic determinations the composition of many flowers has been studied in India. The sugars in nectar are expressed as total solids in the liquid. The three major sugars, sucrose, fructose and glucose which are used as food by the bees, occur in flowers in various combinations of two or three and their proportion in a given plant species is constant. It has been proven by analyzing the nectar of almost 1000 plants (Woykes, 1953; Percival, 1961).

In the total solids of *Thunbergia grandiflora* nectar in India, the three sugars, sucrose, fructose, glucose occur in the ratio 28 : 1 : 1 and in *Moringa pterigosperma* nectar there are 11.81% non-reducing sugars and 0.9% reducing sugars (Nair *et al*., 1962; Nair and Singh, 1974). Soapnut nectar also has 85.5% sucrose and 7.25% each fructose and glucose (Reddi *et al*., 1980), in that proportion of the total solids. The combination of sugars in various flowers is not the same. For example, nectar of *Carvia callosa* contains fructose, glucose, maltose, raffinose whereas nectars of *Schefflera roxburghii, Grevillia robusta* and *Woodfordia floribunda* contain only fructose and glucose (Wakhle *et al*., 1981; Mishra and Sihag, 1987). Sugars in various proportions and combinations have been recorded in nectars of 44 plants by Sihag and Kapil (1983) and of 103 plants by Bahadur *et al*. (1986).

Nectar sugar concentration is an important attribute of flowers to be attractive to bees, although, depending upon the situation bees may collect nectar having sugar as low as 6% and as high as 79% (Deodikar *et al.,* 1957). The concentration in most bee flora ranges between 20-50% and Mishra (1995) has reviewed the studies made in India.

It may be noted that nectar sugar concentration in the plants studied in India varies from 10-79% and honeybees avail of a wide variety of flora available in a locality. It is not enough to have a few plants of a very rich source like *Grevillia* (silver oak), but the bees in order to collect surplus honey need extensive pasturage like a crop of *Brassica* or large plantations of *Eucalyptus*. The sequence of flowering in a locality is very important so that bees can shift from one crop to the other and thus exploit various flora sequentially.

Richness of nectar not only varies with plant species but also with varieties. It should be an important consideration while developing new cultivars. In a given cultivar sugar concentration may be further changed with cultural practices, growth hormones, under changing weather conditions, with the aging of flowers and may even be stimulated by the visits of bees (Woykes, 1953; Percival, 1965; Mishra *et al.,* 1989).

The honeybee activity is basically influenced by the availability of quality nectar. The higher the sugar content of a nectar the greater its attractiveness. When concentration becomes less than 20% its value is not commensurate with the energy spent for collecting it. However, under stress during the dearth period or in tropical areas where nectar gets diluted bees may exploit poor sources like mango which has only 5-6% sugar concentration. For carrying extra volume a honeybee always keeps a spare stomach capacity. For example, the full capacity is 70 mg but an average nectar load is 20-40 mg as determined by quality of nectar, temperature and past experience of the nectar gatherer.

12.5.3 Nectar Flow

In a forest or the edge of a forest there is a great variety of flora which may or may not be sufficient for the collection of surplus honey but may be desirable for the subsistence of colonies. With urbanization and the adoption of intensive agriculture there is a sudden disappearance of the natural flora. It is replaced by crops, orchards and tree plantations which substitution may even be better for commercial beekeeping as it has been observed in North India particularly in Punjab and Haryana. Under most favourable conditions, cruciferous crops bloom from end of September to the beginning of March. Before the flowers fade away, *Eucalyptus,* which has been planted all over North India, starts blooming. Extensive roadside plantations of shisham and fruit trees and orchards provide the supporting flora. Those are followed by barseem which is a wide-spread crop and by sunflower that has been introduced more recently. Bees are busy upto May-June on these crops and collect surplus honey. From then on till August-September it is the dearth period and bees find subsistence food from a variety of trees and crops including *Acacia, Khair,* cotton, cucurbits, avenue plantations, kharif crops, weeds etc.

In western Himalayas during the spring and summer months bees avail flora of rabi crops, pome fruits and a variety of trees like soapnut, tun, wild pear, puna, puja etc. The rainy season is not only a dearth period but a lot of bees are also washed away, reducing the colony strength further. In certain valleys of Himachal Pradesh and Jammu and Kashmir there is autumn nectar flow of *Plectranthus* and *Rabdosia* in October-November from which the bees collect surplus honey. The winter months are again a dearth period.

In the coastal and tropical areas of South India there is dearth of flora during the pre-monsoon drought period and during the heavy down pours from May to September. From October to April-May there is medium to major honeyflow from crops, fruits plantations and trees like jamun, herda, carvi, soapnut, rubber, Schefflera, erythrina etc. Depending upon the area the surplus period may range from 1-3 months in a year.

Although major flora in a region or a state may have a familiar pattern yet the variety and abundance may vary from locality to locality. A serious beekeeper has to make his own observations and record the important flora and prepare a complete chart of the incoming pollen and nectar. Fortnightly records of colonies would give a more realistic picture. When a colony is opened, each and every frame is examined and data is noted in a register. The observations recorded would indicate not only the condition of the queen and brood area of eggs, larvae, and sealed pupae, but also the amount of honey and pollen present in measures of square centimeters. One average colony is placed on a balance and its weight gain or loss is recorded every day. The major or minor honey flow or the dearth of food in the field is thus monitored and the progress of a colony from winter upto honey extraction and further is recorded.

The foraging behaviour of bees is quite complex and the basic principles have already been discussed. The time at which foraging activity starts depends not only on the season but more pertinently on the flora and nectar available. It was observed at Ludhiana that on *Eucalyptus* flowers, *Apis mellifera* started working at 5.45 am when the sun rose at 6.18 am. On barseem the bees started foraging at 8.30 am when the sunrise was at 5.37 am, and on lucern, a closely related crop, they were observed at 10.00 am, when the sunrise was at 5.44 am (Atwal *et al.*, 1970). The starting time for nectar foraging is directly related to nectar secretion, and foraging continues till the secretion goes on during the day. Bees were observed working on this crop early in May from 8.30 am to 6.30 pm and the largest number visited from 10.00 am to 12.30 pm. The comparative foraging activity of the four species of *Apis*, of solitary bees, and other insects, have been discussed earlier in a greater detail. Factors influencing the secretion of nectar and release of pollen grains have also been discussed. Special plant characteristic that attract honeybees are mentioned under each crop.

12.5.4 Variability in Nectar Secretion

The appearance and opening of flowers on a given plant species is a seasonal function. Inflorescence appear on a plant at about the same time of the year but not necessarily on the same date; the variation of a few days depends upon weather which is variable from year to year. When a flower opens nectar is secreted coinciding with the receptibility of the stigma. In nature this mechanism is evolved so that insects attracted to nectar may pollinate the stigma. In other words, the sweet liquid is a reward for the pollination services rendered. After pollination takes place a feed back mechanism comes into operation and nectar secretion is switched off. The sugars already present are reabsorbed and their concentration in nectar is reduced. In dry weather the thin liquid dries up and nectar is no more. In certain flowers nectar secretion is prolonged somewhat if insects keep removing it. It was observed by Mishra and Sihag (1987) that flowers of *Woodfordia floribunda* secreted 126.4 ml nectar in three days; the sugar concentration ranged from 10.6 to 12.2%; it was the minimum on the first day and was the maximum on the second day. In general the duration of nectar secretion varies from one plant to the other species.

Within a species of plant there are varieties and cultivars which have their own characteristic periods of nectar secretion due to genetics or due to a greater heat and drought tolerance. Some new strains of barseem have been selected that flower till late in the season upto the 2nd or 3rd week of June; however, due to extremely hot and dry weather at that time the duration of nectar secretion during the day is shortened very much. There are some species of plants that bear flowers on growing stems or panicules. The plants continue to flower for a longer period thus providing nectar to the bees for a number of days. This succession of flowering in *Brassica* crops lasts for a month and in cardimom forc a number of months (Siddappaji and Channa-Basavanna, 1980).

The other factors that affect the secretion of nectar are sunlight, temperature, relative humidity and soil. Sunlight affects photosynthesis for the manufacture of carbohydrates, (sugars) which appear in nectar. The enzymes that secret nectar are activated at an optimum temperature; higher or lower temperature beyond the optimum adversely affect the secretion of nectar. Since nectar is hygroscopic, moisture is absorbed at high relative humidity, thus lowering the sugar concentration. In case relative humidity is low nectar loses moisture, increasing the concentration; but in case of nectar in tubular flowers the surface of liquid is lowered where the bees may not be able to reach it. Under hot and dry conditions the duration of nectar is reduced significantly and the condition can be improved by irrigating multicut crops like berseem and alfalfa. For producing plentiful flowers and good nectar secretion optimum soil moisture is essential and conditions of excessive water should be avoided. It is also important to keep a right balance of soil nutrients; excessive nitrogen stimulates vegetative growth producing less flowers. Phosphorus and potassium in balanced proportion increase nectar production, whereas high doses of phosphorus adversely affect nectar secretion (Shuel, 1975, 1992).

12.5.5 Pollen Availability

Pollen contains proteins, fats, minerals vitamins, enzymes which are essential nutrients for the growth of developing larvae and adults. To meet these requirements bees collect mixed pollen from more than one source. Pollen grains are released from anthers when dehiscence takes place at the appropriate time of the day which is the characteristic of a plant species. Accordingly, the honeybees regulate their activity for pollen collection. In various species of plants pollen grains differ a great deal in their size, colour, surface, armature, dryness or stickiness. The pollen of maize is very dry and easily falls off whereas that of silver oak is so wet and sticky that bees have difficulty collecting it. From the colour of pollen stored in combs the source can be identified. But a sure test is by recognising pollen grains under a microscope. When honey is extracted small quantities of pollen always remains mixed in it. Individual beekeepers or the testing laboratories interested in determining the source and quality of honey keep slides of pollen of that area for reference. It is a regular test for honey that is exported outside. In commerce it is important to label correctly whether the honey is from a single source or is from mixed sources.

A pollen load of a honeybee is 14-20 mg in weight. Bees prefer pollen from certain species of plants and not from others, even when the species are closely related. The selection is not influenced by age, colour, moisture, protein, or any other known nutritive value of pollen. When it becomes known what makes a pollen attractive the chemicals involved can be sprayed on a desired crop to attract bees for pollination (Levin and Bohart, 1955; Doull, 1966; Free, 1993).

The time of the day and frequency of bee visits for pollen gathering varies with plant species. Bees visit a flower when it opens and dehiscence of the anthers has taken place. There is a great variability in the nature and performance of flowers, and as a result, in the foraging habits of the bees. In some species anthers dehisce in the flower buds and in others after the flowers open. Pollen may be available all day but the peak varies with species. In vast majority of flowers dehiscence takes place in the afternoon when sun is the brightest.

In some flowers all anthers dehisce at the same time and in others spread over a number of days. Weather, particularly temperature, determines bee activity directly and also indirectly through the availability of pollen and nectar. Need of the colony is the basic determining factor (Synge, 1947; Percival, 1955; Ribbands, 1953; Free, 1968a, b). As the bees fly their bodies acquire a positive charge which makes it easier for pollen grains to adhere to body hair (Corbet *et al.*, 1982). After unloading pollen in their hive, as the bees leave, they still have thousands of pollen grains (6,700 were counted) on their body representing 3.4 plant species (Free and Williams, 1972).

12.5.6 Floral Cycle

For successful beekeeping it is desirable to prepare a floral calendar for a locality where apiaries are established. The floral cycle round the year should show major and minor honey flow in the appropriate months and the availability of lesser flora useful for the subsistence of a colony during the dearth period. Intensive observations are made every week and month of the year when a particular source of nectar and pollen starts with its inflorescence, reach a peak, and then decline. In good beekeeping localities there is generally a new flower source coinciding with the decline of the earlier one. The subsistence flora are invariably weeds, trees in the countryside or a leasser crop that is not so popular among the farmers. In Europe and North America beekeepers have long experience of working with *Apis mellifera*. They are generally very serious people and observing the local flora they make their own charts for guidance.

In North India the dry and hot summer and the monsoon seasons are generally the dearth periods. There are some crops and trees that can be planted to serve as the subsistence flora. The foremost among them is sunflower whose planting can be so staggered that flowers continue opening during the dearth period. Sesame and cotton are the other crops. In north-western India among the trees, the promising ones include, bottle brush, siris, jamun, ber, guava, khair, *Eucalyptus,* Acacia, cosmos, cucurbits etc.

Even though it is not possible to prepare floral calendars for a whole state or a region attempts have been by some authors in various parts of India to indicate floral cycles for certain localities or areas (Thakar *et al.,* 1962; Atwal, 1987; Mishra and Kaushik, 1992). The exercise can be fruitful for local conditions by keeping records of a balance-colony.

12.6 BEEKEEPING : A STATUS REPORT

12.6.1 Introduction

This report is a survey on India's export potential of honey with respect to the major importing and exporting countries and a survey of the domestic scenario. The report then goes on to give the strategy needed for establishing India as an exporter of Quality Honey. The purpose of the study is to view the current situation and the potential for expansion in the domestic as well as export markets.

The report has been compiled for The Agricultural and Processed Foods Export Development Authority (APEDA) and has been researched and written by Sarparveen Kaur with the domestic survey of the 12 Indian States and the retail market in the four metros being done by The Greenworld People

12.6.2 Methodology

The objective of the study was to survey the export market and the domestic status and evolve a strategy for developing the Indian industry. To achieve this the major importers, packers, retailers and quality testing laboratory personnel in the importing countries were interviewed and the distribution channels and market preferences studied as also the quality control methods employed and the access to the market for honey from India. A study of the major honey producing countries and the reasons for their success studied. Data were compiled from the relevant trade information, journals, statistics and from the internet.

For the domestic study in each of the 12 States, the relevant organisations, beekeepers and trade were contacted personally and a study made of the status of beekeeping and the honey industry in the State, its beeflora, problems and potential. A brief study of the retail honey market was also conducted.

On the basis of all the above information a strategy for development of the Indian Beekeeping and Honey industry has been prepared.

12.6.3 Beekeeping

This is a brief overview of' the honey bee colony, its organisation and working, the different bee species, the parts of the beehive and a note on how bees collect honey, correct beekeeping practices and techniques for quality honey production, followed by a series of photographs of beekeeping equipment needed to set up a double chamber bee colony.

12.6.4 Honey

Honey has been treasured through the ages by men as a natural sweetening agent. Honey is a valuable food which is appreciated all over the world. There are many types of honey and it is used not only as a spread on bread but also in the bakery, confectionary and pharmaceutical industry. The composition and characteristics of honey are explained.

12.6.5 Bee Products

The other bee products that are extracted from the beehive are royal jelly, pollen, propolis, beeswax and bee venom. These products have a demand on the world market and there is a good potential for India to develop production of these products as supplementary income from beekeeping besides honey. Royal jelly which is a secretion of the bees fed to the queen bee larva is taken internally as a rejuvenating tonic and applied externally as natural cosmetic creams and lotions. Propolis is a natural resinous substance collected by bees from trees and plants which is increasingly being used in natural medication. Pollen collected from flowers is a rich source of protein and is a valuable dietary supplement. It is taken in the form of pollen grains, powder, tablets and capsules. It is also used in natural cosmetics. Bee venom secreted by bees has some medicinal uses for treatment of rheumatism and joint pains and for treating allergies to stings. Beeswax produced by bees to construct the honey comb has many uses and demand exceeds supply on the world market. It is used in the manufacture of cosmetics, to make beeswax, foundation sheets for the beekeeping industry, for candles and in the furniture and paint industry.

12.6.6 Honey Extraction and Processing

The best processing is the least processing but to prepare honey for the market it requires suitable processing. Production of quality honey starts from the hive so the correct techniques to be used for production of honey are given and honey extraction methods both for hobby and commercial extraction of honey along with the list of extraction equipment. A detail of processing methods and machinery for handling honey is included both for processing honey for export and for bottling including a note on moisture removal of honey.

12.6.7 A Model Honey Processing Unit

A short note on a model processing unit for packing honey for export is given with the flow of work and the equipment required.

12.6.8 World Honey Market

The world market for honey was 2,79,000 tons in 1997 and has hovered around the same figure for the last 7 years. The major players in the world honey market are China, Argentina and Mexico. The international market for honey has always been a buyer's market controlled by the major importing countries who set the prices. There has normally been a glut of honey on the world market and prices were more or less steady in the last few decades. It is only between 1995-1997 that world import demand grew faster than available honey due to crop failures in the major producing countries. This led to an unprecedented increase in the prices of honey by importers, but prices have already come down to earlier levels.

A. World's Major Honey Importing Countries

The major importing countries are Germany, U.S.A., U.K., Japan and The Middle East. A detailed analysis of each country is given in the following pages.

Germany

Traditionally Germany has been the largest importer of honey in the world but in 1997 its place was taken by U.S.A. Germany imports around 80,000 to 90,000 tons annually. It also has a well developed domestic beekeeping industry producing around 15,000-30,000 tons annually. Consumption of honey is highest in Germany being around 1.5 kg per capita and the German market has a wide variety of liquid and creamed honeys from all over the world. This is a very quality conscious market and very strict quality control norms are enforced on imported honeys.

United States of America

The U.S.A. has become the world's largest importer of honey replacing Germany in 1997. The U.S.A. is also the second largest honey producing country in the world, producing around 90,000 tons annually. With imports of around 80,000 to 85,000 tons annually the United States is a major honey consuming country, this has been further supplemented by the efforts of The National Honey Board as a natural food and the market for honey and bee products are growing.

United Kingdom

The U.K. till a few years ago was the second largest importer of honey but now it holds the fourth place with annual imports of around 20,000 tons. Honey consumption in U.K. is more or less stagnant and not much growth is foreseen in this market.

Japan

Japan is the third largest honey importing country with honey imports of around 35,000 tons annually. The Japanese market has grown considerably in the last three decades from imports of 6,000 tons of honey in 1965. An interesting aspect of the Japanese market is it has the largest consumption of other beehive products like royal jelly, pollen, propolis and beeswax with annual imports of around 400 tons of royal jelly, 500 tons of beeswax and around 30-40 tons of propolis.

The Middle East

The use of honey is intrinsic to the culture of the Middle East since the Koran appreciates the healing and medicinal properties of honey. The Middle East as a block consumes a reasonable amount of honey and the countries of Saudi Arabia, Kuwait, Yemen, U.A.E., Oman, Iran, Bahrain and Doha together import around 11,000 tons of honey annually. Most of the honey imported is prepacked branded honey from U.S.A., Germany, Australia, U.K., Canada and New Zealand. The Middle East is not a large market for honey.

B. World's Major Honey Exporting Countries

The world's major honey producing and exporting countries are China, Argentina and Mexico. These countries dominate world exports with a share of 54%. We have considered in detail the analysis of the production, exports, reasons for increase and decrease of these countries' market shares, the Argentinian success story and the failure of China to establish themselves as a supplier of quality honey. Mexico has slipped down dramatically in the world market as it is plagued by bee disease problems and antibiotic residues in honey.

12.6.9 Indian Beekeeping and Honey Industry

Beekeeping has been practised in a traditional manner since the time of the Vedas but it is only in the mid 1950's that the KVIC gave an impetus to the beekeeping industry and promoted *Apis cerana* over most of India. This bee was decimated by the Thai Sac Brood disease which gave a great setback to beekeeping especially in States like Kerala. It is only with the introduction of *Apis mellifera*, the commercially used bee over most of the world, that beekeeping in India expanded rapidly from the

Northern States and over most of India to the present production of around 10,000 tons honey and the export of honey from India began in 1995. The current status of honey production and beekeeping both from *Apis mellifera* and *Apis cerana* in India of each State is given. The existing acreages of beeflora allover India namely mustard and rapeseed, sunflower, citrus, litchi, niger/sesamum and rubber are taken and potential bee colonies and honey production figures are given along with the income generation, rural employment and investment required.

Beekeeping and Honey Industries in 12 Indian States

This chapter deals with the current status of the honey and beekeeping industry in the 12 Indian States of Punjab, Haryana, Uttar Pradesh, Himachal Pradesh, Bihar, Assam, Bengal, Kerala, Karnataka, Andhra Pradesh, Maharashtra, and Madhya Pradesh is given including the total production of honey in India which is around 10,000 tons.

An overview of the production, types of bees, beeflora available district wise and the flora calendar are given for each State. The migratory pattern and route where applicable, the organisations involved in beekeeping in the State, the processing units and marketing channels are all detailed. The problems faced by the industry and the potential of each state are also outlined

Retail Market for Honey in India

This is a study of the Indian retail market for honey in the four major metropolis cities, namely Delhi, Calcutta, Mumbai and Chennai, giving the market profile, quality standards, packing and prices, consumer preferences and the distribution network.

Indian Honey Export Scenario

India is a new entrant in the world honey export market since 1995 and quantities exported are still small, being a total of around 5350 tons in three years. In the last two years, there have been crop failures due to adverse weather conditions so the growth has slowed down, it is expected that production will grow in the current year. The few companies that have exported honey so far or been major suppliers to exporters are listed along with a brief note on each company, its capacity and infrastructure.

Package of Practices for the Beekeeping Industry in India

The package of practices for Beekeeping as given by Punjab Agricultural University, Ludhiana, Punjab for handling *Apis mellifera* is given which includes the life cycle of the bee, the work distribution in the hive, list of beekeeping equipment, points to consider in selecting a site for location of bee colonies, the best time to begin beekeeping, movement and placing of beehives. The seasonal management of the bees in each season is given along with the major pests and diseases which the bees are afflicted by along with a note on prevention of loss of bees from the effects of pesticides.

Incentives Available to Honey and Beekeeping Industry

In India as agriculture is the mainstay of the economy, agriculture and other agro based industries were developed but in the case of beekeeping except for the efforts of the KVIC beekeeping and its importance was largely neglected. The different schemes offered by Ministry of Agriculture, KVIC, APEDA giving incentives for development of the honey and beekeeping industry are outlined, along with the India brand Equity fund.

Realising the importance of honeybees in improving crop productivity by pollination, the Ministry of Agriculture took the initiative of setting up a National Beekeeping Development Board in 1994 to promote and develop beekeeping on an all India basis with a major stress on R&D, for expansion of bee colonies, training and development of infrastructure for handling marketing and promotion of honey. An amount of Rs. 18.87 crores was earmarked for the scheme during the VIIth Five Year Plan.

APEDA offers various incentives and financial assistance to the exporters of agricultural food products including honey-these schemes include feasibility studies and surveys, export promotion and market development, packaging development, assistance for promoting quality and quality control and R&D.

KVIC has different schemes for promotion of beekeeping with *Apis mellifera* and with *Apis cerana* being for 30/60 bee colonies of *Apis cerana* and for 25/50 colonies of *Apis mellifera* which are for individual units below ₹ 1.00 lakh, the detailed economics of the schemes as per KVIC are given.

12.6.10 Swot Analysis of the Indian Beekeeping and Honey Industry

An analysis of the strengths, weaknesses, opportunities and threats for the Indian Honey Industry are outlined.

Strategy for Establishing India as an Exporter of Quality Honey

APEDA, MFPI, and other concerned Government organizations should monitor all aspects of honey production to help develop exports of honey from India by setting up a coordination committee so that India can achieve its potential for honey production and exports.

This must be done by making one single agency responsible for monitoring all aspects related to the honey and beekeeping industry, introduction of quality control standards through all stages of honey collection, processing, testing, packaging, marketing and R & D,

Promotion of single source honeys for export ensures quality standards as per the EU monitoring system and gives financial incentives for the development of the honey export industry.

Argentina's and China's strategies for becoming the worlds largest exporters of honey are tabulated.

Strategy for Developing the Beekeeping Industry and Honey Production in India for the Domestic Market and Exports. – A Five Year Plan

India has vast resources of untapped bee flora for honey production. A simple strategy is given as a Five Year Plan for doubling the present bee colonies every year, giving the strategy needed for achieving this by development of various facets of beekeeping.

Bee breeding and production of bee colonies, bee disease testing facilities and R&D, support of migratory beekeeping training, promotion of beekeeping as an aid to agriculture, protecting bees from pesticide damage, tie up of afforestation with beekeeping thus taking into consideration the overall development of the industry.

12.7 HONEY AS A FOOD AND MEDICINE

12.7.1 Introduction

For more than thirty centuries honey was known to man as a food sweetener. It was observed that honeybees made it from plant cell sap, or, the nectar of certain flowers and from honeydew (the excretion of insects like aphids). As time passed, the human mind tried to find out the components it was made of and the uses to which it could be put. In ancient times it was used as a cure for certain diseases like asthma, eczema, psoriasis and open wounds and cuts. It was interesting that upon the excavation of the tombs of Egyptian Pharaohs three thousand year old earthen jars full of honey were found for use in the after-life. There are also records like the Vedas written between 2500 and 1500 years ago in India in which honey is mentioned for specific purposes including the management of chronic wounds. Some of these Ayurvedic formulations are still being used to this day and honey is an important base component of many other traditional medicines.

In addition, Aristotle (384-322 BC) has left a written description of the various colours of honey derived from different sources and considered pale honey the best remedy for sore eyes and open wounds. Until the middle of the twentieth century, Dal Lake in Kashmir was not polluted and lotus honey produced there by bees was exported for diseases of the eyes. Currently, honey has been shown to be effective in the control of certain pathogenic microbes like *Salmonella, Shigella, Escherichia coli, and Vibro cholerae.*

12.7.2 Composition

Honey is primarily composed of sugars (83%) dissolved in water (17-18%). The sugar concentration is homogeneous and fairly stable and in the dry matter of honey ranges from 95-99%. Since sugars are hygroscopic, they easily absorb moisture from the air (60% relative humidity) and can, thus, change the water content of honey particularly during monsoons. Basically, all these components of honey are derived from the nectar of flowers.

After honeybees have ingested the nectar, they add some nitrogenous compounds secreted in specific glands into it. These chemicals are mixed with the nectar as it passes from the proboscis tongue to the oesophagus to the stomach. Upon reaching their hive, the forager bees are helped by the house bees. The nectar is regurgitated and is stored in clean and polished cells. One by one, the house bees suck up the liquid, and work it up and down on their tongues to stimulate chemical reactions and to hasten evaporation. In the honey flow season, there is always an air current produced by fanning inside a hive. The stream of air passes over the frames and goes out of the exhaust hole. When the appropriate consistency of the honey is achieved, it is deposited in clean cells and sealed with wax following further evaporation. Honey, ripened in this way, and, flower pollen, collected by similar foraging bees, are stored in honey comb cells as food reserves for the colony. Honey, nectar, and pollen constitute the main diet of the bees.

The chemical reactions that take place from the stage of nectar collection to that of ripened honey storage have been studied. Various sugars comprise the largest components of honey. These include simple sugars of 6-carbon molecules and sugars of more complex molecules. The simple sugars are fructose or levulose (38.2%) and glucose (31.3%). Among the complex sugars, sucrose (cane sugar) is the most common in the nectar of flowers and is broken down to fructose and glucose while the remainder remains as sucrose (1.3%). It also contains some minor sugars in very small quantities including maltose, isomaltose, and some disaccharides and trisaccharides.

The actual production of fructose and glucose is the result of key chemical reactions initiated by enzymes secreted in the body of the honeybee. Invertase is secreted in the salivary glands of the honeybee and, in addition to causing the conversion of sucrose into simple sugars, it also initiates the production of hydrogen peroxide and organic acids. These oxidize any impurities in the honey and act as preservatives. Another enzyme glucose oxidase (diastase/amylase) secreted in the hypopharyngeal glands in the head also converts sucrose and other sugars into fructose and glucose. In addition, it produces glyconic acid and hydrogen peroxide, the natural preservatives found in honey. The primary oxidants and purifiers of honey constitute 0.5% of honey and they kill practically all the microorganisms and other organic contaminants. Minute quantities of minerals (0.17%) are also present. Those identified include calcium, copper, iron, manganese and phosphorus. Their source is also the nectar from various species of flowers. Depending upon the properties of cell sap of the mother plants, some active phytochemicals may be found in the nectar as well. Some other exudates are active and are collectively described as flavonoids (phenols). Currently, more attention is being paid to this group for studies as a source of non-peroxide anti-microbial compounds with a longer active life.

In pure honey, there are small quantities of vitamin C, B_2 (riboflavin), and B_6 (pyridoxine). Jointly, all the acids maintain the pH value of honey between 3.2 and 4.5. In such an acidic environment hardly any microorganisms can survive. However, certain phytochemicals may alter the level of acidity with the result that some bacteria, yeast, spores of moulds, fungi and their by-products known as mycotoxins may survive. Therefore, the pasteurization of honey after extraction is the recommended scientific procedure; the combined effects of heat, decantation, followed by high and low temperatures are very effective. There is little evidence that microbes causing diseases like cholera, typhoid, and anthrax are found in processed honey.

Poor handling, infected machinery and defective storage conditions are the agents for the spoilage of good, raw honey. If the glucose content is high, it tends to form crystals more easily and to give off water which makes it a favourable substrate for fermentation. This is initiated by yeast left in the honey or new entry of yeast due to the defective sealing of containers.

Among other honey bee hive products, wax and propolis have specific importance. Bee propolis, which contains caffeic acid phenyl esters with anti-proliferative activity towards cancer cells, may also serve a role in mycotoxin detoxification.

12.7.3 Microflora

Microflora are not really one of the components of honey but are contaminants. They originate from flowers probably left there by foraging bees and wind only to be picked up by the new foragers and carry them to their hive. They leave the microbes on frames and everywhere inside the hive. Bees ingest them while feeding themselves or while feeding their larvae. When honey is extracted and handled with infected tools and machinery, the contamination of honey can occur. These places of abode of the microorganisms have been detected by thorough search and microscopic examination of bees, their intestines, inside their hives, and of tools and machines used to handle raw honey. If the honey is bottled and stored, the microbes are killed in time by the action of high sugar concentrations, dehydration due to osmosis, and the reaction of active chemicals naturally present as components of the honey. These oxidants are acids, organic acids, glucose oxidase, hydrogen peroxide and phytochemicals. Most of the microbes are killed but the spores of some species may escape. However, none of the pathogenic organisms have been detected. The continued presence of hydrogen peroxide, organic acids, and phytochemicals (flavonoids) ensures that honey remains disinfected. Among these agents, flavonoids remain active for a long period and research on these non-peroxide agents is under progress.

Scientific studies on the species of microorganisms and their distribution at different loci have been made to determine their spread in the environment of honeybees. Samples were collected from flowers and nectar. Many species turned out to be the same as isolated from the intestines of honey bees. Similar samples were taken in hives frames, combs, and walls. A general distribution in the intestine of a honey bee was 1.0% yeast and 27.0% Gram-positive bacteria including *Bacillus, Bacteriolium, Streptococcus,* and *Clostridium*. The most abundant (70%) were Gram-negative or Gram-variable bacteria including *Achromobactor, Citrobacter, Ervinia, Echerichia coli,*

Flavobacterium, Klebsiella proteus, Helicobacterpylori, and *Pseudomonas*. The intestine of the adult honey bee is actually dominated by the presence of Gram-variable bacteria like various *Bacillus* species, *Enterobacteriaceae, Penicillium, Aspergillis,* and *Torulopsis*.

Other microbes in the atmosphere are fungi (both spore and non-spore forming) and numerous strains of yeasts involved in the process of fermenting sugary liquids and prebaking eatables. In other words, we and our foods are always exposed to contamination by microorganisms most of which are harmless but some of which cause disease. The worst part is that they produce very harmful by-products during growth on their hosts which are now termed *mycotoxins*. These are poisons and, until recently, their properties were largely unknown. It is easy to understand that their entry into foods and hosts is incidental and they have evolved to live in the atmosphere. Another quality of bacteria, yeast, and fungi is that some are evolved to evade extermination by forming spores with a protective layer to survive adverse conditions which would otherwise be lethal to them.

The entry of spore or non-spore forming microbes in exposed honey is accidental and this can be avoided. The remarkable quality of honey is that it has a unique property of killing microbes and can protect humans and animals when taken in diluted form. It is due to the protective elements that the medicinal value of honey is now being studied scientifically. In addition to its hygroscopity and oxidation components already mentioned, certain phytochemicals have also been isolated. Almost 50 plants have been identified from which honey bees collect nectar and other juices which impart curative value to honey. In this respect manuka honey is well known. Some other plants proven to have similar qualities in their sap extracts include *Eucalyptus globules, Acacia nilotica, Azadirachata indica, Prosops juliflora* and many others.

12.7.4 Medicinal Value of Honey

Both *in vivo* and *in vitro* experiments have indicated that honey has a lethal effect on bacteria, yeast, mould/fungi, and mycotoxins. Some of the pathogenic species are also sensitive. The degree of effectiveness depends upon the sources, region, and climate of nectar and honey. The active components are sugars, organic acids, hydrogen peroxide, phytochemicals and other compounds which are grouped together as flavonoids and phenols. Some studies with these chemicals have shown promising results.

Much has been published on the medicinal value of manuka honey of New Zealand. It contains high sugar and good acidity, hydrogen peroxide, and some active phytochemicals which are antiseptic. Many other plants have been identified which produce similar good quality honey including *Eucalyptus globules, Azadirachata indica,* and others. Some honeys have proven bactericidal effects on various *Salmonella* species, *Shigella, Escherichia coli, and Vibro cholera* and have been used therapeutically.

Common honeys have been used for centuries to heal wounds and cuts. Initially, there were some adverse reports of wounds becoming septic instead of healing after the application of honey. However, it has been shown more recently that spores of some resistant strains of bacteria were the cause since they moved to the periphery of the wound and prolonged the healing process. Therefore, thorough analysis of honey is now recommended prior to use in treating open wounds.

Food contamination is a health hazard throughout the world. Babies, children and the infirm suffer the most. Among a number of contaminants, fungal growths on food are the most persistent from season to season. The most well known disorders in human populations are caused by mould growths on food items consumed by humans and animals. The common disease gastroenteritis caused by aflatoxin (a mycotoxin) is produced during the growth of *Aspergillus niger*. Quite often this disease acquires epidemic proportions and children are the greatest sufferers. It has been proven experimentally that the growth of fungi and mycotoxins can be inhibited by adding 10% honey to baby food during its commercial production. In general for the protection of the general population, it has been recommended that all processed food should not contain more than 1μg/kg of aflatoxin.

In the last few decades, much interest has been taken by scientists in determining the role of non-peroxide antibacterial or antifungal substances of plant origin. Their role as antimicrobial agents is being determined since the life of hydrogen peroxide in honey is cut short by other chemicals. The phytochemicals of plant origin (flavonoids), organic acids, and complex phenols are currently under investigation in *in vitro* studies. One group screened 46 aqueous plant extracts for their antifungal potentiality against *Fusarium* species and found 12 of them more effective than blitox, captan, dithane, and thirum of comparable strengths. The synthetic fungicides and extracts of plants were compared on 8 species of *Fusarium*. It is important to note that a significant quantity of human food, particularly staple grains, can be contaminated with mycotoxins produced by *Aspergillus, Fusarium,* and *Penicillium*. As a result, much of that food is unfit for human consumption by scientific standards. Therefore, testing has to be developed for these mycotoxins at every stage from harvesting to storage. In the meantime, processed foods can be protected by the addition of agents like honey and phytochemicals. In the final analysis, it would be fair to say that the infection and damage caused to human food by fungi and bacteria are at least visible to the naked eye. However, the ailments caused by the invisible mycotoxins, which are poisons, are distressing and require greater attention throughout the world.

Aflatoxins in foods are produced during the growth of *Aspergillus niger* and other toxins like fumonisin B_1 are produced by *Fusarium moniliforme*. Chronic fungal, viral, and bacterial infection of healthy human cells can result in an elevated rate of apoptosis [programmed cell death] which is increasingly viewed as a key step in carcinogenesis. Moreover, certain mould metabolites or mycotoxins, like aflatoxins, can also cause apoptosis and necrosis in specific target organs associated with tumours. In mice, the oral administration of honey can reduce the histopathologic changes caused by aflatoxins, possibly due to its proliferative effect on probiotic colonic bacteria which help to detoxify mycotoxins. In another interesting study, aflatoxin B_1 toxicity was decreased in honeybees by propolis extracts which stimulated cytochrome P-450-mediated metabolic detoxification. Thus, propolis may act by enhancing the activity of P-450 enzymes involved in mycotoxin detoxification. P-450 enzymes are also involved in the detoxification of various drugs and carcinogens like aflatoxins in humans which are metabolized mainly in the liver by them. In addition, honey has been demonstrated to stimulate monocytes in cell culture to release cytokines like tumour necrosis factor [TNF] and interleukin-6 [IL-6] which are involved in the immune response to infection and can promote programmed death in surrounding infected tissue by stimulating apoptotic pathways in target cells. This phenomenon may provide a potential mechanism for the host to help rid itself of invading pathogens and may explain some of the antibacterial and antifungal properties of honey.

More recently, much interest has been taken by scientists in determining the role of non-peroxide antibacterial and antifungal chemicals in honey. These substances of plant origin are phytochemicals (flavonoids) or complex phenols and organic acids. Their independent activity was demonstrated experimentally by destroying the effect of hydrogen peroxide with the use of catalase in the laboratory.

Among these non-peroxide elements, flavonoids have been shown to display antifungal activity *in vitro* and they can induce apoptosis in cultured cells. The addition of polyphenols and certain phytochemicals can induce apoptosis in some types of cancer cells and can give protection against chemically-induced carcinogenesis. It has been shown in one study that aqueous extracts of *Acacia nilotica* and *Eucalyptus globules* exhibited antifungal activities against several *Fusarium* species and in equal strength proved to be more potent than the commercial synthetic fungicides. In another study, the essential oil of *Eucalyptus largiflorens* showed antifungal activity against yeast and mould including the mycotoxin-producing species *Aspergillus niger*. Some fractions of manuka oil appear to show varied activity in inhibiting the growth of moulds on panel products used in building materials and furnishings. In addition, manuka honey inhibits the growth of dermatophytes suggesting an antifungal activity. Therefore, it seems that some of these antifungal properties may be transmitted to the honey collected and produced from these plant species by honey bees.

Fumonisin B_1 is a mycotoxin produced by *Fusarium moniliforme* that has been associated epidemiologically with human esophageal cancer. It induces apoptosis in cultured human cells (keratinocytes) and in rat kidney and liver. Furthermore, studies have shown that fumonisin induces apoptosis via the Tumour Necrosis Factor Pathway and caspase activation.

Ochratoxin A is a mycotoxin produced by storage moulds such as *Aspergillus* and *Penicillium* and it has been linked in humans to Balkan endemic nephropathy and the associated urinary system tumours. It has also been demonstrated to induce apoptosis in cultured human proximal tubule cells and in hamster and rat kidney cells.

Aflatoxins B_1, B_2, G_1, and G_2 are produced *by Aspergillus flavus* and *Aspergillus parasiticus,* while aflatoxins M_1 and M_2 (which are found in milk and milk products) are the hydroxylation products of aflatoxin B_1 and B_2, respectively. However, aflatoxin B_1 is the most toxic and carcinogenic mycotoxin among these and it has been associated with liver cancer in humans. Epidemiological studies show a correlation between aflatoxin consumption and liver necrosis. Experimental studies with mice and rats have demonstrated that treatment with a diet contaminated with aflatoxins can cause hepatic necrosis. Moreover, the degenerative and necrotic changes in rat hepatocytes have also been associated with apoptosis.

In mice, the oral administration of honey can reduce these histopathologic changes caused by aflatoxins, possibly due to its proliferative effect on colonic bacteria which help to detoxify mycotoxins. This suggests that aflatoxin can cause an elevation in liver cell death while its negative effect can be lessened by the addition of mycotoxin detoxifying agents. In another study, the effect of this toxin was ameliorated in honeybees by propolis extracts. In addition, two compounds [benzyl caffeate and phenethyl caffeate] isolated from Chinese propolis were shown to have antiproliferative activity towards cancer cell lines, particularly those of cultured colon carcinoma. This suggests a potential anti-cancer role for specific components of bee propolis.

Traditionally, experimental carcinogenesis is a complex, multi-stage process including initiation, promotion, and malignant progression in which the failure of DNA repair mechanisms and the subsequent clonal expansion of damaged cells play a pivotal role. However, more recently, it has become apparent that the pathogenesis of cancer is closely connected with aberrantly regulated apoptotic cell death and the resulting deregulation of cell proliferation. Apoptosis is involved in the homeostasis of cell number in tissues and, although, increased cell proliferation is necessary, it is certainly not sufficient for cell transformation to take place. The increased proliferation in preneoplastic lesions is often accompanied by a parallel increase in apoptosis. Comparative studies with the rat bladder have suggested that apoptosis is closely linked to chemically-induced carcinogenesis. In fact, scientific animal studies have shown that simply increasing the basal frequency of apoptosis in murine skin cells can be linked to the development of squamous cell carcinomas in transgenic mice. Certain human cancer models like Ataxia-telangiectasia also indicate that an elevation in apoptotic activity in specific tissues may be linked to carcinogenesis. Thus, recent emphasis has shifted to inhibition or resistance to apoptosis as the pivotal determinant in cell transformation from the initial focus on increased proliferation and the associated risk of mutation as the major risk factor. In conclusion, there is now evidence to suggest that mycotoxin carcinogenesis may be modelled on this mechanism as well.

SELECT BIBLIOGRAPHY

Al Somal N, Cooley KE, Molan PC, Hancock BM (1994). Susceptibility of *Helicobacter pylosi* to the antibacterial activity of Manuka honey. Russian Med. J.; 87: 9-12.

Amla I, Murty VS, Jayaraj AP, Parpia HAB (1974). Afltatoxin B and Indian childhood cirrhosis – a Review. J. Trop. Pediatrics; 20: 28-33.

Brady NF, Molan PC, Harfoot CG (1996). The sensitivity of Dermatophytes to the antimicrobial activity of Manuka honey. Pharmacy Pharmacol. Comm.; 2: 471-473.

Crane E (1980). A Book of Honey. Oxford, Oxford University Press, 198 pp.

Dadant, CG Ed. (1992). Hive and the Honey Bee. Hamilton, Dadant and Sons, Ill. USA.

El-Arab Ezz et al (2006). Effect of dietary honey on intestinal microflora and toxicity of mycotoxins in mice. BMC Altern. Med., 6: 6 (Full Text 1-13).

El-Letty MA, El-Skabael KB (1992). Role of micro-organisms isolated from bees, its ripening and fermentation of honey. Egyptian J. Microbiol.; 75: 679-681.

Fotedor NR and Fotedor SN (1945). Lotus honey. Indian Bee J., 7: 102.

Galeotti E, Barjle E, Curir P, Dolci M; Lanzotta V (2008). Flavonoids from carnation (*Dianthus caryophyllus* and their antifungal activity). (Important flavonoids: quercetin, epicatechin) Phytochem. Lett.; 1:44.

Gilliam, M, Prest DB (1987). Microbiology of honey bee (*Apis mellifera*). J. Invert. Pathol.; 49: 70-75.

Haffejee IE, Moosa A (1985). Honey in the treatment of infantile gastroenteritis. British Med. J.; 290: 1866-67.

James SJ, Muskhelli L, Gaylor DW, Turturro A, Hart R (1998). Environ. Health Persp.; 106: 307-312.

Java Safaei-Ghomi and Atefeh Abbas Ahd (2000). Antimicrobial and antifungal properties of the essential oil and methanol extracts of *Eucalyptus largiflorens* and E *intertexta*. Pharmacosy Mag.; 6 (23): 172-175.

Lee E, Min H, Park H, Chung H, Kim S, Han Y, Lee S (2004). Life Sci.; 75: 2829-39.

Manriquez A (2010). Toxicity of mycotoxins to honey bees and its amelioration by propolis. Apidology.; 17: 52.

Marin A, Lopez de Cerain A, Hamilton E, Lewis AD, Martinez-Penuela JM, Idoate MA, Bello J (1997). NQ01-based activation as an anticancer strategy. British J. Cancer.; 76: 923-29.

Mizrahi A and Dorin R (1982). Antimicrobial effect of hive products. Israel J. Med. Sci.; 18:23.

Molan PC (1992). The antibacterial activity of honey; I the nature of antibacterial activity; II Variation in the potency of antibacterial activity. Department of Biological Sciences, University of Waikato, Private Bag 3015, Hamilton, New Zealand.

Molan PC and Russell KM (1988). Nonperoxide antibacterial activity in some New Zealand honeys. J. Apic. Rese.; 27: 62-67.

Niu G, Johanson RM, Berenlaum MR (2010). Toxicity of mycotoxins to honey bees and its amelioration by propolis, Apidologie; 10. 1051, pp 1-7.

Olaitan PB, Olufemi E, Adeleke and Lybo O Ola (2007). Honey: a reservoir for microorganisms and an inhibitory agent for microbes. African Health Sci.; 7: 159-165.

Petkova-Bocharova T, Chernozensky IN (1988). Ochratoxin A in human blood in relation to Balkan endemic nephropathy and urinary system tumours in Bulgaria; In Food Additives and Contaminants, A5: 299-301.

Radwan SS, El Essawy AA, Sarhan MM (1984). Experimental evidence for the occurrence in honey of specific substances active against micro-organisms. Zen. Fur Microbiol.; l 39: 245-255.

Safaei-Ghomi, J and Atefeh Abbasi Ahd (2010). Antimicrobial and antifungal properties of the essential oil and methanol extracts of *Eucalyptus largiflorens* and *Eucalyptus intertexta*. Pharmacogn Mag.; 6: 172-175.

Satish S, Raghavendra MP, Raveesha KA (2009). Anti-fungal potentiality of some plant extracts against *Fusarium sp*. Archi. Phytopathol. Plant Prot.; 42: 618-625

Schade E, Marsh GL, Eckert JE (1958). Diastase activity and hydroxy-methyl-furfural in honey and their usefulness in detecting heat alteration. Food Rese.;, 23: 446-63.

Siddoo-Atwal C, Rosin MP, Haas AL (1996). Elevation of interferon B-inducible proteins in Ataxia telangiectasia cells. Cancer Res.; 56: 443-447.

Siddoo-Atwal C (2009). At apoptosis and cancer: a viewpoint. Indian J. Ecol.; 36: 103-110.

Singh T (2010). Efficacy of essential oil extracts in inhibiting mould growth on panels. Scion Res. Private bag 3020; Rotonud, H.Z.

Snodgrass RE, Erickson ER (1992). Anatomy of the honey bee – from Hive and the Honey Bee: 103-169.

Subramanyam M (1991). Topical application of honey in treatment of burns. Br. J. Surg.; 78: 497-98.

Tolleson WH, Dooley KL, Sheldon WG, Thurman JD, Bucci TJ, Howard PC (1996). The mycotoxin fumonisin induces apoptosis in cultured human cells and in liver and kidney of rat. Adv. Exp. Med. Biol.; 392: 237-50.

Tyseet C and De Ratine de la Durand CJ (1991). Contribution to the study of intestinal microbial infection of healthy honey bees. Philadelphia Dep. Agri., Sea-AR, Eastern region Res. Centre pp 21-55.

Usia T, Banskota AH, Tezuka Y, Midorikawa K, Matsushige K, Kadota S (2002). Constituents of Chinese propolis and their antiproliferative activities. J. Nat. Prod.; 65: 673-676.

Wellfor TET, Edie T, Llewellyn GC (1978). Evaluating the inhibitory action of honey on fungal growth, sporulation and aflatoxin production. Zeitschrift-forschung, 166: 280-283.

Wigglesworth, VB (1984). The Principles of Insect Physiology. London, Methuen & Co.

Willex DJ, Molan PC, Harfoot CJ (1992). Comparison of the sensitivity of wound infecting species of bacteria to the antibacterial activity of Manuka honey. J. Appl. Bacteriol.; 73: 388-394.

Wild CP and Cong YY (2009). Mycotoxins and human disease: largely ignored global health issue. Caseinogenesis; 31: 71-82.

13 The Enemies and Diseases of Honeybees

A. HONEYBEE DISEASES

A honeybee colony suffers from various maladies. These maladies include contagious diseases, noncontagious diseases and malfunctions, pests, parasites, and predators. Whether an individual honeybee gets a disease or whether the colony is attacked as a whole the end result is the ill-health of the entire colony that affects its honey production capacity or even its survival.

The infections of contagious and noncontagious diseases are caused by living organisms that include viruses, rickettsiae, bacteria, protozoa and fungi. Not all pathogens are equally destructive to the bees. The viral, bacterial and protozoan diseases are comparatively more virile. Among the pests, parasites and predators, are such diverse groups of animals as, mites, spiders, wasps, hornets, dragon flies, other honeybees, wax moths, ants, termites, lizards, badgers, bears and men.

The noncontagious diseases have many origins. They include hormonal imbalance, nutritional deficiencies, masking of communication pheromones, poisons from plants through nectar and pollen, from pesticides, industrial wastes and affluents. Dysfunctions may be caused by genetic defects as in Africanized bees, or as a result of attack by diseases, pests, and predators. The ectoparasitic mites *Varroa jacobsoni,* and *Tropilaelaps* cause deformities in adults when they feed on bees in the pupal stage. The various maladies are discussed under relevant groups of pathogens and pests. Available literature has been reviewed by Bradbear (1988), Morse (1978), Bailey and Ball (1991), Abrol (1977), and others mentioned in bibliography at the end.

13.1 VIRUSES

The study of virus diseases in bees has been difficult and costly. The laboratories generally do not have a high resolution electron microscope, high speed centrifuge and other ancillary equipment needed to prepare concentrated pure viral suspensions required for photography and antisera studies. The characteristics used for bee viruses include, shape, size and symmetry of the particle; the presence of RNA or DNA in the particle; the location of the particles in the nucleus or the cytoplasm of the infected cell; the presence or absence of crystals or granules of protein in which the particle is embedded as seen by phase microscopy; and the serological reaction of virus suspension and its specific antiserum. Most of the bee viruses are RNA viruses with an icosahedral shape. The incidence and world distribution of honey bee viruses have been described by Allen and Ball (1996).

13.1.1 Sacbrood Virus (SBV)

The sacbrood virus of *Apis mellifera* bees was first identified in USA (White, 1917) but studies on the characteristics of the virus, its multiplication and spread have been made only in recent years (Bailey, 1975, 1982; Anderson and Gibbs, 1989; Dall, 1985). The disease is widely distributed throughout the world and is primarily that of the brood, although it is also found in the adult bees.

Apparently healthy larvae that have been sealed, instead of pupating die of the disease. The larvae remain stretched on their back with their heads towards the cell capping. They acquire sac-like appearance and the fluids that ooze out get dried up within the last larval skin. The colour changes from pearly white to pale yellow and a few days later to dark brown. The head and the thoracic regions darken first. The young larvae injest virus alongwith food and those that are 2 days old are the most susceptible. The food is provided by the nurse bees and therefore they are the vectors of the disease.

Multiplication and Spread

The sacbrood virus develops and multiplies in several body tissues of the young larva which appears normal from the outside. When larva is sealed in the cell for pupation, within 4 days, instead of developing further and transforming into prepupa, it remains confined within the larval skin and dies. A dead larva contains about one milligram of virus, which is sufficient to infect brood of 1000 colonies. In actual practice the disease does not become that virulent and late in summer it suddenly declines. One reason for this is that adult bees detect the diseased larvae at an early stage and remove them from the cells. Moreover, the dried remains of the dead larvae left over quickly lose infectivity.

The disease is endemic in the adult honeybees and virus multiplies in their body without causing a serious problem. The very young workers are most susceptible and they get infected when they remove the dead larvae. Some adults carry the virus from early age of development when infected in the larval stage at two days of age. The hypopharyngeal glands of infected adults degenerate and bees stop eating pollen; they also cease to feed the young larvae. The bees fly and forage but do not collect pollen and die at young age. However, before it happens they might have done much damage and spread the disease among the brood. The virus, whether endemic or ingested by the adults, reaches the hypopharyngeal glands where royal jelly is secreted. The pollen collected is mixed with the gland secretions and the virus is fed alongwith food.

It appears the infected bees, lose appetite for pollen; their metabolic rate is diminished and their lives are shortened by the same level as of the workers totally deprived of pollen. The diseased adults are easily chilled in winter and they die : the colony thus gets a relief. The colony tides over the winter and makes a new start in the spring. The drones do not eat pollen and remain unaffected by the virus.

13.1.2 Thai Sacbrood Virus (TSBV)

This virus is a strain of Sacbrood Virus (SBV) of *Apis mellifera* but has distinctive properties. It was isolated in a very virulent form on *Apis cerana* from Thailand (Bailey *et al.*, 1982). It is closely related to SBV and both viruses have the same dimensions (30 nm) and both are RNA type. The molecular weight of DNA is the same but that of three bands of proteins in TSBV are heavier, in the range 30200, 34000, 38,700 whereas that of SBV is in single band and ranges from 29000 to 31000. Proline content of diseased larvae is 2-3 times higher than that in healthy larvae (Singh and Singh, 1988).

The virulence of TSBV in *Apis cerana indica* is extremely serious The disease started in seventies in Thailand and spread along the Malay archipelago, the Burma, throughout Himalayas and, then across the Indo-Gangetic plains to South India and Ceylon. It had destroyed thousands of colonies by the end of eighties. Beekeeping with *Apis cerana* totally failed in these countries. It was surprising to note that *Apis mellifera* colonies lying in the same apiary were apparently unaffected and in fact beekeepers shifted to the introduced species for relief.

The cause of high susceptibility of *Apis cerana* honeybee to TSBV are not known so far. Kshirsagar *et al*. (1981, 1982, 1983) reported the occurrence of this disease in Bihar and Uttar Pradesh. The possible immune mechanism against the disease is being studied (Abrol, 1993).

13.1.3 Chronic Paralysis Disease of Adult Bees

This virus disease is sometimes known as the "hairless or black bee syndrome". The bees that suffer from this disease often loose hair and become shiny, although they may die of the disease even before loosing hair. A small number of colonies may die and the disease disappear as suddenly as it appeared. The virus in the filterate prepared from diseased bees transmits the disease in healthy ones (Kulnicevic *et al*., 1970). The infected bees die within one week. The virus particles are irregularly shaped and become visible with the aid of electron microscope. The particles are found lodged in cytoplasm of the nerve ganglia.

The enzyme ribonuclease found in normal commercial sugar destroys ribonucleic acid (RNA), the main constituent of chronic bee paralysis virus. Possibly the enzyme attacks naked virus in the gut of the bee. It is not expected to attack virus lodged in the tissues. Therefore, a regular availability of the enzyme in sugar syrup is expected to give protection to the bees.

Among honeybee colonies there seems to be some natural resistance against this disease. Selective breeding is expected to yield good results in raising disease resistant colonies. In general, the disease is more prevalent in the warmer areas. Colonies suffering from the disease early in spring recover with the flow of nectar.

The diseased bees have disjoined wings, carried in an unnatural manner. They exhibit trembling jerky movements and are often pushed around by healthy bees. The sick bees gather on top of frames or other parts of a hive or may crawl out of the hive to die in the front. In severe cases a large number of dead bees will collect in front of the enterance. Others who die inside are pushed out of the entrance and are thrown on the ground. A heap of dead or dying bees is formed in front of the hive. The disease has been recorded in USA, Canada, Mexico, UK, Germany, Switzerland, Ukraine, Russia, China, Australia. The chronic bee paralysis is to be distinguished from other allied diseases caused by viruses, particularly the sacbrood, and many a times the symptoms are similar.

13.1.4 Other Viruses

Bailey and Woods (1974) have isolated from diseased bees at least four viruses which cause various levels of paralysis. These are called Arkansas bee virus, "s" a "slow" bee paralysis virus and a "satellite" virus found in association with the chronic bee virus. Bailey (1975) also found an "iridescent" DNA virus of large size (160 nanometer) and he reported that a combination of sacbrood virus and "X" virus was highly lethal to bees.The iridescent virus is associated with clustering disease of *Apis cerana* in India distinct from the symptoms of accrine disease. The tissues in which it multiplies appear blue-violet or green when seen under bright white light (Morse and Flottum, 1997).

13.2 RICKETTSIAL DISEASES

Rickettsiae are bacteria like organisms that are intracellular parasites in living tissue. They are gram-negative and can be cultured in non-specific cells. For example, Poutiers *et al*. (1969), were able to grow *Rickettsia grylli* of crickets in bee embryonic tissue.

Normally the organism attacks fat body of the adult bee but haemolyph can also be affected in acute infection and it becomes milky owing to the presence of high number of organisms. Since these organisms are affected by antibiotics, the anti-gram-negative drugs are expected to be curative. The symptoms of disease caused by rickettsiae are similar to the paralysis in adult bees.

13.3 BACTERIAL DISEASES

The two most important diseases of young brood of honeybees are the European foulbrood and the American foulbrood caused by two different bacteria. Their name does not indicate the geographical distribution of the diseases. Although extensive surveys have not been made, yet from all field accounts it can be said with certainty, that the diseases are not present in India and Nepal either on *Apis cerana* or on *Apis mellifera*. There is always a danger of their accidental introduction and hence an account of the two diseases is given in some detail.

13.3.1 American Foulbrood

This disease of the honeybee larvae is caused by *Bacillus larvae*. It is the most dreaded brood disease in the world and is highly contagious. If it is not checked in an apiary it not only kills the colony and other colonies in the apiary but also spreads to other apiaries causing a full scale epidemic. The spores of the causative bacteria remain viable on beekeeping equipment indefinitely ; and may cause the disease after remaining dormant for years. The disease was observed in USA by numerous beekeepers for a long time but it was demonstrated conclusively by White (1907) that the causative agent was *Paenibacillus larvae (= Bacillus larvae).*

He cultured it, reinfected the healthy larvae and produced symptoms of the disease. Only the spore state of the *Bacillus* is capable of producing the disease. The bacteria is microaerophilic, gram positive and in the laboratory can be grown on a number of culture media. In a test the spores were shown to remain viable for 35 years (Haseman, 1961) and probably they can survive for hundreds of years.

Woodrow (1941, 1942) observed that the brood larvae one day after hatching could be infected and as they grew older the infectivity was reduced; larvae older than 53 hours in age could not be infected. Inside the larval body the spores germinate in one day and after the rod-like bacteria multiply in the midgut they penetrate the gut wall and reach body cavity. The host larvae ultimately die of septicemia in an upright position and sometimes are transformed into pupae in the cells which may be capped. Scattered brood and sunken cappings are typical symptoms of the disease. If a matchstick or a piece of wood splinter is dug into the dead mass and is pulled out, a thread like or ropy mass is drawn out upto 2.5 centimeters in length. Whereas a healthy larva is pearly white, the diseased one is dark brown and in a month or so it dries out to form brittle chocolate coloured scales, tightly sticking to the cell wall. In some cases the pupal tongue protrudes from the scale to the centre of the cell. The disease is found normally among the worker brood but occasionally the drone and queen larvae are also affected, the queen cells appearing normal from the outside. In advanced stage of putrifaction a characteristic odour comes off the frames; it resembles that of a decaying fish or a heated animal-glue (suresh). Collapsed cappings and patchiness of the brood is a characteristic sign of the disease. A number of sure laboratory tests of this disease are available. The Holst milk test (Holst 1946) is made by preparing one per cent powdered skim milk in water and placing 3-4 milliliters in a tube. A scale or smear from the mass of dead larvae is placed in the fluid and incubated at 37°C. The fluid becomes clear if disease spores are present in the smear. In another test gram stain or a simple stain is used for differentiating the bacteria by its morphology. The bacillus can also be identified biochemically; it reduces a nitrate to nitrite.

The transmission of the disease in an apiary is by various ways. The bees often store honey in the cells containing scales thus contaminating the honey with spores. When a diseased colony becomes weak the robber bees from other hives carry the contaminated honey to their colonies. The disease may be spread by the beekeeper inadvertently through interchange of combs from one hive to the other. At an early stage when the disease is not suspected in the apiary the super combs or brood combs might be given to healthy colonies unknowingly. Swarms from diseased colonies may issue and settle as feral colonies acting as source of infection, or the swarms carrying infected honey may be hived in good boxes. A beekeeper should, therefore, be fully familiar with the symptoms.

For controlling the disease internal quarantine is most effective by burning and destroying the infected bees, hive and all suspected equipment. The bees are killed by putting one or two table spoons of calcium cyanide from the entrance, or pouring a pint of petrol, securing shut the hive and burning it at a safe place inside a hole in the ground dug for this purpose. The ashes are then covered over with 18 inches of soil.

Some gram positive antibiotics are effective against the disease. The most commonly used chemicals are sulphathiazole and oxytetracycline. Genetic methods have also been proven of benefit. By a regular inspection of colonies in apiaries, and taking preventive and control measures the disease in certain areas like Florida has been reduced to one per cent of the colonies. A general survey in USA indicates that even with full precautions in apiaries 1-4% of the colonies remain infested. The bacteria are acquiring resistance against the commonly used chemicals (Shimanuki 1999).

13.3.2 European Foulbrood

The European foulbrood disease reported from almost every continent is not considered a serious disease and is seasonal in nature; colonies used for pollination are more susceptible to the disease. Aleksandrova (1949) was the first to prove that *Bacillus pluton* was the causal agent. She grew the organism in laboratory and reinfected the larvae with it. Bailey in 1956 renamed it *Streptococcus pluton*. It is a small bacteria which is non-spore forming and causes disease of the young larvae, less than 48 hours old. It appears late in the spring (May, June) in Europe and USA, when the colonies are building up strength. The infected colonies do not develop fast enough to gather surplus honey, therefore, honey yield in apiaries is reduced.

The brood larvae are infected by the consumption of contaminated brood food. The bacteria multiplies in the midgut and by the time the larva is five days old the midgut is full of microbes rather than the larval food. The bacteria at first destroy the peritrophic membrane and later on invade the intestinal epithilium (Tarr 1937a, b). The food demand of the larvae increases tremendously and consequently the nurse bees eject such larvae. In this process the colony infection is contained. The nurse bees are also instrumental in spreading the disease within a colony.

The diseased larva first turns yellow, then brown, and its tracheal system becomes visible. It dies in the coiled stage at the bottom of the cell. When the dead mass is tested for its ropiness, the thread is slight much smaller than 2.5 centimeter. Moreover, the remains in the cell are granular. Quite often the larvae die at a late stage and the cells are poorly capped mixed among the normal capped cells. The cells of diseased larvae have concave cappings with punctures. Sour odour emanates from the diseased cells. After some days the dead larvae dry out and form rubbery scales which can be removed more easily than those of the American foulbrood.

Generally the colonies cure themselves of the disease particularly if the nectar flow is good. In weak colonies the disease can be lethal. In good colonies, if there is a break in the brood cycle the nurse bees clean up the cells and disease is suppressed. If it is followed by re-queening with a young queen the colony recovers completely.

The disease can be controlled by feeding the bees oxytetracycline in sugar or by fumigation with ethylene oxide. Pollen from unknown sources should not be fed to bees.

13.3.3 Septicemia

It is a disease of the adult bees which is not serious but is distributed worldwide. Burnside (1928) identified the causal pathogen as *Bacillus apisepticus* Burnside. However, the bacteria were reclassified as *Pseudomonas apiseptica* Burnside. It is a gram-negative, non-spore-forming bacterium which is transmitted with the aid of moisture. Once infection takes place the bees die within 20-36 hours. The haemolymph of infected bees in colony varies from apple brown to chalky white. There is a rapid degeneration of the muscles and it is not possible to pick up whole a dead bee as the wings, legs, head, thorax, fall off. The affected bees in colonies appear restless, do not feed and are unable to fly.

The bacteria probably infect the bees through spiracles. The disease may be found mixed with nosema, acarine, and other abnormal conditions. To control the disease streptomycin has been used successfully but its frequent use is not recommended for fear of bacteria acquiring resistance.

13.3.4 Powdery Scale Disease

It is a bacterial disease of the honeybee larvae caused by *Bacillus pulvifaciens* (Katznelson, 1950). The becterium is gram-positive and produces spores; its maximum growth occurs at 45°C. The infected larvae die in their cells and the scales that result from the dead larvae are powdery, light brown to yellow in colour, and extend from base to the top of the cells. The scales can be distinguished from the brittle scales of the American foulbrood and rubbery scales of the European foulbrood. The cappings of the diseased larvae are punctured and collapsed as in the foulbroods. The disease is not serious and is not well studied.

The causes of brood diseases may be complex and in a given colony or an apiary there may be mixed infections caused by more than one pathogen. Under hot and humid conditions the brood may be affected by certain molds. Dead brood may be observed even when no pathogens can be isolated. Such maladies may be caused by chilling of the brood when colony strength is not great and the young queen is prolific. The brood may also be killed by chemical poisoning or by feeding pollen and nectar from poisonous plants.

13.4 PROTOZOAN DISEASES

The protozoa that are parasites of the honeybee belong to the subphylum Plasmodroma which move by pseudopodia or by flagella. The parasites of honeybee belong to three classes.

(*i*) Sporozoa, includes spore forming species that have not special means of locomotion; *e.g. Nosema apis,* and the gregarines.

(*ii*) Sarcodina are amoeba-like organisms and they move by pseudopodia *e.g. Malpighamoeba mellificae*.

(*iii*) Mastigophora are flagellates.

The protozoa are single celled animals which form cysts or spores to overcome dehydration and death. They multiply by sexual and asexual methods. The honeybees are infected by a number of species and not all have been identified. The infected bees do not show symptoms of any particular disease except that they lose vitality and have a short life. Some of the diseases caused by protozoa, however, are quite characteristic and they have been studied in detail. Unlike the viruses and bacteria that multiply fast and cause mortality, the protozoa are near perfect parasites and do not kill their hosts immediately ; rather they sap vitality, reduce fecundity, shorten life span, decrease movements and responses to stimuli.

13.4.1 Nosema Disease

Zander (1909) first identified *Nosema apis* and located the spores of this parasite in the epithelial cells of the ventriculus of adult honeybee. White in 1919 made extensive studies on the Norsema disease, found throughout the world, more virulent at some and less at other places but not a serious disorder of the bees. He established that the disease was infectious. Recent surveys and studies indicate that the disease is more serious than was thought to be earlier and may be present in more than 60 per cent of the colonies in USA.

The parasite is found in worker, drone and queen adult honeybees. Bees get infected when they eat food contaminated with spores of the protozoa. The spores germinate soon after entering the ventriculus (Bailey, 1955). The vegetative forms then enter the epithelial cells, increase in size, multiply in number and cause concurrent reduction in RNA synthesis in the host cells. Within a week these are full of spores which are shed into the lumen and are voided through the rectum. Some cells burst in the lumen and the spores attack more epithelial cells. The total number of spores estimated in a bee is 180 million. As a result of the disease the hypopharyngeal glands of the nurse bees become atropheid, the production of royal jelly in a colony is reduced and the wintered colonies having diseases dwindle in strength. Naturally the disease causes winter losses and initiates queen supersedure. The queens initially are infected in mating nuclei. In winter the infected workers defecate on the combs from where the infection spreads, further.

A beekeeper is not able to make microscopic examinations and has to rely on outward symptoms of the disease. The infected bees that are strong enough to make field trips for nectar and pollen often drop from fatigue on their return journey. An increasing number of bees collect in front of the hive upto some distance. The bees crawl up the blades of grass in an attempt to fly. Such disabled bees gather in depressions in the ground. The bees, may also be found on the bottom or on frames, in un-natural positions with their abdomens distended with fecaes. Many bees lose their body hair and become shiny. On dissection of the infected bees the mid-intestine is seen swollen and dull greyish-white in colour as it is full of spores.

For the control of nosema, a few general precautions in apiary and bee management are useful. Fresh running water should be made available and pools of stagnant water be drained off. Since queens get infected in mailing cages from the attendants their selection be done carefully from colonies free from the disease. Conditions of cold and humid in winter and hot and humid in summer be avoided. Upward ventilation of the hives is recommended in general.

A new drug "fumagillin" in concentrated syrup at a dose of 2 mg is highly effective in suppressing nosema disease. Hartwig and Prezelecka (1971) concluded that the drug inhibited DNA replication of the spores without affecting DNA replication in the host cells. The drug is also effective in preventing infection of queens in the mating nuclei and in the queen cages. Empty hives can be disinfected by ethylene oxide or by acetic acid fumigation at 120 ml per hive of the latter. After one week the hives can be aired.

13.4.2 Gregarine Disease

The gregarines, the large protozoa, have been observed in honeybees in Switzerland, Italy, Canada, France, USA, Venezuela. For the first time Morgenthaler (1926, 1930) observed it in Switzerland. A number of species of gregarines that normally infect other insects invade the hives (Stejskal, 1955). The honeybees probably ingest the spores while drinking contaminated water or while cleaning the hive. On germination of the spores, the gregarines develop and attach themselves to the ventriculus. The organisms have two distinct parts of the oblong body and the spores are formed in the larger nucleic part. Since the spores are killed on freezing the bee infections are generally prevalent in tropical and sub-tropical parts of the world. Their pathogenic status is not clear and according to Steinhaus (1949) they are nonpathological commensals in the gut lumen. It is observed that the ventriculus of infected bees becomes light coloured similar to that observed in Nosema disease. The affected colonies function abnormally and dwindle in strength.

13.4.3 Amoeba Disease

The amoeba, *Malpighamoeba mellificae* causes a disease of the adult worker bees (Prell, 1927). The bees get infected by ingesting the cysts which accumulate in the posterior end of the midgut (Bailey, 1963) or in the rectum. After germination the amoebae migrate to the malpighian tubules where they feed by pseudopodia between or within the epithelial cells. They multiply by binary fission and within 18-28 days of ingestion by the bee they form cysts. The cysts get loose from the swollen and glassy malpighian tubules, enter the intestine and are voided, alongwith the faecal matter. The disease has a spring peak in April and May. In certain cases infection is serious enough to kill the colony.

By a microscopic examination of the malpighian tubules the cysts, measuring 5-8 microns in diameter, can be seen through the walls of the infected tubules. In vegetative stage the amoebae can be observed by staining.

The disease weakens a colony showing such symptoms as "Spring dwindling", causing economic losses. If queens are confined with diseased workers in cages they can get infected. Normally the queens are free from this disease. The only way to control the disease is by proper hygiene and by disinfecting the equipment. The cysts are found in faecal matter on the frames and on the bottom board, which should be scraped off and disinfected with 2% carbolic acid. In the spring the bees can be transferred to boxes that have been fumigated for one week with acetic acid.

13.4.4 Flagellates

The flagellates are oval or elongate from 5 to 30 microns in length and have flagella on their body. These protozoa 6-12 days after emergence in honeybees are seen moving freely in the pyloric end of the intestine and in rectum. Soon they unite in rosettes adhering to the intestinal wall, and scales develop in epithelial lining at the point of attachment. It is not known what damage the parasites cause to bees or the colonies.

13.5 FUNGAL DISEASES

Generally fungi are saprophytes of bees and combs. However, two fungal diseases of the brood, namely chalkbrood and stonebrood are getting quite common.

13.5.1 Chalkbrood

This disease of the honeybee brood is caused by *Ascosphaera apis*. The disease has been recorded in France, Switzerland, Austria, USA, Newzealand. Another species, *Arrhenosphaera cranei* apparently causes chalkbrood in Venezuela.

The infected brood die and turn into whitish mummified mass full of mycelium of the fungus. The spores are formed within dark brownish-green fruiting bodies after the mycelia of the opposite sex come together. At that stage the mummified larvae turn dark grey or black. The endemic infection can be persistent and damaging, otherwise the disease is transient and is not considered serious.

The honeybee larvae when 3-4 days old are most susceptible to the ingestion of chalkbrood spores. Those on the periphery of brood area that are chilled and sealed become diseased readility. The spores germinate in the gut of the larvae and since the mycelia are aerobic they survive and multiply better in 3-4 day old larvae. Within two days of sealing of the cells the larvae die in upright position and mycilia break out of the hind end of the body (Bailey, 1967). The larvae remain white chalk like so long as mycilium of one strain (+ or –) infects larvae. When mycilia of both strains grow in a larva and fruiting bodies are formed the larval mummy becomes grey or black. Sometimes the mummies remain inside the sealed cells and are heard to rattle when the comb is shaken. The worker bees remove the mummified larvae which are found on the bottom board or the entrance. In an infected colony the spores remain viable on combs and they germinate under favourable conditions, even on the larval cuticle. The spores can remain viable for 15 years. The disease is carried with honey, therefore, giving food frames to healthy colonies from the diseased ones should be avoided. The colonies weakened from other diseases also became susceptible to the disease.

No control measures by the use of chemicals have been suggested. Since the bees remove mummies from the combs, there is a natural decline in the disease after some time. The mummies should be collected and burnt to remove the source of infection. Ventilation in hive should be improved.

13.5.2 Stonebrood

Both brood and adult bees are affected by the stonebrood disease caused by *Aspergillus flavis* Link and sometimes also by *Aspergillus fumigatus* Fresenius. The infected larvae and pupae when dead turn into hard, stone like mummies. The abdomens of the dead adult bees also become mummified. The disease has been recorded in Germany, Denmark, France, Scotland, USA, Venezuela.

Some soil dwelling pathogens that cause diseases in many insects also multiply in bees although it is doubtful whether they are primary pathogens. When the fungal spores are ingested they germinate in the alimentary canal and the mycelia attack all the soft tissues. The spores can also germinate on the cuticle and mycelia penetrate inside. Fungus develops rapidly and as it passes through the cuticle it forms a whitish-yellow ring or a collar behind the head. It then covers the entire cuticle and forms spores outside which fill the entire cells with greenish mass. The mummified bodies stick to the cell walls by aerial mycelia and it is difficult for the bees to remove them without breaking the cell walls.

The symptoms of the disease in adult bees include restlessness, feebleness and paralysis. The abdomen becomes dilated and then mummified. The spores appear near the head and then at the junction of thorax and abdomen. The younger bees die earlier than the older ones.

The disease is probably transmitted through contaminated honey and pollen, therefore, interchange of frames between colonies be avoided. Conditions of high humidity in hives should be minimized.

13.5.3 Pollen Mold

The whitish mold caused by *Bettsia alvei* (Betts) develops on pollen only and it does not affect the brood. It is sometimes mistaken as chalkbrood. Betts (1912a) first described it and later observed that it was similar to the chalkbrood fungus but its spores and fruiting bodies were smaller and they germinated at lower temperature. The fungus mold is white at first and with the formation of cysts it turns grey.

The fungus growth on pollen in hives is common in winter or early spring. The spores grow below 30°C and do not die in the summer heat. They are taken to the new hives by swarming bees.

The fungus dries in layers and is stuck fast in the cells making it difficult for the bees to clean the combs. The fungal growth in hives is thus persistent. A beekeeper has to clean the hives himself by soaking them in water for 24 hours and cleaning out the cells with a needle or a forcep.

13.6 MITES AND INSECTS

A number of mites have been recorded as parasites on the brood and adults of honeybees. The mites live both as ectoparasites on the body and as endoparasites in the tracheae, causing diseases and ailments. Insects live as pests in hives and some of the large ones like dragon flies eat bees as their prey. Work on mite pests is reviewed by DeJong *et al.* (1982) and by other authors referred to earlier. The wingless flies *Braula* spp. are widespread in the world, and are commonly known as bee lice.

13.6.1 Acarine Disease

The 'Isle of Wight' or the Acarine disease of adult honeybees is caused by a microscopic mite that attacks the prothoracic tracheae by entering through the spiracles. Rennie discovered in 1920 that the cause of disease was the parasitic mite, now named *Acarapis woodi* Rennie (Rennie *et al.*, 1921). No less than four *Acarapis* mites have been recorded on the body of honeybees; these are, *Acarapis dorsalis* that lives in the groove of the scutellum on the thorax, *A. externus* living in the ventral side of the neck, *A. vagans* on the face of the anterior border of abdomen and *A. woodie* that lives inside the prothoracic tracheae and is an endoparasitic mite (Morgenthaler, 1930).

The disease can be very serious and may cause dwindling and ultimate death of the colony. It has been recorded in most of western Europe, India, Argentina, Uruguay, Brazil and more recently in Mexico, USA and Canada, Belgian Congo. The outward symptoms of the disease include, 'K' winged condition when the disjoined wings are held at unusual angles, bees are unable to fly and are seen crawling in front of the hive, and the abdomen is swollen with dysentry. The confirmation is made by examining the diseased tracheae under a binocular microscope. For making this study, a bee to be examined is pinned on a wax base with ventral side upwards. With the help of forceps the head and forelegs are removed and the prothoracic segment is teased loose. The silvery white tracheae in the normal bees are exposed free, whereas in the diseased bees the large tracheae are seen congested and are dark brown or black in colour. On further teasing of the diseased tracheae with a needle various developmental stages of the mite are made loose on a glass slide and are distinguished from the entangled mass of the coagulated haemolymph.

Studies on the biology of the mite have been made in Europe, Britain and India (Rennie *et al.*, 1921; Morgenthaler, 1930; Atwal, 1967, 1971; Atwal and Sharma, 1970; Singh and Adlakha, 1958, 1960). Variations in the biological observations have been noticed at various places. Works have been reviewed by Hirschfelder (1952), Bailey (1961), Kshirsagar *et al.* (1980). The mite is generally found in the prothoracic tracheae, unilaterally or bilaterally and, in advanced stage of the disease, in all the thoracic tracheae or even in the head air sacs but not in abdomen. All stages of development of the mite are found together indicating early deposition of eggs by fertilized female mites on entry. Within one day of emergence of the bee female mites have been observed and generally bees older than 12 days are not infested. The mite eggs hatch in about 5 days and the nymphs develop into adults mites in 9 days. There are the larval, protonymph, deutonymph and the nymphal stages. The adult mites mate inside the body of the bee and the fertilized females move out. The male mites mature in 11-12 days whereas females take 14-15 days. When the diseased bee reaches an advanced stage of damage mites come out and infest other younger bees, although bees as old as 2 weeks may sometimes be infested. The mites on reaching thorax of another bee are attached to the wing roots from where they enter the tracheae. In Europe the disease is prominent in the spring after long winter confinement, but in India it is more common during the hot and humid summer, particularly after their confinement during the monsoon rains (Dinabandhoo and Dogra, 1980). As far as possible a good strength of the colonies be maintained to offset the disadvantages of dying bees.

Control. For the control of the mite without harming bees, the chemicals that are effective include, menthol, methyl salicylate, cholorobenzilate and formic acid, the last two being the most effective. Methyl salicylate is placed on a cotton wool pad for its slow fumigation action. For convenience the pad may be placed in a flat tin having perforations in its lid. Nitrobenzine and methyl salicylate in equal quantities can be applied when there is no brood in the colony as the former is harmful to the brood.

Smoke fumigation with bromopropylate or chlorobenzilate (Folbex) is the most effective in killing the mites without causing damage to the bees. Two or three treatments with 350-450 mg given at 7-10 days interval give complete cure. To prepare the paper strips strong blotting paper is soaked in 5 per cent potassium nitrate solution. After drying, it is soaked in 15 per cent solution of chlorobenzilate in benzene, and is dried again. A strip 11 × 4 cm contains the approximate dose. It is convenient to make a number of narrow strips and to light them in various parts of the hive in the evening and to keep the entrance shut for 20-30 minutes. The smoldering strips should not touch the frame lest the burning stops. The strips can also be made to smolder at the entrance. In summer season the hive should not be closed for too long (Atwal, 1967, 1971; Atwal & Sharma, 1970c).

Formic acid and menthol are applied on a swab of cotton at the end of a iron wire which is held in position inside the entrance. The treatment is repeated at weekly interval 3-4 times. In India formic acid treatment is recommended during the summer season as a routine, in order to keep colonies free of the widespread disease (Sharma *et al.*, 1982). Light sulphur dusting on top of the frame for the control of ectoparasitic mite *Tropilaelaps clareae* also keeps acarine under check.

13.6.2 *Tropilaelaps clareae* Delfinado and Baker

The ectoparasitic mite *Tropilaelaps clareae* (Laelaptidae; Acarina) was identified as a new genus from the specimens of mites obtained from among dead *Apis mellifera* bees from the Philippines (Delfinado and Baker, 1961, 1982; Atwal, 1969).

This parasitic mite is very active and is shiny rust red in colour. It is 0.95 mm long and 0.55 mm wide and both the sexes seem to look alike. The dorsal and ventral surfaces of their body are setaceous and in both sexes the labrum is fringed. Their palpal claw is simple and the holoventral plate in the male is separated from the anal plate and in the female the sternal, metastarnal and genital setae are long and subequal.

The mite was recorded on live honeybees, *Apis mellifera* adults, brood and on combs, in the Philippines, Hong Kong, Malaysia and Java (Michael, 1962; Delfinado, 1963), in Punjab (India) (Atwal, 1969). Morse (1978) reported that the mite was a serious pest problem in the Philippines and Stephen (1968) reported the same from Vietnam and India. On *Apis dorsata* the mite was first observed in the wild colonies in Philippines (Laigo and Morse, 1968) and it was also recorded on this bee in India (Bhardwaj, 1968). On *Apis cerana* the mite has not been reported as a serious problem from anywhere. It is now believed that the mite is probably a native of South-East Asia and the Indian subcontinent. Since *Apis mellifera* is a new introduction to this part of the world the mite seems to have a longer biological association with *Apis dorsata* which is a wild indigenous species in this region. It has also been recorded on *Apis cerana* (Delfinado and Baker, 1982).

It was observed at an early stage in the Philippines that the active mites and grown up nymphs attach themselves to the *Apis mellifera* honeybees as well as to the larvae and pupae (Michael, 1962; Delfinado, 1963). It was noticed that the pupae attacked by the mite when emerged as adults had damaged wings. The mites were found to be very active but could be easily picked up with a wet camel hair brush (Atwal & Goyal, 1971).

Apis mellifera was introduced in India in a small way at Nagrota in Himachal Pradesh in 1962-64 and six colonies were later on brought to Ludhiana, Punjab (India) in 1966. In the Himalayas *T. clareae* was not observed and in earlier stages even at Ludhiana in the plains it escaped notice. However, in 1970 a severe attack of this mite was observed and preliminary studies on its biology and the method of its control were conducted (Atwal and Goyal 1971). The immature stages and adult mites were observed on honeybee larvae and pupae in the brood cells. They fed on larval and pupal brood and were also seen entrapped in sealed cells. Although the eggs were not observed the larvae protonymphs, deutonymphs and adults were seen in the brood cells (Hoppe *et al.*, 1989).

As a result of the mite attack of brood at an early stage, the larvae died and lost their shine and showed irregular body outline. The immature stages of mite were whitish, flabby and sedantry at first and then they turned smoky and brownish. They killed the larval brood and anterior end of some dead larvae was seen protruding out of the cells. In heavily diseased colonies 50% of the larvae died. Since young developmental stages of mite were seen on old larvae and pupae it appeared the mite was quite prolific and laid eggs (or larvae) even on the developed brood. The brood attacked at a later stage completed development and the bees sealed the cells. The pupae taken out of such cells harboured numerous mite nymphs in various stages of development and 36 were counted on one pupa. As the emerging adults unsealed the cells the nymphs and adult mites were released and they were seen roaming over the combs entering and coming out of cells actively. Some pupae failed to emerge and died in cells. The adults that emerged showed various body deformities such as distorted abdomen, short and stubby wings, and deformed or missing legs. The brood cells in general showed many perforations. The deformed bees were evicted out of the colony and were seen crawling about on the ground in front of the hive. Unless carefully examined they could be mistaken as "crawlers" suffering from acarine disease (Atwal and Goyal, 1971; Delifinado *et al.*, 1985; Akratanakul, 1987; Burgett *et al.,* 1983).

The severity of mite infestation varies in various colonies and those having lighter attack maintain the colony strength but in case of a severe attack in a colony bees move around lethargically and then abscond, apparently carrying with them some of the mites who stick on to the bees. In a survey made around Ludhiana it was observed that 25 to 50 per cent of the colonies in apiaries carried the infestation. In most of Punjab (India) the mite is noticed as a serious pest of bees and regular control measures have to be taken.

Light sulphur dust passed through a double muslin cloth on frame tops keeps the mite under control. The operation should be repeated two or three times at fortnightly intervals. Formic acid treatment is equally effective (Garg *et al.*, 1984) and perhaps more convenient because it is also controls the Varroa mite and acarine disease of honeybees. Formic acid is diluted in water to 65%. strength and 20 ml. of the fluid is poured on an absorbant cardboard piece (30 × 20 cm) that is placed at the back of the bottom board. The entrance is opened to its full size. Almost complete mortality of all stages of the mite is observed within 24 hours. Formic acid can also be dispensed on cotton wool wrapped at the tip of an iron wire which is pushed inside through the entrance and the other end is bent outwards. At low temperatures 3-4 applications at 10-day intervals will control all the mite parasites (also see Woyke, 1984). For the control of this disease in autumn and winter menthol is recommended. Fine crystals are mixed with hydrogenated vegetable oil and the liquid is spread on a corrugated card board which is placed in the bottom board for a slow release of menthol.

13.6.3 *Varroa jacobsoni* Oudemans

Varroa Jacobsoni Oudemans was described in 1904 and the name, *Myrmozercon reidi* Gunther, is accepted as a synonym. It is considered the most widely distributed Asian mite. It has been reported on *A. mellifera* from Indonesia, Singapore, the Phillippines, Hong Kong, Russia, China, India, Pakistan, Vietnam, Japan, Bulgaria, South Korea, Thailand, Paraguay, and now in a greater part of Europe, North Africa and North and South America. It has also been reported on *A. cerana* from Indonesia, Singapore and Thailand. The mite has spread from Asia to Europe overland through the migration of *A. mellifera* bees and to South America from Japan. Originally the mite was recorded on *Apis cerana* (Pandey, 1967; Delfinado and Baker, 1974) but with the introduction of *A. mellifera* into Asia, the mite adapted to the new species (Ritter, 1981; Kumar *et al.*, 1988). Maps of its world distribution have been prepared by Nixon (1983) and Graffeths and Bowman (1981). It is even suspected that mites on *A.c.* and *A.m.* may be two different strains.

Its biology on *A.m.* was studied by Akratanakul (1987) and by Delfinado and Baker (1984). The mite is brown, leathery in appearance and is 1.05-1.09 mm long and 1.5 to 1.58 mm wide. It is the size of a common pin head and is observed attached to the soft intersegmental sclerites on the abdomen and on thorax. The life cycle is started by a fertilized female living on adult bee and entering an uncapped cell containing 5 day old worker or drone larva. She submerges herself in the leftover royal jelly and feeds on it, keeping her ventral side towards the cell opening. When the cell is capped and the royal jelly has been consumed, the mite is free to feed on the developing larva and prepupa. She lays her first eggs 60 hours after the cell is capped and more eggs at various intervals of one day or so. On hatching, the larva, protonymph, and deutonumph grow into adults in 5.5 days and into females in 7.5 days. A capped cell contains immature stages of mite as well as full grown male and female mites. Mating takes place inside the cell apparently causing lot of inbreeding. More than one mated female has been located in a cell. As a result of the parasitic attack on bee pupa deformities appear in the adult bees. When the adult emerges from the cell the female fertilized mites attach themselves to the body of the emerging bee. The male mites and the immature stages left behind in the empty cell ultimately die. In the colony when bees come in close contact the mites are transferred from one to the other. Feeding on adult workers and drones gives the mites nourishment for the maturity of eggs. One mite parasitizing on a bee reduces its life to half. It is not known whether the mother mite goes back to an adult bee for another bite or she simply enters another cell to raise a second generation of progeny (Shimanuki *et. al.*, 1992).

Varroa is a serious parasite on *A. mellifera* and it causes depletion of strength or even death of the colony. The *Apis cerana* colonies on the other hand are more resistant and the mite is rarely the cause of death of a colony. The fast spread of *Varroa* in *A. mellifera* colonies is due to drifting of workers, brood exchange between colonies, shipping of bees through mail and migratory beekeeping. It is interesting that in Punjab (India) and the surrounding states where *A. mellifera* is in the hands of the beekeepers since mid 1970 this mite is not as serious a problem as the active mite *Tropilaelaps clareae*. The reason might be that light dusting of sulphur universally applied for the control of the latter disease also checks the comparatively less agile former species.

For the control of *Varroa* disease diagnosis is important. One method to collect and identify the mite is to place 200-300 bees in a plastic bag or vial and kill the bees by pouring inside ether or ethanol. As the bees die, mites detach from their body and fall into the liquid from where they can be picked, placed on glass slide and examined under magnification. The mites can also be collected from a hive without killing the bees. A sheet of paper or a cardboard having a circle of grease on it is place on the bottom board. Apistan under pressure, is sprayed from the top of frames or a smoker containing tobacco is used to fume the bees. The mites detach themselves and fall on the sheat of paper from where these can be picked up for counting or studying under a binocular microscope. This mite on *Apis mellifera* has been studied extensively in Europe particularly in Germany where the density of colonies is high and infestation is extensive.

In Europe more than ten chemicals have been found effective and have been recommended for its control. In USA on the other hand public health laws are very strict and so far they have not come up with an exhaustive list of safe chemicals to be used. It is recommended that some characteristic bee behaviour may be used for breeding resistant strains. It has been observed, for example, that *Apis cerana* is good at grooming and gets rid of the mite from its body; the mite is caught in legs and crushed with the mandibles. The physical removal of the mites probably gives *Apis cerana* the power of resistance.

For the control of ectoparasitic mites in India the standard practice remains the use of sulphur and formic acid as described above. In Europe synthetic pyrethroids, Fluvalinate (Apistan), Flumethrin, are in common use; their fumes are released from impregnated strips of plywood or plastic. High mortality of mites is achieved within 6-8 weeks and the treatment is repeated once or twice in a season. Bromopropylate as a fumigant in impregnated paper strips (Folbex) is released as a smoke in empty but enclosed space above the brood chambers. It is effective against all ecto-and endo-parasitic mites (Delaplane, 1997; Atwal, 1967, 1971; Garg *et al.*, 1984). Coumaphos and flumethrin spread on strips of thick paper or plastic placed on bottom board are also recommended. As a regular practice sticky boards covered with a fine plastic screen are used for entrapping mites that fall from above on the bottom board (Delaplane, 1999). In U.S.A., Europe, Australia and Newzealand the infested colonies are effectively burnt to keep the mite population as low as possible.

B. PESTS AND PREDATORS

13.6.4 Insect Pests and Predators

A large number of insects act as enemies of the honeybee, but their attack is not serious in all cases. Some, like the wax moths, are extremely damaging and can cause absconding and death of a colony. Others like praying mantids and dragon flies merely fit in the food chain, forming honey bees as part of their diet. Wasps and hornets, near relatives of the honeybees, actively prey upon them and also rob them of their brood, depleting colony strength to such an extent that beekeeping cannot be practiced in the area of their abundance. The total number of insect species having association with honeybees is very large and their records in India are incomplete, therefore, only those insects will be discussed here that play a significant role in the practice of beekeeping. For a fuller account reader has to refer to Imms General Textbook of Entomology and the original papers.

The important lepidopterous pests include; the greater Wax moth, *Galleria mellonella* (L.) ; the lesser Wax moth, *Achroia grisella* (Fabricius); the Indian Meal-Moth, *Plodia interpunctella* (Hubner); the Mediterranean Flour Moth, *Anagasta kuehniella* (Zeller), the dried fruit moth, *Vitula edmandsae* (Ragonot); the Deaths Head Moth, *Acherontia* spp. A number of other moths live in association with bees in hives and not all of them have been identified (Williams, 1978).

13.6.5 The Greater Wax Moth

It is the most important pest of honeybee products and causes serious losses in commercial beekeeping. The larvae destroy raised combs in storage as well as in hives by tunnelling through near the midrib of a comb. They feed on pollen, wax and protein of the pupal skins. The insects breed on combs and in tropical and subtropical areas they often cause death of the colony. Observations on the seasonal infestation in *A.m.* epidemics in Punjab showed that it was most devastating during and after the monsoons when colony strength is low due to shortage of bee flora (Brar *et al.*, 1985).

Moths appear in the spring from over wintered larvae and pupae found in silver white cocoons sticking to the inner board and the top cover or on top of the frames and the cloth cover (**Plate 11**). The moths are 20 mm or 3/4 inch long and have a wing expanse of 1.25 inches. The forewings are mottled brown to ashy grey and the hind wings are creamy white. The males have two short palps in front of the head and their forewings are fringed on the outer margin. They hide in dark places and become active at dusk and are nocturnal in habit. They mate within 2-3 days of emergence and the females start laying eggs soon after. Within a period of 15 days they lay 400 - 1800 eggs in cracks or crevices of the hive or even on the combs. The eggs are creamy white and hatch in 5-8 days. The young larvae are creamy white and are active. They soon burrow into combs in search of food and make their way in towards midrib of the comb, making silk lined tunnels. They feed on honey, wax, pollen and pupal skins found in the cells. In warm climates they are full grown in 4 weeks, are one inch in length, and acquire grey to dark grey colour on the dorsal and lateral surfaces of the body (Beck, 1960).

In cold climates larvae grow slowly and undergo hibernation for 3-4 months in the prepupal stage and are found inside silken cocoons with an opening in the front, through which the larvae can be teased out. They are found in the spring and early summer alongwith the pupae that are formed at the end of hibernation. The cocoons are generally found in groups, often inseparably touching one another. The pupal stage lasts 7 days. In warm climates more than one generations are completed in a year and the pest multiplies very fast. In case of *Apis cerana,* it is the main cause of colony dwindling in the post-monsoon season.

The damage caused to combs is typical. In early stages ramifications of galleries are seen when a frame is held against light. In the late stage the comb looks like a tangled mass of silk galleries, frass, and hardly a sign of cells. All attacked combs have to be destroyed, and the cocoons scraped off with hive tool and killed.

The wax moth is a pest of all honeybees in India (Singh, 1940, 1962) *viz. Apis cerana, Apis dorsata, Apis florea;* the latter two being wild are a constant source of infestation for the former. Under conditions of severe attack during the monsoon season colonies may abscond (Brar *et al.*, 1985). The moth is a common resident of active *Apis dorsata* colonies in the Philippines (Morse, 1975, 1978). In India the pest also attacks newly introduced *Apis mellifera* but the bee being strong in colony strength does not suffer as much as Apis cerana. Even so *A. mellifera* colonies neglected during the monsoon may be damaged or killed and hundreds of cocoons may be found in the hive.

The wax moth larvae are reared artificially in Europe and USA as fish bait and for laboratory experiments. A race has been developed in which the grown up larva is white and is preferred for fishing.

For the control of wax moth the colony health and good food reserves are important. The bottom board should be cleaned periodically. Light dusting of *Bacillus thuringiensis* spores on combs is effective. The most effective method of control is to fumigate the infested frames in storage with one of the recommended poisonous gases. In an average size apiary it is practicable to store raised frames in 4-5 ft. tall polyethylene bags of thick plastic sheet and to put 1/4 tablet of aluminum phosphide, before tightly closing the open end with a string. Since the gas generated is highly poisonous the storage room should be sway from the main house. Some of the new methods being tried for control are the sex pheromone attractants and bacterial insecticides (Battu & Singh, 1977 ; Shimanaki, 1981).

13.6.6 The Lesser Wax Moth

The larvae and adults of this wax moth are smaller in body size than those of the greater wax moth and it is generally less common and abundant. The adult *Achroia grisella* are slender bodied and silvery-grey to buff in colour with conspicuous yellow head (**Plate 11**). The male moth is 10 mm and the female is 13 mm and they weigh 1/6 - 1/10 th of their counter parts of Greater Wax Moth. When disturbed the moth rapidly run over the combs for a few seconds and fly down over the edge to the ground. The larvae have a narrow white body with a brown head and pronotal shield, and they reach the maximum length of 20 mm. According to Singh (1962) the larvae tend to live isolated in silken tunnels covered with frass and when full grown they tend to congregate. Quite often the larvae are seen on the bottom board feeding on debris, but their natural diet is met in dark combs containing pollen or brood. Both the species of moths may be found together in a colony.

13.6.7 Indian Meal Moth

Plodia interpunctella (Hubner) is a serious pest of cereals and other products. The larvae also feed on dry-fruits, nuts, dried milk and on bee collected pollen meant for commercial sale. The pest is well spread throughout the world.

The moth is 9 mm in length and its head, thorax and distal half of forewings is redish brown. The larvae are white and spin webbings while feeding on pollen and dead brood in stored combs. Before pupation they migrate to other objects or reach cracks and crevices in the ground.

13.6.8 The Mediterranean Flour Moth

Ephestia (= *Anagasta*) *kuehniella* has been widely studied as a pest of flour in granaries and flour mills where the developing larvae make webbings and clamp together the material on which they feed. It also reaches stored honeybee combs and lays eggs on them. The pinkish white larvae feed on pollen in cells, filling them with frass and loose silken webbs. The larvae spin silken cocoons and pupate in the cells. Sometimes the larvae bore through the cell walls and reach the midrib, thus spoiling the stored combs. In temperate climate the larvae of this and other moths can be frozen to death by exposing them to sub-zero temperatures (Atwal, 1960).

13.6.9 Wasps as Predators of Honeybees (Vespidae)

Kshirsagar and Mahindre (1975) gave the distribution of various predatory wasps recorded near experimental apiaries in different parts of India. The species identified are mentioned here.

1. *Vespa auraria* Smith, the golden wasp, was recorded in Kashmir valley.
2. *Vespa velutina* Smith was also recorded in Kashmir valley.
3. *Vespa magnifica* Smith, the giant wasp, is found throughout the Himalayas and in North-Eastern India.
4. *Vespa basalis* Smith, the black wasp, is distributed throughout the Himalayan region and also in North-Eastern India.
5. *Vespa ducalis* Smith is found throughout the Himalayas and in the North-Eastern region of India.
6. *Vespa cincta* F., the yellow banded wasp, is distributed all over India.
7. *Vespa orientalis* Smith, the Yellow-Banded Oriental Hornet is distributed from the Middle-East to the East coast of India. It is a common sight at the fruit and sweet meat shops in the country.

8. *Vespa tropica* L., the Yellow-Banded Black Wasp, is widely distributed in India.
9. *Lianthrena kohlii* Bingh is common in Punjab and Bihar, alongwith other widely distributed species.

Wasps and their control were also studied at other places in India (Dave, 1943; Gandheker, 1959; Sharma *et al.,* 1979; Subbiah and Mahadevan, 1957). The wasps are most active in early hours of the day from 6-8 am but continue predation all day. Their harmful activity continues through the summer months but they are most devastating in the post-monsoon months of August and September (Brar *et al.,* 1985). The predatory wasps catch the incoming loaded bees near the entrance. Generally they cannot be prevented by the guard bees but sometimes they are resisted en masse and bees ball the intruders. Various species of wasps have different habits. The yellow banded wasps alight near the entrance and catch the incoming bees. Sometimes they venture inside the hive where they are killed by the bees and many dead bodies are seen on the bottom board. There is no easy way to control the wasps. Swatting, poison baiting, trapping, nest destruction are the methods suggested for controlling them. None of these are easily practicable but good wasp traps can be used with some success. For further studies on the biology and habits of wasps see (Evans and Jane, 1970; Spacznska, 1975; Sharma *et al.,* 1979).

13.6.10 Black Ants

They belong to the superfamily Formicoidea (Hym.) and all are placed in a single family Formicidae. There are numerous species in India and their size ranges from tiny slow moving creatures to large long-legged fast moving species. Practically all ants are fond of feeding on honey, brood and wax from honeybee colonies. In order to control their attack the most effective and practical method is to dip the legs of hive stands in cups full of water. The ants are more troublesome in the subtropical and tropical areas than in the temperate region.

13.7 SPIDERS AS ENEMIES

The knowledge of spider fauna in India and South Asia is very sketchy and the role of various spider species in killing the honeybees has been studied very little. It is well known that the common genera of spiders have a universal distribution and the living habits of the important groups are similar to those studied in Europe and USA. Langstroth believed that the common domestic spider *Theridion* that makes nests inside the buildings plays a useful role in killing moth worms of combs in storage. However, the orb weavers, crab spiders, and the funnel web spiders are known to catch or ensnare honeybees and eat them up. There is a difference of opinion on the amount of damage caused to the colonies. On the whole spiders are believed to play a minor role as enemies of honeybees.

13.7.1 Orb Weavers

Agriopidae, or the orb weavers, make large beautiful webs in gardens, and not infrequently in the apiaries. The spiders are quite large and have black and yellow or silvery and grey markings on their body. Many drones and honeybee workers are caught in webs and the spiders suck their body juices. Some authors believe they prefer honeybees as prey, but quite often large Diptera, flies and blow flies, are seen caught in their snares.

13.7.2 Crab Spiders

The crab spiders belong to the family Thomisidae and their body is short and broad, somewhat resembling a crab, Most of the time they move sideways and backwards and lie in a bush near flowers.

They do not make webs; and they catch flies and bees much bigger than their own size. The prey is quickly insensitized by a bite and its body juices are sucked.

The jumping spider *Salticus* can catch honeybee twice their size and hold onto them till the bee is dead. The spider may build a silken case as nest inside the hive covers. For keeping a check on them it is recommended to scrape off the nests periodically. The golden rod crab spiders feed actively on honeybees and prefer them over other prey.

13.7.3 Funnel-Web Spiders

The funnel-web spiders belong to the family Agelenidae. They make funnel shaped webs in neglected buildings and godowns. Some species prefer to make their nests in grass and scrub. These webs are very effective in ensnaring honeybees as prey.

13.8 TOADS, FROGS AND LIZARDS

Gandheker (1959) and Singh (1962) mentioned that toads and frogs quite frequently are enemies of bees. Genus Bufo of toads and Rana of frogs are wide-spread in India and are seen near the hives or water holes. The extent of damage to hives caused by them has not been studied. Eckert (1934) in Davis, California, examined faeces of toads and observed that on an average 23.3 honeybee heads were present in excerement.

At Chandigarh (India) the Speckled Garden Lizard was observed to make a tremendous feast of the honeybees. Upto eight immature lizards were counted on the cloth covering of the frames and hundreds of head capsules of the bees were seen litered on the cloth. It appeared the lizards went through holes in the cloth, brought bees from underneath and ate them at leisure. The adult lizards sat at the hive enterance and caught honeybees at ease. At least one of the colonies was eliminated and hardly 100 bees were left when the lizard was driven away. The wall lizard, Gecko sp., is active at night and snaps at stray honeybees who may have lost their way to the colonies.

13.9 BIRDS

Apart from the predatory flying insects the birds on wings are a formidable group of bee enemies in flight. For birds there is no discrimination between solitary bees and honeybees and those adapted to catch flies in flight can also come nearer to the hives for their meals. Some even discriminate between the type of bee they catch or can easily catch. It is reported that in eastern USA some flycatchers wait near the hive for incoming juicy honeybees full of nectar and pollen thus they pick a meal rich in carbohydrates in addition to the body proteins and vitamins. In Europe birds are a problem in the queen mating yards, but in Asia and Africa they cause serious damage to the bees, particularly during the dearth period when colony strength is already depleted. The Green Bee-Eaters (*Meropidae*) are quite devastating in apiaries. The honeyguides (*Indicatoridae*) in the Himalayan region actually guide large animals and man to the bee colonies in jungles so that they can have their own bite of honey and wax from the disturbed colonies.

The birds and their families of interest to beekeepers can be placed in two groups; the major predators and the minor predators. The major predators are placed in family Meropidae, Dicrusidae, and Indicatoridae. The minor or occasional predators are placed in families, Picidae (*Woodpeckers*), Paridae (Titmice), Laniidae (Shrikes), Apodidae (*Swifts*), Tyranidae (Flycatcher or King-bird). Muscicapidae (Swallow), Ploceidae (Subfamily Passerinae—House sparrow and yellow throated sparrow) (Friedman, 1955; Salim Ali, 1977, 1979; Fleming *et al.*, 1976; Fry, 1983, 1984).

13.9.1 The Bee-Eaters

It is believed that there are 24 species in the family Meropidae which are distributed throughout the temperate and tropical area of the world, from Azerbaijan to South Asia, Africa and Queensland. In Savannas of Africa there are 14 species and all of them eat bees. Almost all the bee-eaters prey upon bees and honeybees-range of the individuals is 4-96 per cent of diet (Fry 1983, 1984). Of all the insects caught by these birds 30 per cent constitute honeybees.

The various species and races of Bee-Eaters are fairly uniformly distributed in Central and South Asia, Africa and the East. *Merops supersiliosus* and *Merops apiaster* are found in Azerbaijan and *Merops ornatus* (the Rainbow bird) in Queensland, *Merops apidater* is common in Europe and Africa. The small Green Bee-Eater, *Merops orientalis* Latham and its races are widely distributed in India, Pakistan, Burma, Ceylon from 5000 ft in the Himalayas downwards. In local language the Bee-Eaters are called *Sohanchiri (beautiful sparrow) or Patringa (leaf-coloured).*

13.9.2 The Small Green Bee-Eater

It is a grass-green bird the size of a House Sparrow. There is a tinge of reddish brown on head and neck with a prominent black necklace. It has a long and slightly curved bill and the central pair of feathers are prolonged. The sexes are alike and they are seen in the open sitting in pairs or in groups on telegraph wires, fence-posts and other vantage points. The bird is distributed throughout India, Pakistan, Bangladesh, Ceylon, Burma, Thailand.

The Bee-Eater lives in the neighbourhood of cultivation, fallow land, gardens, golf links, jungle clearings, on sandy beaches or on raised land. When not sitting or roosting they fly around with motionless wings, gliding in the air, looking for bees and other insects. On snapping a prey they fly back gracefully with outstretched motion of wings to their open perch where the prey is killed and swallowed. This hunting game is played all day and is intensified towards the evening when flocks of birds descend near the apiaries. Before roosting in the favourite leafy trees they make much noise, a pleasant jingling-tit, tit or tree, tree, tree.

Their nesting season is February to May before the monsoons. The nests are made in embankments, sides of raised ground and principally in the soft sandy earth of Shivalik Hills and the Southern Ghats. A horizontal or oblique tunnel is dug by both the sexes and at the end is made a widened egg chamber. The clutch consists of 4-7 pure white, rounded oval eggs. Both the sexes take part in feeding the young.

The Bee-Eaters are resident or local migratory birds. In North India they breed in the Shivalik or smaller hills and hillocks. They seem to congregate in certain localities in the breeding season and are seen making nests in thousands, perched on the side of cut earth typical of eroded soil. In the coastal areas also they make nests in the ground. The favourite breeding areas are unsuitable for keeping bee colonies and the migratory beekeepers avoid such localities. Pathankot in Punjab is one of them.

The other allied species of Bee-Eaters have similar habits. They make, their nests on the hill-side, perch and fly around intermixed with small Green Bee-Eater and feed on solitary bees and honeybees on sunny days when the temperatures are moderate and the bees are most active.

13.9.3 The Bluetailed Bee-Eater

Merops philippinus Linnaeus is bigger than the Green Bee-Eater and is the size of a bulbul. It prefers wooded areas near streams and lakes. It has an overlapping range with the large Blue-Checked Bee-Eater. Its tail is bronze-green and the chestnut patch on breast is paler. Both these species are found throughout India; they are resident and local migratory.

13.9.4 The Chestnut-headed Bee-Eater

Merops leschenaulti Vieillot, is of the same size as the above. The sexes are similar and their head and upper back is bright chestnut; the chin and throat are yellow. They live in small flocks in wooded country and perch on exposed branches of trees. Their area of distribution is Western Ghats, Ceylon, Terai area near the Himalayas, upto 4000 ft elevation, Assam, Orissa, Bangladesh. The food and nesting habits are similar to those of the Green Bee-Eater.

13.9.5 Himalayan Honeyguide

Indicator ocanthonotus, is a scarce bird, found in the Himalayas ranging from 2000 to 10,000 ft above the mean sea level (Fleming *et al.*, 1976; Hussain & Ali, 1983). It is 15 cm long about the size and appearance of a house sparrow. It is greyish green in hue and has orange yellow markings on rump and has a yellow forehead. It is solitary or is sometimes seen in loose parties in remote gorges. It glides around like a Bee-Eater or a fly-catcher and frequently returns to its favourite perch, where it sites motionless for long periods. The sexes are similar in size but the female is duller in appearance. Their nesting and breeding habits are not known. They feed upside down on honey and wax from the exposed combs of *Apis dorssata and feral Apis cerana* F. When stomach contents were examined it was found crammed with bees wax.

The large animals and local men in Himalayas are guided by these birds to the hidden *Apis cerana* colonies in the rock hollows. Their areas of distribution include Pakistan and Indian Himalayas, Gharwal, Nepal, Sikkim, North Burma.

There are 11 known species of family Indicatoridae grouped in four genera. Except the two Asian species, the rest are found in Africa, South of the Sahara Desert. Apart from the Himalayan species others are distributed in Burma, Thailand, Sumatra, Borneo, Malaya. Friedman (1955) described the typical behaviour of a honeyguide that has developed symbiotic relationships with certain mammals like honey badgers, baboons and man. The bird makes characteristic sounds like that of a shaking match box and some body movements, flying around with fanning tail, leading a person to the nest site. The honey harvester opens the entrance for honey combs. During this operation the bird hides at some inconspicuous place nearby for a few minutes or even an hour and a half till the person is gone. It then feeds on the left over bits and pieces of the comb. The bird behaviour is quite elaborate so as to lead the mammals, the intelligent lower animals, and man, to an exact tree or the spot on the side of a rock where *Apis cerana* colony might be hidden. They can eat beewax from the combs of *Apis dorsata* colonies on their own without the help of a symbiont.

13.9.6 Woodpeckers

The woodpeckers belong to the family Picidae and they are world-wide, except that they are not found in the extreme north of Asia, Australia, Madagascar and Oceanic Islands. Most of the 208 species are migratory and are considered only a nuisance near the apiaries and can be easily scared away. The green woodpecker (*Picus viridis*) is quite a serious pest of bees in Greece, Russia and Europe during the winter season when no other food is available. It makes holes with its sharp bill near the hive entrance to reach the bees. The holes are further widened by mice and rats.

Three species of woodpeckers are widely distributed in India, Ceylon, Pakistan, Bangladesh and Burma; the Chestnut-Rufous Woodpecker, having black bars on wings is, *Micropternus brachyurus*; the Golden-Backed Woodpecker, *Dinopium benghalense* has golden-yellow upper plumage with margins of black and a red crest; and the Yellow Fronted Pied Woodpecker, *Picoides mahrattensis,* is smaller in size and is spotted black and white above with red spots on vent and abdomen. The woodpeckers breed from March to August and make nests in hollowed out holes in tree stems or branches. They generally lay three glossy china-white eggs and both sexes take part in domestic duties. They are generally seen pecking at bark of trees, for the hidden grubs and insects and only occasionally peck at beehives for catching bees.

13.9.7 The Titmice or the Grey Tit

The size of a sparrow, the Grey Tit, *Parus major* Linnaeus, belongs to the family Paridae, and its earliest written record as a pest of honeybee dates from the days of Virgil in Greece. More recently it has been observed as a bee predator in Russia and Great Britain. The birds in winter months lure the honeybees out and catch them; moreover they carry the dead bees to their nest as food for the young.

The tits in general and this species in particular, is widely distributed in the old world and is migratory in nature. It is a common bird in the Indian country side. The Grey Tit has a glossy black head and white cheeckpatches, gray on top and white below with a broad black band down the centre. It keeps moving the tail up and down characteristically. Single, in pairs, or in parties, they move around with other birds picking flying insects from near the ground. From February to November they raise two broods, laying 4-6 eggs. The eggs are pinkish white, speckled with brown markings. Both the sexes take part in domestic duties.

13.9.8 The Shrikes

Shrikes belong to family Laniidae that is distributed through Africa, Europe, Asia. In Ethiopia, Greece, Russia and Europe shrikes are considered minor pest of honeybees during the active period of April-August. They are seen near the hives, diving at honeybees and catching them.

Two species are common in Ceylon, Bangladesh and India. The Black-headed Cuckoo-Shrike, *Coracina melanoptera,* is the size of a bulbul, is ashy grey with black head, wings, tail, and is whitish with bars on the underside. It is distributed throughout the plains and mountains of India upto 6000 ft, hunting around in parties mixed with other insectivorous birds. In breeding season the males make many sounds ending in quick repeated pit-pit-pit. The large Cuckoo-Shrike, *Coracina novaehollandiae,* is the size of a pigeon but is slimmer, grey in colour with wings and tail black, and grey-white stripes on the underside. It lives near groves in the cultivated areas, making shrill but pleasant double noted call ti-eee. Their movement from one tree to the other tree-top is in pairs or family groups. They feed on large insects including the bees. They nest from May to October in shallow cups made by binding twigs and other materials like lichen, fine bark etc. They lay 3 pale green eggs with brown blotches on them.

13.9.9 The Swifts

They belong to the family Apodidae which has a worldwide distribution with the exception of north-most Eurasia, and the Americas. The spine-tailed swift, *Chaetura dubia,* is an important predator of honeybees in the Philippines and may cause limitations to beekeeping. On cool, cloudy, and windy days flocks of birds upto 300 descent on apiaries during the day time and destroy large number of honeybees whose sting has no affect on them.

In India, Ceylon, Bangladesh and Burma two species of swifts the size of a sparrow are common. The palm swift, *Cypsiurus parvus,* is sooty-grey with a narrow deeply forked tail. The birds fly about gregariously in open country dotted with palmyra palms, with which they are inseparably linked. The birds roost and nest in these palms and also feed on nectar-loving insects, including honeybees that are attracted to the palm sap.

The house swift, *Apus affinis,* has long narrow wings, smokey-black with white throat and rump. They fly about swiftly and gregariously near ancient forts, ruined buildings, and human dwellings. Owing to the peculiar structure of their feet they cannot perch in ordinary way but cling to rough surfaces, angles of walls, edges of thatched roofs etc. They fly all day feeding on bees, flies, midges and other insects. Their presence near apiaries is dangerous. They breed from February to September and make small cuplike nests by cementing feathers and straw with the help of their saliva, usually in colonies, in angles of walls and ceilings. They lay 2-4 white, elongated eggs. Both sexes take part in feeding the young.

13.9.10 The Kingcrows or Drongos

The Kingcrows or Drongos belong to the family Dicruridae which is distributed throughout Asia. They are primarily carnivorous birds and feed on medium and large insects. They generally live in open countryside at the edge of groves where they perch on bamboos, reeds and tall grasses with broad open fields all around. Their value as insectivorous birds is well known but their harmful role in picking drones and queens near mating yards is sometimes disputed.

There are three common species in India, Ceylon, Bangladesh, Burma, and the distribution of the kingcrow extends upto Pakistan. The Racket-tailed Drongo, *Dicrurus paradiseus,* is the size of a myna, glossy black, tufted on the forehead with two long spatula-tipped feathers at the tail-end. It is scarce in north-western India and Pakistan. It makes loud metallic calls, often mimicking sounds of other birds, It hunts on large insects and flies. It is prized as a pet. The White-bellied Drongo, *Dicrurus caerulescens,* is glossy indigo above and white on belly and root of the tail. It is not found west of the line from Kutch to Gharwal and is distributed upto 6000 ft up in the Himalayas. It prefers the bamboo and thin tree jungle, and makes graceful, agile dives after winged insects. It hunts mixed with other insectivorous birds. The Black Drongo, *D. adsimilis* is most widely distributed in the Indian subcontinent.

There are a number of species of drongos distributed in Himalayas (Fleming *et al.,* 1976; Salim Ali, 1977). They include the Black Drongo (*Dicrurus admimilis*), the Grey Drongo (*D. leucophaeus*), the Crowbilled Drongo (*D. annectans*), the Bronzed Drongo (*D. aeneus*), the Lesser Racket-tailed Drongo (*D. remifer),* the Haircrested Drongo (*D. hottentottus*) and the Large Racket-tailed Drongo (*D. paradiseus*).

13.9.11 The Black Drongo or the King Crow

The king crow, *Dicrurus adsimilis* (Bechstein) is the size of a *bulbul,* slim in body, agile, glossy black with a long deeply forked tail. It is distributed throughout India, Bangladesh, Burma, Ceylon, Pakistan. It is a familiar bird of the open countryside near a forest or a grove of trees. It sits on bamboo and other thin plants like reeds, sarkanda etc. or on telegraph wires. It commonly rides cattle or water buffaloes walking in grass or in fields being ploughed, swooping at insects so disturbed out of the ground. It feeds on medium to large insects like, grasshoppers, crickets, moths, bugs etc. If there is forest fire or grass and straw is burnt in the fields they appear in large numbers out of nowhere and sit on isolated bushes and tall reeds, against the hot wind or smoke watching the insects that fly out. They catch them in the mid air or from the ground, and depending upon the size of the prey devour them whole or break them into bits and pieces on their perch and eat them. The birds are particularly active at the harvest time and snap at small and medium insects not easily noticed by man. Occasionally they appear near apiaries and feed on honeybees. While active in the field they utter a variety of challenging sounds not unlike those of the Shikra-hawk. By and large they are friends of the farmers but the beekeepers should scare them away especially during the queen mating season.

They breed from April to August and make nests at the extremity of tree branches of isolated tall trees, 15-39 ft. above the ground. There are 3-5 eggs, white with brownish spots. Both sexes perform parental duties and aggressively guard their nest.

General. Among birds, Bee-eaters are considered the most serious pests of honeybees. They sit and wait or glide around the apiaries in dozens and devour bees at libidum. Since they are gregarious in nature they have the tendency to congregate in certain localities making it unprofitable to keep honeybees there. At times the situation is such that simple scaring them away is not effective. They have to be killed by sling shots or air guns. Hanging the dead birds tied on a rope gives them sufficient warning and eventually they may shift to another locality, only to return after sometime. Bee-eaters are thus a persistent nuiscence in their preferred areas of nesting. Other birds that feed on drones and queen bees should also be scared away in order to protect honeybee colonies.

13.10 COLONY COLLAPSE DISORDER

Colony Collapse Disorder (CCD) is a phenomenon in which worker bees from a beehive or European honey bee colony abruptly disappear. While such disappearances have occurred throughout the history of apiculture, the term *colony collapse disorder** was first applied to a drastic rise in the number of disappearances of Western honey bee colonies in North America in late 2006. Colony collapse is significant because many agricultural crops worldwide are pollinated by bees.

European beekeepers observed similar phenomena in Belgium, France, the Netherlands, Greece, Italy, Portugal, and Spain, and initial reports have also come in from Switzerland and Germany, albeit to a lesser degree while the Northern Ireland Assembly received reports of a decline greater than 50%. Possible cases of CCD have also been reported in Taiwan since April 2007.

The cause or causes of the syndrome are unknown. In 2007, some authorities attributed the problem to biotic factors such as *Varroa* mites and insect diseases (i.e., pathogens including *Nosema apis* and Israel acute paralysis virus). Other proposed causes include environmental change-related stresses, malnutrition, pesticides (e.g., neonicotinoids such as clothianidin and imidacloprid), and migratory beekeeping. More speculative possibilities have included both cell phone radiation and genetically modified (GM) crops with pest control characteristics though no evidence exists for either assertion.

It has also been suggested that it may be due to a combination of many factors and that no single factor is the cause. The most recent report (USDA – 2010) states that "based on an initial analysis of collected bee samples (CCD-and non-CCD affected), reports have noted the high number of viruses and other pathogens, pesticides, and parasites present in CCD colonies, and lower levels in non-CCD colonies. This work suggests that a combination of environmental stressors may set off a cascade of events and contribute to a colony where weakened worker bees are more susceptible to pests and pathogens."

Applying proteomics-based pathogen screening tools in 2010, researchers announced they had identified a co- infection of invertebrate iridescent virus type 6 (IIV-6) and the fungus *Nosema ceranae* in all CCD colonies sampled. However, subsequent studies have questioned the methodology used in these proteomic experiments.

13.10.1 History

From 1972 to 2006, there was a dramatic reduction in the number of feral honey bees in the U.S. (now almost nonexistent) and a significant though somewhat gradual decline in the number of colonies maintained by beekeepers. This decline includes the cumulative losses from all factors, such as

* Colony collapse disorder—Wikipedia, the free encyclopedia

urbanization, pesticide use, tracheal and *Varroa* mites, and commercial beekeepers' retiring and going out of business. However, in late 2006 and early 2007 the rate of attrition was alleged to have reached new proportions, and the term "colony collapse disorder" began to be used to describe this sudden rash of disappearances, (sometimes referred to as Spontaneous Hive Collapse or the Mary Celeste Syndrome in the United Kingdom).

Losses had remained stable since the 1990s at 17%-20% per year attributable to a variety of factors, such as mites, diseases, and management stress. The first report of CCD was in mid-November 2006 by a Pennsylvania beekeeper overwintering in Florida. By February 2007, large commercial migratory beekeepers in several states had reported heavy losses associated with CCD. Their reports of losses varied widely, ranging from 30 to 90% of their bee colonies; in some cases beekeepers reported loss of nearly all of their colonies with surviving colonies so weakened that they might no longer be viable to pollinate or produce honey.

Losses were reported in migratory operations wintering in California, Florida, Oklahoma and Texas. In late February, some larger non-migratory beekeepers in the mid-Atlantic and Pacific Northwest regions also reported significant losses of more than 50%. Colony losses also were reported in five Canadian provinces, several European countries and countries in South and Central America and Asia. In 2010 the USDA reported that data on overall honey bee losses for 2010 indicated an estimated 34 per cent loss, which is statistically similar to losses reported in 2007, 2008, and 2009.

Limited occurrences resembling CCD have been documented as early as 1869 and this set of symptoms has in the past several decades been given many different names (*disappearing disease, spring dwindle, May disease, autumn collapse,* and *fall dwindle disease*). Most recently, a similar phenomenon in the winter of 2004/2005 occurred, and was attributed to *Varroa* mites (the "Vampire Mite" scare), though this was never ultimately confirmed. Nobody has been able to determine the cause of any past appearances of this syndrome. Upon recognition that the syndrome does not seem to be seasonally restricted, and that it may not be a "disease" in the standard sense—that there may not be a specific causative agent—the syndrome was renamed.

13.10.2 Signs and Symptoms

A colony which has collapsed from CCD is generally characterized by all of these conditions occurring simultaneously:

(*a*) Presence of capped brood in abandoned colonies. Bees normally will not abandon a hive until the capped brood have all hatched.

(*b*) Presence of food stores, both honey and bee pollen:

- (*i*) which are not immediately robbed by other bees
- (*ii*) which when attacked by hive pests such as wax moth and small hive beetle, the attack is noticeably delayed.

(*c*) Presence of the queen bee. If the queen is not present, the hive died because it was queenless, which is not considered CCD.

Precursor symptoms that may arise before the final colony collapse are:

- (*i*) Insufficient workforce to maintain the brood that is present
- (*ii*) Workforce seems to be made up of young adult bees
- (*iii*) The colony members are reluctant to consume provided feed, such as sugar syrup and protein supplement.

14 Beekeeping Industry and Trade

14.1 THE INDIAN BEEKEEPING INDUSTRY*

Beekeeping has been practiced in India since ancient times and the earliest references to beekeeping are found in the Vedas and the Ramayana. Traditionally bees were kept in log and mud pot hives and in wall recesses but these methods were crude and wasteful. Scientific beekeeping was introduced in the mid 19th century with *Apis cerana* being kept in moveable wooden hives.

14.1.1 Beekeeping In India

The beekeeping industry in India was given an initial impetus in the early 1950's by the Khadi and Village Industries Commission, (KVIC) which took up beekeeping scientifically using the Indian honeybee *Apis cerana*. *Apis cerana* was propagated by the KVIC who did pioneering work in this field all over India. Beekeeping with *Apis cerana indica* was done in most Indian states but KVIC was particularly active in U.P., Bihar, the Southern states, the North Eastern states and Bengal.

The major production of honey in India was initially from *Apis cerana* and *Apis dorsata* but by the attack of the Thai Sac Brood Virus in India, in the early 1990's, *Apis cerana* was practically wiped out and beekeeping received a serious setback all over India. Particularly the North Eastern and Southern states which had a large population of *Apis cerana* and many beekeepers involved in beekeeping, were seriously affected and the whole industry suffered due to the decimation of the population of *Apis cerana* because of Thai Sac Brood.

There was a big controversy all over India about the import of *Apis mellifera* and it was only through the efforts of Dr A.S. Atwal and Dr O.P. Sharma that *Apis mellifera* was introduced to India in 1962-64 through the Punjab Agricultural University at its research station at Kangra, Himachal Pradesh. Since import of colonies of Mellifera was banned, due to the fear of diseases like American foul brood and European foul brood coming into India, this was done by a unique experiment for the first time in the history of apiculture in which the Mellifera queens were imported and given to brood and bees of *Apis cerana* who accepted the queens of Mellifera and started rearing the Mellifera brood. In this way *Apis mellifera* was introduced to India without bringing in any of the brood diseases which

* From an APEDA Project by Sarparveen Kaur and her husband Peter Satwant Singh.

are prevalent all over the world. However, there were still restrictions on the movement of Mellifera out of the Punjab Agricultural University as the KVIC believed that it would bring disease to the rest of India.

However, *Apis cerana* was practically wiped out by the Thai Sac brood virus which came into India from the far east and through no fault of *Apis mellifera*, because *Apis mellifera* was not present in the areas from where the Thai sac brood spread all over India. Beekeeping suffered a serious set back in India because of this and the stage was set for *Apis mellifera* to take over. It is only with the introduction of *Apis mellifera* in 1962 that beekeeping in India took off and developed to its present status of 1,00,000 bee colonies in the Northern States.

Beekeeping with *Apis mellifera* was taken up in India, through the pioneering efforts of the Punjab Agricultural University, (PAU) who introduced this bee into India in 1962. The PAU has done a lot of R&D in beekeeping. After research for a number of years on Mellifera, the PAU decided to distribute these bees to farmers on an experimental basis and start training of farmers with *Apis mellifera*. The Punjab Agricultural University, Ludhiana held its first course on beekeeping with *Apis mellifera* in 1976 which was attended by progressive farmers from all over Punjab and by Sarparveen Kaur and Peter Satwant Singh on whose farm near Ludhiana, the PAU set up the first research and development centre at the farmers field for beekeeping with *Apis mellifera*, with a view to finally distribute the bees to the farmers in the state. The course and the field experiment at the farm was a great success and subsequently in the next two years bees were given to select farmers all over Punjab.

Beekeeping with *Apis mellifera* in India never looked back after that and *Apis mellifera* grew rapidly in Punjab from 1978 onwards. It then caught the interest of beekeepers from Himachal, particularly beekeepers from Simla and Kangra who came to Punjab, bought *Apis mellifera* from Punjab farmers and starting migrating the bee colonies all over Punjab and Haryana. The real impetus to beekeeping in this area came because of the large plantations of *Eucalyptus* and then the start of Sunflower cultivation in these areas. Whereas initially around 1978, the major honey crops were clover, *Eucalyptus* and Arhar, by the change of the cropping pattern to oilseeds like *toria* and Sunflower the bees started producing honey all the year round in these states, starting from *toria* honey in November-December, to mustard honey from February to March, sunflower honey from May to June and then for the dearth period of the monsoons during June-July there were Khair (*Acacia catechu*) plantations along the foothills of the Shiwaliks range of the Himalyas that gave the bees sustenance in the off season.

All this led to the rapid development of Mellifera in the northern states and seeing the success, and production of honey in these states, beekeepers from Bihar, U.P., Bengal, and the southern states started coming to Punjab and Haryana, buying bees and taking them to their home states, so that Mellifera started to spread to other parts of India. One person instrumental in the spread of *Apis mellifera* all over India has been Dr R.C. Mishra who headed the Mellifera coordination committee under the Indian Council of Agricultural Research (ICAR)

Beekeeping in Punjab and Haryana moved in fits and starts as there was no sure market for the honey produced. And by 1992 the situation became problematic as beekeepers in Punjab and Haryana could not sell their honey. The Punjab Government stepped in through Punjab Agro Industries and set up the first integrated joint venture project in India for beekeeping and honey processing, for the development of value added products from the bee hive and for the export of honey. The project under the name of Honeybee Natural Products Ltd – was the first such integrated project in India with forward and backward linkages to the farmers, and was set up in collaboration with Sarparveen Kaur and Peter Satwant Singh who had set up an R&D unit for bee breeding registered with the Department of Science and Technology. The setting up of this project resulted in the export of honey from Punjab in the year 1995. The honey exported was from rapeseed (*toria*) which honey had a marketing problem in India as none of the packers liked to buy this honey for packing because it granulated rapidly and also because of its strong taste which was not liked in the domestic market However, rapeseed honey is liked in Germany and in the USA.

Subsequently, another processing unit was set up by The Punjab Beekeepers Federation at Fatehgarh Sahib in Punjab with the KVIC using the honey processing plant developed by the CBRI at Pune. The Punjab Beekeeping Federation has beekeepers as members and processes and sells honey to bulk packers. Doraha in Punjab became the centre of beekeeping because of its close proximity to the PAU and Kashmir Apiaries in Doraha also set up a processing unit with KVIC assistance for bottling honey and selling through the KVIC outlets in various parts of India.

After the export of honey from Punjab in 1995, other companies also moved into the honey market in Punjab and Haryana. JCT's Agro Division moved into honey exports in 1996. Subsequently, JCT ceased operations in honey export and Pathbreakers, a break off of Employees from JCT has set up a unit at Hoshiarpur for honey processing and exports. Kejriwaal Enterprises came into export of honey in a big way and started buying and exporting honey from Punjab, Bihar and Bengal.

Meanwhile Dabur decided to expand its honey operations in India and started advertising honey on television and the media in a big way. Dabur initially handled honey from its plant at Sahibabad in U.P. but now has set up a modern processing plant at Baddi, Himachal Pradesh with a capacity of 15000 tons per annum for honey processing. Their plant also has facilities for reducing the moisture level in honey after the honey with high moisture comes into the plant. Dabur has rapidly become one of the largest buyers of honey from this area.

Beekeeping rapidly increased in Punjab, Haryana and Himachal from 1995 till 1997 as the demand for honey grew. However, from October 1997 onwards, there have been successive failures of honey crops which have coincided with a fall in international prices, leading to many honey processing units becoming not viable any longer for export. The domestic players like Dabur, however, are rapidly expanding their domestic market as they offer higher prices for the honey which is making the export of honey from India not a viable proposition for the exporters.

Beekeeping has also rapidly developed in Bihar from 1995 onwards as the demand for Litchi and Karanj honey from Bihar started growing steadily after its export by Honeybee Natural Products, JCT, and Kejriwaals from Bihar since 1995. Dabur also is a buyer of the honey from Bihar. Thus beekeeping in Bihar with Mellifera has received a great impetus. The KVIC and the Mirza Nagar Gramudyog Sociey have done a lot of work in state and this Society has exported around 100 tons of Litchi Honey. However, the last two years of bad weather have also taken their toll of the industry in Bihar.

Bengal is another state which has developed beekeeping with *Apis mellifera* and also increased honey production from the Sunderbans area. The West Bengal Beekeepers Association is very active in beekeeping there and supplies exporters and helps set up beekeeping. The KVIC has also aided the 24 Parganas Beekeepers Society and the Rama Krishna Mission is active in beekeeping and offers training to beekeepers.

Beekeeping in Kerala after the loss of *Apis cerana* has started moving forward again. Father Mathews and the Malanadu Society is very active with beekeeping in Kerala and have helped develop beekeeping again in Kerala and also has taken up some beekeeping with *Apis mellifera*.

In Maharashtra the KVIC and the CBRI in Pune are very active though the loss of *Apis cerana* has seriously affected their operations in those areas.

U.P. Agro Industries Corporation has taken up beekeeping in Uttar Pradesh in a big way and has helped develop beekeeping rapidly all over U.P. They have developed various belts of beekeeping in U.P. over the last few years. The KVIC is also extremely active in beekeeping in U.P.

In Andhra Pradesh, the A.P. Girijan Cooperative Society is very active in beekeeping with *Apis dorsata* and deals with wild honey from the forests.

The present scene on the export scenario of honey from India is dismal as the honey production in the major producing states of India has gone down because of the failures of crops in the last two years, existing units are working at a loss because of under utilisation of capacity and the world honey prices have crashed since last year and now come back to the old prices which existed before the boom till 1997. The entry of many players in the honey market has coincided with failure of crops and the present beekeeping and honey production in India is not enough to support all of them.

The domestic demand for honey is now increasing as there are many packers in the domestic market. The domestic market for honey is around 6,000 tons of which Dabur is the major packer handling around 2,000 tons of honey annually. The Khadi and Village Industries Commission handles another 2,000 tons through its many regional packing plants all over India and sells honey through the Khadi Bhavans throughout the country. The balance 2,000 tons or so are handled by other packers like Baidyanath, Zandu and Charak, regional packers like Honeybee Natural Products Ltd., Mehsons, Leksons and Allied in Delhi, Square D in Madras, local packers in Mahabaleshwar and other local packers all over India.

14.1.2 Feasibility

Beekeeping and honey production is a very feasible proposition all over India as we have large acreages of bee flora in the form of oilseds like *toria*, mustard, sunflower, niger, safflower and sesame (Table 12). Pulse crops like pigeon pea (Arhar), gram and soyabean are also good sources of nectar for bees. Large plantations of *Eucalyptus* exist in many areas which are an excellent source of honey. We then have horticultural crops like citrus, litchi and plantations like rubber, cardamom and coffee in the South. Our forests also have many nectar yielding plants all over India, we have Karanj and litchi as a good source of nectar in Bihar and the Himalyan region has an excellent honey from *Plectranthus*.

Beekeeping has the advantage of increasing crop yield through pollination by honeybees and in particular sunflower has an increase of yield by over 100% and other crops from 20 to 50%.

TABLE 12. Districtwise Available Bee Flora in Punjab

Area in 000 Hac (1996-97)

District	*Sunflower*	*Rapeseed & Mustard*	*Arhar*	*Litchi*	*Citrus*
Hoshiarpur	20	3	2	29.2	564.2
Jalandhar	16	2	9	14.4	165.3
Nawansahar	5	2	1	0	0
Ludhiana	3	3	38	1.7	125.2
Ferozepur	7	22	2	0	500.4
Amritsar	10	12	25	14.6	186.3
Gurdaspur	3	4	1	124.3	186.3
Kapurthala	4	1		2.0	75.3
Bhatinda	1	12		0	145.5
Patiala	11	4	4	8.4	99
Sangrur	1	3	18	.1	78.7
Ropar	4	2	18	13.1	112.4
Faridkot	1	1		0	363.7
Moga		1	2	0	0
Mukatsar	10	6		0	0
Mansa	1	7	1	0	25.5
Fatehgarh Sahib.	2	1	8	.4	23.4
Total	99	86	129	208.2	2651.2

Source: Dept. of Agriculture & Horticulture, Punjab

Other than this the other major flora is from eucalyptus which is there in districts of Ludhiana, Amritsar, Hoshiarpur, Patiala and all along the Grand Trunk road.

14.1.3 Beekeeping in Punjab -Areas, Beekeepers, Colonies

Areas in Which Beekeeping is Done

The main beekeeping districts in Punjab are Hoshiarpur, Amritsar, Patiala, Ludhiana. Jalandhar, Nawansahar and Fatehgarh Sahib.

Number of Beekeepers/Colonies and Their Production

The information on beekeepers and their colonies differ from source to source, but on the whole it has been established that Punjab produces about 2500-3000 MT of honey per year. However, in the last two years production was adversely affected on account of bad weather and shortage of flora. But it will be pertinent to say that in Punjab:

Total Number of Beekeepers	8,000-10,000
Total Number of Colonies	1,00,000
Average size of the Beekeeper	30-50 Boxes
Average yield	30-40 kg/Colony/Season
Production of Honey	
(Apiary)	3,000 MT
(Wild)	20-50 MT

Migratory Pattern of Bee Colonies

Since beekeepers in Punjab have a large number of colonies, they need to migrate the bee colonies over the year for increased production and sustenance. Majority of the beekeepers even moves their bee colonies out of the state in search of flora. The major migratory routes are given in Table 13.

TABLE 13. Major Migratory Beekeeping Routes in Punjab

Flora	*Months*	*Areas*
Eucalyptus	Feb.-March	Ludhiana, Amritsar, Hoshiarpur, Patiala, G.T. Road (Amritsar to Ambala)
Sunflower	End April-May	Hoshiarpur, Jalandhar, Kapurthala, Amritsar
Rapeseed	October-Dec.	Rajpura, Amritsar, Ropar, Jalandhar
Mustard	Dec.-Feb.	Mansa, Ganganagar (Rajasthan), Rewari, Hissar (Haryana)

14.1.4 Organisations Involved in Beekeeping

The success of beekeeping is mainly through the efforts of the Punjab Agricultural University which has promoted *Apis mellifera* all over the state and the efforts of Punjab Agro Industries Corp. and Agriculture Department in giving training and subsidy on bee colonies. The main organisations involved are

1. Deptt. Of Agriculture, Govt. Of Punjab—provides subsidy on bee colonies, about 5,000 colonies per year. In 1997-98, 3400 boxes were given in the districts of Hoshiarpur, Ludhiana, Amritsar, and Gurdaspur.
2. Punjab Agro Industries Corpn., Chandigarh – helps in setting up joint ventures in the state. In Punjab PAIC also have a joint venture with Honey Bee Natural Products (P) Ltd. Honey Bee Natural Products (P) Ltd., Lalru—has taken up 750 farmers for beekeeping in Punjab over the last 3 years under a scheme of the Punjab Rural Development Department in which training and follow up is given at site for 1 year and the complete package is provided to the farmers at site including bees, equipment, training at site under a scheme sanctioned by NABARD having over 80% success rate.
3. Punjab Agricultural University, Ludhiana—mainly involved in research and training. PAU in fact played a pioneer role in popularisation of *Apis mellifera*.

4. KVIC—was involved earlier in setting up the co-operatives and even financing the bee keeping through various schemes.

14.1.5 Processing Units in the State

Punjab has a number of honey processing units:

1. Honey Bee Natural Products Pvt. Ltd., Lalru has a processing capacity of 6,000 MT/year. They pack honey both for export and the domestic market. In the domestic market they sell their honey bottled under the brand name Honey Bee and Golden Meadows in Delhi and Northern part of India
2. Path breakers Trading Co. Pvt. Ltd., (Formerly JCT LTD.), Hoshiarpur—has the processing capability of 2000 MT/year. They pack honey for export in bulk.
3. The Punjab State Beekeepers Federation, Bassi Pathana, Fatehgarh Sahib - The Chief Executive is Dr B.S. Chahal (Retd. Head of the Department of Entomology, P.A.U., Ludhiana). They have beekeepers as members of the federation whose honey is collected and processed. They mainly supply in bulk after processing to packers all over India. The federation fixes procurement prices after discussion with the beekeepers. They have the processing plant with the capacity of 4000 MT/annum.
4. Kashmir Apiaries, Doraha, Ludhiana—has a bottling and processing unit at Doraha. They bottle honey under the Agmark license and supply to the KVIC and other private retail outlets under the brand name of Kashmir Honey. They also process and supply bulk honey for export. They handle about 1,500 tons of honey annually.

14.1.6 Marketing

As Punjab is the major honey producing state, all the major players who are on the national map right from the processors/exporters from Delhi to packers in Pallakad, Kerala buy honey from Punjab.

Bulk Honey

Over the last three years Punjab catered to the major share of the export demand. The major producing areas are Hoshiarpur, Amritsar, Patiala, Jalandhar, Ludhiana, and Ganganagar. The following organisations procure honey from all over the state for export in bulk:

Company	*Channel*
Honey Bee Natural Products (P) Ltd.	Directly
Dabur India Ltd.	Agents
Kashmir Apiaries	Directly
Kejriwal	Agents
Pathbreakers Trading	Directly

Retail Packing

The following organisations also bottle honey for the retail market.

Company	*Channel*
Honey Bee Natural Products (P) Ltd.	Directly
Dabur India Ltd.	Agents
Kashmir Apiaries	Directly
Punjab Bee Fed	Directly

14.2 PROSPECTS OF HONEY PRODUCTION IN INDIA

As can be seen from the production figures of 10 major Indian states (Table 14), Karnataka had a production of 582 tons, Kerala 1963 tons and Tamil Nadu 2379 tons. These figures are no longer valid due to the failure of beekeeping in the southern states of India due to Thai sac brood affecting *Apis cerana*. Production of honey in these states has reduced drastically while production in the Northern states has increased dramatically due to the successful expansion of *Apis mellifera*.

TABLE 14. Status of Beekeeping in 10 Indian States

States	*Colonies (Number)*	*Earning (Lakh Rs.)*	*Production (kg) Honey*	*Production Wax*	*Production Value (₹ Lakh)*
Andhra Pradesh	22006	45.02	98099	878	59.62
Assam	57065	106.93	388411	776	116.91
Bihar	29682	194.35	207414	747	242.95
Himachal Pradesh	10735	16.16	64410	350	17.0.3
Karnataka	136526	172.98	582126	4077	176.66
Kerala	242541	398.95	1963823	1184	498.69
Punjab	303200	19.55	107688	–	27.93
Tamil Nadu	303364	398.4	2379547	4445	554.26
Uttar Pradesh	26018	46.04	105853	–	57.56
West Bengal	63085	208.8	488529	5050	264.67
Total:	1194222	1607.18	6385900	28163	2016.26

Source : KVIC Annual Report 1991 – 1992

The KVIC data are the data on record but in actual fact let us take the example of Punjab for verfication of the data. In the year 1992, Punjab had a maximum production of 500 tons of honey when Punjab Agro Industries set up its joint venture in the state for honey processing and Haryana had around 200 tons of honey. Taking an average production of 30 kg per colony, this means that Punjab had 16,666 or say a maximum of 20,000 bee colonies as opposed to the figures of 3 lakh bee colonies given above.

From the survey done by the Greenworld people in India, it has been found that figures given by Government Agencies did not tally with figures on the ground as according to the agencies a certain number of bee colonies had been distributed in a particular area over a number of years—which did not exist in reality (Table 15).

TABLE 15. Status of Beekeeping and Honey Production in India for the year 1996-97 Collected from Government Agencies and KVIC

State	*No. of Bee Colonies*	*No. of Beekeepers*	*Avg. Yield (Kg.)*	*Honey (Tons)*	
				Apiary	*Wild*
Punjab	75000	10000	30-40	3000	0
Haryana	15000	—	30	2000	50
Andhra Pradesh	40000	5000	5	100	300
Karnataka	21000	13128	10-15	250	150
Uttar Pradesh	30000	—	30	500	500
Bihar	20000	20000	—	600	200
West Bengal	27000	4250	30-40	500	700
Assam	48425	—	10	500	120
Maharashtra	9000	2413	5-10	50	100
Himachal Pradesh	15000	4000	—	50	50
Madhya Pradesh	—	—	—	0	100
Kerala	62640	18052	12	300	100
North East (5 states)	—	—	—	200	200
Gujarat	—	—	—	0	50
Tamil Nadu	—	—	—	100	0
J. & K.	—	—	—	200	0
Rajasthan	—	—	—	350	50
Total:	363065	76843	—	8700	2670

Blanks (—) in the above represent figures that were not available.

This shows a total honey production of around 9000 tons of apiary honey and 3000 tons, of wild honey in India, rounding off to the nearest 1000 tons, and a total bee colonies' population of around 3.63 lakhs. However, even here by the figures collected from the individual states the number of bee colonies does not tally with the honey produced, as when we calculate the number of bee colonies from the average production of honey per colony in these areas then we come to a different set of figures (Table 16), where we have broken up the states into those having mainly *Apis mellifera* and others having *Apis cerana*.

The figures for the bee colonies for *Apis cerana* are not certain as it is not perfectly clear whether the honey in those states having *Apis cerana* is only from *Apis cerana* or may be honey from *Apis dorsata* which is clubbed with the production of *Apis cerana* in the area.

TALE 16. Present Estimated Status of Bee Colonies in India

(Bee Colonies of *Apis mellifera* calculated by taking production averages of honey produced in each state assuming figures of production as obtained from field visits and trade estimates)

State	*No. of Bee Colonies*	*Estimated beekeepers assuming each owns*			*Yield (Kg.)*	*Honey (Tons)*	
		5 colonies	*10 colonies*	*50 colonies*		*Apiary*	*Wild*
Punjab	75000	15000	7500	1500	40	3000	—
Haryana	50000	10000	5000	1000	40	2000	50
Bihar	20000	4000	2000	400	30	600	200
Uttar Pradesh	16667	3333	1667	333	30	500	500
West Bengal	16667	3333	1667	333	30	500	700
Himachal Pradesh	5000	1000	500	100	10	50	50
Total	183334	36666	18334	3666		6650	1500

Thus the major 7 honey producing states of India have a maximum of 1.83 lakh bee colonies of *Apis mellifera* producing about 6700 tons of Apiary honey.Around 1500 tons of honey is collected from *Apis dorsata* from the forest areas in these states.

Table 16 gives a reasonably realistic estimate of the bee colonies of Mellifera in India. The figures of Punjab, Haryana and Himachal are misleading as they include the number of bee colonies that are migrated into Punjab and Haryana from Himachal. Around 50% of these bee colonies belong to beekeepers from Himachal who migrate to Punjab and Haryana in the months of October to June.

These figures seem to be more realistic and tally with the survey conducted by the Greenworld People in the above states. The difference as seen by the survey of the Grenworld people is that the bee colonies exist in the Government records but during the field visits done for the survey, many areas had actually none or a minimal number of bee colonies which did not tally with the government estimates given.

According to the available figures (Table 17), there are 1.23 lakhs bee colonies of *Apis cerana* in five states. However, the bee colony figures seem to be inflated even here because particularly Assam does not seem to have 50,000 bee colonies of *A. cerana*, Karnataka does not seem to have 21,000 bee colonies and Andhra Pradesh also does not seem to have 20,000 bee colonies as seen from the field survey done by the Greenworld people. Thus even calculating backwards from the honey production taking averages of honey produced per colony, we do not seem to bee getting the correct figures for these states. It may be that the honey produced in these states is being clubbed with honey from *Apis dorsata* or otherwise is being totalled including the honey imported from the honey producing states of the north. During the survey, it was brought out that even in areas like Mahabaleshwar and Coorg, which had their own honey industry, the honey which was shown as honey from these states included honey imported from Punjab, Haryana and Bengal.

TABLE 17. Bee colonies of *Apis cerana* calculated by taking production averages of honey produced in each state assuming figures of production as obtained from field visits and trade estimates

State	*No. of Bee Colonies*	*Estimated beekeepers assuming each owns*		*Yield*	*Honey (Tons)*	
		5 colonies	*10 colonies*	*(Kg.)*	*Apiary*	*Wild*
Kerala	25000	5000	2500	12	300	100
Assam	50000	10000	5000	10	500	0
Maharashtra	7143	1429	714	7	50	100
Andhra Pradesh	20000	4000	2000	5	100	300
Karnataka	20834	4167	2083	12	250	150
Madhya Pradesh					0	100
Total	122977	24596	12297		1200	750

These states which had a previous history of beekeeping and a developed honey industry, have had problems because of the failure of *Apis cerana* and as yet not so wide-spread use of *Apis mellifera*, which needs to be propagated in a big way in these areas as there is plenty of bee flora available (Table 18).

TABLE 18. Potential of Beekeeping in Different States in India

State	Acreage of Bee flora '000 hectares					
	Rapeseed-Mustard	Sunflower	Citrus	Litchi	Niger/Sesamum	Rubber
Punjab	86	50	40	2	0	—
Haryana	579	40	4	0	0	—
Bihar	97	16	18	23	24	—
Uttar Pradesh	854	17	22	9	0	—
West Bengal	378	0	6	3	6	—
Assam	266	0	13	4	24	—
Maharashtra	8	497	92	0	111	—
Himachal Pradesh	9	0	38	0	7	—
Madhya Pradesh	506	0	10	0	225	—
Kerala	10	0	0	0	10	443
Karnataka	4	886	36	0	43	—
Andhra Pradesh	3	160	62	0	18	—
Total	2800	1666	341	41	468	443
Colonies / hectare	5	5	5	5	5	5
Bee colonies (millions)	14.00	8.33	1.71	0.21	2.34	2.22

14.3 BEEKEEPING AS A VILLAGE INDUSTRY

Beekeeping as a cottage or a village industry has been promoted in India since the beginning of this century. Many State Government and Non–Government Organizations (NGO) particularly the Khadi and Village Industries Commission of India, prepared and executed development programmes for beekeeping. Apart from giving training to the prospective beekeepers, subsidy was provided for the purchase of honeybees and bee equipment. When honey was extracted, it was purchased by the Commission for cash so that beekeepers did not have to go far to sell their honey. Cash was handy for the poor farmers who produced very little else on their farms during the year that brought cash. The seasonal crops barely met their food requirements. Therefore, Khadi products and beekeeping provided some income to the rural people. Gradually, beekeeping with *Apis cerana* was established as a cottage industry in various parts of the Himalayan region and in South India. The tradition of keeping bees in skeps and hollowed house walls was much older but modern beekeeping was started with the introduction of various types and sizes of wooden hives having movable frames. Pure and clear honey was extracted from honey frames with the help of centrifugal machines.

With the above background, beekeeping with *Apis mellifera* in Himalayan region was only one step forward and prospects for commercial beekeeping seemed promising. As a result of many years of research at Nagrota (H.P.) and at Ludhiana in Punjab, a few hundred colonies were developed. In the training programs, *A.m.* was introduced to the farmers, who were impressed by its performance. They hoped that the new imported bees would be given to them. The beekeepers in Himachal Pradesh and J & K had a good tradition for beekeeping and they were alive to the value of exotic honey bees. Over the years, a great deal of this research was published in the newspapers and beekeepers became anxious to work with these bees. Under this public pressure, *A.m.* was given to selected and trained beekeepers in Himachal Pradesh in 1974 and in the Punjab plains in 1976. The enthusiasm with which they worked with the bees and the speed with which they multiplied the colonies was beyond all expectations (see Table 2). This extraordinary interest shown by the farmers for the honey production corroborated that of the seed of the new Mexican varieties of wheat, in the mid-sixties.

People in Himachal Pradesh, Uttar Pradesh, and Jammu Kashmir were familiar with honeybees and the art of beekeeping was passed on from father to the son. These farmers adopted *A.m.* beekeeping with ease. It was surprising how people of Punjab not familiar with beekeeping suddenly fell in love with the bees. These practical and innovative people adopted new farm practices whenever they saw the profit involved. Once honeybees came into their hands they multiplied them at a tremendous speed and sold the nucleus colonies to their neighbours. There was a great demand for training courses at the Agriculture Universities and the State Department of Agriculture. Many farmers took training from their neighbours who were already keeping bees. They learned the basics of bee management very quickly and became proficient in dividing colonies and grafting queen cells. Under good floral conditions, they also extracted surplus honey and competed with one another for obtaining higher yields per colony. The performance of *Apis cerana* (Annonymous, 1973) and that of *Apis mellifera* in India has been reviewed (Annonymous, 1989) and the prospects of the latter in Indo-Gangetic plains is described by Atwal (1987).

During the eighties, this industry saw many ups and downs. Compared with the earlier years, much greater quantities of honey were produced which could not be sold. In Himachal Pradesh, the Government purchased honey at remunerative prices ranging from rupees 20-30 per kg in various years, which they could sell further with difficulty. In Punjab, the vanaspati tins full of honey accumulated with beekeepers and Cooperative Societies. The stocks kept accumulating for 3 or 4 years. In the absence of retail marketing of bottled honey, the main purchasers were the firms like Dabur who manufactured indigenous medicines but they would not pay the remunerative price for machine extracted honey.

The Punjab State Marketing Federation (Mandi Board) then stepped in and started purchasing and retail–marketing of honey, which gave some relief to the beekeepers. A few individuals in Delhi and Bombay started retail marketing and also bulk exportation of honey to the Middle East. This gave further relief to the beekeepers who were becoming disheartened and had started selling their bee colonies to people in other parts of India such as Bihar, Tamil Nadu, Kerala etc. However, the committed beekeepers continued to produce honey. The price stabilized around rupees 25 per kg. It appeared that finally the concept of scientific beekeeping had caught on in the plains of Punjab, Haryana, and the adjoining states. While beehives became a familiar sight in the country side signboards selling pure honey appeared along the highways from 1980 onwards.

In the nineties, there appeared many hopeful signs for the progress of beekeeping industry. Many Beekeeper's Cooperative Societies organized into Federations and mobilized bigger funds for the purchase, processing bottling and marketing of honey. The Punjab Beekeeper's Federation set up a honey processing plant in 1993 at Bassi Pathana with a capacity of 30 quintals per day. Dr. Baldev Singh Chahal, after retirement as Extension Entomologist from Punjab Agricultural University, continued to be the guiding force. It was on the model of Khadi Commission plant at Mahabaleshwar. The same year Punjab Agro-Industries Corporation finalized a 2 Crore project as a Joint Venture Company in collaboration with Honey Bee Natural Products Ltd. The promoters are a husband-wife team Peter Satwant Singh and Mrs. Sarparveen Kaur. Apart from promotion of beekeeping and supplying hives and bees to the farmers on loan, the company also purchased, processed, and sold honey in bulk and in retail. The processing plant in their factory at Lalru, near Ambala, had a capacity of 3,000 quintals per day. For the first time in India, a sizeable quantity of 800 tons of honey was exported by them in 1995–96 season to USA. However, international trade is a tricky business and you have to compete in the world market.

In order to procure honey the traders now had to go to the field in apiaries and make payments in cash. The beekeepers were in a position to bargain and were anxious to increase the number of colonies for honey production rather than for the sale of nucleus colonies. The bulk price of honey increased to rupees 30 per kg. With this new trend, there seemed to be greater scope for the expansion of beekeeping industry in India.

The initial investment of approximately 1000 rupees per colony, with very little recurrent expenditure, gave one of the highest investment : employment ratio. It was hoped that the bulk sale of honey was only a temporary feature which ultimately would give place to bottled honey for sale in India and abroad. Honey from single floral sources with special aromas, could be produced with proper management practices so that this honey could be sold at premium. For the present, beekeepers, once again, had started multiplying colonies for the production of honey, which could be sold readily.

There appeared to be great prospects for honey industry in India. The rich flora available in the subtropical and other parts of the subcontinent, could be exploited by the bees and there was a huge unexplored domestic market for the retail sale of bottled honey and an ever expanding world market for export. Apart from the variations in the yearly cycles of honey production in the world, the beekeepers in many of the large producing countries had to keep many thousands of colonies in order to make a family living. Therefore many Asian countries with abundance of bee flora and manpower were prepared to enter this world market and trade of honey.

14.4 COMMERCIAL BEEKEEPING

The above background would indicate that beekeeping from a home or a cottage industry had to go a long way in order to make it a successful commercial venture with large number of colonies. Like any other business venture the first need was that of finances. The banks and other financial Institutions had to come forward to loan money at concessional rate of interest, since bees are a live material and high risk is involved as against manufacture of consumer goods. Moreover bees had to be moved from season to season and could not be installed like a factory. Since the introduction of *Apis mellifera*, China has made a tremendous progress in beekeeping which has been well documented by those who visited that country (Fang, 1984; FAO, 1984, 1986; Gong, 1983; Guanghung Liu, 1985; Deh-Feng, 1981).

14.4.1 Enterprise Budgeting

Beekeeping generally means honey production, but a beekeepers can produce many other commodities for sale and thus generate income. These include, queen rearing, package bees, pollen and propolis collection, wax, royal jelly, bee venom, and renting colonies for pollination. The honey itself can be sold in bulk or can be bottled for retail selling and, further, it can be produced as comb honey or creamed honey which are sold at a higher price. All these sub–activities can be broken down to separate enterprise budgeting and from the income and cost the weak and strong points of the industry can be determined.

For keeping accounts both the fixed and variable costs have to be recorded. The fixed costs include capital, depreciation of equipment, taxes, insurance or rent. The variable costs include labour, fuel, chemicals, supplies and bee food. The fixed costs depend upon the size of enterprise, *i.e.*, whether the colonies number is 100 or 1000. Leaving aside the hives the cost of other equipment and its depreciation is going to be near about the same. In variable costs the investment on labour or bee feed is going to be much different in the two cases. The accounts of the income and cost on the enterprise maintained would give the final profit or loss.

The production and marketing strategies would depend upon the type of operation or the kind of colony to be used. The operation can be for honey production only or for pollination in addition. The colonies used may be over-wintered or the new package bees. Management in all these cases would vary and labour, chemicals, feed etc. would also be different. The hives and supers used determine the winter management and cut on maintenance and losses. Timely replacement of old frames is desirable and the additional cost of production of new frames should be taken into consideration. New comb foundations can be prepared from recycled wax. The foundations can also be made of plastic which are coated with a thin layer of wax. Such combs are not attacked by wax moth but their acceptability has to be studied carefully.

The cost of extracting honey depends upon proper sequence and movement of operations. Manual removal of bees from the supers can be slow and more costly whereas the use of a blower can facilitate and expedite the operation, reducing cost of the operation. The expenditure can be compared with that of the use of fume boards whoes effectiveness also depends upon the prevailing temperatures.

The honey extractors are available in various sizes and the selection of their size and number would depend upon the size of the apiary and the number of hives. For 1000 colonies the equivalent of two 50–frame extractors are required. If those are of inadequate total capacity the cost of extracting honey would shoot up for want of proper handling and flow of the operations. The same applies to settling tanks and uncapping knives. Proper capacity and efficiency would limit the costs. In brief, each and every operation, should be according to the economy of scales and should be cost effective.

14.5 MARKETING

A person who starts beekeeping as a hobby will find that he is left with surplus honey in excess of his need. He can dispose off the produce by giving it to friends and relatives or try to sell it in the neighbourhood. After the initial start a beekeeper might like to expand his apiary and to increase production. The sizeable quantities are handled in a different way. He can pack honey and sell it in retail or sell it to wholesale units which would have a bigger outlet. Honey can also be sold in bulk to Government agencies, to packers, or to large traders who may sell it in bulk or export it in large drums of 200 litre capacity, food lacquered, with two holes. A drum of this size holds 290 kg honey. The drums are sent abroad in 20 ft long containers which carry 18.85 tonnes of net honey.

Honey, whether given as a gift or sold in retail, should be in attractive jars and containers possibly having catching labels. Honey sold in recycled bottles does not receive the appreciation that it deserves. If sold in narrow mouth squash bottles, it granulates in winter and is difficult to take out. Moreover, it may give a lingering feeling that granules are perhaps crystals of sugar. A beekeeper having a small business should process the honey in improvised equipment and properly filter it, removing the air bubbles and wax that come up on top when honey is made to stand for sometime.

For retail selling the best strategy is to meet the needs and desires of a special group of people who always purchase honey with great curiosity for its good flavour. Creamed honey which can be produced at a small scale in winter months attracts lot of people. Some people may find comb honey attractive although it would be something new for India and the taste has to be developed. The new products should be introduced gradually, in step with the likings of the people. In contrast to the sale of honey in shopping malls the retail sale of honey should meet the fancy and requirements of individuals. There is a lot of health consciousness these days and some persons prefer to eat raw unprocessed and unfiltered honey produced in areas where crops are not sprayed. Such a health food is sold at a premium price.

14.5.1 Outlets for Retail Sales

There are many situations from where honey can be sold, starting from home to places of work, factories, in fairs and males or festivals.

(*i*) *Sale from Home*. It is the least expensive method and honey can be kept in honey house or a side room approachable from outside. Apart from the packed honey it can also be sold in a container of the purchaser.

(*ii*) Honey can be sold outside large offices factories or other places of work, with the permission of seniors incharge.

(*iii*) Roadside stalls or juice and cold drink bars are quite suitable places. In fact honey bottles always catch the eye of the people. For giving them assurance of purity the name of well known farm can be displayed alongwith honey bottles.

(*iv*) Farmer's Markets where fresh vegetables are sold are also attractive places for the sale of honey.

(*v*) Country fairs arranged near temples are particularly popular among women and children. Honey is considered good for suppressing cough and can be sold easily.

(*vi*) The annual festivals like Basant, Baisakhi, Id, Janamashtmi, Dussehra, Diwali and other Saint's days are celebrated throughout the country. Generally there are large gatherings and that is a good occasion for the sale of honey and other products. The wholesale traders may also visit there and make enquiries for bulk purchase.

(*vii*) Annual beekeeping fairs at National, State or District levels supported by Beekeeper's Societies are already becoming a regular feature. Such exhibitions are particularly visited by traders and wholesalers who want to purchase honey, wax, pollen etc. in bulk. The well reputed beekeepers can make contacts there.

(*viii*) Honey open houses are organized by some well established beekeepers during the extracting season. Honey extraction is a mysterious and interesting operation for many people to watch. Alongside other sales like pumpkins, sugarcane juice, fruit juice and tody or rice wine, can make the event interesting. The honey harvesting demonstrations can be quite educative, so they should be neat and clean free from the robbing honeybees and wasps in the compound where arranged.

14.5.2 Wholesale or Packaged Honey

Instead of selling packaged honey yourself other people can join in and sell it in larger stores, super markets, health food stores etc. The sales of specific brands can be supported by advertising through flyers, by advertisements in local and regional news papers with signs of the labels. Sale promotion can be handled by giving insertions on yellow pages in a phone book, and through broadcasts on radios or telecasts on TV. Naturally brands that have sufficient stocks and supplies should be advertised on mass media. In due course some of the brands become well known nationally and are universally accepted. It has been noticed in Punjab that there is a natural growth from a small beekeeper, to an enterprenure in honey sale, and even a bulk purchaser and seller within a matter of 2–3 decades.

14.5.3 Export of Honey

Export of honey to USA and possibly Germany and other countries is a new phenomenon observed since 1995. Earlier, small consignments were sent to the Middle-Eastern Countries by small firms but the major break through was achieved by Honey Bee Natural Products and Punjab Agro-Industries Corporation Ltd. who negotiated with firms in USA and Germany and obtained an order for the export of 1000 tonnes of honey in the season 1995-96 but they could supply only 800 tonnes. Other large scale exporters had also appeared in the market who probably procured 200 tonnes from Punjab and Haryana.

The international trade is governed by a number of factors. The first and formost is the identification of trading partners from among the importing countries. These countries include Germany, the Netherlands, France, Belgium, Saudi Arabia, Kuwait, Hong Kong, Japan etc.

It is not easy for individual honey producers to make contacts with agencies abroad who would buy their product. Government agencies of export have long procedures and lot of formalities. Intermediate companies who can handle export include management companies, export trading companies or by utilizing the services of export department of large corporations who normally handle other commodities.

The foreign buyers firstly want quality control of honey free from contaminants like sulphur which is used for controlling mite diseases. They generally like to import single source honey with specific flavours. The mixed source of exported honey can be detected from the pollen slides (Nair, 1964). In international trade the tests of quality are rigorous. Then long–term commitments of supply are required as the foreign firms once they launch a product want to continue the same supply of the standards fixed. Proper contacts are needed with a reliable freight company who can advise on freight rates, documentation, export packing and other requirements. For maintaining quality of honey their route through the tropical seas and the time of transport are other significant considerations.

14.6 QUALITY CONTROL

The primary goal of beekeeping is to obtain honey from the colonies. Other products obtained are wax, propolis, pollen, bee venom. It is estimated that annual world production of honey is around one million tonnes valued at 1000 million US dollars.

According to Codex Alimentarius Commission (1983/84) honey is defined as : "Honey is the unfermented, sweet substance produced by honeybees from the nectar of blossoms or from secretions of or on living parts of plants which they collect, transform and combine with specific substances, and store (and ripen or mature) in honey combs. Honey shall not have any objectionable flavour, aroma or taint absorbed from foreign matter during its processing and storage and shall not contain natural plant toxins in an amount which may constitute a hazard to health".

TABLE 19. ***The main constituents of honey (%)***

Constituents	*U.S.A. (Standard)*
Water	17.2
Fructose	38.2
Glucose	31.3
Sucrose	1.3
Disaccharides (maltose)	7.3
Higher sugars	1.5
Total acids (gluconic)	0.57
Minerals	0.17
Proteins and amino acids	0.04
Enzymes	traces
Aroma constituents	traces
Other substances including vitamins	traces

14.6.1 Chemical Analysis of Honey

Honey produced in various parts of the country and climates varies in colour from water white to almost black and has a great variety of aromas depending upon the flowers from which it is collected. The main constituents in all honeys are the same which are given in Table 19.

If honey is stored at temperatures much below the hive temperature (92–95°F) some of the honey sugars crystalize and it is said to have granulated. Most honeys contain more fructose than glucose but there are exceptions in which it is the reverse. A honey in which glucose is in excess granulates more readily, the example being *Brassica* honey. Composition of *A.c.* honey from unifloral and multiforal sources has been studied by Phadke (1962a; 1967a; 1967b). He also analysed honey collected and produced by *Apis dorsata, Apis florea, Trigona* sp. at Mahabaleshwar. Honey of *A.d.* collected from Punjab shows some differences (Mann *et al.*, 1970), which are on the positive side and constituents are close to the machine extracted *A.m.* honey.

For details on acids, minerals, enzymes, aroma constituents and other minor sugars see Crane 1990).

14.6.2 Bees Wax

Bees wax is secreted by worker honeybees on the underside of abdomen where there are located our pairs of wax glands. It is a plastic used for making combs and is supple so that at the hive emperature it can support 2–3 kg of honey per frame. The melting point of wax is 63–65°C. The omposition of bees wax does not vary much from one country to the other and therefore, adulteration an be detected fairly easily.

Wax of *Apis mellifera* and the African races is distinct from that of the three Asian honeybees *A. erana, A. florea, A. dorsata*. Wax from the Asian bees is known as Ghedda in international trade. For etails consult Phadke *et al.* (1961, 1969, 1972) and Poncini (1987).

The standard *A.m.* bees wax is composed of nearly 300 fractions of compounds which consist of ydrocarbons (10.5%) mon-esters (35%), diesters (14%), hydroxy polyesters (8%), hydroxy nonoesters (4%), free acids (8%), acid esters and polyesters (3%), triesters (3%), unspecified (13%). The available literature has been reviewed by Callow (1963), Tulloch (1980) and Crane (1990). Tests or its adulteration are described by Tulloch (1973). As compared to other waxes, bees wax has a omparatively low melting point and between 25–35°C a comb can support many kilograms of honey veight.

Beeswax when applied on a surface in thin layers gives protection from water. Wax is an inert naterial which retains its shape and quality for centuries.Wax from the cappings is of the highest quality. That from old frames is cleaned by removing the dark scum that settles at the bottom of a wax nelting tank. It is strained further to clarify it and is then rendered into blocks for sale.

14.6.3 Propolis

It is a dark solid workable plastic substance collected by bees from the resins and gums of plants and trees. It is a material that honeybees and some other bees collect from living plants and use it as such or mixed with wax for the construction and repair of their hive. When collected out of a hive it generally has a mixture of wax in it. At hive temperature it is soft but at low temperatures it becomes hard and brittle.

In Europe propolis is used for its medicinal value. It is supposed to have antiallergic and anti-bacterial properties. It is effective against herpes virus, and is useful for wound healing and tissue regeneration. Its extracts are a cure for many dermatological disorders. The exact ingradients in propolis that are effective are not known, Walker and Crane (1987) have listed 149 compounds and 22 minerals from the samples of propolis. They considered that a group of 38 flavonoids probably constituted a greater bulk of this material.

The annual production of propolis in China is reported to be 55 tonnes; in Argentina 7–8; Chile 7–8; Canada 3–4 tonnes.

14.6.4 Royal Jelly

Royal jelly is a rich food on which honeybees rear their brood. Queen larva is given abundance of this food and it grows 1,300 times in 6 days. Royal jelly is a secretion of the hypopharyngeal or brood food glands located in the head of young worker honeybees. The worker larvae are given less of this food and they remain smaller in size as compared with queen larva. When they transform to adults the same size difference is manifest in the two.

Since fifties royal jelly is being used as human health food in the form of capsules and is taken first thing in the morning. It is believed to be a general tonic and a cure against arthritis and it imparts immunity against common colds (Stein, 1989). In natural form it is used as face cream and is applied on skin to clear blemishes and eczema. Almost 500 tonnes are produced in the World and it is fast becoming an important item of trade.

Royal jelly is manually harvested from the queen cells which are reared in large numbers by the standard method of queen rearing. Royal jelly is collected by skilled operators and the larvae whoes food is taken away die as they cannot complete development. One man can collect it from 2–3 hive in one hour and from 13–17 hives he can collect 100 g of jelly in one day. For commercial production of royal jelly the procedure has to be programmed so that there is a continuous supply of queen cells in the active season of collection (Smith, 1960, 1964).

The composition of royal jelly as given by Takenaka (1982) is : water 66.9%; proteins, 11.4%; sugars, 9.1%; other extracts, 6.2%; ash 0.94%. Among the enzymes it contains glucose oxidase, phosphatase and the vitamins recorded were B_1 (thiamine); B_2 (riboflavin); niacin; biotin; folic acid; pantothenic acid. It also contains inositol and acetycholine.

For the collection of royal jelly it is imperative that pure hygenic conditions are maintained in the apiary and in the colonies while it is being collected. The greatest care is needed to avoid contamination after it has been collected. It has to be converted to capsules under aseptic conditions and in case it is converted into face cream it has to be handled similarly.

14.6.5 Pollen

Pollen is collected by bees from plants which they pack like pellets in the baskets of hind legs and bring to the hive as food for their brood. If a pollen trap or a narrow grill is placed in front of the hive entrance the bees as they pass through the grill drop off pellets in the tray below. Large quantities of pollen are collected that way and after 2 weeks the pollen trap is removed for one week so that bees can collect sufficient pollen as a food for their own brood also. Generally pollen is a mixture from various plants to meet the nutritional requirements of their brood.

The pollen so collected is shade dried with the help of an air blower in trays placed one on top of another. The pollen remains well preserved and does not germinate as the bees mix certain substances while collecting it in the field. Pollen is used as a human health food and it is believed to have antibiotic and antiallergic properties. Air dried pollen can be kept at cool temperature and it deteriorates after some months. It is eaten raw or in capsules.

Pollen is exported by Spain, Hungary, Argentina, Taiwan, Australia. The annual production of Western Australia is 60–130 tonnes; that of Taiwan is 150 tonnes; and of Vietnam is 15–20 tonnes. Pollen is imported by Germany, France, Italy, U.K and U.S.A. The exact trade figures are not available (Crane, 1990).

The composition of pollen collected by honeybees is : water 11.2%; crude protein 21.6%; other extract 5.0%; carbohydrates as reducing sugars 25.7%; non reducing sugars 2.7%; starch 2.6%; ash 2.7%; undetermined compounds 28.9%. A large number of other substances have been identified from pollen and they include phenolic acids, free amino acids, nucleic acids, terpenes, enzymes, vitamins, carotenoids, flavonoids, growth regulators etc. (Stanley and Linskens, 1974).

Pollen is not supposed to be a staple food for man but considering its high cost of collection and marketing, it has a scope for use as a health food supplement and in certain cases as an antiallergic agent for those who suffer from specific allergies in certain seasons. Naturally it should be collected, processed and preserved under strict hygenic conditions.

14.6.6 Treatment of Bee Venom and Stings

Honeybee sting is a traditional remedy for rheumatism. By making a bee sting at the site the dose of venom injected cannot be controlled. Therefore efforts were made to extract venom and make it available in injectable form which could be administered in standard doses. The bees sitting on a nylon frame were given mild electric shock from below and they embedded their stings in the cloth without losing them. The venom on the screen was washed with water and on dehydration a crystalline powder was formed which could be used for injections in measured concentration (Marcovic and Molnar, 1955).

Many scientists have tried to determine the composition of venom. Shipolini (1984) has compiled the data. The venom contains enzymes, small proteins and peptides, physiologically active amines, and amino acids. It has been reported that; AID, MS and cancer patients have high level of costisol in their blood and bee venom artificially turns off body's ability to overproduce this chemical (Rose, 1999).

14.7 HONEY AND BEESWAX TRADE

According to 1984 FAO figures around 263 thousand tonnes of honey is imported by various countries and 269 thousand tonnes is exported; the respective figures for bees wax are 5,571 tonnes imports and 6,146 tonnes consumption mostly in France, Germany, Netherlands, Switzerland, USA and Japan (Hepburn, 1986). Honey and wax trade in India is reviewed by Joshi (1982). The estimated figures of honey and beeswax production in the world are given in Tables 20 and 21.

14.7.1 World Production

Honey and beeswax production in various continents is given in Table 20.

TABLE 20. *World Honey and Beeswax production*

Continent	*Honey (1000 tonnes)*		*Beeswax (M. Tonnes)*	
	1971	*1984*	*1993*	*1984*
Europe	122	165	169	1,777
USSR	220	193	240 (1991)	3,860
N. America	—	118	—	3,027
C. America	—	87	—	10,528
N. + C. America	158	205	198	13,555
S. America	—	57	85	3,268
Oceania	25	31	31	592
Asia	241	246	352	15,870
Africa	25	95	114	8,682

After FAO Production Figures.

The honey and beeswax productions of individual countries of the World have been compiled by Crane (1990). The production figures of some of the important countries of the world as reported by Food and Agriculture Organization of UN are also given here (Table 20). The statewise production of honey in India is given in the Table 21.

TABLE 21. *Honey and Beeswax Production in Major Countries*

(M. Tonnes, after FAO)

S. No.	*Country*	*Honey*		
		1979–81	*1993*	*Beeswax*
1.	Angola	15,000	15,000	1,300
2.	Argentina	34,533	48,000	1,500
3.	Australia	20,923	22,556	482
4.	Bulgaria	9,334	3,000	290
5.	Canada	32,303	31,000	1,327
6.	China	1,39,110	2,05,200	12,800
7.	Cuba	7,602	10,500	—
8.	Czechoslovakia	6,043	9,225	—
9.	Ethiopia	20,500	23,700	2,100
10.	France	12,321	17,500	—
11.	Germany	16,416	28,000	—
12.	Hungary	13,531	10,700	—
13.	India	43,333*	51,000*	1,800*
14.	Kenya	10,834	19,500	1,050
15.	Mexico	65,758	55,000	9,150
16.	Poland	11,889	12,900	—
17.	Romania	14,279	10,000	—
18.	Spain	12,835	25,000	730
19.	Tanzania	9,500	16,000	1,050
20.	Turkey	27,090	61,000	2,302
21.	USA	94,404	90,000	1,700

* All *Apis* spp. combined; 60–70% Beeswax is from *Apis dorsata*.

14.7.2 Honey Production in India

Reliable data on honey production in various states of India is not easy to get because it is not a part of the Agricultural Statistics collected by the State Departments of Agriculture nor by the Economics and Statistic Cells. The Khadi and Village Industries Commission of India, Bombay, have their net work of development centres throughout India and they have made efforts to collect statistics on the number of hives and beekeepers, and the honey production. It would be seen that when the honey production in various States (Table 22) is totalled it does not add up to the figures sent from India to the FAO world production statistics. In any case the Statewise honey production in the country is given to indicate the comparative beekeeping activity in various parts of India.

TABLE 22. *Statewise Production of Honey in India*

S. No.	*State*	*Production of Honey (Metric Tonnes)*
1.	Andhra Pradesh	104.20
2.	Arunachal Pradesh	0.30
3.	Assam	401.15
4.	Bihar*	360.97
5.	Goa	0.16
6.	Gujarat	2.02
7.	Haryana*	6.96
8.	Himachal Pradesh*	60.00
9.	Jammu & Kashmir*	96.70
10.	Karnataka	662.59
11.	Kerala	3,658.52
12.	Madhya Pradesh	86.32
13.	Maharashtra	78.24
14.	Manipur	207.50
15.	Meghalaya	57.92
16.	Mizoram	2.50
17.	Nagaland	19.22
18.	Odisha	590.00
19.	Punjab*	540.00
20.	Rajasthan	13.96
21.	Sikkim	1.46
22.	Tamilnadu	1,946.00
23.	Tripura	2.72
24.	Uttar Pradesh*	97.33
25.	West Bengal	496.96
	Total apiary honey	9,254.33
	Total forest honey (approx.)	18,000.00
	Grand Total	27,254.00

* Includes honey from *A. mellifera*. **Source**: beekeeping KVI (Khadi Village Institutes) Report-1992.

The most significant feature of Beekeeping Industry in India is the export of honey to other countries which started at a reasonable scale in 1995–96. Under a project of Punjab Agro–Industries Corporation, 800 tonnes of honey was exported to USA in 1996. Another 200-300 tonnes of honey was exported by other private firms to the Middle-East countries. There is already a demand from Europe for litchi and other flavours of honey from India.

It is apparent now that there is a great potential of Mellifera beekeeping in the country and the World market is ready to purchase any amount of honey from India. There is a sizeable local market also for sale of honey and for its use in indigenous medicines. One well known firm in India uses at least 2,000 tonnes of honey for these purposes. Then there are other smaller packers and retail sellers. It is clear there are great prospects of honey production and sale in India in the years to come.

Book Two

POLLINATION OF CROPS BY BEES

Pollination of Crops by Bees

Plants play a vital ecological role in the biosphere but a vast majority of them are not of direct economic value to man in providing timber, food, fiber and firewood. In this section on pollination we are basically concerned with flowering plants (angiosperms) which vary in size from some of the tallest trees to the tiniest weeds growing on land or as parasites on other plants. Among them the economic plants including crops, fruits, vegetables, number around one hundred. Their cultivation and propagation is through seeds which are produced by the pollination of flowers. Agricultural grasses (wheat, barley, oats etc.) are pollinated mostly by wind but many other crops including horticultural crops are adapted for insect pollination and they have conspicuous coloured and scented flowers that attract insect and other pollinators (Free, 1993).

Pollination is caused by a simple process of transfer of a compatible pollen grain (male part) to the stigma of ovary (female part). If there is no pollination the flowers wither and produce no seed. The factors that govern pollination are so diverse and the environmental influences are so complex, that the ecological relationships have not been fully understood even in case of the most commonly cultivated crops. Firstly, the diversity, mechanism, and physiology of flowers varies a great deal and, secondly, the agents of pollination are influenced by the environmental conditions so much that the end results are unpredictable and certainly not consistent from year to year.

Among the agents of pollination, wind, gravity, dew drops and rain give the greatest coverage. Next come animals including mammals, birds, insects, the last mentioned being the most effective group. Among insects those known to play this role are beetles, moths, butterflies, bugs, plant hoppers, plant lice, thrips, ants, wasps, flies (particularly Diptera) and Apoidea. The important genera of Diptera include *Eristalis, Syrphus, Platycheirus, Rhingia, Calliphora, Lucilia, Sarcophaga, Bibio, Dilophus, Bombylius*. Except for Apoidea most of the insects are supplementary pollinators as they are basically visitors to flowers and are not totally dependant on them for their food. It is different with Apoidea comprising the solitary bees, subsocial bees, bumblebees and honey bees.

Bees are dependent on flowers for their food. There exists a close, association and mutually beneficial relationships through their evolution together since 60-100 million years ago. Bees not only aid in self pollination within a flower they also bring pollen from other plants for cross pollination thus widening the scope of gene pool and increasing biodiversity among the plants visited by them. It is estimated that there are 15-20 thousand species of bees in the world and no less than 4,000 exist in USA and Canada (Hoopingarner and Waller, 1992). In the Indian subcontinent there are perhaps fewer number of species of bees visiting flowers.

15.1 SCIENTIFIC TERMS

In the study of pollination by bees we come across a number of scientific terms and to understand their meaning is important. These terms include the following :

1. Flower. In the centre of a simple flower is an ovate ovary with an elongated style terminating in a sticky stigma. Growing around the base of an ovary are elongated male sex organs, the stamens, which vary in number from species to species. Each stamen elongates into a stalk which supports at the end a sac-like *anther* which produces the pollen grains, generally yellow in colour. The base of the flower on which those structures are located is the inner rounded structure which carries varying numbers, forms and colours of *petals* and the outermost part of the flower which originally sheathed the bud is called *calyx*. The calyx and the corolla together form the perianth which is the colourful circus of the flower, giving protection to the inner parts and at the same time attracting pollinators by the colour, odour, and by nectar secreted in the nectaries. The bees see all colours except the red which is perceived by them as black or shades of dark grey.

2. Pollen. The yellow powdery stuff generally seen on flowers is pollen which is a collection of pollen grains. When seen under a microscope the pollen grains of different species of flowers show their own shape and distinctive armature. From the identification of pollen on bees or in honey the species of plants visited by them can be determined.

3. Pollination. The deposition of right pollen on the stigmas of flowers of the same species is known as pollination.

4. Fertilization. Once a pollen grain is deposited on the stigma it germinates and grows a long pollen tube through the stalk and reaches an ovule inside the ovary. The male and female gametes unite and fertilization takes place.

5. Self-Pollination. A plant is considered self-pollinated when pollen from an anther reaches stigma of a flower of the same plant or another plant of the same clone. If viable seed is produced plant is self-fertile and in case the seed produced is not viable it is self-sterile. Auto-self-pollination takes place with the growth and movement of flower parts and no external agent is involved in the transfer of pollen.

6. Cross-Pollination. It takes place with the transfer of pollen from an anther to a stigma of a flower of a different plant of the same species.

7. Pollinator. Agents like wind, insects etc. that bring about pollination are called pollinators.

8. Pollinizer. Pollinizer is the plant from which pollen originates.

9. Self-incompatible. In a self-incompatible crop pollinizer varieties are planted in between the rows so that bees can transfer pollen from one to the other. In fruits like apple and almond most of the varieties are self-incompatible. In orchards of plum, the indigenous variety Amritsari is planted as a pollinizer.

10. Self-compatible. In these varieties there is no need of plant pollinizer varieties as pollen is compatible within the variety. Such varieties can be planted in large blocks of land but the role of pollinators is essential to harvest a good crop.

11. Polyphilic. It is a blossom pollinated by many species of pollinators.

12. Oligophylic. A blossom pollinated by a few closely related species of pollinators.

13. Monophylic. It is a blossom pollinated by one or some closely related species of a pollinator, *e.g.*, honeybees.

14. Polytropic. It is a pollinator that visits many different species of flowers.

15. Oligotropic. A pollinator that visits only a few species of plants, *e.g.*, many solitary bees.

16. Monotropic. It is a pollinator that visits a single or some closely related species of flowers, *e.g.*, orchid bees.

17. Constancy of Bees. In the presence or abundance of blossoms of the same kind (Brassica field) an individual bee may work over an area not more that 30-40 feet square.

18. Fidelity. It is observed that in the presence of many species of flowers bees at a time stick to one species for the collection of pollen which is of advantage to the plant. If bees visit many species of plant their value as pollinators will be reduced.

19. Solitary Bees. The females individually make their nests in burrows, place a ball of honeyed pollen in each of the cells, lay an egg and seal the chamber. The larva grows on its own into an adult who cuts open the cell and escapes to start next generation by making a new burrow.

20. Sub-social Bees. The female visits the nest in a burrow repeatedly and sequentially bring in provisions till the larvae are fully developed and emerge as adults. The nest is then abandoned.

21. Semi-social Bees. More than one female live and work in a nest and any one of them may lay eggs on pollen balls, the other bringing in provisions. After raising a few generations in a season the nest may or may not be abandoned.

22. Social Bees. They live in colonies and make multiple cells in their nest or comb and raise the brood by feeding them repeatedly and visiting them frequently. The food consists of pollen and honey and their own glandular secretions. Food is stored in the nest and caste system is well developed and reproductive queen is the centre of the colony.

15.2 POLLINATION NEEDS OF PLANTS

The crops commonly grown are either self-fertilized with their own pollen or they are self-infertile and need to be cross pollinated with pollen from other plants of the same species. The self-fertile plants are further of many kinds; some are pollinated with pollen of their own flowers, some have peculiar structure of their flowers and have to be pollinated by wind or insects and others will produce seeds and fruits by self-pollination but the yield and quality is higher when cross pollinated.

The self-fertile wheat and many other agricultural grasses are aided by wind as the pollinator, whereas, most other agricultural and horticultural crops that have brightly coloured and scented flowers are so evolved that they attract insect pollinators. The successful functioning of insects will depend upon whether the flower is self-infertile or it is self fertile but not fully self-pollinating. Many self-fertile and self-pollinating plants also benefit from insects. Some studies have been made on crop plants but the extent to which they need insect pollinators is not clear, particularly, in the self-fertile crops. There are various ways of determining it.

Bees are free moving insects and they cannot be directed to visit more a certain field and to pollinate less another field, although, field cages can be constructed on a crop before flowering. In one of them bees are released, in the second no bees are released, and the third is the open field where bees visit naturally. The three yields can be compared by measuring the weight of seeds and the percentage of flowers that set seed. In such experiments the cage effect cannot be avoided and the results are not altogether natural but they do show difference it makes to have bee pollinators and not to have them. Inside a cage with bees the crop would be over saturated with the pollinators which again is not a natural situation.

Another method of determining the effect of bee pollination is to place colonies in the centre of a large crop field and measure the yield at various distances going outwards. Depending upon the availability of manpower a number of repeats can be set up in the same year in that area or experiments can be carried out again next year.

For more close observations individual plants or parts of a plant can be caged in cloth or paper bags. Visits of pollinators have to be watched. Pollination can also be done by hand using a suitable technique and the flower shoot can then be caged. The seed or fruit set inside can be compared with controls outside. Other techniques suiting various crops can be devised for experimentation.

Bee population per unit area is crucial for the successful pollination of a crop. In the first place it should be known what is the concentration required for that crop and secondly how to count the number of pollinators in a field. The concentration required depends upon the effectiveness of pollinators which is quite difficult to note in the field as bees keep moving and the flowers visited cannot be tagged easily. Attempts can be made with radio tracer techniques for specific studies. The actual counting of bees is equality tricky. The common method suggested by (Free, 1970) is to hold a stick one meter long, to extend it on the crop and count insects under it as one walks at a steady pace along the edge of a field for a set distance. One acre-length (220 yards) is convenient as the fields are well demarcated in North-Western India. Another method is to mark (with reeds or poles) squares in the field distributed at random and to count bees at the required interval (Vansell and Todd, 1946). A portable frame of one meter square can be carried inside a field and insects counted randomly at selected spots for the required length of time (Smith and Townsend, 1952; Free, 1993).

If honeybees are to be used as pollinators it is possible to make some numerical counts but we must know what is the optimum number required for a given field of fruit trees. Based on the experience of beekeepers and growers estimates were made that 2.5 colonies per hactare were required to pollinate fruit trees (Hendrickson, 1916; Tufts, 1919). A colony of average size having two-third foragers would have bee force of 20,000 per acre. Naturally the number of bees required for pollination will depend upon the type of crop, the concentration of flowers and their attractiveness. In the beginning when flowers start opening in an orchard their concentrations is low and they are placed far apart. So colonies are moved to an orchard gradually, bringing in the maximum number when the crop is in full bloom.

In order to facilitate pollination and to increase the yield of crops the new plant types produced by breeding should have such characteristics as abundance of honey and pollen so that pollinators are attracted to them. Any defects in flower structure should also be removed. Other factors to be taken into consideration are suitable cultural practices to make plants attractive; efficient use of honeybee colonies and encouraging wild bees by increasing nesting sites; protecting pollinators from large-stretch spraying of pesticides; introducing new promissing insect pollinators from abroad.

15.2.1 Crops Benefiting from Insect Pollinators

Hambelton (1944) prepared a list of 50 agricultural crops which are either dependant upon honeybees for pollination and production or they yield more when bees are abundant in that area. Under Indian situation some of the crops that benefit from pollination by honey bees (four species) are different and even though the list cannot be considered complete an attempt has been made to do so as based on the literature available.

TABLE 23. *List of Crops and Fruits that benefit from Pollination by Honeybees*

Alfalfa	Cashewnut	Litchi
Almond	Cauliflower	Loquot
Apple	Cardamom	Mango
Apricot	Coffee	Nectarine
Anaseed	Coriander	Niger
Asparagus	Coconut	Okra
Barseem	Citrus	Onion
Blackberry	Cucurbits	Peach
Blueberry	Cucumber	Pear
Brassica	Ghia	Pepper
Toria	Pumpkin	Green Chilli
Sarson	Melon	Guava
Raya	Tori	Jujube
Lahi	Tinda	Persimon
Kale	Watermelon	Plum
Radish	Celery	Prune
Turnip	Clovers	Raspberry
Brinjal	Cherry	Red oil Palm
Broccoli	Chickpea	Sesame
Brussels sprout	Cotton	Safflower
Buckwheat	Eruca	Soybean
Cabbage	Fennel	Sunflower
Carrot	Linseed	

15.3 BEE POLLINATORS AND THEIR FIELD ACTIVITY

Under natural conditions there was originally mixed forest vegetation with a few cultivated fields near human habitations. It would appear the wild bees were adapted to meet the pollination requirements of the plants around them. With the extension of agriculture and its modernization, there came clean culture, double cropping constant stirring of the soil, and great concentration of single flower species at places. The worst was insecticidal spraying of crops for killing some pests but at the same time eliminating pollinators. With intensive agriculture and urbanization the natural nesting sites of solitary, subsocial and semisocial bees that make their nests in the ground were further truncated. One has to simply scan the countryside where agriculture is based on irrigation and machines. Where is the land for bees to make nests in the ground ? After the ground is flooded in monsoon where do the bees go ? They make a tenacious survival by making nests along the canals, roads, rail road tracks, village shamlats, fields of tree farming and the forest edges that have receded to the rivers, streams, ravines and unproductive lands. Suitable nesting sites have become scarce and the population of bees seems to have dwindled accordingly. The biology of some of the wild bees has been studied (Atwal, 1970, Kapil and Jain, 1980) which indicate the varied nesting and living habits.

The wild bees work on flowers very intensively and are thus efficient pollinators. They collect nectar and pollen on the same trip and are bound to transfer pollen to a stigma. Some of them are specific pollinators of certain plants and when introduced to new areas the crops are not a success due to lack of pollination (Bohart, 1966). The wild bees essentially work as solitary insects and they cannot be manipulated to pollinate crops as we desire. Some efforts have been made to artificially increase their population by providing nesting sites. The well known examples are those of the leaf-cutter bee, *Megachile rotundata,* the alkali bee, *Nomia melanderi,* and the bumblebee, *Bombus* (Fig. 21). They have been used for improving the pollination of alfalfa in USA.

M. rotundata was inadvertently introduced into Eastern USA around 1930 from Eastern Europe or West Asia. It spread westwards and was found to be an efficient pollinator of alfalfa. Efforts were made to artificially increase its population and it was found that man made nesting sites were accepted by the bees. They made nests with cut leaf pieces as linings inside holes in wooden blocks; semicircular grooves 3 mm wide were made in wood strips and when two faces were matched together holes 6 mm in diameter were formed. The length of the tunnels was 11.5 cm. The blocks of wood were arranged one above the other forming many rows of tunnels. The structure of artificial domiciles was placed on a stand and was provided with a shade as protection from sun, wind and rain. It was screened in the front to prevent birds from attacking the nests. Instead of wood plastic castings could also be used (Bohart, 1962; Stephen, 1962; Hobbs, 1964).

Nomia makes its nests in the ground and prefers certain sites where soil is alkaline and is constantly moist. The main vertical shaft of the nest tunnel is 15-20 cm long and at the end there are 3-4 branches, 5-7 cm in depth. The nests are provisioned and brood develops on its own. Attempts to increase the number of nests artificially have been successful. The required type of sites are created by compacting the ground below or placing a polyethylene sheet and covering the area with suitable soil. Since water does not percolate below, soil remains moist, and it can be made more attractive to the bees by mixing sodium or potassium salts (Stephen, 1965; Bohart, 1967).

Fig. 21. Top—a stingless bee ; **Below**—a bumblebee.

In search of more solitary bees for the pollination of alfalfa a US-PL-480 grant was given for work at Ludhiana (Punjab). Author was the principal investigator and Bohart (1965-70) was the collaborating scientist. Insect pollinators of this and other crops were surveyed and their biology was studied (Atwal, 1970; Batra, 1977; Kapil and Dhaliwal, 1968, 1969). A bright green carpenter bee, *Pithitis smaragdula* which makes nests in the pith of reeds was found promising and was later on introduced into USA. In the experiments at Ludhiana it was possible to attract the bees to make their nests in one-meter long dry reeds fixed in various positions. The acceptability of those fixed vertically as individual stalks in the ground around alfalfa fields was 50.8% and of those fixed on poles just above the crop (0.25 m) horizontal or at 30° was 30.9 per cent. The bees made 1-11 cells in the reed lengths (Atwal, 1970; Kapil & Kumar, 1969). The bees easily made nests in roofs of huts close to alfalfa fields at various distances. By releasing marked bees and observing them at the nests it was determined that from 100 meters all bees returned but from 250 m none of them returned to their nests. In plots of 500 m^2, nests were provided at the rate of 100, 150, 200, 250 and population per 25 sweeps was recorded 10 days later; it was observed to be 11.7, 14.4, 16.3, 19.3, whereas the initial population was 3.0, 4.5, 1.5, 0. It indicated progressive rise in Pithitis population (Dhaliwal and Atwal, 1980).

By improving the artificial nesting technique the solitary bees can be used for pollination under special situations such as confined spaces where hybrid seed of flowers and vegetables is produced.

Regardless of the isolated successes achieved in artificially increasing the population of solitary bees they cannot be relied upon for the pollination of valuable crops or those grown in large fields commercially. The honeybees are more dependable. The two wild species, *Apis florea* and *Apis dorsata*, cannot be manouvered to the desired crops but their natural population is quite large and on some of the crops they are the most common visitors (Kapil and Brar, 1971). The two hive-species, *Apis cerana* and *Apis mellifera*, are the most useful bees for directed pollination needed for obtaining economic yields. It is, therefore, meaningful to study the field behaviour, foraging habits, pollination efficiency, fidelity and other characteristics of the hive bees, particularly *Apis mellifera* which has been the subject of much interest in the world.

The comparative abundance of the four species of *Apis* on some of the flowering crops was studied in North-Western India in the late sixties. In the higher mountains of Kulu valley *Apis cerana* formed 85 per cent of the insects collected on *Brassica* crops and cruciferous vegetables (Kapil *et al*., 1971). It was the most abundant insect on pome fruits. In the plains this honeybee was experimentally released in large field screen cages on *Brassica campestris* var toria, and on American cotton; the increase in yield was recorded on both these crops. It ranged from 100-151 per cent in the former and 17.4-19.0 per cent in the latter crop (Grewal, 1963; Sidhu, 1959; Sidhu and Singh, 1961, 1962).

In the Himalayas, the position of *Apis cerana* as pollinator has changed very much since 1985 because 95% of the colonies have perished by the attack of TSBV, and its place has been taken over by *Apis mellifera* which was introduced in the field in mid seventies. The role of the wild honey bees, *Apis florea* and *Apis dorsata* on *Brassica* crops is also very interesting (**Plates 13, 14, 15**). They are not found on the high mountains but at the lower altitudes the latter is the most abundant among insect visitors. In the southern districts of Punjab, Haryana and part of Rajasthan the former is the most abundant and forms 73-84% of total insects collected (Atwal, 1970; Kapil *et al*., 1971; Kumar *et al*., 1988). It may, however, be noted that all the *Apis* species gather nectar from the side of the corolla of *Brassica* flowers, hence they are not as efficient pollinators of flowers as their abundance would suggest. On open flowers like cotton, sunflower, pome fruits etc. they are efficient pollinators and in fact the hived species are a necessity for the apple orchardists in Uttar Pradesh, Himachal Pradesh, Jammu and Kashmir.

15.4 HONEYBEES AS POLLINATORS

Extensive studies on the foraging habits of *Apis mellifera* have been made in Europe and North America. Through dances bees convey to new recruits the foraging sites in a field. Once the bees reach that field individual bees concentrate on their own site which may be only a few feet square. They recognize the plants through odour of the nectar and it is also likely that they demark their territory by leaving behind their odour of pheromone (Free, 1987). Bees get conditioned to that source and exhibit constancy ranging from 40-60% (Free, 1963). In a field of *Brassica juncea* bees were marked and were observed to work within an area of 10 m^2. They were then marked in a smaller area of 5 × 7 m and it was observed after 24 hours that 90% of them still worked there (Minderhoud, 1931; Butler, 1941). Foraging bees were marked on a number of plants and it was found that foraging area was quite small during calm and sunny weather when nectar and pollen were abundant. When the density of flowers was low or when the flowers had less nectar and pollen bees foraged in a larger area even though they landed in the same restricted area. The departure points varied in different trips (Singh, 1950). They can locate their foraging site more easily if there are some prominent land marks like a row or grove of trees, a building etc. (Weavers, 1957; Free, 1966).

In a season bees forage on a number of plant species but collect their nectar and pollen from a few only and stick to them as long as pollen and nectar is available (**Plates 13, 14, 15**). Colonies lying side by side in an apiary may exploit different sources altogether, and once they start using that source they stick to it as if they become conditioned to it. However, if there is rich source in abundance bees from all colonies may converge on it which is the case during a main honey flow (Todd and Bishop, 1941; Synge, 1947).

Foraging range of a colony depends upon flora available. Although bees have been noticed foraging at a distance of 11.3 km, they generally forage within 4-6 km under conditions of sparce flora but on a rich crops they concentrate within 0.4-0.8 km. They prefer forage near the hive and can recruit more foragers for short distances than for long distances. Gradually they spread outwards but density remains higher near the hive. It was observed in a field that bees foraged within 137 m on the first day, upto 549 m on the second day and 686 m on the third day (Karmo, 1958). In general the young inexperienced workers tend to remain near the hive and older ones venture further away. In an experiment bees were marked with radio active gold and it was observed in the field that their proportion to the total bees decreased with the increasing distance from the colonies; those were 48% at 100-200 m; 38% at 300-400 m; 28% at 400-500 m (Haragsim *et al*., 1965).

Gary *et al*. (1973) made extensive studies in an alfalfa field of 65 ha where colonies were placed in small 29 groups; the mean foraging distance from the base group was 198 m for nectar gatherers and 240 m for pollen gatherers. It has been shown by many workers that for proper pollination in a crops colonies should be placed single or in small groups of 4 with their entrances facing outwards so that bees spread equally in all directions. Levin (1961a, b) observed, that foragers from a group of colonies decreased in numbers towards a neighbouring apiary probably due to competition.

Under natural conditions in India honeybees may be thinly distributed in a locality and wild honeybees or solitary bees may be quite abundant and play an important role in the pollination of a crop. The element of competition among various species may be quite manifest. Further on it is mentioned under individual crops that one or the other wild honey bee may visit a crop in great prepondence depending upon the general population in that region but the solitary bees never out-number the wild honeybees or the hive honeybees. Moreover, there is foraging competition between honeybees, wild honeybees and solitary bees (Dhaliwal and Atwal, 1976; Free, 1993; Sugden, *et al*., 1996).

It was observed on *Brassica campestris* Vastoria that when more than one species of insects (wild honeybees, syrphid flies, bumble bees etc.) are working on a flower and a forager like a hive honeybee (*Apis mellifera*) comes to that crop in large and threatening numbers the earlier visitors tend to fly away and sort of vacate the source for the new (may be powerful) comer. This phenomenon was observed most spectacularly at Ludhiana (Punjab) when a good size apiary of *A.m.* was established there for the first time and there was no general population of this species in that area, the number of insects caught in 25 net sweeps by the hour were counted at 1 km distance. *A.m.* ranged from 4.3 to 18.7 whereas all the other bees combined (*A. florea, A. dorsata, Andrena ilerda*) ranged from 0.5 to 5.2. At a distance of 2 km, however, *A.m.* ranged from 1.5 to 6.5 and the other bees combined numbered from 4.0 to 11.3. It was indicated that upto 1 km from the apiary it was the area of dominance by *A.m.* and at 2 Km the other bees foraged at the normal population level (Dhaliwal and Atwal, 1976).

Even though bees tend to forage in a small area of their choice (4-5 m^2) they tend to drift to a wider area as the resources become exhausted. It inevitably leads to the contamination of seed. In order to maintain purity of seed a breeder has to isolate the seed plots. Certain isolation distances have been worked out for some crops. For the production of registered seed the distances recommended are : clovers 90 m; cotton 400 m; onion 800 m; sesame 360 m; sunflower 800 m; tobacco 500 m; tomato 300 m; water melon 800 m (Frankel and Galun, 1977).

Colony management techniques for pollination have been reviewed by Seeley (1985); Free (1987); Winston (1987); Gould and Gould (1988); Roubik (1989). Recent advances in the behaviour of pollinators and nectar gatherers indicate they can switch from one duty to the other according to the needs of the colony. The presence of queen and egg laying is directly correlated with forage activity. Pheromones play an important role. These chemicals are mixed in nectar and as the load is transferred to house bees inside the entrance they pass on odour to other bees; pollen gatherers mix these chemicals while depositing the loads direct in the cells. The amount depo-sited depends upon the number of cells available or the number of brood to be fed.

Some of the newer techniques being developed for improved pollination include: breeding special races of bees for more efficient pollination; synthetic pheromones to attract bees to flowers or to the brood for increased collection of pollen; use of solitary bees for certain crops (*e.g., Megachile, Pithitis, Osmia*); rearing and maintaining bumble bees, stingless bees, blow flies etc. for controlled pollination in enclosures, particularly for the production of hybrid seed.

The use of synthetic pheromones is on the increase for the pollination of some precious crops grown commercially. In many advanced countries synthetic, Queen Mandibular Pheromone (QMP) is sprayed on crops like kiwi fruit, strawberry, blueberry, apple, almond etc. to attract honeybees for greater pollination and high fruit production (Kevan, 1999). The commercial products are sold under various trade names such as Bee Boost, to attract a queenless swarm, and Fruit Boost, to attract honeybees (*A.m.*) to a given field for maximum pollination (McCarthy, 1999).

Apart from the reduction of population of wild bees under intensive agriculture there are other factors that influence pollination, requirements of crops and fruit trees.

(*a*) **Pollinizers for cross-pollination**. The plants that are not self-fertile need pollen from other plants for cross-pollination by the bees. At the time of planning an orchard of apples, almonds and pome fruits the pollinizing varieties should be planted in alternate rows or at regular intervals mixed with the main variety to be pollinated. Pollinators while working tend to move along the rows of fruit trees and they do not criss-cross.

(*b*) **Competing pollinizers**. There are weeds and plants that bloom earlier than the main crop to be pollinated. The bees get diverted to these uneconomical plants. In case those blossom much earlier, the bees can make use of them for brood rearing. If on the other hand they bloom coinciding with the main crop the competition should be eliminated, *e.g.*, weed dandelion competes with apple blossom. Some repellants can be applied or the area can be cleared of such weeds.

(*c*) **Climate**. Even in the presence of pollinators a crop may not benefit fully and there may not be proper seed and fruit set. Untimely frost, rain, wind, or storm may interfere in pollination. If it is a regular feature from year to year it would mean the locality is not suitable for that crop and another one should be tried in its place. In case the condition is due to lack of soil moisture and poor nectar in flowers arrangements should be made for irrigation, coinciding with the blooming period.

(*d*) **Wintering conditions**. The honeybee colonies before winter are packed with sufficient food and good insulation to pass the cold weather without heavy losses. The colonies before the spring are expected to start brood rearing, making use of wild pears, dandelions and other weeds in the mountainous areas where winter is cold and spring flowers are late. The colonies should be ready with sufficient field workers who can pollinate apple and other pome fruits in time. If the colony strength is poor there would not be sufficient number of foraging bees to pollinate the fruit trees. In case the winter is too severe there is expected to be heavy mortality among the wild bees also and there would be further paucity of pollinators. In order to meet these situations migrating bee colonies to lower altitudes should be practiced for building colony strength during winter and bringing them back to apple orchards in the spring.

(*e*) **Hazards to Bees**. Apart from cold and starvation, bees can experience other hazards which may reduce the colony strength. Diseases like Acarine and Varroa attack the colonies in winter when bees are confined inside in a congested place. The protozoan disease Nosema can also be damaging. Pesticides can cause heavy mortality. In the mountains although dormant sprays on fruit trees are safe, yet their drift on weeds and early flowering plants can kill the bees in large numbers. The colonies migrated to the plains can be exposed to spraying on *Brassica* crops causing losses among the bees and if protection is not given to the colonies the gains expected might turn in to losses of bee workers of the foraging age.

15.4.1 Comparative Efficiency of Bees

It has been observed that bumble bees work more rapidly and can cover more blossom in a given time than the hive bees. In studies on pollination of cultivated blueberries it was found that at 54-69°F, honeybee visits averaged 5.6, bumblebees 13.0, and solitary bees 2.7 blossoms per minute. On alfalfa it was observed that on an average 7 *solitary bees* (*Megachile*) and bumblebees both visited flowers per minute and they tripped 17 flowers each, whereas only one pollen collecting honeybee visited and tripped 7 flowers. In general bumblebees visit blossoms 2-3 times the rate honeybees visit but in terms of the total fruit set the latter owing to their large population play a greater role (Eckert and Shaw, 1960).

In order to identify foraging honeybees from a colony they can be marked by various ways; marking with dyes; by feeding radio-isotopes mixed with food; by sprinkling florescent dyes on them; by using different strains of bees having distinct body colour; and by catching bees in the field (chilled or anaesthized) and tagging iron discs of various colours and with the help of a magnet collecting those at the hive entrances (Gary, 1971).

In earlier work from USA it was reported that heavy concentration of *A.m.* honeybees decreased the attractiveness of alfalfa fields to bumblebees, alkali bees (Nomia) but not the leaf-cutter bees

Megachile rotundata. In fact even in the presence of moderate population of *M.r.* the honeybees stayed away from the field (Bohart, 1967). It was concluded that both honeybees and bumblebees were quite zealous and competed with each other. Ultimately the honeybees with their higher population excelled. It would seem the behaviour of *A.m.* honeybees observed on *Brassica* crops is governed by similar factors.

The daily foraging rhythm of *Apis cerana* and *A mellifera* was studied at various altitudes in the Himalayas (Atwal and Sharma, 1968; Verma and Dulta, 1986). The foraging activity was observed at the gates of the respective hives and it was noticed that the former species worked for longer hours in the cold weather and was active between 60-95°F becoming inactive at 50°F, whereas, the latter was active between 70-95°F and the activity remained poor below 70°F. The foraging behaviour of the two species was then studied at three altitudes, 1350, 1875, 2400 meters and it was observed that worker bees of the former started earlier in the morning and worked till later in the evening, the starting times for the former were 0603, 0606, 0618 hours and the closing times were 1913, 1902, 1825; for *A.m.* the starting times were 0627, 0641, 0648 and the foraging stopped at 1855, 1838, 1804 hours at the respective altitudes. In both honeybees the hours of foraging on apple were reduced as the altitude increased. It was observed that during the abundance of blossom both species worked zealously and there was no dominance exerted by one on the other species and it was suggested that in order to avail of longer hours both the species could be employed for apple pollination (Verma, 1991).

Further studies on apple pollination indicated that although the number of flowers visited by the two honeybees per minute was the same (6.4) the number of flowers visited per foraging trip was 172 by the former and 193 by the latter. It was further observed that in the morning hours among both bees more foragers worked on flowers from the top and in the afternoon hours more worked from the side, apparently for collecting nectar. It was also recorded that both the species foraging at a distance of 100 m were twice in number than those at 500 m (Rana, 1989; cited by verma, 1991).

15.4.2 Constancy of Bees

Bonnier (1906) was the first to make detailed studies on the constancy in honeybee behaviour. It was observed that bees once find a rich source of pollen and nectar they stick to that plant till the source is exhausted. In a hive there are two kinds of field bees, the searchers or scouts and the collectors. The scouts locate good sources of nectar and pollen and through the communication dances, direct the collectors to that source. He observed that certain bees confined their foraging to limited areas of the bloom of a tree or a crop. This phenomenon was further studied and it was shown that an individual bee may work in an area not more than 30-40 feet square (Singh, 1950; Hoopingarner and Waller, 1992).

The comparative studies on hive bees and wild bees Halictinae, Andreninae, Bremini, have shown that no group is particularly superior than the other in respect to constancy and they all collect mixed load of pollen and the Andrenids are the most variable. On apple blossom 80% of the *A.m.* bees carried pure load, *Halictus* carried 72%, Bremus 65%, and Andrena 57%. Thus honeybees appeared to show greater fidelity than the others (Britain *et al*., 1933). Some other authors have found that there were fewer mixed loads, for example, Betts (1935) found 28 mixed loads out of 915 examined, Vansell and Griggs (1952) recorded only a few out of 1000, and Eckert and Shaw (1960) found only 60 mixed loads among 1200 examined. In general it is believed honeybees do show fidelity to plants and when the source is near exhaustion they shift to others. It should also be remembered that a single type of pollen may not meet the nutritional requirements of the brood, therefore, a certain amount of mixed pollen is required by the colony.

Taking advantage of the bee behaviour of constancy some practical field trials have been carried out to attract bees to the desired crop. Bees were fed sugar syrups containing extracts of flowers of a crop and expected them to reach there. It has not been found practicable. Sugar syrup was also sprayed on plants and it was found that bees picked up sugar and neglected the flowers. A commercial product "Bee-Scent" has been formulated with chemicals relating to the natural Nasanov pheromone but it has given mixed results. Bees are known to "tag" flowers with certain scents and in future some chemicals might become available for attracting bees to the desired crop. Work on queen pheromones and semiochemicals to attracts honeybees is being persued vigorously.

15.5 FACTORS INFLUENCING POLLINATION

Apart from the factors discussed above there are others that determined the effectiveness of bees in pollination. The goal of bee management is to guide the bees towards the patterns of activity desired by the beekeepers whether it is for honey production or pollination. A full understanding of bee behaviour is essential. Bees communicate direction, distance and odour of a new source of food through dances. They are soon distributed on the nectar and pollen sources within their flight range of 2-3 miles. They are also savers of energy and they like to find food as close to their hive as possible which behaviour can be put to much use for the pollination of attractive crops. Unless there is some cause of unattractiveness bees easily go to plants that are rich in pollen and have nectar rich in sugar. From the farmers' point of view good bee population is important for obtaining effective pollination of his crops.

15.5.1 Population of Foragers

The number of honeybee foragers available for the pollination of a crop will depend upon the stage of colony development, the time of year and the type of crop to be pollinated, *viz.*, self-pollinated with pollen from other plants of the same species. As discussed in earlier chapters the good condition of a colony in winter determines the early brood rearing in spring and development of the colony strength in time for taking advantage of the early bloom of plants, particularly the orchard plants like, citrus, litchi, guava, mango in the plains or apple and other pome fruits in the mountains at higher altitudes. As the colony grows from spring to summer the proportion of bees that are potential foragers increases; the large colonies have proportionately more foragers than the small ones. During the development of a colony the bee to brood ratio changes from 1 : 1 to 3 : 1 and in the end more foragers are available which is desirable for the pollination work.

In the spring and early summer there are four kinds of foragers in a colony : (1) nectar collectors, (2) Pollen collectors (3) bees that collect both nectar and pollen ; (4) water collectors. The nectar foragers in proportion to the pollen foragers are generally in the ratio 3 : 1, but the ratio changes all the time in a season or even from hour to hour in a day according to the needs of the colony. Within the hours of a day a colony has more pollen foragers in the morning than in the afternoon and the nectar foragers are slow in the morning but predominant in the afternoon. It is a perfect relationship between the plants and the bees as the nectar secretion is more profuse in the afternoon when the sun is high up in the sky, temperature is warm, some bees who might have already visited the flowers might have stimulated them to secrete more nectar of a higher sugar concentration (Shuel, 1975, 1992).

For the flowers to benefit from pollination from both type of foragers there has to be a proper synchronization among the two types. On alfalfa honeybee pollon foragers were observed tripping seven flowers per minute but the nectar collectors tripped only 0.1% flowers. On apple blossom the predominant number of afternoon nectar foragers tried to gather honey from the sides and not from the top which would be more efficient for pollination. However, it is only a comparative behaviour and a certain percentage of both types of foragers do pollinate apple flowers at all hours of the day during their activity.

At 1350 m altitude in the North-Western Himalayas it was observed that among *Apis cerana,* there were 46.0% of pollen foragers at 9 am and only 20.0% at 3 pm and at the same timings there were 51.0% and 76.0% nectar foragers. Among *Apis mellifera* workers there were 18.0% pollen foragers at 9 am and 22.0% at 3 pm and during the same timings there were 73.0% and 70.0% nectar collectors (Verma and Dulta, 1986). It would appear that pollination, to a lesser or a greater degree, continued all day on apple blossom.

The reason for greater pollen gathering in late morning or near noon is that as soon as the flowers open there is dehiscence of anthers, and pollen is available. There are also dew drops and humidity is high in the forenoon and some of the dry pollen like that of corn or cotton can be managed by the bees more easily. It has been observed that pollen gatherers increase their activity if a obstruction like a pollen trap is placed at the gate of a colony; 50-70% of the pollen pellets are retained in the trap and honeybees try to make up for this loss by increasing their foraging activity, and hence pollination of crop (Free, 1970).

A colony produces upto 2000 new bees every day and they start foraging on the 10th day, but sometimes they wait a few days more. Once they start foraging they work for about three weeks upto the end of their life. There is a great deal of difference in the foraging habits of bees at various ages. The young bees are novices and at first they forage close to the hive. Thus a colony brought to an unattractive crop like onion or carrot will serve the purpose of pollination for a few days. Then the bees become experienced and they spread out to other attractive crops. Even on an attractive crop the bees after sometime develop such a pattern of flight that they concentrate on the richest section, neglecting other areas. It is desirable to exchange colonies between fields 2-3 miles apart. The bees will start foraging closer to the hives and cover neglected areas of the crop.

Crops generally become attractive to the bees when 10-25% of the flowers bloom. Colonies should be brought at that time for proper coverage. In case of certain high value crops like apple, strawberry, cucumber, blueberry, onion, carrot etc. it is economical to bring in bee colonies even when the flower bloom is less than 10%. It is to ensure proper pollination and high yield.

15.5.2 Weather and Pollination

Inside the hive grows brood and the strength of a colony builds up but once the foragers go out as individual bees they are on their own and have to face the weather. The elements of weather not only influence their foraging efficiency but their very survival. Extreme heat may cause exhaustion and a bee might die during the flight. A heavy shower of rain may drown her. In extremely cold weather she might get chilled and never return to the hive.

The elements of weather not only affect the bees but also the condition of plants they visit for gathering food. Bees would go to flowers if food is available. The daily or even hourly secretion of nectar and pollen depends upon weather; temperature, light intensity, and humidity have a direct influence on the opening of flowers as if to be in readiness for the bees to visit. Bees are adapted to gather nectar and pollen and the flowers are adapted to receive pollen grains on their stigma for fertilization.

When various species of bees are seen working in the field it is observed that solitary bees work for shorter hours than the bumblebees or hive bees under similar temperature and light conditions. In fact the hive bees work at temperatures lower than the optimum under low light intensities. The environmental factors that influence bee activity are temperature, hour of sunshine, light intensity, relative humidity and wind.

Temperature

Based on observations there is a general statement that *Apis mellifera* field workers make short flights at 60°, free flights at 65° and full flights from 70°F and above (Phillips, 1930). Further it was observed that in strong colonies the workers made full flights from 60° and above whereas in weak colonies workers did not make such flights until the temperature was 72°F (Woodrow, 1933). The bumblebees fly at lower temperature and visit flowers but their number is inadequate for the desired pollination

The foraging activity of workers of *Apis cerana* and *Apis mellifera* was observed in the Himalayas at various temperature from morning to the evening. The former worked at a longer range of temperature from 60° to 95°F and the latter between 70° to 95°F (Atwal and Sharma, 1968). Their activity on apple blossoms at Shimla showed that peak activity of *Apis cerana* was between 0900 and 1100 hours when the temperature ranged form 13.5 to 21.0°C and that of *A.m.* workers leaving the hive was between 1100 and 1300 hours when the temperature ranged from 21 to 25°C. The respective pollen loads ranged 9.06-9.26 mg and 9.24-12.22, the morning load being always smaller and the mid-day load the heaviest. It apparently was related to temperature and the abundance of pollen. The *A.m.* foragers carried heavier pollen loads then the *A.c.* bees, throughout the day (Verma and Dulta, 1986).

Light

The effect of light and temperature are inseparable. In winter bees work for shorter hours than in summer. Although the length of the day must have some influence, it is temperature that apparently is the dominant factor. The availability of nectar and pollen is also important. It is generally observed that within a narrow range of temperature bee activity increases with higher light values and decreases in the afternoon when values are lower. The effect of clouds, haze or fog is quite dramatic and bees immediately react to those conditions. According to Britain *et al.* (1933) light apparently has a more important influence than slight variations in temperature.

Moisture

Honeybees do not fly in rain and remain confined to the hive. Soil moisture and relative humidity influence bees through their effect on nectar and pollen. The opening of anthers is directly related to atmospheric humidity; and lack of soil moisture stops nectar secretion in a number of crops. The effect is more dramatic when temperature is also high. It is very noticeable in barseem fodder which provides main honey flow in April and May and every new cutting is followed by a bloom of flowers. In June when the daily temperatures range from 22° to 45°C the crop suffers from low soil moisture and the flowers that appear do not have nectar in them. Honeybees stop visiting the flowers and move on to those fields which are irrigated more frequently. The life of the crop and availability of nectar can be extended by 2-3 weeks by irrigation. In fact during this extension period colonies can collect sufficient honey to tide over the subsequent dearth period in the monsoon.

Wind

It is generally stated that bees stop foraging when wind velocity is more than 25 miles an hour (Eckert and Shaw, 1960). Probably much lower wind velocities influence various species of bees. Some studies were made in Punjab on the effect of air temperature, relative humidity and wind velocity on bees visiting Egyptian clover (Dhaliwal and Atwal, 1976b) (Table 24).

Comparative abundance and flower visits were studied among solitary bees, *Megachile flavipes, Pithitis smaragdula,* the wild honeybees, *Apis florea, Apis dorsata,* and the hive honeybee *Apis mellifera*. The bees were counted for 5 min. in 10 m^2 areas and there were 12 repeats at temperatures 26.5–35°C, when humidity ranged from 46-70%. Wind was the principal controlling factor for the bee visits. In descending order of abundance the number of various bees observed were *A.f.* 38; M.f. 26.5; P.s. 22.5; A.m. 19; A.d. 16 and they visited 80.9, 70.8, 83.5, 64.7, 107.2 flowers, respectively (Dhaliwal and Atwal, 1976b). The maximum number of all bees was observed when wind velocity was less than 1 km per hour and the minimum was observed when velocity was above 16 kmph.

The dwarf honeybee was the most tenacious and visited flowers even when the wind velocity was 22.5 mph and no other bee was seen in the field. The giant honeybee, perhaps owing to its large size, was the first to stop working at 11.4 kmph wind when all other bees were working. During the period of observations the wind velocity varied from 0.2 to 22.5 and it was interesting to note that the maximum bee activity was at wind velocity less than 1.0 kmph. All species of bees reacted very sharply to the prevailing speed of wind and stopped working when it was too windy for them. Considering the temperature and humidity range at which the dwarf honeybees worked (see chapter 2) and even the strong winds that they resisted the insects appear to be pollinators par excellence. In surveys made from 1965 to 1970 in North-Western India this wild honeybee was by far the most abundant pollinator recorded on all varieties of *Brassica, Eruca,* alfalfa and barseen (Atwal, 1970).

TABLE 24. *Effect of Air Temperature, Relative Humidity and Wind Velocity on Bees visiting Berseem at Ludhiana (April-May 1971).*

Avg. Tempe-rature (°C)	*Relative Humidity (%)*	*Wind Velocity (Kmph)*	*Number of bees/10 m² (average of 4 plots)* A. florea	*A. dorsata*	*A. mellifera*	*P. smaragdula*	*M. flavipes*	*Total*
26.5	64.0	6.10	2.00 (7.5)	1.00 (8.5)	1.00 (7.5)	1.00 (6.5)	1.50 (9.20)	6.50
32.0	46.0	3.28	3.50 (8.0)	1.50 (7.7)	2.50 (7.0)	2.50 (7.0)	3.50 (9.0)	13.50
31.0	66.0	4.66	4.50 (7.5)	1.50 (6.5)	2.50 (7.5)	2.00 (7.5)	2.00 (8.5)	12.00
31.5	65.0	10.02	2.00 (5.7)	1.50 (4.5)	0.50 (6.5)	1.00 (4.5)	1.00 (7.5)	6.00
32.0	58.0	11.36	1.50 (6.0)	0 (0)	2.00 (6.0)	1.50 (4.0)	1.50 (7.5)	6.50
35.0	52.0	16.68	1.00 (4.0)	0 (0)	0 (0)	0 (0)	1.00 (7.0)	2.00
31.0	54.0	0.20	6.50 (8.5)	3.00 (7.5)	2.00 (10.0)	3.00 (6.5)	4.00 (12.5)	18.50
31.0	49.5	0.30	5.00 (8.5)	2.50 (10.5)	3.00 (10.0)	2.50 (8.0)	4.00 (14.0)	17.00
32.0	49.0	22.49	1.50 (3.2)	0 (0)	0 (0)	0 (0)	0 (0)	1.50
31.0	64.0	2.00	4.50 (7.5)	2.00 (10.0)	2.50 (9.5)	3.00 (6.5)	2.50 (11.0)	14.50
32.0	68.0	1.16	4.00 (8.0)	2.00 (9.5)	3.00 (10.0)	4.50 (7.5)	4.00 (11.5)	17.50
31.0	70.0	9.86	2.00 (6.5)	1.00 (6.0)	0.50 (7.5)	1.00 (5.5)	1.50 (7.5)	6.00
Total			38	16	19.50	22.50	26.50	

1. Value of r for wind velocity and the number of *A. florea* = – 0.8358*; *A. dorsata* = – 0.8868*; *P. smaragdula* = – 0.8353*; *M. flavipes* = – 0.9063*
2. In the parentheses are given the number of flowers visited / min. Significant at 1% level

15.6 POLLINATION, YIELD AND QUALITY

15.6.1 Pollination and Yield

It is well known that yields of many crops in the groups of oilseeds, clovers, and pome or other fruits increase with pollination activity of solitary bees and honeybees. The increase in yield is recorded in many self-pollinated but self-infertile crops and also those which are cross-pollinated. It may be brought about by a single species of pollinator, a group of pollinators, or by more then one agent of pollination. The yield increases recorded will be discussed under each crop at the end of the chapter. At this stage the outstanding examples will be reviewed to establish the principles involved.

In alfalfa flowers there is a tripping mechanism activated by solitary bees and the hidden male and female parts of the flower are released. Only then pollination can take place. These bees are effective in the North-Western states of USA and they trip the flowers much more efficiently than the honeybees who avoid approaching the male parts of flowers because of the physical impact on their body. After the solitary bees, *Megachile* and *Nomia,* have been active in the field, honeybees are brought in to supplement the work of wild bees (Bohart, 1967; Hobbs, 1964; Hoopingarner and Waller, 1992). In the hot and dry States of Nevada, California and Arizona the flowers get tripped accidentally and there is poor pollination giving no seed crop or production of seed is not economical.

Among cruciferous crops there are vegetables and oilseed crops and they have different pollination requirements. In order to produce pure seed in the former cross pollination is avoided and in the latter cross-pollination is beneficial and it increases yield. Among the *Brassica* oilseed crops the yellow seeded sarson has such a flower structure that it is adapted for autofertilization, but other varieties of *Brassica campestris* (toria, sarson, raya) have adapted for cross-pollination by wind or by insects. It was observed that yield was higher in bee pollinated plants than in the wind or self-pollinated plants (Bisht *et al.*, 1980; Mishra *et al.*, 1987). It was further noticed on raya that heavy aphid infestation reduced the number of open flowers and hence the bee visits. Very low yield was obtained as a result of aphid attack as well as poor pollination (Dhaliwal and Atwal, 1975). In good weather and low aphid infestation honeybees are very effective pollinators. In two experiments reported by Rehman (1940) and Singh (1954) five colonies of *Apis cerana* honeybees per hectare of sarson crop increased the yield by 30% and 10-25%, respectively.

In sunflower 2-4 rows of florets on the periphery open every day and the structure of flower prevents self-pollination. Fertilization takes place through cross pollination with pollen from other florets of the same flower or preferably from another flower head. Free and Simpson (1964) covered with muslin bags one head separately and two heads together and released bumblebees inside. The seed set in double head was 45% and in the single head it was 24%. Rudnev (1941b) recorded yields of 53, 76, 133, 210 kg of seed per million florets from plantations in which average bee visits per floret were 1.0, 1.4, 3.4, 6.1, respectively. In another experiment he observed that starting from an apiary with increase in distance from 500 m to 1250 m the number of bee visits per 100 m^2 fell from 100 to 61 and the yield of seed was reduced from 6.0 kg to 3.7 kg per plot. The role of honeybees in sunflower pollination is so important that 50-100% increase in yield is expected simply by bringing to the field honeybees at the rate of 2.5 colonies per hectare.

Most apple varieties are self-incompatible and some are also cross incompatible. Although a certain amount of pollen is carried by wind it has been shown conclusively that wind pollination plays no role in fruit production (Free, 1964). Primarily honeybees provide pollination for this crop. In an orchard of self-infertile variety, rows or individual trees of pollinizer varieties are planted for bees to transfer pollen. For getting a full commercial crop only 5-10% of the total bloom need be fully pollinated, but in actual practice there should be more fruit set to start with as many fruit drops are expected in the growing period. Generally one honeybee colony per acre is considered adequate to obtain the desired fruit set. In case of improper pollination all the seeds inside are not set and that part of the carpel does not develop, giving rise to a misshapen apple. Such fruits drop early and are no good for commercial purposes. In case of branches of 10 apple varieties it was observed in Hungary that if the duration of bee visits was limited to the extent that there was insufficient pollination the yield was greatly reduced but there was no effect on quality of fruit (Benedek and Nyeki, 1996).

Excessive fruit set is also not desirable as fruit size remains small and in that case thinning is resorted to through chemicals. The role of bees, particularly honeybees, in having a successful apple crop is absolute and without their contribution in pollination no fruits can be harvested.

15.6.2 Pollination and Quality of Seeds and Fruits

The seeds and fruits produced by cross pollination have a greater weight and flavour quality than that produced by self-pollination. In oilseed crops the size, viability and oil content is more and the plants grown from such seeds have greater vigour, flowering capacity and yield potential. It has been observed for a number of crops in India and, literature has been reviewed (Deodikar and Suryanarayana, 1977). The following advantages of cross-pollination have been reported :

(*a*) The germination of pollen on stigma is improved and fertilization is assured.

(*b*) The number, size weight and quality of seeds produced is improved.

(*c*) Fruit set is better and as the seeds grow, carpels develop uniformly producing well formed fruits. The fruit drop is reduced to the minimum.

(*d*) The fruits formed are more nutritious and have a better flavour.

(*e*) The seeds produced have more oil content and nutrients and when sown there is a greater germination and healthier growth.

(*f*) The plants grown from such seeds have more vigorous growth, producing more flowers and increasing the crop yield.

(*g*) The plants grown from cross-pollinated seeds are observed to have greater resistance to disease and other adverse environmental conditions.

Apple is a very valuable commercial crop and the role of honeybee in its pollination has been studied a great deal and the literature has been reviewed by Free (1993) and McGregor (1976). The European and American literature covers a period of many decades resulting in very precise recommendations for growing commercial apple crop. Some observations on fruit quality were also made in western Himalayas on the self-compatible, variety Golden Delicious. Honeybee pollinated crop was compared with open pollinated crop and it was found that fruit weight and number of seeds inside was 22% and 17% higher under the former conditions. In the self-incompatible variety Royal Delicious the fruit weight was 33% higher and there were 49% more number of seeds in crop pollinated by honeybees as compared with the open pollinated. Similar differences were observed in fruit drop. In the self- compatible variety Red Gold the fruit drop in the control, where there were no pollinators, was 38.0% and it was 25.0% in crop pollinated by honeybees (Dulta and Verma, 1987).

Honeybees do a thorough job of pollination on apple and produce not only higher fruit set but also the quality of fruit produced is better and the fruit drop is also less. As compared with the self-compatible varieties, the incompatible ones and cross-pollinated show very marked differences in their flavour and the role of honeybees is essential in their case.

15.7 POLLINATION SERVICES

The services of honeybees for the pollination of certain valuable crops are essential and the farmers generally rent the colonies from experienced beekeepers who know how to prepare colonies for this specialized work. Moreover the number and strength of colonies for pollination work are different than those required for honey production. For pollination colonies are brought in from outside for a short period of time when the orchard or a crop is in bloom. Although bees collect food for rearing new brood, the older worker bees exhaust themselves and there is a temporary set back to a colony. The beekeeper is justified in asking for compensation. The orchardist sometimes asks for a share of the honey gathered and the terms and conditions are mutually worked out before hand. In the practice of pollination service there are many considerations for the beekeeper and the farmer.

15.7.1 Type of Colony

In USA 3lb packages of bees are available for pollination work. Those are brought to the orchard 4 weeks before the bloom and are installed in a hive on drawn combs. Within that time they settle down, rear brood and become equal in strength to an average colony. Without this start the package by itself is not good enough for pollination work. An over-wintered colony with the same weight of bees is considered superior. Comparing the two, Farrar (1931) found that 2-4 times more bees fly out per minute from the over wintered colony than from the package.

The over wintered colony has to be prepared in the winter specially. By stimulation feeding brood rearing is initiated early and before pollination begins the ratio of worker to brood becomes 3 : 1. Earlier, during the year also the brood rearing activity is watched so that it can be timed when to expect the field bees.

15.7.2 Strength of Colony

The beekeepers that rent colonies have to agree to certain standards of strength and bee population. Two weak colonies do not give out the same number of pollinators as a strong one of the same strength. The colonies are considered standard, substandard, and above standard. In an apiary the colonies can be equalized by giving them brood frames so that they come upto the desired standard.

In USA a colony is considered good for pollination if it is on two chambers and all the 18-20 frames are densely covered with bees and there is 750-1000 square inches of brood area. In the view of some, this degree of strength is difficult to achieve for the pollination of early bloom, so, they accept one chamber colonies with all frames, full of bees.

For apple pollination in the Himalayas package bees as commercial venture are not available. The overwintered colonies have to be prepared. The weeds and wild flowers for early build up are not available near the apple orchards. Beekeepers take bees to lower altitudes to avail of *Eucalyptus* or take them to the plains where in addition to this source *Brassica* crops are plentiful. Through this migratory effort the beekeepers not only extract honey but also build up colony strength early in spring and bring back the colonies for the pollination of pome fruits. In case the colony strength is good the pollination activity is effective. The experienced beekeepers can even extract apple honey which is prized for its flavour. The Kashmir honey also has a characteristic flavour which has a ready market abroad.

15.7.3 Number of Colonies Per Acre

The right number of colonies required for optimum pollination of various crops has not been worked out scientifically. By trial and error certain norms have been suggested. Since the beginning of this century in USA it is a common practice to provide one standard colony per acre for apple orchards. For the non-preferred crops like onion and carrot a large population of bees is required to concentrate on these crops, till the bees as a normal habit drift away in a few days. Two colonies per acre of these crops is considered necessary, the same number is considered desirable or the seed crops of broccoli and kohl, mustard and cabbage. For cucumber, strawberry and blueberry fruit crops, one colony per acre is considered sufficient. Pollination efficiency can be further increased by fitting pollen traps at the entrance of the colonies. As explained earlier the pollination beekeepers receive incentive for working harder as they lose 50-70% of their pollen load in the traps and there is a set back to brood rearing in their colonies, hence delay in population build up.

For the oilseed crops like sunflower, *Brassica,* niger 1.5 colonies per acre are considered the optimum. For *Brassica* crops the minimum number should be considered as half colony per acre considering the presence of a large population of solitary bees and wild honeybees on this crop. However, these crops are rich in nectar and pollen and for gathering surplus honey in North India more than 1.5 colony per acre may be provided.

As discussed above a pollination colony may be on two chambers having a strength of 50,000 bees or on one chamber having a strength of 25,000 bees. So, the number of colonies per acre is not an absolute figure. A farmer has to judge for himself the concentration of optimum number of honeybees required on his crop or the orchard. Considering the variability in weather from place to place and that of climate from year to year, a farmer has to make his own judgement of the concentration of honeybees per unit area in his crop or per tree in his orchard. Based on his judgement from year to year on the number of bees pollinating and the yield of apple obtained (for example) he can draw his own conclusions as to the number and size of bee colonies required by him that year.

The concentration of honeybees per unit area can be measured by various ways, some of which have been described earlier. By experience of observations and the yield obtained by him a farmer can draw his own conclusions on the population of honeybees required per unit area, which would determine the number of colonies per acre. The number of bees on a tree or a crop can be counted directly while walking in a strip of 1 meter or 2 meters wide. On a dwarf tree bees on the entire tree can be counted but on a large tree half side can be counted at one time. After making estimates of population a farmer can decide the number of colonies required. It is estimated that for sunflower crop there should be one bee per plant and since only a few florets open in a day the bees would be busy for 10-15 days.

15.7.4 Pollination Contracts

From experience it is considered desirable to draw out a written contract to be signed by the beekeeper and the grower. A sample document can be obtained from field inspectors. Essentially it should cover (*a*) the rental price and date of payment; (*b*) the dates of bringing in colonies and taking them away; (*c*) the number of colonies to be rented and the route by which those are be brought in to the field; (*d*) prior information on the likely pesticide use in the area; (*e*) providing water to the bees; (*f*) the standard and size of the colony; (*g*) since colonies are placed at night, whether mechanical equipment could be used or those have to be taken to the site manually.

While agreeing upon the rent per colony many things should be taken into consideration. The size of the field and the type of crop will determine hardship to bees. If the crop is rich in nectar and pollen the beekeeper can benefit from the collection of honey and the condition of the colonies would improve. Lower rent can be fixed in that situation. In case the field is small or isolated from where bees cannot gather much honey the rent should be more. The distance to which bees are to be moved will also determine the rent. Whether the colonies are to be placed in groups in a self-incompatible crop should also be taken into consideration.

A beekeeper has to make good estimates of his expenditure in transport, extra night time work, reduced honey yield, out-of-pocket expenses, hazards from pesticides and losses to colonies. All these considerations go into the reasonable demand for rent to be charged so that it is an economical proposition both for the grower and the beekeeper.

15.7.5 Hazards to Honeybees

For pollination services honeybees colonies have to be transported from a distance, have to be packed properly, loaded on a truck, then unloaded at the destination and taken to the site of placement. Honeybees face many hazards. For transport in the migratory beekeeping the colonies on top are given screens for ventilation and water is also sprayed. When colonies are transported for pollination early in the spring, temperature is not high and water requirement is not a critical factor, as the bees are to be transported over night for 8-10 hours. In the plains a truck can move them from 350-450 km and in the mountains they can be taken 200-250 km away. The colonies are loaded on a truck with the top covers on, but the hive must be tied up tightly with ropes or with stapled steel strips individually so that they reach destination in one piece. The placement of hives is most important; the frames of colonies must be along the long axis of the truck so that frequent brakes on the way do not damage the combs. It is particularly important on the narrow and winding roads in the mountains.

In USA providing honeybee colonies for pollination is a professional job with some beekeepers, but in India this degree of sophistication has not been achieved so far. The movement of honeybee colonies from lower altitudes or the plains to the apple orchards is already being practiced at a reasonable scale with very promising results. The mortality among honeybees in transit is a reality that has to be guarded against.

The requests for pollinating colonies is on an individual basis and no one monitors the concentration of honeybees in a given locality after bringing in the colonies. The beekeepers supplying the colonies are also not in communication with one another. Situations can arise where too many colonies are brought in one locality. The honeybees foraging there naturally would have over-competition which would be a set back to the individual colonies. Since the number of colonies available is much less than the requirements in western Himalayas for apple pollination some organized effort is required for an equitable distribution of colonies for the benefit of growers and beekeepers so that there are not too many colonies at one place and too few at another.

It may be remembered that in western Himalayas the size of apple orchards is small, generally not more than one hectare per family. It is not feasible for them to hire colonies from outside all the time and therefore they raise their own honeybees. They must learn the winter management and early spring colony build up in order to provide their own pollination service for the orchard.

When honeybee colonies are brought for pollination to new areas the beekeeper is not familiar with the local crops other than the one to be pollinated and does not know the farm practices prevailing, particularly the spraying of crops. Pesticides are one of the biggest hazards to bees and mortality can be extremely high. In USA there are laws for the protection of bees and a farmer who wants to spray his crop when honeybees are around is required to inform the beekeepers in that area. However, nothing is as good as good-neighbour relations and by the word of mouth prior information can be collected so that the bees are shut inside their hives for the period required to off-set the poisonous effects.

When colonies are moved from the plains to mountains, sometimes an overnight, or one or two days stop becomes necessary in the Shivalik areas. These clay mountains are the breeding places of the Green Bee-Eaters who make their nests on the mountain sides and congregate to breed in the spring. There is lack of food for them at that time and the foraging bees are the obvious prey. The colonies have to be protected from the birds and if possible to avoid stopping in those areas.

There can be many other hazards like accidents, delays, misinformation on the arrival dates, breaches on mountains, untimely weather changes, misdirections and loosing the way at night on mountain roads. In a country like USA the pollinating beekeepers have the problems of big movements, say 2000 colonies at a time, and in a developing country like India the problems centre around lack of transport and communication.

Winter time is ideal for the development and spread of tracheal (Acarine) and ectoparasitic mites. Fumigation of colonies in autumn is expected to keep the diseases suppressed. In the spring, before taking colonies for pollination work, it is even more important to ensure that the colonies are free from these diseases. Bees suffering from the diseases and also from the paralysis viruses have shorter life and their performance in the orchards is certainly not going to be to the level of expectations. The diseases should either be controlled prior to the pollination service well in time or such colonies should be isolated and not employed for pollination work.

15.8 WILD BEES AS POLLINATORS

On the Indian subcontinent there are about four thousand species of bees in Apoidea and not more than a few hundred are identified. Some of the groups of social bees like the honeybees, bumblebees, stingless bees are well known and their account is given in chapter 2. Apart from these groups there are a large number of solitary bees, subsocial, semisocial bees and parasitic bees who visit flowers very diligently and bring about the pollination of fruits, vegetables, field crops annual flowers and even weeds. Their population at places can be as large as that of the wild honeybees and the hive honeybees and they do play an important role in pollination.

These bees are distributed in nine families of Apoidea and the biology and field activity of some of them has been studied which is given under the respective families. The species studied were not necessarily typical of the family concerned, but were picked up from the samples of insect visitors on crops of economic importance. The work on their biology was mainly carried out at two centres *viz.* Ludhiana (Punjab) (Atwal, 1970) and Hissar (Haryana) (Kapil and Jain, 1980).

The solitary bees are considered to be more efficient pollinators because the same individual bee collects both pollen and nectar from a flower and by this very habit they pollinate more flowers in a shorter time as compared with honeybees who have separate pollen and nectar foragers. However, the latter out-do the former by virtue of their sheer numbers. The population of some of the solitary bees can be increased artificially and an outstanding example is that of *Megachile rotundata*. In India *Pithitis smaragdula* = (*Ceratina bhinghami CKll*) can also be manipulated to increase its population artificially by providing nesting materials. Their service can be very useful in producing hybrid seeds during the summer.

In Northwestern India studies on insect pollinators were conducted on various crops namely *Brassica, Cucurbits* and *alfalfa* from 1965-70 (Atwal, 1970) and from 1973-79 (Kapil & Jain, 1980). Their comparative abundance and activity on these crops were studied and the biology and behaviour of some of the social and solitary bees were recorded. The major findings are summarised here under various families.

A. Family Apidae

Apidae includes both truly solitary and highly social bees. The females have pollen collection baskets whereas the males do not. The family includes bumblebees and stinglees bees both of which are socially evolved, the latter have large colonies and their strength may reach 1,00,000. They do bite but do not have the sting. The pollination activity of the wild honeybees of the subfamily Apinae is discussed first.

1. *Apis florea* F.

The nests of these wild social bees are found in bushes, hedges, stacks of sticks etc. throughout the plains of North India and they are more abundant in dry and hot areas of this region. The *Brassica* crops abundantly visited by this bee are mostly grown under irrigated conditions except taramira (*Eruca sativa*) which is almost always grown under unirrigated conditions in the states of Punjab, Haryana, Himachal Pradesh, Uttar Pradesh and Rajasthan.

Apis florea was found to be active on all the varieties of *Brassica* (*i.e.*, toria, sarson, raya) and also on *Eruca*. Estimates of absolute population per unit area were not made but their comparative abundance alongwith other pollinators was studied. The measure used was the average number of bees caught in 100 sweeps of insect nets at various localities and the data of three year (1965-66-67) were confounded for obtaining the overall averages. On *toria* the population of *Apis florea* formed 84 per cent of the total insects collected (also see Kapil and Brar, 1971).

On *sarson, Apis florea* formed 73% of the total insect visitors. On raya, it constituted 75% of the total insects collected. Both sarson and raya grow side by side and flower at about the same time, but the latter attract more insect pollinators than the former. On taramira also this wild honeybee was the most abundant insect and it formed 50.8% of all the insects collected.

On alfalfa crop again *Apis florea* was the most abundant insect visitor although it did not always trip the flowers successfully and was not an efficient insect pollinator. It formed 46.3% of the insects collected on this crop. On the other hand *Pithitis smaragdula* and *Megachile flavipes,* which constituted 8.6% and 8.1% of the population respectively, were efficient insect pollinators. The mere abundance of *Apis florea* on this crop was of no benefit for the purposes of pollination.

On cucurbits *Apis florea* was always present but it did not have a dominant position. On some cucurbit species it was most abundant whereas on others it was only second or third or at a lesser place. *Apis dorsata* was generally in a more dominant position on these crops. On various cucurbits different insects were the most abundant species (also see Kapil *et al.*, 1971).

2. *Apis dorsata* F.

The Giant honeybee owing to its size is quite a prominent insect visitor on the *Brassica crops*, although in numbers it stands second or sometime third in position. In the submountainous districts of Punjab and Haryana it was the second most abundant insect but in the dry and hot zone of southern districts it held only third position on toria (4%), sarson (7.2%) and raya (3.9%). On taramira it was the second most common visitor and formed 20.0% of the total catches. Considering its abundance in absolute numbers its population was less in southern districts but was more further north. Its natural population seemed to be related to the number of colonies in a locality as determined by the luxuriance of crops.

Apis dorsata is not attracted much to clovers and on alfalfa it formed less than 1% of the total insects. On cucurbits on the other hand, it made 15% of the population and was the third most common insects particularly on *C. pepo* and on *luffa*.

3 & 4. *Apis cerana F.* and *Apis mellifera* L. (Apinae)

The abundance of these bees depends upon the number of colonies in a locality. They visit a wide range of crops and pollinate open flowers. The strategies to be followed for their proper employment as crop pollinators have already been discussed.

5. *Pithitis smaragdula* Ckll (Ceratinini)

Pithitis smaragdula Ckll are metallic green bees, slightly smaller than *Apis florea*. They are wide spread in the plains of Punjab and Haryana and even in the Himalayas upto Kufri (Shimla) at an altitude of 2500 m. The bee forages on clovers (*Medicago sativa* and *Trifolium alexandrinum*), cucurbits, and late summer crops like arhar (*Cajanus cajan*) and San-hamp (*Crotalaria juncea*). These crops are grown throughout the area of distribution of this bee. Studies on its biology and field activity were carried out at Ludhiana from 1967-69 (Atwal, 1970; Kapil and Kumar, 1969; Kumar *et al.*, 1968).

Pithitis make their nests in the pith of sarkanda (*Erianthus munja*) and sometimes in the dry sticks of *Brassica* and castor. Sarkanda is commonly used in the roof thatchings of huts and the insect selects those sites for its nest building. The bees are active from February to September and during the winter adults are occasionally seen on *Brassica* crops, otherwise they remain in dormacy from mid October to mid February. They start building nests in February in the roof thatching or stalks of reeds used as support for growing tomatoes, garden peas etc. A series of 1-11 cells are made separated by partitions of pith but sometimes a gap of 2-cell length are left between the cells. The entrance of the tunnel is 0.24 cm and that of the main tunnel is 0.30 cm. The concave partitions are made of a mixture of pith and some other material which imparts iodine colour.

The cells are provisioned by female bees though males are seen coming in and out in some nests. The females bring pollen and nectar, prepare a boat-shape mass with a longitudinal groove. It is completed in 28 trips made in 578 working minutes. The pollen mass is 0.5 cm long and 0.3 cm wide, ranging from smoky yellow to orange in colour. In weight it is 0.03 gm.

The eggs are elongated, curved, translucent about 0.3 cm long and are laid in the furrow of the pollen mass. The larva develops in 9-10 days at 27.5°C. Young larva is translucent and its head, thoracic and abdominal segments are distinctly marked, and the intestine is visible through the cuticle. The spiracles are prominent. As it grows it takes a C shape and develops fat bodies. The prepupal stage lasts about 3 days and the pigment develops in pupal stage. The bright green colour is visible at first in the thorax and spreads to head and posteriorly and then descends to legs; full colour appears in 3 days at 25°C.

The bee completes 6-7 generations in a year and the two peak periods of egg laying are early in April and in September. The acceptability of reeds fixed vertically and horizontally were studied in the clover fields; the former were preferred when 25 cm above the crop height. In various experiments acceptability ranged from 7 to 47%. On alfalfa and cucurbits, Pithits formed 9.0% and 4.6% of the total insect visitors.

In February and March, when *Pithitis* starts nesting actively *Brassica* inflorescence is on the wane, but this bee does exploit the pollen and nectar sources. However, their number is always less than 1.0 per cent of the insect visitors. Alfalfa and *Trifolium* are its forage crops and by the time flowers appear in the end of March and April it builds population in the field and forms 9% of the total insect collection on alfalfa, as against 46.4% of *Apis florea,* 5% of the Halictine bees, 7.9% *Megachile flavipes* and 24.2% of the *Nomioides* spp. On cucurbit crops *Pithitis* forms only 4.6% of the total insects collected.

In Utah its biology was studied in greenhouse based on the earlier field observations made at Ludhiana (Bohart 1965-70; Atwal, 1970). Insects were provided with American pithy reeds caged in a large (6.1 × 3.4 × 1.8 m) 10 mesh nylon screen. In pollinating alfalfa these bees were as effective as *M. rotundata* and *Nomia melandari* (Batra, 1976). Further, P.s. were released in the field at Davis, California but they could not establish under natural conditions there (Daly *et al.*, 1971). Temperature alone might not have been the limiting factor because the insect can servive in snow bound Kufri beyond Simla in the Himalayas (Atwal, 1970). There is scope for making repeated attempts.

The other allied species collected by Batra (1960) at Ludhiana were *Allodarula pervula* Smith and *Ceratina sexmaculata* Smith.

B. Family Halictidae

They are burrowing bees, many thousand in number are distributed throughout the world. They are small bees 3-11 mm in length and have pollen collecting brushes. The various bees belonging to this family have all transitions from solitary to social life. Halictine bees are also known as alkali bees. They are important pollinators of *Brassica, Trifolium,* cucurbits in Punjab, Haryana and Himachal Pradesh. Among *Brassica* they were common on all the four crops *viz.* toria, sarson, raya and taramira and formed 5%, 6.8%, 3.7% and 4.1% of the total insects collected respectively. On these crops they held second or third or fourth position : *Apis florea, Apis dorsata* and in some cases *Andrena ilerda* were more abundant. The *Halictine* bees are quite a large group and only some of the species were identified in the studies and the rest were lumped together as a group.

Halictine bees make nests in open level ground exposed to the sun, in naked soil or sometimes in grass cover (or on embankments). They generally have tumuli and make nest straight down ending in variable number of lateral cells with pollen balls as provision. The cell is closed with earth. The burrows of *Nomioides variegata* go from 15 to 45 cm. deep and have 0-13 cells at the end of laterals. The diameter of burrows in various species ranges from 2.5 to 4.0 mm. The bees are active throughout the year on a number of crops and more than two generations are completed in a year.

The genera of Halictidae recorded in the surveys includes *Halictus, Nomia, Nomioides, Thrinchostoma, Sphecoides, Eupetersia, Homolictus, Lasioglossum, Systropha.* Some species of the first three genera were identified (Atwal, 1970; Batra, 1966, 1977). They include *Halictus lucidipennis, Nomia callichlora, N. canitata, N. ebrneigera, N. oxybeloides, Nomioidea variegata, N. pusilla, N. divisus.*

The Halictine bees are an important group of pollinators as they remain active throughout the year and cover a number of crops. In the surveys as a group they formed important insect visitors on various crops; toria 9.7%; sarson 8.2%; raya 4.1%; taramira 6.9%; alfalfa 31.4%; cucurbits 41.3% and among them *Halictus* spp. alone formed 5.1, 7.0, 3.7, 4.7, 5.0, 6.5 per cent on respective crops. On alfalfa and cucurbits *Nomioidea* spp. among Halictines were the most prominent bees.

C. Family Andrenidae

There are 1000 species of bees in this family distributed all over the world. They are mining bees and the genus *Nomada* consists of highly coloured members. In Northwestern India throughout the plains and also in the Himalayan valleys *Andrena* spp. are common on *Brassica* crops. The three species commonly observed were *Andrena ilerda, Andrena leaena, Andrena levilabris,* the first named being the most abundant of the three. Their activity is synchronized with the *Brassica* crops in winter and early spring, after which their activity comes to an end in the summer.

1. *Andrena ilerda*

The adult female is 11-13 mm long, stoutly built, with darker head and thorax, having four light brown bands on the abdomen. The general body colour is reddish brown, and there are white hair on head and thorax; the male is somewhat darker in colour and is more slander with head narrower than the thorax.

On the first crop of the season, toria, which flowers in October-November, the insect was the fourth most abundant, *Apis florea, Apis dorsata* and *Halictine* bees being in premier position. On the next crop, sarson, it gained the second position, and maintained it on raya also. On taramira, the bees conceded this position to *Apis dorsata*. Their activity on these crops is from November to the end of February.

Considering the role played by *Andrena* group on *Brassica* crops studies were conducted on the biology and nesting behaviour of these bees (Sidhu & Atwal 1970; Grewal *et al.*, 1969). In this genus 25 species were recorded in India (Sharma, 1959). In Punjab, Batra (1967) collected 12 species. The insects are widely distributed on *Brassica* crops and *Andrena ilerda* is by far the most abundant and in certain places may form 40-54% population of all insects visitors. A close species *Andrena leaena* has the peculiar habit of congregation of nests (Grewal *et al.,* 1969). On February 11, 1967 near Ferozepur at a locality it comprised 203 bees per 100 sweeps as compared to 63 of *Apis florea* which normally forms 75% of the population in general. *Andrena levilabris* is another species that visits *Brassica* crops but is quite scarce.

Andrena ilerda is active in nest building and brood rearing from November to February, and remains in hibernation in adult stage in the old burrows from March to the end of October. There are two generations; one is completed in November-December and the other in January-February. The population increases substantially in the latter period particularly on taramira which remains in bloom in February-March.

This solitary bee makes burrows in the sunbaked open light soils on the side of sand dunes or embankments, near the *Brassica* fields and it avoids clay-loam soils. Bees emerge from last year's burrows in the last week of October, first the males and then the females. The life of the male is one month and that of the females about two months. Mating must take place soon after emergence because the females start digging nests on the day of emergence. The emergence continues for a number of days.

It takes a female 8-10 days to dig a burrow of appropriate depth containing 3-6 brood cells. While digging nest in a depression of a furrow or a hoof mark, they excavate soil and make a hillock or a tumulus 4.2 to 7.5 cm in height and 5.4-8.2 mm in diameter at the base. If it rains the bees temporarily make turrets which are extensions of the burrow outward; they have 1-2 mm thick walls and are conical in shape. The burrows are 44-78 cm deep and at the above 16-25 cm portion they slant at an angle of 45° and then go downwards. There are laterals at the 3-4 cm basal portion of the shaft arranged at right angles. At the end of a lateral there is a horizontal brood cell made of compact soil. The bees are seen bringing pollen on the 9th or the 10 day of start of work; loads of pollen are brought and made into a pollen ball 3.5-4.5 mm in diameter. It takes a bee 18-34 days to complete the nest and to provision the brood cells. After completing one nest, marked bees were seen to start work on a second burrow which they completed in 16-30 days; these nests had fewer cells. By the end of December the bees were exhausted and their mandibles were worn out. Adults of the first brood were seen emerging in the last week of December and soon they started work on rearing the second brood during January-February.

The eggs of *Andrena ilerda* are white, smooth and slightly curved in the middle. The young larva is white and straight and is ridged on the sides, below the spiracles. The full grown larva loses the ridges and is creamy white and C shaped. The prepupa is white and pupa is darker in colour and lies in the brood cell with its head towards the neck of the lateral. Life cycle is completed in 38 days; egg 7-9 days; larva 16 days, prepupa 4-5 days, pupa 10.5 days. Exact life of the adults could not be determined, but those emerging out of hibernation were observed in the field.

The males fly around on flowers and feed on nectar, whereas the females collect both nectar and pollen Males are not seen near the nesting sites and instead they hover over the flowers in search of females. Mating is quite spectacular and female carries a male to some distance, and the act lasts 5-8 seconds. While mating is going on other males also come closeby and try to mate, indicating that a strong sex attractant, plays an important role and males can discern it from 1 meter distance. The mated males do not live for more than 24 hours. The males, in general, are not allowed to enter the burrows and for a repose at night they find cracks and crevices in the ground or simply dig shallow holes for the night. They appear on the flowers one hour before the females do and then their activity concentrates on finding mates.

The females close the nest entrance for the night with the help of moist soil brought from below. The bee starts its daily activity at about 9 a.m. when the air temperature is around 10°C. Before leaving they hover around the nest as if making orientation flight, but in subsequent flights they come straight to the nest. If some landmark is removed they get disoriented and eventually find the burrow. In a day they make 3-4 trips of variable duration and quit for the day about 1 hour before sunset. In the winter months they bask in the sun for about 15 min before starting the day at 10 am and they quit work at 4 pm. Both males and females end days work at the same time.

D. Family Colletidae

There are 500 species of these bees distributed all over the world. They are commonly known as Plasterer or Yellow Face bees and they line their burrows with a glandular secretion. *Colletes nursei* is an oligolectic bee the size of *Apis cerana.* Its dorsal side of the abdomen has alternating black and creamy white stripers. Its activity is associated with raya and taramira and in the field it is found from the middle of February to the end of March uptill flowers in the late crops fade away. Studies on this bee were made in February-March 1969 at Ludhiana in Punjab (Atwal, 1970; Grewal *et al.*, 1970).

In field studies nests were located in sunny alkaline soils, on uneven ground on the sides of sand-dunes, mostly in unirrigated fields but some nests were also located in irrigated fields of raya. The opening of the burrow is 0.8-1.1 cm in diameter and it leads to a slanting tunnel which further down goes straight and again slopes at the end. The 3-5 laterals go straight down. The brood cells are noticed at depth ranging from 150 to 256 cm and are in the form of oval chambers measuring 0.9 × 1.8 cm, and are lined with papery material imparting smoothness. The cells are provisioned with pollen balls and only the females dig the nests. From the duration of field activity it would appear that only one generation is completed in a year.

E. Family Xylocopidae

This family of carpenter bees is widely distributed in India. The bees are largish, wasp-like, jet-black and are commonly seen on cucurbit crops which they pollinate (Fig. 21). Their body size is 3.5-4.2 cm. There are two allied species commonly met with in Punjab and Haryana belonging to the sub-family Xylocopinae. *Xylocopa pubescens* (Spinola) is dark grey and the females have a conspicuous yellow mark on the thorax. *Xylocopa fenestrata* Fab., is uniformly black with shining wings and a purple hue. Both species are active in the field from February to the middle of November and they make nests in similar situations. The common nesting sites are in the hollow of dry bamboos used for making roofs of huts in the countryside. Studies on their biology and behaviour were conducted at Ludhiana in Punjab (Kapil and Dhaliwal, 1968, 1969). Of the two species *Xylocopa fenestrata* was found to be more abundant on flowers.

It was observed in February on *Brassica* crops, radish, citrus, and later on it moved to cucurbit crops in May. After the rains when cucurbit flowers fade away, it moves to *Solanum, Sesbania,* sanhemp, cosmos, lantana, etc. It was active upto the middle of November. It remained in hibernatum in winter in the adult stage inside the bamboo nests. The females remained single but 2-7 males were congregated together.

It was studied that carpenter bees completed four generations in a year and were more active in brood rearing from March to May and comparatively less active the rest of the year. The adults that emerged in the end of October or early November did not come out of their nests and remained inside at a height of 5-7 feet from the ground.

The nest on an average contained 6 cells, each measuring 1.9 cm in length. The cells were placed end to end, separated by uniconcave partitions made of bamboo-scraps. The brood cells were always constructed and provisioned by the female and the male played no part and lived a solitary life elsewhere. The female started its activity at 7 a.m., suspended it during mid-day when it was hot, and resumed it in the evening around 4.30 to 5.00 pm. In order to complete one cell a female made 25 foraging trips of 10-17 min each, bringing 0.06 gm of pollen in a trip. The males also visited flowers, but in a more leisurely manner. For example on Solanum the females visited 16.2 flowers per minute whereas the males visited 5.7 flowers; on luffa the visits were 13.4 and 4.2 respectively.

The pollen balls made by the females are 1.3 gm in weight and are made by mixing nectar. One egg is placed on a pollen mass. When fresh, the egg is translucent-white and is 1.2 cm in length. The eggs hatch in about 3 days and as the larva grows it becomes opaque-white and later on creamy-white. It has a rather square head and large eyes. The pupa is also creamy white. At 25°C the prepupal and pupal stages are completed in 14.8 days and the life cycle upto the emergence of adult is completed in 30.8 days.

Although *Xylocopa* spp. visit many crops their main activity is on cucurbit crops which they pollinate. On cucurbits they form 7.4% of the total catches, as against *Apis florea,* 18%; *Nomioidea* spp. 25%; *Apis dorsata* 15%; Halictine bees 9.8%; Pithitis 4.6%. Owing to their large size *Xylocopa* and *Apis dorsata* appear to be the more prominent insect visitors on cucurbit crops.

F. Family Megachilidae

It is a very large family with its natural distribution in all the continents of the world. The genus *Megachile* includes the leaf-cutter bees and the mason bees. They construct their nest in rotten wood or in the soil and cells are made out of cut pieces of green leaves which are rounded or oblong in shape. The mason bees make their nests with mud which are quite hard structures. Some of them occupy tunnels made by wood borers and they line them with cut leaves or with human or cattle excrement. Among the solitary bees they are highly evolved. They use the same sites for making nests, and in time, nests look like apartment buildings. *Megachile rotundata* is a well known Eurasian species which has been introduced in many parts of the world and is now present in North and South Americas. This solitary bee is an efficient pollinator of alfalfa owing to its capacity to trip the keel. It can be encouraged to accept hollow drink straws or tunneled blocks of wood with appropriate size holes, for lining their cells with cut leaves. Such boxes are placed near the alfalfa fields obtaining increased pollination and higher yield of seed crop.

Work on Indian Megachilids was carried out in Ludhiana (1965-70) and at Hissar (1973-79) and some publications originated from these two centres (Atwal, 1970; Kapil and Jain, 1980; Kapil *et al.*, 1969; Grewal *et al.*, 1970; Batra, 1967, 1977). Those collected at Ludhiana in central Punjab were got identified by Batra (1966) and are listed here.

1. *Megachile lanata* Fabricius, on legume, 27 March, 1965.
2. *Megachile fasciculata* Smith, on *Crotalaria juncea,* 14 September, 1964.
3. *Megachile flavipes* Spinola, on legume, Sep. - Oct., 1964.
4. *Megachile rotundata* Fabricious, on *Luffa aegyptica,* 14 September, 1964.
5. *Megachile femorata* Smith, on *Trifolium pratense,* 20 April, 1965.
6. *Megachile vera* Nurse, on marigold, 1 December, 1964.

7. *Megachile disjuncta* Fabricious, 1 May, 1965.
8. *Megachile bicolor* Fabricious, on legume, 28 October, 1964.
9. *Coelioxys apicata* Smith, on Cicer sp., 15 September, 1964.

At Hissar (arid Zone) Kapil and Jain (1980) in addition to the ones at 3, 5, 8 above, also collected *Megachile nana* Bingh, and *Megachile cephalotes* Smith. The biology of some of the *Megachile* was studied and is described here.

1. *Megachile lanata* Fab.

They are dark brown bees of the size of *Apis cerana*. Their thorax is orange red and on the dorsal side of their abdomen they have dark brown and orange red stripes. The bees visit flowers of san-hemp (*Crotaliaria juncea*), arhar (*Cajanus indicus*), and ghia-tori (*Luffa aegyptica*) and other crops. They are widely distributed in Punjab, Haryana, and in Himachal Pradesh upto 1500 meters.

In fields around Ludhiana the bees are most active from August to October coinciding with the flowering of san-hemp and arhar crops. They select a suitable nesting site which is generally below some projection of earth or a pre-existing cavity in the soil, with some vegetation around it, but at some distance from the host plants. They dig the soil further to make a tunnel of 17 cm long and 1.0 cm in diameter. While digging they employ all the legs and push the earth 3-5 cm away from the hole and push out almost 90 earth loads. It takes about 2 hours to make a tunnel, it is then lined with elongated pieces of leaves of host plants measuring 2.5 cm in length and 1.4 cm in width and these are arranged diagonally opposite to make a broad chamber at the terminal end. The pieces of leaves used are not necessarily from the hosts mentioned above. The bees also use leaf pieces from *guara* (*Cyamopsis tetragonoloba*) and moong (*Phaseolus aureus*), but the pollen balls are from the main host plants and most of the leaf pieces are also from those plants (Kapil *et al*., 1969; Grewal *et al*., 1970).

It was studied that a nest either had a single chamber or two chambers, connected from the main tunnel. The females made 2-6 cells in one nest; a cell was cylindrical 1.2-2.0 cm long and 0.7-1.0 cm in diameter. After making a cell pollen ball weighing 0.23 gm, cylindrical in shape was placed in it and an egg was laid on the slope of this pollen mass. The smooth surface of the leaf pieces faced inside the cells, and their overlapping edges were glued with a substance which on drying looked papery in appearance. After making provisions in a cell and laying an egg the cell was sealed and the next one was constructed in the same manner, in series, one next to the other. Each cell was capped with circular pieces of cut leaves (0.9 cm), smaller than those used for lining the tunnel.

After completing and providing for the cells in a chamber it was sealed off with leaf pieces. The larvae and pupa developed on their own and on transformation into adults, emerged one by one, the males completing their development about 2 days earlier than the females. At least two generation were completed in these type of active-period cells.

In the second half of October the female bees start making earthen cells and houses for the winter-generation that lasts many months. There may be a single cell or there may be 3-4 cells built end to end. The cells measure 2 cm in length and 0.7 cm in diameter and are provisioned with pollen mixed with nectar, each weighing 0.23 gm. The eggs are laid on the pollen masses and every cell is sealed separately. An egg is a curved, smooth translucent structure about 3.65 mm in length and 0.76 mm in width. The larva on hatching starts eating vigorously and its intestine full of yellow pollen is seen through the body wall. A larva starts defecating after 5 days and when full grown it is 1.7 cm long and remains in hibernation all winter, early spring and most of the summer. Before entering hibernation larvae spin brown silken cocoons in layers, with a caplike structure at one end. Pupation starts in the middle of August and adults emerge by the end of this month.

The bees primarily visit san-hemp and arhar flowers but their comparative abundance was not recorded on these crops. On lucerne all *Megachile* spp. formed 11% of the total visitors and among them *M. lanata* was only 0.2%. On toria their number formed an insignificant 0.2% and none on sarson and raya in the winter. On cucurbits they formed 5.3% of the total insect visitors and *M. lanata* was only 1.6%.

2. *Megachile fasciculata* Smith

They are jet-black shining bees, the females measuring 1.9 cm and the males 1.5 cm in length. They are not so common but are widely distributed in the plains and submountainous regions of Punjab and Haryana where san-hemp (a fiber crop) and arhar (a pulse crop) are grown. Their activity from 3rd week of August to October-end coincides with the flowering of these crops. However, the bees end their activity in the 1st week of November when the crops are still flowering, indicating that they enter hibernation. They also visit other crops mentioned above.

The nests of the bees were located in small or large hollows in the stems and trunks of shisham (*Dilbergia sisoo*) trees, situated at a height of 0.2-7.0 m above the ground level. Shisham is the most common tree around and bees select suitable nesting sites amongst them in the area where host crops are grown, using cut leaves for nests from both the sources. Some nests are also made in the cracks of loosely placed bricks, in vertical, horizontal, or slanting plane according to the space available (Grewal *et al*., 1970b).

It was noticed that the females constructed 2-11 brood cells in 1-12 series, arranged more or less parallel to one another in a nest. The series were 4.2-18.0 cm long and were held in position by extra packing of leaves and in addition an outer covering enclosing the cells. The length of leaf pieces ranged from 1.8 to 3.9 cm and the width ranged from 1.1 to 2.2 cm. The smooth and shiny surface of leaves was on the inside. The pieces used for making cells were glued together and the cappings were made of circular pieces. On completing a cell in about 4 hours (4-6 leaf pieces per hour) a female provisioned it with pollen ball weighing 0.76 gm; it was completed in 15-32 field trips. The bees worked from 7 am to 7 pm on every bright and sunny day. The full grown larvae pupated in August and bees emerged at the end of that month.

The biology of four *Megachile* species *viz*. *M. nana* Bingh; *M. flavipes* Spinola, *M. fumorata* Smith, and *M. cephalotes* Smith was studied at Hissar (Kapil and Jain, 1980). The location is in the midst of zerophytic arid zone but the irrigation canals on the sides present some mesophytic conditions also.

3. *Megachile nana* Bingh

It is a leaf cutting bee that makes nests with cut pieces of leaves. The female bee is 8.2 mm in length whereas the male is 7.23 mm. The thorax is covered with bright yellow hairs on top and white hairs on the sides. The forewings are hyaline and the abdomen is black with yellowish and white spots. The legs are black, covered with thin white hairs. Activity of the bee coincides with the flowering of lucerne and *Parkinsonia*.

During a year this bee completes three distinct life cycles (*i*) the winter development and hibernation ; (*ii*) the spring and summer life cycle ; (*iii*) the post-summer life cycle. In autumn the post-summer females have staggered emergence. They lay eggs in cells on pollen masses and the larvae develop normally to their full size. They spin cocoons around themselves and enter hibernation for the winter, lasting 3-4 months.

In March pupation takes place and adults emerge in the end of the month in successive waves till the end of May. The females make nests from cut leaves in natural holes in live or dead wood. They rear the last brood from the end of March to the middle of June using flowers of lucern. The bees stop activity in the end of May or early June when the flowers dwindle. The larvae that complete development spin cocoons and enter aestivation taking 3-5 weeks to pupate and then emerge as adults.

The post summer adults in June-July come from the aestivating brood of the summer and also from the winter brood that did not emerge in the spring. The females raise a number of overlapping generations upto the middle of November when further emergence stops. It is interesting to note that a part of the winter hibernating generation does not emerge in spring and has to pass hot summer before the larvae pupate and transform into adults which takes almost 9 months.

It was observed that *M. nana* made nests in many types of existing holes in dry wooden materials such as holes made by wood borers, or in dry hollow sticks of bamboo or castor having an appropriate diameter. The nests and cells were made with oblong or round pieces of leaves of alfalfa, rose, pomegranate, guava etc. The oblong pieces were 9.4 × 5.2 mm in size and were used for lining the tunnel. The rounded pieces 4.9 × 4.2 mm were used for making the cell walls and for closing the mouth of a cell. A finished cell was 7.1 mm in length and 4.5 mm in diameter.

Since the bees accepted pre-tunneled cavities for nest making many artificial materials having holes were offered to bees and they accepted them. The most accepted diameter was 5-6 mm and the length 5 cm. The types of materials accepted were dry caster sticks, soft-drink straw pipes, holes dug into wooden blocks and wooden strips having grooves joined together to form holes. By placing these devices near the alfalfa fields population of the bee was increased indicating that in nature there was dearth of pre-existing tunnels of the right diameter.

M. nana started foraging activity at 8 a.m. in the morning and continued collecting pollen till 6 p.m. On a visit they approached 6-12 flowers and there was no definite pattern of high or low activity at various hours. A bee spent 5-19 seconds to search and trip an alfalfa flower and the tripping efficiency ranged from 87 to 100%.

4. *Megachile flavipes* Spinola

The male bees are 9.0 mm long and the females are 11.5 mm long. Their thorax and abdomen is grey with white hairs laterally. The legs and antennae are reddish brown with whitish hair. The fore wings are fuscous and annal area is slightly smoky. The bee has two clear life cycles. The summer cycle extends from March-April to September-October and the autumn winter cycle extends from August-September to March-April including the period of winter hibernation.

In the summer the life cycle is completed in 72-181 days and in the post summer period it is completed in 133-209 days. When tried experimentally the bees did not accept artificial nesting sites so readily. These mason bees made their nests with mud available nearby; a bee would dig out from the same spot making a little pit. They also made nests in suitable natural tunnels or among folds of clothes or roof thatchings. The bees made 1-5 cells in a nest. In the absence of suitable sites the bees reluctantly accepted tunnels in dry castor stems but they did not colonize as abundantly as did *M. nana*.

The bees were particularly active on alfalfa and formed 7.9% of the total insects recorded in the central districts of Punjab. They tripped the flowers effectively.

5. *Megachile femorata* Smith

In body colour they are very much like *M. flavipes*. The male is 8.7 mm long and the female is 8.9 mm long, and their life cycle and biology is very similar. Their nests are made of mud and 1-5 cells are made in a nest. Like other solitary bees *M. femorata* also locate their nests instantly on return from a pollen trip; if interrupted by an obstruction or another bee they may take 2-7 seconds to search the nest.

Their field trip lasts 20-22 minutes each. Once inside the nest the bees unload in 2.6-3.7 minutes and 10-50 seconds later they fly off again. It takes a bee 6-8 hours to provision a cell.

For provisioning a cell they make 13-18 trips. Their greatest activity is between 10 am to 2 pm which coincides with the dehiscence of flowers. The bees are also associated with alfalfa and to some extent cucurbits and other flowers.

6. *Megachile rotundata* Fabricious

A great deal of research work on this species has been carried out in USA, Canada, Europe, South America and Austrialia (also see section 15.3). Its population near alfalfa fields has been increased by providing artificial nesting sites for effective tripping and pollination. The flowers are not tripped by the honeybees who avoid the impact of the keel and instead gather honey from the sides of the petals, thus reaching nectaries without tripping.

Other genera of Magachilidae recorded in Punjab were *Anthidium, Lithurgus, Coelioxys, Heriades, Osmia, Anthocopa, Anthidiellum, Stelis* (Atwal, 1970; Batra, 1977).

G. Family Anthrophoridae

The *Anthrophorids* were earlier placed in the sub-family Anthrophorinae (Apidae). They are either parasites of other solitary bees or become soil-nesting industrious bees, often living in large colonies near canal banks. *Anthrophora* is a large genus of stout hairy bees. They are one of the earliest spring bees and at Ludhiana have been observed on *Brassica* flowers including toria, sarson, raya, and on cucurbits, gram and Compositae. The other genera of the family recorded are *Amegilla, Tetralonia,* and *Nomada*. The *Nomada* are generally hairless wasp like bees usually black and yellow in colour and parasitize other bees.

Batra (1977) made a more extensive survey of Anthrophoridae and in addition recorded : *Ammobates, Morgaria, Thyreus, Melecta, Habropedoa, Epeolus*. In the net sweep catches Anthrophora were found in insignificant numbers during a larger part of the year on winter, spring and summer crops including *Brassica,* cicer, clovers and cucurbits. Since their population was small their role in pollination of crops was expected to be minor.

As mentioned earlier some Diptera, particularly the hoverflies (Syrphids, Aristalis) (**Plate 16**) play an important role in the pollination of certain crops. Early in spring in North India their population is more than 95% of all insect visitors and method of collecting and keeping them in captivity for the pollination of valuable crops has been described by Glands (1996).

15.9 POLLINATION OF CROPS BY BEES

Many of the cultivated crops and commercial plantations provide nectar and pollen to the honeybees. Those are generally grown in large strectches of land and their contiguity is desirable for commercial beekeeping. The crops that need cross pollination also benefit from the foraging activity of honeybees. The evolution of this symbiotic relationship is over 60 million years old. Honeybees may also forage on certain crops which are not dependant on their services for pollination. For example, maize is wind, rain, or gravity pollinated and yet honeybees exploit its rich source of pollen.

In this section are discussed some crops and plantations which provide forage to bees and in return they benefit from cross pollination. Invariably cross pollination improves crop yield and thus man is doubly rewarded for keeping honeybees by way of obtaining a bumper crop and at the same time extracting honey from the bee colonies. Some other crops that provide only bee forage but do not benefit from bees have also been discussed.

1. Onion (*Allium cepa* L.; Alliaceae)

The inflorescence stalk of onion is 1-1.5 m tall and on the top is a ball shaped umbel of 50-200 flowers (Fig. 23). The individual flowers are arranged in groups of 5-10 (cymes). Flowers continue opening for 30 days but stigma remains receptive for a few days only. The anthers dehisce at irregular intervals and pollen has a greater ability to germinate in the morning and it declines after the first day (Nye *et al.*, 1971).

Nectar is secreted at the base of stamens and sugar concentration may rise from 40 to 75% when it is difficult for bees to collect it (nectar contains 53-56% fructose; 40-43% glucose; 4% sucrose (Walker, 1975). The crop is highly self compatible but wind and gravity play a minor role in pollination and seed setting (Walker, 1975). It has been shown experimentally that as compared with natural pollination the caged plants have less than 10% seed set and moreover the seeds weigh less and their germination is poor (Singh & Dharamwal, 1970; Mohanrao and Lazar, 1980; Jadhav, 1981; Kumar *et al.*, 1985; Rao & Suryanarayana, 1989).

In most parts of India there is rich fauna and the important insect pollinators are bees, Diptera, Neuroptera and some moths. The above mentioned workers recorded that in various parts of India bees, *Apis cerana, Apis dorsata, Apis florea, Trigona iridipennis,* comprised 70-96% of the visitors on flowers. It was recorded that *A.c, A.f, T.i,* visited 2, 3, 1.3 umbels and 5.9, 6.5, 2.5 flowers per minutes indicating that the first species of bee was more active in the field, but the second was more thorough in exploiting the source. The overall effect of natural pollinators was observed in an experiment; the field caged without bees gave 61% seed set, that caged with bees give 90% and the open field give 72% seed set. Thus a good crop is possible without honeybees provided the fields are not sprayed unwisely against thrips or aphids, which are serious pests.

When honeybee colonies are taken to onion crop the bees drift to other more attractive crops like *Brassica, Citrus, sunflower,* carrot etc. Probably honeybees avoid onion nectar as it contains too high a potassium ion concentration (1800-7400 ppm).

Kumar *et al.* (1985) observed that umbels of Japanese bunching onion (*Allium fistulosum* L.) when bagged, caged without insect pollinators or left exposed to natural pollinators gave 15%, 29%, 46% seed set and weight of 100 seeds was 269, 426, 476 mg respectively. They noted that *Apis cerana* was the most common visitor from 9-10 am and comprised 47% of the total. The moth *Helicoverpa armigera* became most abundant later on in the day. It was observed that *Apis dorsata* collected nectar only and played very little role in pollination. Only a few *Apis mellifera* colonies were available in that area and the bees were probably too thinly distributed at the time experiments were conducted.

2. Cashew Nut (*Anacardium occidentale* L.; Anacardiaceae)

Cashew nut trees are medium tall in height grown in subtropical and tropical parts of the world. The pink fragrant flowers appear in February-March and are a good source of nectar. A tree produces two types of flowers, the abundant male and the fewer hermaphrodite; both have one tall stamen, and nine short ones the latter have an ovary in the centre with a single ovule. Pollination is a serious problem and fruit set is very poor (3-10%). Insects, most importantly honeybees, play the crucial role, although hand pollination has a limited success especially where there are few insect pollinators (Rao and Hasson, 1957; Reddi, 1987). However, Pillai and Pillai (1975) who achieved 85% successful pollination by hand could harvest only 4% mature fruit, as majority of them fell off.

To obtain a good crop of cashew nuts fundamental studies need to be made on flower behaviour, physiology, genetic variability and utilization of honeybees and other insects for pollination. Some interesting studies were initiated in Malaysia (Phoon, 1983); and in Tanzania (Northwood, 1966) where 33% fruit set was obtained.

Sugar concentration in flowers varies from 30-55% and the white flowers after one day turn red and also become fragrant that attract honeybees and other insects.

3. Mango (*Mangifera indica* L.; Anacardiaceae)

It is the most common fruit tree grown from the southern tip of India upto 3,000 ft in the Himalayas. Half the area of fruit acreage in India is under mango. Polyploidy is most pronounced and there are countless varieties with varied characteristics, fruiting habits, fruit taste. The grafted cultivars are medium tall and the seeded varieties are tall with a large spread.

The predominantly male flowers and fewer hermaphrodite flowers (1-35%) are borne on panicles which are upto 60 cm long. There may be 200 to 6000 flowers on a panicle and there may be a few hundred to thousands of panicles on a tree. A perfect flower has ovary on disc containing a single ovule and one fertile stamen; these flowers are borne on the terminal portion of a panicle. Considering the huge number of flowers on a mango tree only a small number of fruits are produced; a cultivar bearing more perfect flowers bear more fruit (Naik and Rao, 1943; Mukherjee, 1953; Singh, 1978).

The flowers open all day but the peak is from 8-12 h. Soon after the flower opens dehiscence takes place. Stigma remains receptive for at least one day. Self fertilization is observed but cross pollination takes place in some varieties. When panicles are bagged fruit set is 25% as compared with that in open flowers (Malik, 1957; Naik & Rao, 1943; Singh, 1960). However, Sharma and Singh (1970) observed in four cultivars that self-pollination gave only 0-2% fruit set and those also dropped within 4 weeks, whereas, cross pollination gave 6-23% set and the fruit drop was much less.

The insects recorded on mango flowers belonged to Coleoptera, Lepidoptera, Hymenoptera and Diptra. Honeybees seem to play a small role in the pollination of mango. It was reported by Singh (1960) that bees did not visit flowers but Rao and Nair (1985) reported that *Apis cerana* and *Apis florea* visited flowers all day. There is only 5% sugar in nectar and the bees prefer to visit other flowers. During dearth periods, however, honeybees forage on jassid honeydew from the mango leaves. Syrphid flies are the most predominent visitors.

It seems, owing to a great variability in mango tree and the wide range of climate in which is is grown, no set pattern of pollination requirements has emerged. Studies will be relevent for a given cultivar in a well-defined climatic zone. However, at Ludhiana where mango blooms from the 4th week of February to the 4th week of March all the four species of *Apis* were observed foraging for nectar on mango (Atwal *et al.*, 1970).

4. Sunflower (*Helianthus annuus* L.; Compositae)

It is a photo-insensitive and short duration crop that is grown in many parts of the world. It is a new introduction to India and fits in very well in many crop rotations practically throughout the country. In India it covers an area of 1.5 m ha. In some countries abroad intensive studies have been made on its pollination requirements and increase of yield; literature has been reviewed by (Free, 1933), Rudnev (1941a, b), Goyal and Atwal (1973a, b), Parker (1981a, b, c).

Helianthus annus L. is an important oilseed crop prized for its oil of low cholestrol quality and is grown throughout the world. It is photo-insensitive and can be grown over a wide range of climate throughout India. It can cover the unsown fields after the rabi and kharif sowings. It particularly suits the fields after potato harvest in December or early spring and the crop favourably utilizes the left-over nutrients in the soil. In recent years it has become very popular in Central and Northern India and its cultivation is being encouraged by price support policy.

Like other compositeae, the sunflower head is surrounded by green bracts and when it opens has a single row of ray golden yellow florets which are very conspicuous (**Plate 17**). From the outer periphery inwards the rings of concentric disc florets open at the rate of 2-4 per day and continue for 6-10 days and the crop flowers for 3-5 weeks. In a flower head the number of florets many range from 1000 to 2000.

Each disc floret has a single ovary containing one ovule which on fertilization produces one sunflower seed. A floret on opening develops in two stages, first the male and then the female. Even though there is mechanical possibility for the achievement of self-pollination intra-floral pollination does not occur in nature. However some of the varieties may be partially self-fertile. The role of honeybees and other insects is essential for inter-floral cross pollination. It was demostrated that bumblebees achieved 35% seed set whereas there was only 1% seed set in flower heads covered with cloth bags. Cross pollination takes place by the transfer of pollen (by insects) from another floret in the same flower head or from florests on other flower heads (Free and Simpson, 1964). There are important advantages of cross pollination: high percentage of seed set and crop yield; as a result of hybrid vigour seeds are more stout and raise healthy plants; greater percentage of oil is obtained from hybrid seed (Furgala, 1954a, b).

Sunflower crops are visited by a number of insects including Hymenoptera, Lepidoptera, Coleoptera and Diptera. Honeybees, wild honeybees, bumblebees and solitary bees are the most important groups. All-day observations on sunflowers showed that *Apis mellifera* visited a flower 261 times, *A. dorsata* visited 11 times and a moth, a butterfly, a housefly and a beetle each visited one only (Goyal & Atwal 1973a, b). Honeybees among all the insects aquire pre-eminent position because the colonies can be moved to the desired fields for pollination and at the same time surplus honey can be gathered. Some detailed studies have been made in Europe and North America on various aspects relating to cross-pollination and crop yield.

In an attempt to determine the effect of honeybee visits on sunflower seed production in Manitoba, Furgala (1954b) placed 2.5 colonies per hectare along the East border of sunflower fields. The seed production per hectare in these two fields was as much as recorded in four fields located 4.8 km away from the honeybee colonies. Furthermore, he noted that seed yield in the farmers fields decreased progressively as the distance increased from the East to the West. Some closer observations were made by Rudnev (1941) in Russia in plots 100 m^2 at distances of 500-1,250 m from the honeybee colonies. The number of bees per day were counted on the crop and later on seed yield was recorded. Within that distance the number of bees decreased from 100 to 61 and the yield decreased from 6,000 gm to 3,700 gm.

In general, the greater the number of bee visits per floret, the higher the seed set. Rudnev (1941a, b) observed that when the bee visits per floret were 1.0, 1.4, 3.4, 6.1 the average seed harvest per million florets was 53, 76, 133, 210 kg, respectively. He further calculated that for efficiently pollinating all florets in one hectare of sunflowers, 2,000 bees per day are needed for 3 weeks of flowering period which number can be provided by a strong colony of *Apis mellifera* bees. More recently, in central Bohemia, nectar secretion of 15 sunflower cultivars was determined at an average of 18.1 mg/ 50 flowers. The higher the sugar production the greater the number of bee visits. It resulted in an average increase in seed yield at 26.0% but the highest increase was upto 101% (Kalmer, 1996).

It has been estimated that, depending upon weather conditions, 21-75 kg of honey can be extracted from one hectare (Montgomery, 1958). In Manitoba, an actual gain of 47 kg was recorded from colonies by placing them in sunflower fields for 15 days (Mitchener, 1950). Thus honeybees can play a vital role in sunflower seed production and also in collecting surplus honey.

There was variation in sugar concentration in different cultivars and places; the range was 49-51% (Lasik, 1953) and 13-26 (Parker, 1981). Total sugar produced per flower ranged from 0.4 to 0.6 mg. In

USA honeybee colonies taken to sunflower fields produced on an average 36 kg of honey and in USSR from 14-47 kg/colony. It is generally believed that honeybees visit this crop for nectar only but there are records indicating that 60-85% of the pollen collected was from sunflower. Availability of alternative sources of pollen is the deciding factor.

Panchabhavi *et al*. (1976) in Karnataka kept 67 *Apis cerana* colonies in a sunflower field of 20 ha and observed that there was 72% seed set whereas in 2 other fields open to natural pollinators but without colonies there was 55 and 57% seed set.

In hybrid seed production using cytoplasmic male sterile lines a large population of pollinators is required to transfer pollen from male lines to the female (male sterile) lines. Honeybees can play a crucial role. Insect pollination of sunflower has also been studied in Pakistan (Manzoor-ul-Haq, and Faiz, 1980).

5. Niger (*Guizotia abyssinica* Cass; Compositae)

It is a comparatively new crop in South India and provides forage to honeybees during the dearth period. It covers an area of 4.8 lac ha and seed production is 1.2 lac tonnes. It is grown is Madhya Pradesh, Maharashtra, Bihar, Orissa, and Tamil Nadu.

The composite head has 40-60 tubular hermaphrodite flowers and a row of marginal florets. Spread over 7-8 days the flowers in the outer rows open and dehisce in the morning and the style lobes of stigma grow during the day and curl backwards but do not touch their own pollen, therefore, cross pollination is common through external agencies like insects. In an experiment in Maharashtra 40 flower heads in two plots were caged, one with *Apis cerana* and the other without bees; in the former on an average 40 seeds were produced but in the latter there were only 15 (Bhambure, 1958). In another experiment yield was compared under different situations; in measured plots without *A.c*. bees it was 0.9 kg, those caged with bees 1.2 kg and in the open with natural pollinators was 1.1 kg (Pande *et al*., 1988). In the open, bees were observed at 1.8 to 8.1/m^2 per minute. It indicates that there is some self-fertilization but the crop benefits from pollinators that bring about cross pollination.

6. Flax or Linseed (*Linum usitatissimum* L; Linaceae)

Linum usitatissimum L. is grown for fiber from the stems, and oil from the seeds, throughout the countries of Asia and Europe. In India the crop is mainly sown as linseed, predominantly in Madhya Pradesh, Uttar Pradesh Bihar and Maharashtra. In many other states it is also grown in solid fields or more often in lines in crops like wheat, cicer, and barley. The total acreage in India is 1.9 m hectares and production is 5 lac tonnes of seed.

The flower has five bright blue or white petals and the colour is red in ornamental varieties. The ovary has five chambers which have five erect styles. On warm clear mornings when the flowers open anthers dehisce and self-pollination takes place; cross-pollination is less than 1%. The nectar glands located at the base of stamens are visited by insects including *Bombus* spp., Diptera, butterflies and most importantly, the honeybees. The visits of honeybees last 7-8 seconds per flower and they avoid standing on stalks outside the flowers. Only modest quantities of honey are gathered by the colonies kept near linseed fields. The role of honeybees in pollination and increase of yield is not clear. When bees were confined in caged plants 22-24% increase was observed in certain varieties (Gubin, 1945a).

7. Sarson and Toria (*Brassica campestris* L. *sarson*; var. *toria*; Cruciferae) (Plate 18)

Like other cruciferae these plants have four sepals, four petals and four long stamens around the ovary and two small ones in an outer whorl. An ovary contains two fused carpels. In Indo-Gangetic plains both are important oilseed crops and their leaves and soft shoots are eaten as vegetable. The two crops have distinct growing periods under irrigated conditions. Toria is sown in September and produces flowers from early October to the end of November, the peak of blooming being from mid-October to mid-November. Sarson and raya (*Brassica juncea*) are sown, often side by side, in October and November. They produce flowers from December to February. In favourable weather of cold nights and sunny warm days flowering continues sequentially and the spikelets gain considerable length and practically all the pods are full of seed. Out of 21 m ha, under all oilseed crops 5 m is under *Brassica* alone.

The crops are grown in multiples of half or full acres and quite often as distant rows in wheat or chickpea crops. The bright yellow fragrant flowers attract a large number of insects which come for nectar as well as for pollen. The average sugar concentration in the afternoon ranges between 40-45% whereas in the morning it is 16% only (Sharma, 1958; Murrel & Nash, 1981). Mohammed (1935) in West Punjab at Lyallpur recorded 117 visiting species of insects belonging to seven orders; *Andrena ilerda, Apis florea, Halictus* spp. comprised 82% of insects and were the most important in this order. He also observed that stigma of the crops remained receptive for two days after opening and in 5 days their fertility was lost whereas pollen remained viable for 7 days. It was observed in Australia that among *Apis mellifera* foragers 72% collected nectar only 25% both nectar and pollen, and 3% collected pollen only (Hangridge & Goodman, 1975). The pollen was collected between 10 am to 2 pm.

Auto-fertilization of these crops was studied by Mohammed (1935) for 6 years by enclosing individual plants in muslin bags; in toria, seed-set was 12.3% in brown sarson, was 20.3% and in yellow seeded sarson 91.0%. Then he compared self-pollination to cross pollination in toria. He bagged flowers for cross pollination and hand pollinated them; there was 100% pod formation. Sihag (1986) studied seed set in toria plots caged and not gaged and left open for natural cross pollination; there developed 25 and 1060 pods per plant, 6.3 and 18.2 seeds per pod, and 4 and 385 seeds per plant. Similar observations on sarson gave 59 and 95% pods containing 3.5 and 12.7 seeds per pod (Mishra *et al*., 1988). Observations made in field plots of both sarson and toria showed that when crop was caged to exclude insects yield was 68 g; when caged with *Apis cerana* gave 219 g and in open field near honeybee colonies where other insect pollinators were also present yield was 244 g, (Latif *et al*., 1960b). In general, there is quite a high degree of self incompatibility in these crops and good yields are obtained by cross pollination with the aid of natural fauna in most places where good weather prevails and probably yield can be further increased with the help of honey bee colonies to the extent of 10-30% (Rahman, 1940; Singh, 1954) covering for inclement weather. In Uttar Pradesh it was observed that bee pollinated plants in field cages as compared with self-pollinated plants in individual bags produced 3 times heavier pods, 4 times more seed per pod, 50 times more seeds and 84 times more seed yield per plant (Singh and Singh,1992).

In continued good weather an inflorescence is visited by insect pollinators 71 times and a flower 26 times and these is hardly any chance of a flower escaping pollination. In prolonged cloudy weather pollination may be inadequate and podless gaps may appear on seed branches (Mohammed, 1935). In bad weather yield may be affected for two reasons; fewer insects may be active or there may be heavy infestation of aphids, reducing sugar concentration making the plants less attractive to bees. Observations were made at Ludhiana in the month of January between 12 noon & 1 pm and the total number of *Apis florea*., *Apis dorsata* and *Andrena ilerda* visiting per 10 m^2 of raya (*Brassica juncea Coss)* were counted for a duration of 10 min. When there were 283 aphids per 10 cm of twig, there were 7.7 flowers per inflorescence, sugar concentration was 42.1% and only 8.7 bees visited. However, when there were 25 aphids the number of flowers was 14.8, sugar concentration was 49.5%, the number of bees visited was 68.2. On the first day, it was reported that only 0.2 bees visited an inflorescence per 5 mim even though wind velocity was 1.3 km/h and temperature was 12.9°C. On the second day 1.7 bees visited and the temperature was 10.5°C and wind velocity was 7.4 km/h (Dhaliwal and Atwal, 1975).

It was recorded in West Punjab (Pakistan) that 105 species of insects belonging to 9 orders visited toria and sarson, the prodominant orders being Hymenoptera and Diptera. Among the former *Apis florea* comprised 16.5% of visitors on toria and 50.5% on sarson, whereas *Andrena ilerda* comprised 42% on toria and 4.0% on sarson. Before 10 am and after 4 pm Diptera (Syrphidae) were the most abundant and in between these hours bees were the predominant visitors (Rahman, 1940; Bhalla *et al.*. 1983). The speed of working of these insects, was also studied by these authors. It was observed that *A cerana* and *A. mellifera* visited 17.7 and 23.4 flowers per minute.

In East Punjab including Haryana, and Himachal Pradesh in India extensive surveys of insect pollinators of *Brassica* were made in 1965, 1966, 1967 during the mid-day and the comparative abundance of those species was recorded when they were repeatedly caught in nets (Atwal, 1970; Kapil *et al.*, 1969b, 1971)* . In all 157 samples were collected 69 on toria and 88 from sarson in different agroclimatic zones including the southern districts where annual rainfall was less than 50 cm, central zone with 100 cm rainfall, and northern zone including Himalayan slopes with rainfall 100 to 150 cm. In all zones *Apis florea* F. was by far the most abundant insect on toria (84%) and sarson (73%) and it was indicated in each and every sample (157 in all) of 30 sweeps each. Population density was the highest in central zone, followed by southern and northern areas; annual rainfall upto 100 cm seemed suitable but not so above that. On these crops as well as on raya the only other species represented in most of the samples were *Apis dorsata, Andrena ilerda* and *Halictus*. On toria, *Halictus* bees (9.7%) were the next most abundant, followed by *A.d.* (3.7%), *Andrena* (1.6%) and all the rest of insects comprised less than 1% (also see Dhaliwal & Atwal, 1975, 1976a, b). On sarson, these bees in the same order, comprised 8.2%, 7.8%, 5.9% and all the rest 5.0%. Since *Andrena adults* emerge from their burrows in middle of November their population significantly increases on this crop, but toria bloom at that time is on the decline. The most glaring difference between the West and East Punjab is that *Andrena* in the former area is as abundant as *A.f.* (84%) in the latter region. The density of population of all visitors combined was 2.8 insects per net for toria and 6.7 for sarson. Between the two observations there is a gap of many years and agroclimatic conditions probably changed significantly at places in the two Punjabs.

It was observed that while foraging on *B. campestris* the majority of honeybees (*Apis mellifera*) collected nectar (72%) and fewer collected nectar and pollen (25%) and only 3% collected pollen only. The proportion of bees in these studies changed from year to year. Mohr and Jay (1988) observed that in one year 89% of bees collected nectar only and 11% both nectar and pollen, but in another year the respective percentages were 64 and 36. Most nector gatherers approached flowers from the top and 24-53% passed over stigmas. In Punjab (India) it was observed that *Apis florea* and *Apis dorsata* landed on flowers from the top, passed through the body of rather small and delicate flowers, and then searched for nectaries. They collected pollen also but their proportion to the total was not recorded. They tended to collect nectar from the side between petals but their weight was borne by the main body of flower. The two wild honeybees together constituted 80.8 to 87.7% of the total population of insect visitors and their role in pollination of these crops seemed all important. More recently with the introduction of *Apis mellifera* pollination has improved further and surplus honey to the tune of 10-15 kg/colony is commonly gathered from *Brassica*. Colonies properly managed gave an average yield of 50 kg/colony in a year.

Apart from yield of these oil seed crops there are problems of seed production and maintaining genetic purity. Owing to wide-spread cross pollination there is intermixing of varieties and even

* Other bees recorded on *toria* and *sarson* were reported by Batra (1977).

species. For preventing contamination, crops have to be isolated, but in areas where *Brassica* is sown all over it becomes difficult and a seed crop can be easily ruined. The distance of isolation of a seed crop vary from 0.4 to 3.2 km depending upon the purity required. According to Gill and Vear (1958) most contamination occurs within 90 m of pollen source and if a crop is isolated at a distance of 900 m, it is quite safe to produce a seed crop. It also depends upon the cultivars involved those—having high sugar are more intensively pollinated (Koltowski, 1999).

8. Raya (*Brassica juncea* L.; Cruciferae)

It is primarily a self pollinated crop. Howerd *et al.* (1916) described how a honeybee works in a flower. It lands on a flower and pushes its tongue between the long and short stamens to reach nectar that has sugar content of 22-65% (Sharma, 1958). Its head touches long anthers and stigma and thorax touches anthers of short stamens. As it moves over to the other side of flower its thorax touches stigma, transfering pollen. The exact role of bees in pollination has not been determined.

It was observed in UK that plots of India mustard (*B.j.*) caged without bees or with bees produced the same yield as obtained in the open (Free and Spencer-Booth, 1963). In India, plants inside cages devoid of pollinating insects produced 14.4 seeds per pod, 581 pods per plant and average yield of 193 g seeds per plant as compared with the figures 17.2, 1090, and 349 from the exposed plants visited by all insects (Sihag, 1986). It was suggested by Free (1993) that in order to eliminate caging effect foraging bees should be released inside another cage and yield of plants be observed.

Surveys carried out from 1966-68 in Punjab (including Haryana and Himachal Pradesh) showed that in 59 samples of 30 nets each the average density of insect visitors (2.3/net) was about the same as recorded on toria and sarson. *Apis florea* was by far the most common bee (75.4% of total) and the other common bees were *Andrena ilerda* (9.3%), *Andrena leaena* (4.1% and *Andrena* spp. (2.1%), *Apis dorsata* (3-8%) *Halictus* spp. and all *Halictines* (3.7 & 4.1%). All the rest of visitors comprised 2.1% of the population and almost all the groups were represented as recorded on toria and sarson. It was interesting to note that after *A.f.*, *Andrena* was the conspicious group (15.5%) and bees of family *Halictidae* (7.8%) were also active on this crop as on toria and sarson (Atwal, 1970; Kapil *et al.*, 1969b, 1971). *Megachile* even though a small group was observed on toria and sarson but not on raya.

It was observed that among *Apis dorsata* foragers 18% collected pollen. The nectar collectors also touched stamens and stigmas. They started foraging when the temperature was at least 15°C (Kapil and Kumar, 1975). The reason for *Andrena* group prefering raya above sarson was not clear. Perhaps tallness of this crop stimulated mating races (see biology of Andrena).

9. Taramira (*Eurica sativa* Lam; Cruciferae)

This oilseed crop is grown under unirrigated conditions by itself or mixed with chickpeas. Its flowers open in the morning and remain so for three days. Anthers dehisce as soon as the flower opens and nectar is available sometime later on; therefore, pollen gatherers visit first and nectar gatherers follow them. It has been reported that taramera is self-sterile and seed is produced with the help of outside agencies (Cobley, 1956).

Its pollination requirements were studed by Sihag (1986). He caged plants to exclude insects and compared them with those exposed to natural pollinators; in the two cases he recorded 79 and 1075 pods, 2.2 and 15.5 seeds per pod, 4 and 321g seeds per plant.

Insect visitors were surveyed (58 samples of 30 sweeps each) in Punjab (including Haryana and Himachal Pradesh) for there seasons during 1965-68. Of all the visitors *Apis florea* (50.8%) was the most numerous, followed by *Apis dorsata* (21.2%). *Andrena ilerda* and allied species (18.5%),

Colletes spp. (2.2%), and all the rest of bees (6.9%). Halictidae were also recorded but the presence of *Colletes* on this crop was unique. On the whole population density was less than half of that recorded on other oilseed Cruciferae. In the drier southern districts, *A.f.* held first position and *Andrena* spp. the second position. However, in central districts *A.d.*, was the most abundant, followed by *A.f.* and *Andrena* (Atwal, 1970; Kapil *et al.*, 1971). It was reported that *Andrena* in West Punjab had comparatively higher population perhaps owing to the drier climate there. The overall density of population on taramera was 0.9 insect per net sweep in East Punjab.

10. Seed Cauliflower (*Brassica oleracea* L., var. *botrytis*; Cruciferae)

Other vegetables that belong to this species are cabbage, Brussels sprout, broccoli etc. Cauliflower seed is produced from plants of the previous season, grown in isolated fields at higher altitudes. Flowers are self-pollinated also but for seed production, cross-pollination is considered desirable. Among the insect pollinators, honeybees (*Apis cerana* and *Apis mellifera*) are considered most useful. The wild honeybees, Andrena, Syrphidae and other Diptera also play an important role (Kakkar, 1980, 1981).

The pollination requirements of cauliflower were studied in Himachal Pradesh (Rauala, 1972). It was observed that plants in cages with honeybees (*Apis cerana*) had 68% seed set, 9 seeds per pod, and 627 seeds per 100 flowers, whereas in the cages without honeybees there was 9% seed set, 2.3 seeds per pod and 28 seeds per 100 flowers. In another experiment it was found that when plants were bagged there was 19% seed set and 2.3 seeds per pod as against 45% and 4.0 for plants in cages of wire netting that excluded large insects; the figures for exposed plants in the open were 93% and 6.4 (Verma and Joshi, 1983). Sihag (1986) also observed big differences in seed set and yield in plants caged and not caged.

Insect pollinators including honeybees are active from morning till evening but the highest seed set is obtained by their activity between 11.00 and 13.00 h (Dhaliwal & Bhalla, 1981). It was observed that *A.c.* colonies enhanced pollination in the open fields; upto 50 m seed set was 94.5%, number of seeds per pod was 21, and there were 84 seeds per plant; at 200 m where honeybees were not visiting, the figures were 81%, 15, 41. At 200 m *Apis dorsata, A. florea, Andrena* and other insects were active in the field (Tewari & Singh, 1983). It appears foraging range of *A.c.* on cauliflower bloom is limited. Within 200 m 53% and within 400 m 86% of the honeybees foraged and maximum distance at which they were seen on this crop was 900 m. It was estimated that if fields were 2 km apart purity of seed could be maintained (Dhaliwal and Sharma, 1973; Adlakha and Dhaliwal, 1979; Kumar *et al.*, 1988). Radish seeds are produced in the plains of North India and honeybees play an important role in the pollination of this crop.

11. Cucurbit Vegetables (Cucurbitaceae)

There are a large number of species of cucurbits grown throughout the world. Most of them are monoecious and have separate staminate and pistillate flowers. However, some like American varieties of *Cucumis melo* L. have staminate and hermaphrodite separate flowers. In certain localities and climates perfect pistilate flowers may also be produced. Ovary in a female flower may be three or five chambered with corresponding number of stigmas; the seeds in a fruit indicated the number of ovules fertilized (Free, 1993).

In various parts of the world different species of cucurbits are preferred and some are universally liked everywhere, namely musk melons and canteloups (*Cucumis melo*) and water melon (*Citrullus vulgaris* or *C. lanatus Mansf.*). In Europe and Americas, Cucumber (*Cucumis sativus* L.) and pumpkin or squash (*Cucurbita pepo L.*) are grown most commonly, whereas in India apart from these vegetables many others are liked; those include ghiyatori (*Luffa cylindrica; L. aegyptiaca; L. acutangula*) chapan or veg. marrow (*Cucurbita pepo*), tar (*Cucumis melo var. ultilissimus*), karela (*Momordica charantia*), petha or ashgourd (*Benincasa cerifera or B. hispida*), on which insects visitors were recorded.

In central Punjab (Ludhiana) these cucurbits were surveyed for insect activity in 1966 and 1967 by collecting 78 samples of 30 sweeps each. The average population of all insects on ghiyatori was 13% of the total, on chapan 36%, on tar 17%, on karela 16%, on petha 7%, on musk melon 10%. The presence of various insect species was recorded on each crop. On all crops as a whole Halictidae comprised 42% (15.1% *Halictus* spp; 4% *Nomia* spp; *Nomioides* spp. 23%), *Apis florea* 28%, *Apis dorsata* 14%, *Xylocopa* spp. 6.5% and *Pithitis smargdula* 6.0%. The bees that visited all crops studied and were caught in most of the samples were, *A.f.*, *P.s.*, *Halictus* spp. and *Xylocopa fenestrata*. *Apis dorsata* was not observed on petha and muskmelon. Some unidentified *Nomioides spp.* in 1966 were observed in unusually large numbers on tar and karela. It is likely that there were agregations of nests near the crops from where only 5 samples were collected (Atwal*, 1970; Batra,* 1977). Luffa flowers when bagged do not set fruit as they are monoecious and need pollination services of insects. Pollination studies were made in south India near Bombay (Bhambure, 1957, 1958). In smooth ghiyatori (*Luffa aegyptiaca*) the male flowers start appearing one week earlier than female flowers and are always in preponderance. They are visited by *Apis cerana, A. florea, A. dorsata, Melipona* spp., *Bombus* spp. At Ludhiana the most common insects were *A.f.* and *Halictidae* in equal numbers followed by *A.d., Pithitis smaragdula, Xylocopa fenestrata,* all the three in equal numbers. Among Halictidae, *Halictus* spp. and *Nomioides* spp. were as common as the last three mentioned. The other species recorded have been mentioned in the foot note. In Pakistan most numerous pollinator on cucurbits in general was *A.f.*, followed by *Anthrophora confusa, A. zonata, Pithitis* sp., syrphid fly, *Megachile monticola,* 3 species of *Xylocopa* and *Thyreus histerio,* (Saddiqui and Mecci, 1974).

Kalitori, ridge or spong loofah, also have monoecious flowers which open between 0900 to 10.30 h. The nectar chamber in staminal base opens 2-3 h after anthesis and the bees by that time are covered with pollen. Near Bombay the most common insects visiting flowers were *A.c.*, *A.f.* and *Melipona* spp. and they were active during daytime. Moths and butterflies were also important pollinators (Shrivastava, 1991).

Karela or bittergourd, is dioecious and cross-pollination takes place. Anthesis takes place at 2200 h and nectar near ovaries is secreted a few hours later. In the absence of male flowers and pollination fruit can develop as a result of parthenocarpy (Shrivastava, 1991). In Punjab (Ludhiana) insect pollinators of this crop were surveyed in 1966-67 and it was recorded that Halictidae comprised 75.4% (Halictines 32.9% and *Nomioides* spp. 42.5%) of all the insects, followed by *A.f.* 10%; *P. smaragdula* 9% and all the rest reported on cucurbits 6%. *A.d.*, and *Xylocopa* spp. were barely noticed in our samples (Atwal, 1970). In another survey after 10 years. spread over a larger area Grewal and Sidhu (1978) recorded *A.f.* 28%, *Halictidae* 52%, and *Anthrophoridae* 10%.

Pumpkin, squash, gourd and chapan have separate male and female flowers borne singly in the exils of leaves. The male flowers, 12 times the number of female flowers, have five stamens bearing anthers. The female flowers have three chambers in the ovary with a ring of nectar around it. Male

* The species of bees recorded were: *Allodapes; Anthophora* spp. ; *Megachile bicolor; M. disjuncta; M. fasciculata; M. femorata; M. flavipes; M. lanata; Nomia oxybeloides; Nomia* sp.; *Nomioides* spp. *N. divisor; N. minutissima; N. pusilla; N. variegata; Xylocopa* spp., *X. fenestrata; X. pubescens*. Batra (1966, 1967, 1977) further identified some more species and also studied their nesting habits; *Megachile coclioxyoides; M. monticola; Nomia eburneigera, N. curvipes; Lasioglossum cathulum, L. albescens; Tetralonia ovata; Ceratina* sp.; *Braunsapis* sp.; *Thyreus* sp.

flowers are produced first and then the female flowers appear but the habit can be reversed in certain seasons and nectar is found in both types. Sugar concentration is around 30% but it changes according to place, time of day, fertility and moisture of the soil (Girish, 1981). It was observed near Bombay that among seven species of gourd and pumpkins flowers opened from 6-8 am and closed between 12-6 pm and in two species flowers opened at 5 pm and closed sometimes at night. Every species has its own timings (Bhambure, 1958). It was observed both in Banglore and Punjab that *C. pepo* flowers that were pollinated by insects in early hours of the day had a greater fruit set than those pollinated later on. Flowers receiving from 3 to 10 bees progressively set higher percentage of fruits ranging from 30 to 100%. Caged plants devoid of all insect pollinators produced no fruit and fruits of self pollinated plants were of less weight than those of cross pollinated plants (Grewal and Sidhu, 1978, 1979; Girish, 1981).

The *Cucurbita pepo* plots without insects and without fruit to start with were compared with caged plots with *A.c. bees inside* and the exposed fields with natural pollinators; the respective fruit set was 46% and 57% (Girish, 1981). Similarly in Punjab it was observed that plots caged to exclude insects produced 10% fruit set, those caged with females and nests of *Pithitis smargdula* set 42% , and in the exposed fields fruit set was 62%. It was also noted that *A.d.* (45%), *A.f.* (35%), *Halictine* bees (15%) were the principal pollinators and other bees like *Xylocopa* sp., *Megachile* sp., and Anthrophids were also observed; (Grewal and Sidhu, 1978; Atwal, 1970). At Banglore in the month of February *A.c.* was seen foraging at 6.00 h and *A.d.* and *A.f.* two hours later; all of them stopped foraging at 1200 h. Very few solitary bees were seen foraging. The time spent by honeybees on flowers was observed; nectar was collected both from male and female flowers. Their body gets covered with pollen and pollination is facilitated. *A.c.* is seen on this crop when grown near an apiary otherwise the wild honeybees play the most important role. The solitary bees were scarce (Girish, 1981).

Cucumber (*Cucumis sativus* L.) is perhaps the most important crop of Cucurbitaceae in Europe and America but on the Indian subcontinent there are other more important crops. It is primarily a monoecious crop bearing at first staminate flowers and later on pistillate flowers. The female flower has a three chambered ovary and rudimentary stamens. The male flowers have three stamens and dehiscence of anthers takes place at 18-21°C. Pollination is essential for most of the cultivars but some (in Britain and Holland) develop fruit by parthenocarpy and in fact if pollinated they become misshaped thick at one end with a narrow neck (Free, 1993). Except for these special cultivars, pollination with the help of insects is encouraged both in the glass houses and outside in the fields. After a five year study it was proven by Canada, Department of Agriculture (1961) that mean yield per hectare in plots caged without honeybees was 1.7 tonne, that caged with honeybees was 4.7 and in exposed plot was 5.8 tonnes. Considering the importance of this commercial crop, extensive studies on the role of insect pollination have been made in Europe and North America. In India Bhambure (1958b) observed near Bombay that *A.f.* and *Melipona* spp. and not *A.c.* were active in cucumber fields. In Punjab mostly *A.f.*, some *A.d.*, *P.s.*, *Xylocopa fenestrata* and even *A.m.* were actively foraging on this crop (Grewal and Sidhu, 1978). The absence of *A.c.* in the former case and presence of *A.m.* in the latter case appear to be just by chance depending upon the bee colonies in the vicinity of observation fields. However, both these bees are known to forage on cucumber.

Canteloups and muskmelons are monoecious and some cultivars have staminate and hermaphrodite flowers. In a female flower ovary has three to five chambers and 400-600 ovules. Pollination is by insects even in hermaphrodite flowers whose pollen is too sticky for autofertilization. It was observed that in California, flowers opened at 0700-800 h and anthers dehisced at 09-030 h (18°C), however, in Punjab flowers dehisced 30 min earlier than anthesis that occured between 0530 and 0630 h (Rosa, 1924; Nandpuri and Brar, 1966). Sugar concentration of male and female flowers ranged between 20-40%. In 1967 it was observed in Punjab that bees of Halictidae group formed 50%

of the insect visitors and *Apis florea* formed 40.7% but in subsequent observations after 10 years the two groups of bees formed 23% and 65% and *Xylocopa* was also present (Atwal, 1970; Grewal and Sidhu, 1978). It has been observed in USA that in areas where beekeeping is practiced honeybee, *Apis mellifera* pollinate flowers adequately, if not, 2.5 colonies per hectare are required. For maintaining purity of seed, crop has to be grown in isolation.

Near Bombay it was recorded that male flowers of watermelon (Citrullus lanatus) opened at 0830 h and female flowers at 0930 h. Within 3-4 h anthers and stigma were fully exposed and the flowers closed by 1400 h. Peak pollen collection by honeybees was observed at 0900 h and the species recorded were *A.c. 87%*., alongwith *A.f.* and *Trigona iridipennis*. The Indian honeybee worked upto noon whereas *A.f.* worked for longer hours (Bhambure, 1958; Rao and Suryanarayana, 1988). The flower structure and pollination requirements of watermelon are similar to those of maskmelon. In Karnataka where yield in open pollinated fields was 15.2 t/ha, was enhanced to 26.5 and 20.7 t/ha when natural pollinators were helped by two and one colony/ha, respectively (Lingappa *et al.*, 1999).

12. Alfalfa or Lucerne (*Medicago sativa* L.; Leguminoseae)

It is a very important fodder crop in Europe and North America whereas in India and Pakistan other fast growing clovers are preferred. A great deal of literature has been produced on the pollination requirement of this crop in Europe, Canada, USA and other countries that have been reviewed (Bohart, 1962; Hobbs, 1964; Free, 1993). Since honeybees avoid tripping the flowers, solitary bees are being exploited for its pollination and search has been going on for the introduction of exotic solitary bees into USA. In certain areas of India particularly Gujarat pollination requirements are met through the activity of natural pollinators.

Medicago sativa L. is a biennial rich fodder crop grown throughout the world and is an important source of honey production. Its seed setting is poor in colder parts as the flowers do not trip readily. The tubular flower has five petals joined at the base. In the middle is the large standard petal and two smaller ones are one on each side forming wings. Two other petals form the keel which hold staminal column within and retain it under pressure. Keel petals are interlocked with projections from the wings. When any insect like honeybee or a wild bee presses the keel the staminal column snaps forward and strikes against the standard petal scattering pollen. When tripping takes place the column hits the bee at the lowerside of its head which it does not like. The nectar at the base of tube can be reached from the sides avoiding tripping. The honeybees learn to do that and hence are not good pollinators. Some of the wild bees are more efficient and can take the brunt of tripping. The stigmatic membrane is ruptured during tripping and becomes receptive to pollen in this process. Naturally flowers that do not trip produce no seed.

The capacity to produce seed varies from plant to plant which may be completely self-sterile or totally self-fertile. More seeds are produced from cross-pollination than from self-pollination. As the staminal column strikes a bee there are chances of both self-pollination and cross-pollination the latter being more common in nature. In order to keep purity of seed the varieties should therefore be isolated.

It is possible to cause tripping mechanically by passing a rope over the plants. The tripped flowers are more acceptable to honeybees who come to collect nectar and in this process cause cross-pollination. Seed yield can be increased by moving in honeybee colonies. Right concentration of bees and the distance at which colonies are kept near alfalfa fields is important. Sovoleva (1952) observed that fields at a distance of 500 m and 1250 m from an apiary produced 310 kg and 81 kg of seeds per hectare respectively. High yields were obtained (131 kg/ha) by keeping 2.5 colonies per hectare as compared to 10.5 colonies (50 kg/ha.) (Pankiw *et al.*, 1956). Honeybees are not effective in certain areas where climate is cold or where more attractive crops are grown and bees drift away from alfalfa.

A certain amount of tripping occurs naturally without the role of insects but the percentage is too low for economical seed production. There are localities where climate is more favourable for natural tripping; there may be 30-40% automatic tripping where temperature reaches 32°C or above (Hobbs and Lilly, 1955).

As mentioned earlier some wild bees are very efficient pollinators and their population can be increased by proper management. The two well known examples are *Megachile* spp. and *Pithitis* sp. A survey of insect visitors on alfalfa in Punjab, Haryana and Himachal Pradesh showed that a large number of wild honeybees and solitary bees, visited this crop (Atwal, 1970; Batra, 1977; Kapil and Jain, 1980). In 100 samples of 100 sweeps each collected over a period of 4 years the average percentages of various species per sample recorded were: *Apis florea* (46.3%), unidentified *Nomioides* spp. (16.8%), *Pithitis smaragdula* (8.6%), *Megachile flavipes* (8.1%), *Halictine* bees (5.1%), *Nomioides minutissma* (3.9%), *Nomioides pusilla* (1.9%), *Nomioides oxybeloides* (1.8%), *Nomioides variegata* (1.3%), *Megachile disjuncta* (1.6%) and the rest of the population (4.6%) comprised of *Allodapula, Andrena levilabris, Anthrophora* spp., *Apis dorsata,* Megachile spp., *M. bicolor, M. disjuncta, M. femorata, M. lanata, M. rotudata, Nomia* spp., *Xylocopa* spp., *X. fenestrata, X. pubescens*.

The effect of abundance of insect pollinators on seed production was studied at two places in Punjab; at one place there were 278.2 pollintors per 100 net sweeps and the seed yield recorded was 270 kg/ha., while at the other place the count of pollinators was 153.3 and the yield recorded was 158.0 kg/ha (Goyal and Atwal, 1975). It indicated that in North-Western India the total combine of insect pollinators played an important role in seed production. Seed production could also be improved by encouraging carpenter bees. By providing reeds of *sarkanda*, bees are attracted 80% to make nests in them and increase population (Atwal, 1970; Goyal and Atwal, 1975). Similarly Megachile flavipes could also be encouraged (Kapil *et al.*, 1968, 1980). Nest-building bees can be gainfully employed in enclosed spaces for hybrid seed production of flowers and vegetables.

13. Barseem (*Trifolium alexandrinum* L.; Leguminosae)

Barseem or Egyptian clover is a very important fodder crop in India and covers an area of 800 thousand ha. In the vicinity of towns and villages where milch animals are kept the crop is grown in almost contiguous stretches. Farmers like to keep their own seed and sell the surplus. Therefore yield and purity of seed are important considerations. In Egypt, Hassanein (1953) studied its pollination requirements. He kept plants in screen cages with *A.m.* honeybees, plants caged without bees, plants in the open exposed to natural fauna, and plants in cloth cages without bags to exclude the effect of wind, and observed that the number of seeds per head were 43.8, 1.2, 22.3, and 0.4. It was concluded that the plants were practically self-sterile and needed cross-pollination and that wind played a minimal role. At Faisalabad (Pakistan) it was observed that plants caged with *A.c.* bees produced 26.1 seeds and those caged without bees gave 1.1 seed per head. In open field when *A.c.* bees were present with other insects 53.2 seeds were produced and in the absence of the honeybees 23.2 seeds were formed per head (Latif *et al.*, 1965). At Pusa in Bihar also it was observed that plants caged to exclude bees and plants in the open gave an average of 111 and 4010 seeds per 100 flower heads recorded in two years (Narayanan *et al.*, 1961). It is quite clear that in the absence of pollinating insects, no seed can be produced. In the presence of natural pollinators a fair yield is expected but in order to obtain good yield honeybee colonies should be kept near the crop.

Near Cairo, Wafa and Ibrahim (1959) collected on barseem 4 species of Lepidoptera, 14 of Diptera, 16 of Coleoptera, 35 of Hymenoptera. Musca, forming 38% of the total, was the most common insect visitor. In Bihar insect visitors were observed for two years and it was found that honeybees comprised 90.5% of the total; *A.c.* was the most abundant and *A.f. A.d.* were also present

(Narayanan *et al.*, 1961); probably the field was close to an apiary. In Punjab Ludhiana, Batra (1967) observed that barseem was visited predominantly by *A.f*. and a few *Pithitis smaragdula* and the latter was an efficient pollinator. At that time there were no colonies of *A.c.* or *A.m.* near there, otherwise they are known to forage on this crop. In fact it was observed that in tripping efficiency various bees in descending order were *A.f., Megachile flavipes, P.s., A.m., A.d.* (Dhaliwal and Atwal, 1976b); between the temperatures 26.5–35°C and relative humidity 46–70% they visited 6.7, 9.7, 6.4, 8.4, 7.8 flowers per minute and all the species combined visited in largest number when wind velocity was 0.2 km/h and their count was 18.5 per 10 m^2 5 min; the lowest count of 1.5 was when wind velocity was 22.5 km/h. Bees of all species combined worked on barseem untill the velocity was 10 km/h and above that speed they either decreased in number or disappeared altogather. *A.f.* was the most tenacious and was seen even at 22.5 km/h. It was further observed on this crop that *A. f.* spent 41.9 seconds per flower, *A.d.* 15.7 s, *A.c.* 9.4 s (Atwal *et al.*, 1970).

All the four *Apis* started working between 0800–0830 h and continued upto 1800–1830 h except that *A.d.* started working a little earlier and stopped foraging at 1700 h. There were two peaks of activity, one in the forenoon upto 1200 h and the other after 1500 h for about 1.5 hours. Foraging range of *A.m.* on this crop was also studied (Dhaliwal & Atwal, 1974). Daily observations were made for 25 days in May, the peak flowering month. At distances of 100, 700, 1000 meters from a solitary apiary of *A.m.* the number of bees collected per 25 net sweeps were, on an average, 15.9, 6.7, 1.8 respectively. It was concluded that for proper pollination bee colonies should be within 1 km of the field.

Barseem provides major honey flow after the *Brassica and Eucalyptus* seasons and honeybees collect surplus honey. Chitalla (*Trifolium hybridum* L.) is another clover grown as fodder in India. It is self sterile and in nature many solitary bees aid in its pollination and yield of seed is much increased with the help of honeybees. At Ludhiana, it was pollinated primarily by *A.d., Nomioides variegata, Lasioglossum cattulum;* the other bees that actively foraged on the plants were *A.f., A.c.*, 4 *Megachile* spp., 2 *Xylocopa* spp., *Nomia* sp., *Nomioides* spp., *Halictus* sp., *Sphecoids* sp. and 4 additional *Lasioglossum* spp. Another forage crop senji (*Melilotus indica*) was visited by *A.f.*; 6 *Andrena* spp., Colletes sp., and 3 *Lasioglossum* spp. (Batra, 1967).

14. Groundnut (*Arachis hypogaea* L.; Leguminosae)

Groundnut or peanut plants have automatic self-pollination. The flowers open at sunrise and anthers dehisce at about the same time pollinating the ovary.

15. Arhar (*Cajanus cajanus* (L.) Millsp.); Leguminosae)

Arhar or pigeon pea or red gram is the most popular pulse eaten with the staple food rice or chapati; hence there is a great interest in developing varieties that produce high yield. Before the flowers open stigma becomes receptive and 12 h later the anthers dehisce followed by the opening of flowers 6–8 h later on. It is a self fertilized crop but many insects visit flowers and their role is not fully understood. At Hydrabad insects recorded belonged to the orders Hymenoptera, Diptera, Lepidoptera, Neuroptera, Homoptera and Orthoptera and the most important insect pollinators belonged to the genera *Megachile, Xylocopa, Chalicodoma, A.d.* and *A.f.* When *A.d.* and *A.f.* land on the keel they do not depress it by body weight and they obtain pollen by inserting their probosis between the standard and wing petals; they spend 14.7 s and 25.6 s respectively as compared with 3.8 s and 3.6 s by the efficient *megachile* and *Xylocopa* (William, 1977, 1980). Although at Hydrabad, *A.c.* were unable to obtain nectar from flowers, in Agartala (East India) they were observed to be more active nectar gatherers, than the common Trigona (Pande and Bandyopadhyay, 1985). According to

Chaudhry and Jain (1978) *Magachile lanata* population on this crop can be increased by providing artificial nesting materials. Bhatia *et al*. (1981) observed outcrossing in this crop at four places; Hydrabad 20%, Varanasi 27%, Badnapur 3%, Coimbatore 40%. Moreover, it varied with different cultivars. The crop as a whole may not benefit much from cross-pollination by insects but honeybees foraging on it seem to gather some nectar and even pollen. Since the crop blooming continues up to autumn, the honeybee colonies can build up their strength on this crop, particularly in those areas where toria is not grown.

16. Para Rubber (*Hivea brasiliensis* Muell-Arg.; Euphorbiaceae)

Para rubber is an important source of nectar from leaf buds and the honeybees collect surplus honey in Kerala, particularly in areas where coconut and oil palms grow closeby to supply pollen (Mathew, 1975). The male and female flowers are separate on rubber plants, the former open first and drop after a day and the latter open for 3-5 days and the total flowering period is 2 weeks. In bagged flowers there is no auto-fertilization and in the open 0.3–1.6% set fruit and their number increases from 8.4 to 46% by hand pollination; cross pollination is beneficial in some strains (Morris, 1929; Rao, 1961). Most of the female flowers fall off without setting any fruit.

Rao (1961) collected 36 species of insects belonging to 19 families of the orders Diptera, Hymenoptera, Thysanoptera. In India *Melipona* sp. and *Trigona* sp. collect pollen from rubber flowers whereas the honeybees *A.d., A.c.* do not; in fact they are rarely seen on flowers (Jayarathnam, 1970; Suryanarayana, 1980). In Brazil and Puerto Rico midges play important role in pollination.

17. Banana (*Musa paradisiaca* L. or *M. acuminiata* L.; Musaceae)

There are many cultivars of banana grown along and below the Tropic of Cancer in India and many tropical areas of the world. Separate male and female flowers appear on every subsequent node along a central spindle. Self-pollination cannot occur and fruits are produced parthenogenetically. Both male and female flowers produce nectar as much as 162–423 mg in a day; pollen is also produced. Sunbirds, many bees and Melaponidae visit flowers but play no role in pollination. In diploid varieties pollination is carried out by sunbirds and long–tongued bats between 1700 and 0700 h when the flowers open (Nur, 1976).

Since plants flower throughout the year and nectar has sugar concentration of 25–30% honeybees exploit this source.

18. Guava (*Psidium guajava* L.; Myrtaceae)

Guava covers an area of 30,000 ha throughout the country, the largest area being in Uttar Pradesh and Bihar. It can withstand heat and prolonged drought. The flowers are hermaphrodite which appear twice a year, in April–May and then from September to December. The flowers are a medium source of nectar and pollen and they benefit from the activity of bees. Flowers are white and fragrant and the honey produced from them is amber in colour. Many bees and other insects visit flowers and produce 35% corss–pollination, although the flowers are basically self fertile.

19. Ber (*Zizyphus jujuba* Mill. and *Z. mauritiana* Lam.; Rhamnaceae)

The Chinese and Indian ber have small greenish yellow flowers 0.5–0.6 cm in diameter, borne singly or in groups upto ten in the leaf axils. Anthers dehisce soon after a flower opens and the stigmas become receptive 2 days later. The Indian ber flowers open between 0730–0900 h or from 1200 to 1300 h and are self unfruitful and some are even cross-incompatible (Vashistha and Pareek, 1979;

Teaotia and Chauhan, 1964). Insect pollination is believed to be benefitial and at Jodhpur 7 species of Diptera and 11 of Hymenoptera visited flowers and at Ludhiana. *Polististes hebraeus,* the yellow wasp, was most abundant. If there are honeybee colonies in the vicinity *A.c.* and *A.m.* also visit the flowers (Singh, 1984; Dhaliwal, 1975). The role of *Musca* is not clear as no pollen grains are noticed on their body.

20. Safflower (*Carthamus tinctorius* L.; Compositae)

It is a winter crop grown in dry parts of India, generally under unirrigated conditions. Its oil contains unsaturated fatty acids and it is considered good for dieting. The yellow and orange florets are surrounded by sharp and pointed bracts. Anthers dehisce in the morning and soon after the stigma appears above them catching pollen grains, thus self fertilization is possible. Extensive crossing takes place in certain parts of India and not in others. When seed crop is separated by 30 m with another crop in between there is very little contamination (Howard *et al*., 1916; 1919 ; Kadam and Patankar, 1942). Nectar-sugar concentration ranges from 13-29% at different times of the day and honeybees *A.m.* mostly visit for nectar and very few for pollen. It was observed near Bombay that *A.c., A.d., A.f.* began foraging at 1000 h. and pollen collectors worked upto 1400 h and nectar collector upto 1500 h ; the three species were at these jobs for 24 and 21 s, 17 and 13 s, 32 and 33 s, indicating that *A.f.* worked more thoroughly (Deshmukh *et al*., 1985). In Karnataka honeybees (*A.c.*) increased the yield by 54. 8% over naturally pollinated crop; crop pollinated by two colonies in the open registered a yield of 14.4 q/ha as compared with 12.2 q/ha pollinated by one colony (Lingappa *et al*., 1999).

21. Coffee (*Coffea arabica* L.; Rubiaceae)

Coffee trees are grown in the hilly tracts of Kerala, Tamil Nadu, Karnataka and Andhra Pradesh, covering an area of 155.5 thousand ha. The buds appear on trees from November to February and flowers open in February–March. The crop faces serious competition from Brazil and other South American countries, therefore, its yield should be improved. Its flower has five white petals which are joined at the base and form a cylindrical tube in which develop five stamens; the ovary at the base has two chambers having one ovule each. Nectar contains 38% sugar and flowers are visited by bees, wasps, flies; the honeybee, *A.m.,* were the most abundant at midday, half of them collected nectar and the others pollen. Flowers open at dawn and the anthers dehisce soon after, close to the receptive stigma on a long style. Coffee is tetraploid and self–fertile (McDonald, 1930). In Jamaica it was observed that bushes covered to exclude insects and given honeybees inside cages gave the same fruit set of 84–90% but there were more of medium size berries in the fomer and more of large size in the latter treatment; moreover, the number that matured and their weight was at least 50% more as a result of pollination by honeybees (Raw and Free, 1977). The amount of cross–pollination reported from various countries ranged between 4–93%. There is a general recommendation that colonies of honeybees should be kept in coffee plantations. *C. robusta,* the other species of coffee, is self-sterile and insects play an important part in its pollination. The flowers are a minor source of pollen and a rich source of nectar for honeybees.

22. Coriander, Carrot, Saunf (Umbelliferae) (Plate 19)

Coriander (*Coriander sativum* L.), carrot (*Daucus carota* L.), *saunf* (*Foeniculum vulgare* Mill), sweet saunf or anise (*Pimpinella anisum* L.) and cumin (*Cuminum cyminum* L.) all have small flowers grouped together in umbels on top of the plants. A flower has 5 tiny sepals, 5 petals and 5 stamens. Nectar is on top of ovary which has two chambers, each with one ovule. Although flowers are self-fertile, no self–pollination takes place as the availability of viable pollen and receptivity of stigma do not coincide.

Coriander is an important commercial crop in parts of central India. Localities are selected according to yield potential which probably depends upon the population of insect pollinators available and weather at the time of maturity. It was observed near Bombay that bee pollination increased the yield by 187% and in another experiment it was found that caged plants produced only 56% of the yield recorded in open fields (Deodikar and Suryanarayana, 1977; Baswana, 1984). In Haryana, Sihag (1986) observed that plants caged to exclude insects produced 183 seeds with a weight of 2.9 g per plant and open crop not caged produced 1203 seeds weighing 15.7 g. The effectiveness of honeybee (*A.c.*) pollination in cages varies with varieties also; increase in one was 21% and in another 124%. Perhaps these varieties were developed for growing at different times of the year (Shelar and Suryanarayana, 1981). The crop is visited by a large number of Diptera and Hymenoptera and honeybees are active if colonies are close by. Among the wild fauna, *A.f.* is the most important pollinator (Singh, 1954). At Ludhiana, Batra (1977) collected a number of species on jowain *(Trachyspermum ammi*); *A.f.*, *A.d.*, *Andrena* sp., *Nomioides* sp., *Halictus* sp., *2 Lasioglossum* spp.

Seed carrot is another valuable crop raised from stumps of mature carrots. After flowers open anthers dehisce in one or two days and stigma becomes receptive on third or fourth day, and remains in that condition for a week. Pollen grains when shed, lose viability very soon, thus self-fertilization is not possible. There is nectar on top of the ovary to which insects are attracted but honeybees ignore the flowers and drift to some other crop even if hives are kept in a carrot field. A good seed crop is possible only through cross-pollination brought about by insects belonging to Apoidea, Sphecidae, Syrphidae, Stratiomyidae; Bohart and Nye (1960) in Utah collected 334 species belonging to 71 families.

Bohart's group earlier (1956) observed that the average yield in open fields exposed to all pollinators was 674 kg/ha, those enclosed in nylon cages with *A.m.* bees was 864 kg, enclosed in nylon cages without bees was 367 kg, and enclosed in muslin cages without bees was 112 kg/ha. It was indicated that with natural pollination a fairly good crop was obtained but when honeybees were forced to work in an enclosure produced a higher yield. Muslin stops the flow of wind and hence the yield is very low. It is recommended that attempts should be made to rear some of the flies in decaying vegetable matter and to release them in seed carrot fields. I have observed at Chandigarh that Syrphidae are by far the most common insects on this crop particularly in areas where Brassica crops are grown in winter and the population of aphids is high providing food to the syrphid larvae; subsequently, adults remain active until the ambiant temperature becomes too high.

Fennel benefits tremendously from insect pollination and without their role yield falls by one half or even one quarter in various parts of India. At Ludhiana yield from caged plots was 15.6 g as compared with 31.2 g per 1000 umbels from the uncaged plots exposed to insect pollinators. At Hissar the caged and exposed plots yielded 392 and 1364 seeds weighing 3.4 and 9.6 g per plant, respectively (Sagar, 1981; Sihag 1986). At the former place the insects recorded were *A.f.* 54%, Syrphidae 34%, *A.m.* 5%, *A.c.* 4%, *A.d.* 2%. Near Bombay the insects recorded were *A.f.* 82%, *A.c.* 6%, *A.d.* 1%, Syrphidae 6% and other Diptera 5% and, they were most active from 1100 h to 1300 h (Suryanarayana *et al.*, 1960). In areas where *A.f* and Syrphidae are abundant there is no need of placing honeybee colonies; *A.m.* and *A.c.* probably visit other more attractive crops. In any case, both fennel and anise are grown in small fields open to many insect pollinators. Cumin is comparatively a more important crop dependent on insect pollinators. Fortunately these three crops are not sprayed with insecticides and insect visitors are quite safe.

23. Citrus Fruits (Rutaceae)

The commonly cultivated fruits are : sweet orange (*Citrus sinensis Osceck*); sour orange (*C. aurantium* L.); *santra (C. raticulata Blanco) ; grape fruit (C. paradise Macf.)* ; lime (*C. aurantifolia*

Swing); lemon *(C. lemon* L.); Citron (*C. medica* L.); Sweet lime (*C. limettioides*). They occupy 67,650 ha in various regions of India and flower from February to May. Citrus flowers are a source of nectar and a minor source of pollen for honeybees.

The white scented flowers of most citrus species and cultivars have an average of 5 petals (range 4-8) and 20-60 stamens that form a ring around the style (Fig. 20). Inside this ring nectar is secreted 3–5 days before the flower opens. Sugar concentration inside the flower bud of orange was recorded at 13–17% but as the flower became older it increased to 31% (Vansell *et al.*, 1942). The flowers are hermaphrodite in most species but in lemon and citron ovaries are not functional and seedless fruit are produced. Insects are attracted to nectar and pollen of most citrus species but they do not need their services to set fruit. In same cultivars pollen is not viable and parthenocarpy is common. In others there is abundant fertile pollen and self-pollination takes place but fruit set is more through cross-pollination; in the former case seedless fruit is formed. Some cultivars are wholly or partially self-incompatible and their pollination needs are met by grafting a few branches or a row of pollinizer varieties which are worked on by insects. It was observed at Ludhiana that trees of sweet lime caged with or without honeybees (*A.m.*) produced 6% fruit set and those cross pollinated by hand gave 53–67% fruit set (Nijjar and Sandhu, 1971). In Pakistan, branches of Kinnow mandarin bagged to exclude insects gave less and poor quality fruit as compared with those in the open, pollinated mostly by *A.f.* (Manzoor-ul–Haq, 1978). Citrus pollination, studied in various countries is reviewed by Free (1993) and in India further studies are needed on specific varieties grown in various parts of the country. Luckily insecticide spraying is not recommended on plant blossom and thus natural fauna comprised of Hymenoptera, Diptera, Lepideptera, Neuroptera, and Coleoptira are spared when visiting this crop.

24. Loquat (*Eriobotrya japonica* Lindl.; Rosaceae)

The tree is very attractive and fits in good landscaping. The fruit is a delicacy only. White flowers appear on panicles and each has plenty of nectar around the ovary attracting insects. Some of the cultivars are self–incompetible and the others also benefit from cross-pollination by insects or even by hand. It was observed at Ludhiana that bagged flowers produced 15% less fruit than those in the open. Many bees visited flowers and *A.d.*, was the most common (Mann and Sagar, 1987). In Srinagar *A.c.* and *Bombus* were the most common and both were equally abundant (Abrol, 1988b). A great deal of genetic diversity is available in civil lines of Dehradun and around Saharanpur.

25. Tomato (*Lycopersicon esculentum* Mill.; Solanaceae)

On an inflorescence there are 4–5 pedent flowers, each with 6 stamens and bright yellow anthers and a short style which is receptive at the time of dehiscence. Flowers are self-fertile and self-pollination takes place quite readily. There is no nectar and insects visit for pollen only. Tomatoes grown in a glass house need the services of honeybees or bumblebees but an electric device for vibrating flower clusters produce equally good results. Under field conditions if honeybee colonies are kept the bees drift to other crops. In Haryana *A.f.* and *Halictus* spp. were seen visiting flowers and were most abundant between 0900 to 1100 h; *A.d.* and *A.c.* were also seen (Choudhry *et al.*, 1973). There is natural 2–4% cross-pollination and for hybrid seed production honeybees can be employed fruitfully.

26. Chilli (Solanaceae)

There are two species cultivated; *Capsicum annum* L. and *Capsicum fruitescens* L. There are many cultivars grown in the length and breadth of India under different climates, soil fertility, and insect fauna. Stigma is generally reseptive before anthers dehisce and hence there is high degree of cross-pollination which may range from 58-68%. Sugar concentration in nectar is also quite high (33-69%) and insects belonging to many families visit flowers for nectar and pollen. For hybrid seed production honeybees and flies (*Calliphora and Lucilia*) can be employed. Honeybees forage on these crops during dearth periods. They also forage on tobacco in south India and thus tide over the dearth period.

It was observed in Southern Quebec that under glasshouse conditions the yield and quality of fruit of sweet pepper (*Capsicum annum* L.) was greater when hoverfly (*Eristalis texax* L.) was released as a pollinator (Jarlan *et al.*, 1996).

27. Grapevine (*Vitis vinifera* L.; Vitaceae)

Flowers are hermaphrodite and are in clusters; those at the base open first and within 1–2 days all of them open. In a flower there are 5 stamens with a nectary at the base of each. Ovary is 2 chambered each with 2 ovules. anthers are close to pistil and dehisce to cause self-pollination and, tapping the flower cluster is helpful. Pollen in some cultivars is sterile and cross-pollination takes place, either naturally, or by artificial means. In such cultivars pollinizer vines are planted in between; wind and insects act as pollinating agents. Insects are not attracted to nectar but collect some pollen. The insect visitors include honeybees, *Halictus* sp., *Syrphidae* and some beetles. At Ludhiana, *A.d.* was recorded as the most abundant insect and the bees worked from 0700 to 0930 h (Mann and Tanda, 1984).

28. Cardamom (*Elettarica cardemum* (L.) Maton; Zingiberaceae)

Cardamom is an important spice grown in western parts of South India. The crop is important for beekeeping as colonies build up their strength on its flowers. There are 2–5 flowers on each of the 20 or more lateral branches of the long panicle which may droop or lie flat depending upon the cultivar. In a flower there is one stamen with a large anther that dehisces at the time of flower opening between 0500 and 0800 h and pollen becomes non-viable by 1400 h. At 1000 h stigma is most receptive but if flowers are enclosed in sleeves of mosquito netting the fruit set is only 11% with 2.1 seed pods per panicle as compared with 66% and 27.4 seed pods on the panicles exposed to insect pollinators. Nectar has a sugar concentration of 37–43% and is available from 0700-1400 h when insects visit for sugar as well as pollen (Chandran *et al.*, 1980; Pattanshetti and Prasad, 1973). Fruit set was observed by Verma (1987) for seven years and he reported that when flowers were bagged set was 5–7% whereas that of open pollinated flowers ranged between 37–82%.

The insect visitors recorded on cardamom were; *A.c.*, *A.d.*, *A.f.*, and *Trigona* sp. Beekeeping is traditionally practiced with *A. cerana* and concentration of colonies varies from locality to locality which seems to affect the fruit set. In a field where there were 10 colonies/ha the fruit set was 67% but 2 km away where fewer bees were visiting the set was 60%. Another observation showed that in four plots where there were 0, 2, 4, 6 colonies per hectare the fruit set was 36, 41, 46, 44% indicating that natural population of bees including *A.c.* were effective in pollination activity but the presence of honeybee colonies improved fruit set (Verma, 1987; Madhusoodanan and Dandin, 1981). It was also observed that in the presence of *A.c.* there was 35–45% greater weight of capsules as compared with that in fields where there were no bees. It was observed that *A.c.* collected nectar upto 0900 ha and thereafter worked on pollen; *A.d.* on the other hand foraged on pollen all day (Siddapaji & Basavanna, 1980). Elsewhere *A.m.* is also known to forage on cardamom. In general the wild honeybees seem to play a significant role in the pollination of this crop.

As an agricultural practice it has been recommended to expose the panicles and flowers by removing leaf mulch. A higher fruit set takes place and yield is increased. The presence of honeybee colonies is a help in this process. It appears, when there is abundant natural bee fauna, presence of colonies helps no more although there is some evidence that more the quantity of pollen deposited on stigma, the greater the fruit set and seeds per pod and higher the yield. Well directed studies on the corss–pollination of this valuable crop and the role of honeybees (*A.c.* and *A.m.*) are needed to derive the full benefits.

29. Cotton (*Gossypium hirsutum* L.; *G. arboreum* L.; Malvaceae)

American cotton *G.h.* is medium stapled and has broad leaves which are less hairy and harbour sap sucking insects and a number of worms that bore into buds and bolls. It is grown extensively in India and is sprayed with insecticides 4–10 times during the growing period. *G.a.* is the Asiatic or desi cotton which is hairy and is rarely sprayed. It is of short staple and covers much less area. All cottons combined cover an area of 8 m hectare in India. The annual production ranges between 65 and 80 lac bales.

Cotton flowers have 5 large prominent petals and their colour varies with species and cultivar concerned. Ovary at base has 3–5 united carpels, each with many ovules. Around the style there is a tube like structure above whose length are attached numerous stamens and one anther each producing lot of pollen. The flowers open at dawn and witter by the evening. Anthers dehisce soon after the flower opens and stigma is also receptive at that time. Pollen remains viable for 12 hours and fertilization takes place after 30 h. In *G.h.* the stigma is close to the anthers and self pollination takes place, where as in *G.a.* flower, pollen does not reach the top without the presence of insects and both self and cross-pollination is observed. In the former there is primarily auto-pollination (self-fertile) but insects increase the yield by 5–10%. Wind does not play any role (McGregor and Todd, 1956; Afzal and Khan, 1950; Sidhu and Singh, 1961, 1962).

In cotton, nectar is profuse secreted by one floral and four extra floral sets of nectaries. The floral nectar inside the base of petals is difficult to reach by small insects but the extra floral nectar attracts many insects. Nectar has sugar concentration of 18–24% in various cultivars and owing to evaporation concentration may reach 20–54% in floral and 60–82% in extral floral nectars (Vansell, 1944). Mound (1962) observed that leaf nectaries secrete at different times of the day *viz.* in morning hours, noon hours and in the afternoon. Sugar concentration reaches the maximum between 1400 and 1600 h.

It was recorded at Ludhiana that 41 species belonging to 23 families and 7 orders visited *G. hirsutum* and many of them either did not visit floral nectaries or did not touch the sexual parts. It was observed that each flower was visited by 47 insects and *A.d.*, *A.c.*, *A.f.* were the important pollinators; the bees visited 6.7, 6.6, 5.8 flowers per minute and 141, 128, 120 flowers per visit respectively. Only experimental colonies of *A.c.* were available (Sidhu & Singh, 1961). On *G. arboreum A.c.*, *A.d.*, *A.m.* formed 32% of insect visitors and solitary bees were 27%; the three honeybees visited 104–124, 94–102, 87–99 flowers per trip, respectively. The plants caged with honeybees (*A.c. or A.m.*) gave 10% more bolls and 15% more yield and 17% more seed as compared with those caged without bees. It was observed that *A.c.* did not touch the anthers but brushed past the stigma in 60% of the visits whereas *A.m.* touched anthers and brushed past stigma in 70% of the visits (Tanda, 1983, 1984; Tanda and Goyal, 1979a, b), indicating the possibility of cross-pollination. *A.m.* was present with beekeepers and *A.c.* came only from experimental colonies.

The paradoxical situation in cotton is that beekeeping is not recommended in *G. hirsutum* areas whereas *G. arboreum* can benefit from pollination by *A.m.* honeybees. A perfect understanding is needed between the farmers who spray their crop and the beekeepers in that area. The same applies to hybrid cotton.

30. Bhindi or Okra (*Abelmoschus esculentum* (L.) Moench; Malvaceae)

It is an important summer vegetable and is also called lady's finger. It is cooked when quite supple but the seed crop is allowed to mature till bhindi becomes brown and is picked before it splits and scatters its seeds. The large yellow flowers are borne in leaf axils and open at 0700–0800 h and close between 1200–1700 h; petals and calyx are shed soon after. The long style has 5–9 stigmas and at the base it is surrounded by 50–120 stamens borne on a tube like structure. Anthers dehisce soon after flower opening and continue to do so till noon, with maximum at 0800 – 0900). Pollen is fertile for a few hours and in the afternoon it becomes sterile. Hand pollination in the forenoon gives 80% fruit set (Srivastva, 1969; Sukhija *et al.*, 1981).

It was observed in Punjab that there was 5–20% cross-pollination in a commercial seed crop (Purewal and Randhawa, 1947). Insects were attracted to pollen as well as nectar and the groups of insect visitors included *Apis, Halictus, Ceratina, Megachile, Xylocopa, Bombus, Diptera* (Chaudhry *et al.*, 1973). Honeybees played an important role in cross-pollination and it was observed in Himachal Pradesh that length of bhindi increased by 18%, its weight by 25% and number of seeds by 33% (Mishra *et al.*, 1967). The percentage of *A.c.* collecting nectar, pollen or both was 87, 10, 3 and those of *A.m.* were 88,4,4 respectively. It was observed at Ludhiana that nectar gatherers of both species while visiting a newly opened flower, landed on stigma and anthers then reached for nectar and while leaving they took off from petals only, indicating good chances of cross-pollination (Tanda, 1985).

It appears there is adequate natural pollination for the production of vegetable crop which can be further facilitated by shaking the plants. For obtaining high yield of seed crop intensive pollination by honeybees can increase yield by one-third.

31. Sesame (*Sesamum indicum* L; Pedaliaceae)

Sesame is thinly spread throughout India as a kharif crop in the North and a rabi crop in the South covering an area of 2.3 m ha. Plants have floral and extra floral nectaries at the base. Its nectar has 38% sugar concentration and honeybees exploit it during the dearth period.

Sesame or till is eaten raw and is also a source of valuable oil. Flowers are hermaphrodite and are borne singly or in groups; they have 4 stamens and an ovary with many ovules. When a flower opens, at 0400 h the stigma is receptive and is covered over with pollen. The flowers are shed 12 h later. In many flowers anthers do not shed pollen and insects attracted to extra floral nectaries bring about cross-pollination (Howard, 1919). The plants exposed to insects gave 25% higher yield as compared with those covered with cages to exclude insects. Cages with *A.c.* inside gave 2.6 kg okra per plot as compared with 1.6 kg in cages without bees (Phadke *et al.*, 1967; Pande *et al.*, 1988). Out in the open *A.c.*, *A.d.*, *A.f.* were the most common insects on the crop and they began foraging at 0600, 0700, 0800 h and visited 6.3, 5.0 and 1.9 flowers per minute and flowers pollinated by them gave 51, 53, 4 seeds per pod, respectively. Seed weight was also heavier as compared with the controls (Rao, *et al.*, 1981). While foraging 72% of *A.c.* collected nectar 23% pollen only and 4% both nectar or pollen. At one time or the other they touched anthers to bring about cross-pollination and thus increased the yield (Phadke *et al.*, 1976). *A.m.* is also known to work on flowers very actively. Extensive studies were made in Punjab on the genetics and pollination requirement of this crop (Sikka and Gupta, 1949).

32. Coconut (*Cocos nucifera* L.; Palmae)

Coconut grows along the coastline of India and on the near by islands, covering an area of 108 m ha. of land. With a production of 6000m nuts it stands second in the world. The plant starts flowering when 5–7 years old.

The male flowers are in thousands and only a few or even one female flower may be present at the base of spadix. The male flowers on opening shed pollen in a day or two and all of them dehisce about a month earlier than the female flowers open at the base of spadix. The female flowers open 3–6 days after the male phase. There is nectar both in male and female flowers and insects visit them. In the tall cultivars male flowers dehisce at various intervals on different trees so that pollen is available when female flowers open. With the aid of insects both self–pollination and cross-pollination occurs but wind plays a very little role. Since, pollen remains viable for 2–9 days and with proper techniques it can be stored for several months, hand pollination is successful particularly in the dwarf varieties. In different parts of the world honeybees (*A.c.* and *A.m*), *Trigona* sp., wasps (*Polistes* spp.) and ants forage primarily on pollen throughout the year. It has been observed that wasps tend to drive away solitary bees and honeybees (Free *et al.*, 1975). If female flowers are bagged the fruit set is 3.4% as against the expected 22% of flowers in the open. Sugar concentration in nectar was recorded 17–23% and Jay (1974) observed a honeybee visited 20 female flowers in 3 minutes; 67–99% of the honeybees carried pollen on their bodies. They collected pollen all day and there were two peak periods at 1000–1100 h and again in the afternoon. Coconut pollen plays an important role in successful beekeeping near rubber plantations in Malaysia and South India (Makhdzir *et al.*, 1980).

33. Oil Palm (*Elaeis guineensis* Jacq.; Palmae)

The male and female inflorescences are separate but on the same palm; trees flower at different times, spread over a period of months. Male flowers on a spikelet start opening at the base and complete the process in 2–4 days producing 25–50 g pollen per inflorescence. The female inflorescence consists of about 200 spikelets arranged spirally with 10–30 flowers borne on each. The flowers open in a certain propertion each day and in all it takes one week; the stigma remains receptive for 36–48 h. Syed (1979; 1982) in Malaysia has proven that profuse light pollen is not carried by wind for pollination but insects carry it and deposit it on stigma. It was observed that 88% of the insects on palms belonged to the groups *A.c., A.d., A.f.,* and *Trigona* sp. and only 5% of them were recorded on female flowers. He also observed that thrips played an important role in cross-pollination but the greatest success was achieved by the introduction of the beetle *Elaeidobius kamerunicus* which was successfully introduced from Cameroon (Syed, 1982). The kernal and oil yield has now increased from 20 to 53%. The beetle is perhaps also present in Kerala, India. Oil palm is an important source of pollen for honeybee colonies.

34. Plum, Peach and Apricot (Rosaceae)

The stone fruits Plum (*Prunus domestica* L.; (**Plate 20**), Peach (*P. persica* L.; **Plate 21**), Apricot (*P. armeniaca* L.) have many cultivars for temperate and subtropical climates having high chilling and low chilling requirements.

The flowers of *Prunus* have 5 petals, many stamens, an ovary with two ovules inside a single carpel. The stigma is receptive as soon as the flower opens. The anthers dehisce soon after the flowers open but in warm weather they may do so even before the opening of a flower.

Plum is self–unfruitful but peach and apricot are largely self-fruitful; corss-pollination is essential for the former and is beneficial for the latter. It was observed by Free (1962c) that more honeybees visited plum trees nearer the colonies and gave higher seed set than the trees further away. Apart from honeybees many other groups of insects namely *Bombus, Andrenidae, Megachilidae, Diptera,* also visit the flowers but honeybees, both wild and hived, are the main pollinators. The proportion of various *Apis* species depends upon their local population and the popularity of beekeeping in that area. For example at Ludhiana on plum, *A.d.* formed 62%, *A.f.* 9%, *A.m.* 29% and, on almond they were in the percentage of 45, 14 and 41 (Mann and Singh, 1981, 1983). Near Solan, Shimla and Srinagar although *A.m.* was present *A.c.* was observed as the most important pollinator (Kumar *et al.*, 1985b; Verma and Chauhan, 1985; Abrol, 1988a; Bhalla *et al.*, 1983b). It has been reported that bumblebees though not as abundant as honeybees may be more efficient as they work faster and continue to do so even in bad weather. *Aristalis* spp. Syrphid flies and some moths are also abundant and important pollinators at certain places in Himachal (Mishra *et al.*, 1976; Singh & Mishra, 1986).

In the Himalayan areas of Utter Pradesh pome fruit trees are pollinated primarily by the honeybees, *A.c., A.f., A.d., A.m.,*. At higher altitudes, where the wild honeybees are not found, the trees are pollinated primarily by halictine bees (*Lassioglossum matianense*), bumblebees, and the giant race of *A.d.* (= *Apis laboriosa*) which is adapted to cold climate (*Batra,* 1996). It was observed near an *A.m.* apiary that it foraged dominantly and the other honeybees were driven away (also see Dhaliwal and Atwal, 1974, 1976a). At lower altitudes and in the Shivaliks at places where *A.d.* is predominent the smaller bees stay away. In a forest area it was recorded that among the foragers *A.d.* comprised 80%, *A.c.* 15%, and all other bees 5%. The solitary bees recorded were; *Braunsapis, Amegilla, Pithitis, Trigona, Anthrophora, Tetralonia, Systropha* and other halictids, *Bombus, Xylocopa, Osmia* and *Andrena* (Batra, 1996).

Sugar concentration in nectar of plum, peach and apricot ranges between 10-40%, 20–38%, 5–25% respectively and insects are attracted to flowers. Honeybees collect nectar, pollen or both (Singh, 1954; Sharma, 1958; Mishra *et al.*, 1985). Nectar is more concentrated in older flowers and sugar content is also affected by relative humidity. Apricot flowers are wide open and nectar is diluted by dew and rain. Even without these factors honeybees may ignore apricot and pear and show preference for apple and cherry. In self fertile fruits pollination is probably by gravity and wind. Honeybees are beneficial for pollination and fruit set, but their role is not fully understood. However, in Japan it was demonstrated that *Osmia cornifrous* bee greatly improved fruit set of peach (Kitamura and Maeta, 1968). It was observed at Ludhiana that low chilling varieties of peach benefitted from insect pollinators (Mann and Singh, 1981) but at higher altitudes in Himachal, Kumar *et al.* (1984) studied 14 cultivars of high chilling requirements and concluded that they did not need insect pollination. Plum on the other hand greatly benefits from insect pollination and honeybees play a dominant role.

Mann and Singh (1981) at Ludhiana recorded insects visiting almond (*Prunus dulcis* Miller) which is self-unfruitful and needs cross-pollination through insects. More than one plant are needed for the supply of pollen. The bees visited flowers in the ratio *A.d.* 45%, *A.f.* 14%, *A.m.* 41%. Sugar concentration in nectar was 16–36%. Germination of pollen on bodies of pollen collectors was observed at 65% and from nectar collectors 38% (Klungness *et al.*, 1983). It has been observed in Israel that in order to obtain maximum yield of almond the main and pollinizer varieties should be grown in alternate rows rather than in blocks (Ersikowitch, 1999).

Specific requirements of these fruits and their cultivars should be made at various altitudes since they have different pollination requirements. Moreover, the population of effective insect pollinators varies from place to place.

35. Apple (*Malus domestica* Barkh; Rosaceae)

Apple varieties are either diploid having plenty of pollen and are generally self-fruitful or they are triploid which are generally self-unfruitful and have to be pollinated with, pollen from pollinizer varieties. Even self–fruitful varieties need the interplantation of other varieties for obtaining high yields. Among the fruits grown in temperate India apple covers 65% of the area totalling 1700 ha. There is an annual increase of 10% in area.

Genus Malus has five petals and numerous stamens. there are five styles united at the base and ovary consist of 5 carpels of two ovules each. Since proper fruit growth takes place around the seeds only, adequate fertilization of the flowers is necessary. The full bloom (March–April) of apple is far in excess and if 10–15% flowers are fertilized commercial crop is raised. The flowers that are not pollinated and fertilized drop off. Although stigma is receptive as soon as the flower opens, wind plays no part in pollination. Honeybees, both *Apis cerana* and *Apis mellifera,* exploit pollen and nectar containing 12–18% sugar and, bring about cross-pollination so important for successful production in most cultivars of this fruit crop.

It was reported that the Indian honeybee visited apple for nectar and pollen (Rehman and Singh, 1941). More pertinent studies were made by Mishra *et al.* (1976) before the introduction of *Apis mellifera* in Shimla Hills. The most common insects observed on apple bloom were honeybees, *Bombus, Eristalis, Syrphus* and in the evening some Lepidoptera. Honeybees were the most abundant and they visited in the highest number between 1200 and 1500 h. Diptera were the next in abundance and they visited in the greatest number before noon. Mishra *et al.* (1976) placed bee colonies in a Red Delicious orchard and observed that the highest fruit set was 33.1%, within 10 m of the colonies and beyond that upto 200 m. the fruit set ranged between 13.4 and 17.9%. At longer distances even though no honeybees were seen a good crop was harvested. Apparently wild bees and other insects produced adequate pollination.

In a later survey it was recorded in the Shimla hills that 44 species belonging to 14 families and five insect orders visited apple blossom; *Apis cerana* was the most abundant and formed 24.0–43.0% of the population; Bombus was scarce and comprised 0.2–2.4%; the rest were mostly Diptera of which Muscids were predominent and formed 19.7–38.3% of the population (Verma and Chauhan, 1985). Fruit set in two self compatible varieties, Golden Delicious and Red Gold was studied: in covered plants fruit set was 24.6% and in open and honeybee pollinated plants was between 30.7 and 34.5% in the former variety; and 15.7, 18.3–22.4% in the latter variety under the two situations. Fruit drops in covered plants was 38% whereas in open pollinated and honeybee pollinated plants it ranged from 25–28%. As compared with control the increase in fruit weight in the two varieties was 22 and 18% respectively. In the self-incompatible variety Royal Delicious the increase in weight was 33% (Dulta and Verma, 1987; Verma, 1987b).

By that time *Apis mellifera* had been introduced in the Shimla hills and at a locality where it was found working along with *Apis cerana* on apple their comparative foraging activity was studied: the nectar plus pollen foragers among them were recorded as 6–11% and nil, respectively. Nectar collectors among *A.m.* at 0900 and 1500 h were 73 and 70% whereas pollen collectors were 48 and 22% respectively. The proportion of nectar and pollen gatherers in *A.c.* was the reverse in afternoon. The peak activity among the former species was between 1100–1300 h. when temperature was 21-25°C whereas in *A.c.* the peak activity was from 0900–1100 h when temperature was 13.5–21.0°C. Pollen loads carried by the former bee at 0900, 1200, 1500 h were 9.24, 12.22, 11.12 mg respectively, whereas those carried by the latter at the same hours were 9.06, 9.26, 8.64 mg respectively (Verma and Dulta, 1986).

At the lower altitudes near Solan, apart from the above mentioned insect pollinators of apple, some solitary bees belonging to the families Andrenidae and Megachilidae were also recorded (Bhalla *et al.*, 1983). These bees as well as bumble bees are known to be efficient pollinators whereas the role of Syrphidae and Calliphoridae is not clear. The role of honeybees is well determined, moreover, their population can be controlled by placing the right number of hives in a orchard. The comparative efficiency of various pollinators on apple was studied by Kendall (1970) in U.K. through well planned experiments; the percentage of ovules fertilized was 20% by honeybees, 39% by Andrena spp., 42% by Bombus spp. and 72% by hand pollination.

For the pollination of apple one honeybee colony of good strength per acre is considered sufficient. The colonies should be placed in groups of 4–6 at sunny, wind protected place.

36. Litchi (*Litchi chinesis* Sonn; Sapindaceae)

There are three types of flowers: males with functional anthers, no style but nectar at base; hermaphrodite with functional anthers and rudimentry style; hermaphrodite with non-functional anthers but full style and stigma and those form one third of the total flowers (Das and Choudhary, 1958; Chaturvedi, 1969, 1965). There is abundance of insects on flower panicles but compared to the total flowers fruit set is low; probably as determined by the carrying capacity of the tree the embryos abort. There is some self pollination by wind but most of the cross-pollination is brought about by insects. It was observed that flowers bagged with mosquito netting set 15% and those exposed gave 43%. In another study bagged panicles gave 0.03–0.1% fruit set and exposed ones gave an average of 2.7–4.9% set. In terms of fruit count there were 12 on bagged and 53 on exposed branches (Pandey and Yadava, 1970; Phadke and Nain, 1974).

It was observed in Uttar Pradesh that *A.d., A.c., A f.,* and *Melipona* spp. were the predominant visitors on flowers although other insects like *Syrphidae, Musca, Vespa* spp. black ants were also present. In the southern slopes of western Himalayas *A.c.* comprised 50% and *A.f.* 26% of insect visitors and in Bihar the respective percentages were 18 and 60 (Pandey and Yadava, 1970; Dhaliwal *et al*., 1977). In Chandigarh where there are hardly any *A.c.* colonies *A.m.* was the principal forager alongwith abundant Syrphidae but no *A.f.* or *A.d.* were observed nor their colonies spotted in March 1998 after the severe winter (Atwal, 1998).

Along the Shivalik foothills at Pathankot Hoshiarpur, Dehradun and Ranchi *A.c.* and *A.m.* colonies are kept by beekeepers to collect surplus honey which has a special flavour. In order to achieve that, the late winter honey from *Brassica* and *Eucalyptus* should be extracted first. In Bihar 7–8 kg of honey per *A.c.* colony has been recorded whereas in Punjab and Dehradun 20 kg per *A.m.* colony is the average (Phadke and Naim, 1974; Atwal, 1998).

37. Maize (*Zea mays* L.; Gramineae)

Corn or maize is grown as kharif crop mainly to the north of a line connecting Surat with Calcutta. The area under this crop is 5.8 m ha, apart from the crop raised as fodder. The plants produce large tassels which carry abundant dry pollen in the morning hours. At that time there is dearth of pollen and honeybees utilize it. Sometimes the honeydew found on plants is also used by the bees.

38. Buckwheat (*Fagopyrum esculentum* Moench; Polygonaceae)

Buckwheat is a food grain of Asian origin and has been extensively grown in Russia, Europe and North America. It is a course grain with high fiber content and its cultivation has declined in recent decades. In the Himalayas it is still grown at higher altitudes. The plants are unable to self pollinate and honeybees are important pollinators, increasing the yield by 63% (Deodikar and Suryanarayana, 1977). Crop is a rich source of nectar and the honey produced is very dark in colour and is generally mixed with other types of honey before sale.

To conclude, the symbiotic relationship between honeybees and commercial crops or plantations is a unique feature of biological evolution. But for this biodiversity and co-existence the earth would be by far a barren place. Imagine, for example, a landscape without colourful and scented flowers and without the human and activity of bees and butterflies that dance in the fields. The romance and sweetness of honey would have had no mention in the folklore of mankind which existed since time immemorial.

In the comming decades great progress is expected in the science and practice of beekeeping and *pollination* of crops. New honeybee strains would be produced genetically that would excel in specific behaviour and tasks to be performed by them. Chemistry of pheromones is already being studied in order to find altogether new approaches to the management of bees and crop pollination. Semiochemicals in plants are being isolated for the same purpose. It has been observed in Japan that an oriental orchid, *Cymbidium floribundum* Lindl., has an unusual attraction for the workers, drones and queens of *Apis cerana japonica* but not for those of *Apis mellifera*. Analyses of flower aroma and *A.c.j* Nasonov gland extract showed that 17 compounds were common to the two (Sasagawa *et al.*, 1999). Such approaches are just the beginning. Future plant genetics will no doubt be employed to produce plant types having attributes of quality as well attractiveness for pollinators.

Bibliography

Abrol, D.P. (1988a). Pollination studies in almond (*Prunus amygdalus* L.). Curr. Sci.; *57* : 397-398.

Abrol, D.P. (1988b). Ecology and behaviour of three bee species pollinating loquat (*Erioliotrya japonica* Lindl.). Proc. Indian Nat. Sci. Acad. B; *54* : 161-163.

Abrol, D.P. (1993). Possible immune mechanism in indigenous honeybee, *Apis cerana indica* F. against Thai Sacbrood Virus. J. Anim Morphol. Physiol.; *39* : 165-166.

Abrol, D.P. (1997a). Bees and Beekeeping in India. Kalyani Publishers, Ludhiana, New Delhi ; 449 pp.

Abrol, D.P. (1997b). Honeybee Diseases and their Management. Kalyani Publishers, Ludhiana, New Delhi; 194 pp.

Adam, Brother (1968). 'Isle of Wight' or acarine disease : its historical and practical aspects. Bee Wld.; *49*(1) : 6-18.

Adams, J.F. (1972). Beekeeping the Gentle Craft. New York, Double day & Co. Inc.; 182 pp.

Adams, J.F.; Rathman, F.D.; Kerr, W.E.; Paulino, Z.L. (1977). Estimation of the number of sex allels and queen matings from diploid male frequencies in a population of *Apis mellifera*. Genetics; *86* : 583-596.

Adey, M.; Walker, P.; Walker, P.T. (1986). Pest Control safe for bees : a manual and directory for the tropics and subtropics. London, IBRA; 224 pp.

Adlakha, R.L. (1971). Peculiarities of mating of honeybee queens of *Apis mellifera* and *Apis cerana* and begining of oviposition. Proc. Int. Beekeeping Congr.; *23* : 434-439.

Adlakha, R.L. (1976). Acarine disease of adult honeybees in India. Am. Bee J.; *116*(7) : 324-344.

Adlakha, R.L.; Dhaliwal, H.S. (1979). Insect pollination of seed cauliflower (*Brassica oleracea v. botrytis*) with special reference to the role of honeybees. Indian Bee J.; *41* : 13-16.

Adlakha, R.L.; Sharma O.P. (1971). Introducing honeybee queens of *Apis mellifera* into colonies of *Apis indica*. Proc. Int. Beekeeping Congr.; *23* :.415.

Adlakha, R.L.; Sharma O.P. (1974). Interspecific robbing. *Am*. Bee J.; *114* : 248-249.

Adlakha, R.L.; Sharma, O.P. (1975). *Apis mellifera* vs. *Apis indica*. Glean. Bee Cult; *103*(5) : 160.

Adlakha, R.L.; Sharma, O.P. (1976). Stylops (Strepsiptera) Parasites of honey bees in India. Am. Bee J.; *116*(2) : 66.

Afzal, M.; Khan, N. (1950). Natural crossing in cotton in western Punjab; Agron. J.; *42* : 14-19; 89-93; 202-205; 236-238.

Ahmed, R. (1982). Introduction of *Apis mellifera* and factors affecting its establishment in Pakistan. Rep. Pak. Agric. Res. Comm.

Ahmed, R.; Campher, E.; Ahmed, M. (1983). Factors affecting honey yield of the oriental bee, *Apis cerana* in Pakistan. Pakistan J. Agric. Res.; *4*(3) : 190-197.

Ahmed, S. (1984). Nectar and Pollen plants of Pakistan, J. For.; *34*(2) : 75-78.

Ahmed, S. (1984). Country status report on beekeeping in Pakistan Proc. Expert Consult. Bangkok; 203-210. FAO Rome.

Akratanakul, P. (1984). Beekeeping industry with *Apis mellifera* in Thailand. Proc. Expert Consult. Bangkok; 222-232. FAO, Rome.

Akratanakul, P. (1986). Beekeeping in Asia; Rome, FAO : 112 pp.

Akratanakul, P. (1987). Honeybee Diseases and Enemies in Asia—a Practical Guide; Rome, FAO : 51 pp.

Alfonsus, A. (1933). Zum Pollenverbrauch des Bienenvolkes. Arch. Bienenk. *14* : 220-223 cited by Gary (1992) Activities and behaviour of honeybees pp. 269-372 in Hive and the Honeybee. Dadant & Sons, Hamilton, III.

Allen, M.D. (1960). The honeybee queen and her attendants. Anim. Behav.; *8*(3/4) : 201-208.

Allen, M.; Ball, B. (1996). The incidence and world distribution of honeybee viruses. Bee Wld.; *77*(3) : 141-162.

Alpatov, W.W. (1929). Biometrical studies on variation and races of the honeybee *Apis mellifera* L. Rev. Biol.; *4* : 1-57.

Alpatov, W.W. (1938). Contribution to the study of variation in the honeybee; 5. Number of eggs tubules in ovaries of worker bees in the genus *Apis* as related to the evolution of polymorphism. Zool. J.; *17* : 241-245.

Ambrose, J.T. (1976). Swarms in transit. Bee Wld.; *57*(3) : 101-109.

Ambrose, J.T. (1992). Management for honey production Chapter 14, pp. 602-655, from the Hive and the Honey Bee; Hamilton Dadant & Sons, III., U.S.A.

Anderson, L.D.; Atkins, E.L. (1968). Pesticide usage in relation to beekeeping. Annu. Rev. Entomol.; *13* : 213-238.

Anderson, D.L. and Gibbs, A.J. (1989). Transpuparial transmission of Kashmir bee virus and sac-brood virus in the honeybee (*Apis mellifera*). Annu. of Appl. Biol.; *114* : 1-7.

Andrewartha, H.G. (1961). Introduction to the study of Animal Populations (2nd ed. 1970). Methuen & Co. Ltd. London.

Anonymous (1973). Resume of research activities -1962-1972. Khadi and village Industries Commission, Vile Parle (West), 3 Irla Rd. Bombay 400056.

Anonymous (1989). Report of the Quinquennial Review Team on All India Coordinated Project on Honeybee Research and Training. ICAR Report, New Delhi.

Armon, P.J. (1980). The use of honey in the treatment of infected wounds Trop. Doctor ; *10* : 91.

Atkins, E.L. (1992). Injury to honeybees by poisoning; Chapter 26, pp. 1155-1208, from the Hive and the Honeybee. Himilton, Dadant & Sons, III., U.S.A.

Atwal, A.S. (1960). Influence of temperature, duration of conditioning and age of Anagasta (Ephestia) kuhniella (Zell.) (Lep. Pyralididae) on acclimation to sub-zero temperature. Can. J. Zool.; *38* : 131-141.

Atwal, A.S. (1964). Introduction di api regine di razza italiana in alveari di *Apis indica*. 1' Apicoltore di Italia, Aug. 1964 : 100-101.

Atwal, A.S. (1967). Acarine disease problem of honeybees in the Punjab. Tech. Report, Coll. Agric., PAU, Ludhiana : 39 pp.

Atwal, A.S. (1969). Serious infestation of *Apis mellifera* L. bees by the ectoparasitic mite, *Tropilaelaps clareae* : 22nd Int. Beekeep. Congr., Summary : 108.

Atwal, A.S. (1970). Insect Pollinators of Crops : Report on the Project-Biology, Ecology and Utilization of Insects other than Honeybees in the Pollination of Crops. Punjab Agric. Uni. Press, Ludhiana (India); 116 pp.

Atwal, A.S. (1971). Acarine disease problem of the Indian honeybee, *Apis indica* F. Part I and Part II. Am. Bee. J. *111*(4) : 134-135 and *111*(5) : 186-187.

Atwal, A.S. (1986). Agricultural Pests of India and South-East Asia. Ludhiana, India. Kalyani Publishers : 529 pp.

Atwal, A.S. (1987). Performance and prospects of *Apis mellifera* L. (Apidae, Hym) in the Indo-Gangetic plains. Indian J. Ecolo.; *14*(1) : 92-101.

Atwal, A.S. (1998). Personal observations. March, 1998. Chandigarh, U.T. India.

Atwal, A.S.; Bains, S.S. (1987). Applied Animal Ecology. Ludhiana, India; Kalyani Publishers : 245 pp.

Atwal, A.S.; Balraj Singh (1990). Pest population and Assessment of crop losses. ICAR Publs., New Delhi ; 131 pp.

Atwal, A.S.; Dhaliwal, G.S. (1969a). Robbing between *Apis indica* F. and *Apis mellifera* L. Am. Bee J.; *109*(12) : 462-463.

Atwal, A.S.; Dhaliwal, G.S. (1969b). Some behavioural characteristics of *Apis indica* F. and *Apis mellifera* L. Indian Bee J.; *31*(1) : 1-8.

Atwal, A.S.; Dhaliwal, G.S. (2013). Agricultural Pests of South Asia and their Management. Kalyani Publishers, Ludhiana, New Delhi.

Atwal, A.S.; Goyal, N.P. (1971). Infestation of honeybee colonies with *Tropilaelaps clareae* and its control. J. apic. Res.; *10*(3) : 137-142.

Atwal, A.S.; Goyal, N.P. (1973). Introduction of *Apis mellifera* in Punjab plains. Indian Bee J.; *35*(1/4) : 1-9.

Atwal, A.S.; Goyal, N.P. (1974a) & (1974b). *Apis mellifera* turns indifferent to Shain (*Plectranthus rugosus*). Everyday Science, P.U. Chandigarh ; *19*(2) : 25-26.

Atwal, A.S.; Grewal, G.S. (1968). Role of insect pollinators in the production of horticultural crops. Punjab Hort. J.; *8* : 204-210.

Atwal, A.S. (2000). Essentials of Beekeeping and Pollination. Kalyani Publishers, Ludhiana, New Delhi: 395 pp.

Atwal, A.S.; Sharma, O.P. (1964). Introduction of Italian honeybee in India. Proc. Indian Ent. Soc., Silver Jublee No : New Delhi (IARI) ; May, 1964.

Atwal, A.S.; Sharma, O.P. (1967). Introduction of *Apis mellifera* L. queen into de-queened colonies of *Apis indica* F. XXI Int. Beekeeping Congr. Beltsville; 70-71.

Atwal, A.S.; Sharma, O.P. (1968). Introduction of Apis mellifera queens into *Apis Indica* colonies and the associated behaviour of the two species. Indian Bee J.; *30*(2) : 41-56.

Atwal, A.S.; Sharma, O.P. (1970a). Studies on the performance of five strains of *Apis mellifera* L. as compared with *Apis indica* F. under Nagrota conditions. J. Res. Punjab Agric. Univ.; *7* : 477-86.

Atwal, A.S.; Sharma, O.P. (1970b). Brood rearing activity of the Italian honeybee, *Apis mellifera* L. in the Punjab plains at Ludhiana. Indian Bee. J.; *32* : 62-67.

Atwal, A.S.; Sharma, O.P. (1970c). Acarine disease of adult honeybees prevention and control. Indian Farmg; May 1970.

Atwal, A.S.; Sharma, O.P. (1971). The dominance of *Apis mellifera* over *Apis Indica*. Am. Bee J.; *111*(9) : 343.

Atwal, A.S; Sharma, O.P. (1982). Beekeeping in Punjab, Ludhiana. Punjab Agric. University Press, Ludhiana (India) ; 2nd Edn. : 71 pp.

Atwal, A.S.; Bains, S.S.; Singh, B. (1970). Bee flora for four species of *Apis* at Ludhiana. Indian J. Ent.; *32*(4) : 330-334.

Australasian Beekeeper (1986). Phostoxin (R) registered for wax moth control. Australas Beekpr. *87*(8) : 152-153.

Ayers, G.S.; Harman, J.R. (1992). Bee forage of North America and the potential for planting for bees. Chapter 11, pp. 437 535, from the Hive and the Honey Bee. Hamilton, Dadant & Sons, 111., USA.

Ayer, G.S.; Hoopingarner, R.A.; Howitt A.J. (1987). Testing potential of bee forage for attractiveness to bees. Am. Bee J.; *127*(2) : 91-98.

Bahadur, B.; Chaturvedi, A.; Swami, N.R. (1986). Nectar types in Indian plants. Proc. Indian Acad. Sci.; *96*(1) : 41 - 48.

Bailey, L. (1955). The infection of the ventriculus of the adult honeybee by *Nosema apis* Zander. Parasitology ; *45* : 86-94.

Bailey, L. (1961). The natural incidence of Acarapis woodie (Rennie) and the winter mortality of honey-bee colonies. Bee World; *42* : 96-100.

Bailey, L. (1963). Infectious Diseases of the Honeybee. Land Books, London.

Bailey, L. (1967). The effect of temperature on the pathogenicity of the fungus, *Ascosphaera apis* for the larvae of the honeybee, *Apis mellifera*. In proceedings of the International Colloquium on Insect Pathology and Microbial Control (P. A. Van der Laan Ed.) Amsterdam : 162-167 pp.

Bailey, L. (1975). Recent research on honeybee viruses, Bee World ; *56* : 55-64.

Bailey, L. (1981). Honeybee Pathology. London, Academic Press : 124 pp.

Bailey, L. (1982). Viruses of Honeybees. Bee Wld.; *63*(4) : 165-173.

Bailey, L. (1988). Kashmir bee virus, BBKA News ; *68* : 2.

Bailey, L.; Ball, B.V. (1991). Honey Bee Pathology. London, Academic Press : 193 pp.

Bailey, L. and Woods, R.D. (1974). Three previously undescribed viruses from honeybee. J. of General virology; *25* : 175-186.

Bailey, L.; Carpenter, J.M.; Woods, R.D. (1982). A strain of Sacbrood Virus from *Apis cerana*, J. Invert. Pathol.; *39* : 264-265.

Banks, B.E.C.; Shipolini, R.A. (1986). Chemistry and pharmacology of honey bee venom. Chapter 7, pp. 329-416, from Venoms of the Hymenoptera ed. T. Pick. London, Academic Press.

Barbier, M.; Lederer, E. (1960). Structure chemique de la "substance royal" de la reine d' abeille (*Apis mellifera*). Compt. Rend. Acad. Sci. (Paris); *250* : 4467-69.

Basavanna, G.P.C. (1979). Role of insect pollinators in sunflower. Curr. Sci.; *8*(1) : 1-3.

Baswana, K.S. (1984). Role of insect pollination on seed production in coriander and fennel. South Indian Hort.; *32* : 117-118.

Bastian, J.; Esch, H. (1970). The nervous control of the indirect flight muscles of the honeybee. Z. vergl. Physiol.; *67* : 307-324.

Batra, S.W.T. (1966). "Identified insect pollinators of crops at Ludhiana (India)". Personal letter of April 14, 1966 from Snow Entomological Museum, Department of Ento., University of Kansas.

Batra, S.W.T. (1967). Crop pollination and the flower relationships of the wild bees of Ludhiana. J. Kansas Ent. Soc.; *40* : 164-177.

Batra, S.W.T. (1976). Comparative efficiency of alfalfa pollination by Nomia melandari, Megachile rotundata, Anthidium florentionum and Pithitis smaragdula (Hym). Apoidea. J. Kansa Ent, Soc. *49* : 18-22.

Batra, S.W.T. (1977). Bees of India (Apoidea); their behaviour, management and a key to their genera. Oriental Insects; *11*(3) : 289-324.

Batra, W.T. (1979). *Osmia cornifrons* and *Pithitis smaragdula,* two Asian bees introduced into the United States for crop pollination, Proc. 4 Int. Symp. Poll. : 307-312.

Batra, S.W.T. (1996). Fruit pollinating bees of the Gharwal Himalayas, U.P., India. Proc. 7th Int. Symposium on Pollination, Lethbridge; 325-328.

Battu, G.S.; Singh, K. (1977). On the possibility of using *Bacillus thuringiensis* Berliner against wax moth, *Galleria mellonella (*L.*),* in bee hives. Indian Bee J.; *39* : 27-29.

Beadnell, C.B. (1939). Transporting bees abroad. Brit. Bee J.; *67* : 190.

Beck, S.D. (1960). Growth and development of the greater wax moth *Galleria mellonella (*L.*),* (Lepidoptera). Trans. Wisc. Acad. Sci. Arts Lett.; *49* : 137-148.

Beck, B.F. & Smedley, D. (1966). Honey and your Health. New York, Dodd, Mead and Co.

Bee World ed. (1976). Harvesting pollen from hives, Bee Wld ; *57*(1) : 20-25.

Benedek, P.; Nyeki, J. (1996). Yield of selected apple cultivars as affected by the duration of bee Pollination. Proc. 7th Int. Symposium on pollination, Lethbridge; 207-212.

Bentley, B.L.; Elids, T.S. (eds.) (1983). The Biology of Nectaries. New York, Columbia University Press : 259 pp.

Benton, A.W.; Morse, R.A.; Stewart, J.D. (1963). Venom collection from honey bees Science, N.Y.; *142* : 228-230.

Berry, R. (1987). The arataki tube package : a major breakthrough in air-freighting bees. N.Z. Beekpr ; (195) : 8-10.

Berthold, R. (1987). Building your own beeswax processor. Glean. Bee cult.; *115*(12) : 687, 689.

Betts, A.D. (1912). A bee-hive fungus, *Pericystis alvei,* Gen. et sp. Nov. Annals of Botany; *26* : 795-799.

Bhalla, O.P.; Verma, A.K.; Dhaliwal, H.S. (1983a). Insect visitors of mustard bloom (*Brassica campestris* v. sarson), their number and foraging behaviour under mid-hill conditions. J. Ent. Res.; 7 : 15-17.

Bhalla, O.P.; Verma, A.K.; Dhaliwal, H.S. (1983b). Foragting activity of insect pollinators visiting some fruits. J. Ent. Res.; *7*(2) : 91-94.

Bhambure, C.S. (1957). Importance of honeybees in the pollination of *Luffa aegyptiaca*. Indian Bee J.; *19* : 144-151.

Bhambure, C.S. (1958a). Further studies on the importance of honeybees in pollination of Cucurbitaceae. Indian Bee J.; *20* : 10-12.

Bhambure, C.S. (1958b). Effect of honeybee activity on niger seed production. Indian Bee J.; *20* : 189-191, 195.

Bhardwaj, R.K. (1968). A new record of the mite *Tropilaelaps clareae* from *Apis dorsata* colonies. Bee World (3) : 115.

Bhatia, G.K.; Gupta, S.C.; Green, J.M.; Sharma, D. (1981). Estimates of cross pollination in *Cajanus cajan* (L.) Millsp.; several experimental approaches. In Kumble, V. (ed) Proc. Int. Workshop on Pigeonpeas. Vol. 2, 129-136. Andhra Pradesh, India, Int. Crop Res. Inst. for the semi-arid Tropics, Hyderabad.

Bisht, D.S.; Nain M.; Mehrotra, K.N. (1980). Studies on the role of honeybees in rapeseed production. Proc. 2nd Int. Conf. Agric. Trop. Climate. New Delhi, ICAR Publ.; 1983 : 491-496.

Bisht, D.S.; Naim, M.; Mehrotra, K.N. (1982). Comparative efficiency of Jeolikote villager and Newton beehives for maintaining *Apis cerana indica* Indian J. Ent.; *44*(4) : 368-372.

Blum, M.S. (1992). Honeybee pheromones. Chapter 9, pp. 374-400, from the Hive and the Honeybee. Hamilton, Dadant & Sons, III, USA.

Boch, R.; Shearer, D.A.; Stone, B.C. (1962). Identification of isoamyl acetate an active component in the sting pheromone of the honeybee. Nature; *195* : 1018-20.

Bohart, G.E. (1952). Pollination by native insects Yb. Agric. US. Dep. Agric. : 107-121.

Bohart, G.E. (1962). How to manage the alfalfa leaf-cutting bee (*Megachile rotundata* Fabr.) for alfalfa pollination Circ. Utah. Agric. Exp. Stn., 144.

Bohart, G.E. (1966). Studies on pollination of diploid and tetraploid red clover varieties in Utah. Bee Wld .; *47* (Suppl.) : 43-50.

Bohart, G.E. (1967). Management of wild bees. Agric. Handb., U.S. Dep. Agric. No; *335* : 109-118.

Bohart, G.E. (1965-1970). Personal communications as collaborating scientist : PL480 Project on Biology, Ecology and utilization of Insects other than Honey Bees in the Pollination of crops. Punjab Agric. Uni. Ludhiana (India).

Bohart, G.E. (1972). Management of wild bees for the pollination of crops Annu. Rev. Entomol.; *17* : 287-312.

Bohart, G.E. and Nye, W.P. (1960). Insect pollinators of carrots in Utah. Bull. Utah. Agric. Exp. Stn.; 419.

Borthwick, H.A. and Emsweller, S.L. (1933). Carrot breeding experiments. Proc. Am. Soc. hort. Sci.; *30* : 531-533.

Bradbear, N. (1988). The world distribution of major honeybee diseases and pests. Bee Wld.; *69*(1) : 15-39.

Brar, D.S.; Sharma, S.C.; Brar, H.S. (1985). Role of *Apis dorsata* F. (Hymenoptera; Apidae) in the pollination of *Leucaena leucocephala* (Lamb) de Wit, in Kandi area. National Symposium on Products and Conservation Forestry April 12-13, 1985.

Brar, H.S.; Gatoria, G.S.; Jhaji, H.S; Chahal, B.S. (1985). Seasonal infestation of Galleria mellonella and population of *Vespa orientalis in Apis mellifera* apiaries in Punjab. Indian J. Ecol .; *12*(1) : 109-112.

Breed, M.D., et al. ed. (1982). The Biology of Social Insects. Boulders, Co., USA : Westview Press.

British Columbia, Ministry of Agriculture (1984). Pollination and honey production, Southern interior--estimated costs and returns. Pbl. Econ. Br. Min. Agric. B.C. No. *267* : 16 pp.

Brittain, W.H.; Gooderham C.B.; Herman, F.E. (1933). Apple pollination studies in the Annapolis valley, Nova scotia. Can. Dep. Agr. Bull. *162* : 5-198.

Broadman, J. (1962). Bee Venom : the natural curative for arthritis and rheumatism. New york, Putnam : 220 pp.

Brown, A.G. (1951). Factors affecting production in plum. Fruit Yb. 1950 : 12-18.

Bryant, T.G. (1987). Honey filteration unit. N.Z. Beekeeper ; (196) : 25-26.

Burgett, D.M. (1980). The use of lemon balm (*Melissa officinalis*) for attracting honeybee swarms. Bee Wld.; *61*(2) : 44-46.

Burgett, D.M.; Krantz, G.W. (1984). The future of European honeybee (*Apis mellifera*) in Southeast Asia : constants of parasitism. Proc. Expert. Consult. Bangkok : 34-40, FAO Rome.

Burgett, D.M.; Akratanakul, P.; Morse, R.A. (1983). *Tropilaelaps clareae;* a parasite of honeybees in South-east Asia. Bee Wld.; *64*(1) : 25-28.

Burnside, C.E. (1928). A septicemic condition of adult bees. J. Economic Entomology; *21* : 379-386.

Butler, C.G. (1941). A study of the frequency with which honeybees visit red clover (*Trifolium paratense*) together with the environmental conditions. A Appl. Biol.; *28* : 125-134.

Butler, C.G. (1974). The World of the Honeybee. London, Collins.

Butler, C.G.; Calam, K.H. (1969). Pheromones of the honeybee the secretion of the Nassanoff gland of the worker. J. Insect Physiol ; *15* : 237-244.

Butler, C.G.; Simpson, J. (1967). Pheromones of the queen honeybee (*Apis mellifera* L.) which enable her workers to follow her when swarming. Proc. R. ent. Soc. Lond. A; *42* (10/12) : 149-154.

Butler, C.G.; Calam, D.H.; Callow, R.K. (1967). Attraction of *Apis mellifera* drones by the odour of the queens of two other species of honeybees *Apis cerana indica and Apis florea*. Nature (Lond.) ; *213* : 423-424.

Butler, C.G.; Callow, R.K.; Johanston, N.C. (1961). The isolation and synthesis of queen substance, 9-oxodec-trans-2-enoic acid, a honeybee pheromone, Proc. Roy. Soc. London (B) ; *155* : 417-432.

Cadapan, E.P. (1984). Beekeeping with *Apis mellifera* in the Philippines. Proc. Expert. Consult. Bangk : 211-216, FAO, Rome.

Callow, R.K. (1963). Chemical and biochemical problems of beewax. Bee Wld.; *44* : 95-101.

Canadian Beekeeping (1988). Closing the Canada-US border, a new concept. Can. Beekeep.; *14*(2) : 27.

Candian Department of Agriculture (1961). Effect of honeybees on cucumber production. Charlottetown Exp. Farm P.E.I. Res. Rep. 1958-61.

Carr, W.E. (1957). Introducao de abelhas africanas no Brasil. Brasil apic.; *3*(5) : 211-213.

Casteel, D.B. (1912). The behaviour of honeybees in collecting. USDA Bur. Entomol. Bull. 121. In : Gary (1975). Activities and behaviour of honey bees pp. 185-264 from the Hive and the Honey Bee ed. Dadant & Sons. Hamilton Ill.

Cavalloro, R. (ed.) (1983). *Varroa jacobsoni* Oud. affecting honeybees : present status and needs, Rotterdam; Balkema : 107 pp.

Chahal, B.S.; Gatoria, G.S. (1983). Import of exotic bees (*Apis mellifera*) and spread of disease ? Indian Bee J.; *45*(1) : 11-13.

Chahal, B.S.; Brar, H.S.; Gatoria, G.S.; Jhaji, H.S. (1986). Aggressive behaviour of *Apis florea* towards *Apis mellifera* in hive robbing and foraging. J. Apic. Res.; *25*(2) : 106-108.

Chahal, B.S.; Brar, H.S.; Gatoria, G.S.; Jhajj, H.S. (1995). [Italian Honey bees and their Management]. Punjab Agric. University Press, Ludhiana (India); 116 pp. (In Punjabi).

Chakrabarti, K.; Chaudhuri, A.B. (1972). Wild life biology of the Sundarbans forests : honey production and behaviour pattern of the honeybee. Sci. Cult.; *38*(6) : 269-276.

Chandran, K.; Shah, F.A. (1974). Beekeeping in Kodai hills (Tamilnadu : India). Indian Bee J.; *36*(1/4) : 1-8.

Chandran, K.P.; Rajan, D.J., Suryanarayana, M.C. (1980). Studies on the role of honeybees in the pollination of cardamom. Proc. 2nd Int. Conf. Apic. Trop. Clim, New Delhi. ICAR Publ : 497-504.

Chaturvedi, R.B. (1965). Preliminary studies in the sex distribution, pollination and fruit development in litchi (*Lichi chinensis* Sonn.). The Allahabad Farmer ; *39* : 49-51.

Chaubal, P.D.; Deodikar, G.B. (1965). Morphological characterization of pollen grains of some major honey yielding plants of the Western Ghats (India). Indian Bee J.; *27* : 1-28.

Chaubal, P.D.; Kotmire, S.Y. (1980). Floral calender of bee forage plants at Sagarmal (India). Indian Bee J.; *42*(3) : 65-68.

Chaudhari, R.K. (1977). Bee forage in Punjab plains (India). 1. Pathankot and adjacent villages. Indian Bee J.; *39*(1/4) : 5-20.

Chaudhuri, R.K. (1978). Floral fidelity in the Indian honeybees (*Apis cerana indica F*). Indian Bee J.; *40*(2) : 239.

Chaudhury, B.; Choonsai, M.L.A.; Menon, M.G.R. (1973). Insect pollination in some vegetable crops. Haryana, J. Hort. Sci.; *2*(1/2) : 56-62.

Cherian, M.C.; Mahadevan, V. (1945). The longevity of Indian bee (*Apis indica*). Indian Bee J.; *8*(3/4) : 41-43.

Chonduri, M.M. (1940). Beekeeping in Orissa. Indian Bee J. (cited from Bee World *1941* : 66).

Cobley, L.S. (1956). An Introduction to the Botany of Tropical Crops; London : Longmans.

Codex Alimentarious Commission (1983/84). Proposed draft Codex Standard for Honey (World-wide Standard) (Romone FAO/WHO) CX/PFV 84/13, also revisions ALINORM 85/20 Appendix IX & X.

Commonwealth Secretariat; IBRA (eds.) (1979). Beekeeping in rural development : unexploited beekeeping potential in the tropics with special reference to the Commonwealth. London, Commonwealth Sectt : 196 pp.

Cooper, B.A. (1986). The Honeybees of the British Isles. Derby, U.K.; British Isles Bee Breeders Association.

Crane, E. (1973). Four bees foraging, editorial. Bee Wld.; *54*(4) : 145-146.

Crane, E. (1976). East West-home's best; 57(4) from World Perspective in Apiculture.

Crane, E. (1976a). Honey : A Comprehensive Survey. Heinemann, London.

Crane, E. ed. (1976b). Apiculture in Tropical Climates : Report of the First Conference on Apiculture in Tropical Climate. (Cairo); London, IBRA.

Crane, E. (1980). A Book of honey. Oxford, Oxford University Press : 198 pp.

Crane, E. (1983). The Archaeology of Beekeeping. London, Duckworth : 360 pp.

Crane, E. (1985). World Perspective in Apiculture. London, IBRA .

Crane, E. (1990). Bees and Beekeeping ; Science Practice and World Resources. Oxford, Heinemann Newness. London, Mathuen. Also Comstock publishing Associates (Cornell Uni. Press), Ithaca, N.Y : 634 pp.

Crane, E. (1992). The World's beekeeping past and present. Chapter 1 In the Hive and the Honeybee. Hamilton, Dadant & Sons Inc., III., U.S.A. : 1-22.

Crane, E. (1998). Wall hives and wall beekeeping. Bee Wld.; *79*(1) : 11-22.

Crane, E. (1999). The World History of Beekeeping and Honeyhunting. Gerald Duckworth and Co. Ltd. London; 682 pp.

Crane, E.; Walker, P.; Day, R. (1984a). Dicrectory important world honey sources. London, IBRA : 384 pp.

Crane, E.; Walker, P.; Day, R. (1984b). Pollination Directory for World Crops. London, IBRA.

Dadant, C.C. (1992). Beekeeping equipment. Chapter 12, pp. 539-600, from the Hive and the Honey Bee. Hamilton, Dadant & Sons, III ., USA.

Dade, H.A. (1977). Anatomy and Dissection of the Honeybee. International Bee Research Association; Cardiff (Wales).

Dall, D.J. (1985). Inapparent infection of honeybee pupae by Kashmir and Sacbrood bee viruses in Australia. Ann. Appl. Biol.; *106* : 461-468.

Daly, H.V.; Bohart, G.E.; Thorp, R.W. (1971). Introduction of small carpenter bee in California for pollination. I-release of *Pithitics smaragdula*. J. Econ. Entomol.; *64* : 1145-1150.

Das, C.S.; Choudhury, K.R. (1958). Floral biology of litchi (*Litchi chinensis Sonn*.). South Indian Hort.; *6* : 17-22.

Dave, M.L. (1943). Wasps. Indian Bee J.; *5* : 127-131.

Deh-Feng, M.; Wen-Cheng, H. (1981). Apiculture in the new China. Bee Wld.; *62* : 163-166.

De Jong, D.; Morse, R.A.; Eickwart, G.C. (1982). Mite pests of honeybees. A. Rev. Ent.; *27* : 229-252.

Delaplane, K.S. (1997). Practical Science—research helping beekeepers. 3, Varroa. Bee Wld.; *78*(4) : 155-164.

Delaplane, K.S. (1999). Varroa jacobsoni ; research based answers for a global problem. Proc. 36th Apimondia Congr., Vancouver; 53-57.

Delfinado, M.D. (1963). Mites of the honeybee in South-East Asia. J. Apic. Res.; *2*(2) : 113-114.

Delfinado, M.D.; Baker, E.W. (1961). *Tropilaelaps* a new genus of mites from the Philippines (Laclaptidae; Acarina). Fieldiana Zool.; *44*(7) : 53-56.

Delfinado, M.D.; Baker E.W. (1974). Varroaidae, new family of mites on honeybees. Wash Acad. Sci.; *64* : 4-10.

Delfinado, M.D.; Baker, E.W. (1982). New records for *Tropilaelaps* clareae from colonies of *Apis cerana indica*. Am. Bee J.; *122* : 382.

Delfinado, M.D.; Underwood, B.A.; Baker, E.W. (1985). The occurrence of *Tripilaelaps* mites in brood nests of *Apis dorsata* and *Apis laboriosa* in Nepal, with description of the nymphal stages. Am. Bee J.; *125* : 703-706.

Deodikar, G.B. (1962). Some aspects of bee botany. Indian Bee J.; *24* : 3-22.

Deodikar, G.B. (1970). A preliminary survey of beekeeping in Kashmir. Indian Bee J.; *32* : 12-22.

Deodikar, G.B.; Suryanarayana, M.C. (1977). Pollination in the service of increasing farm production in India. In Advances in Pollen Spore Research Vol. 2 ed. P.K.K. Nair, New Delhi. Today and Tomorrow Printers and Publishers ; 60-82 pp.

Deodikar, G.B.; Thakar, C.V. (1953). A pollen study of major honey yielding plants of Mahabaleshwar hills Poona. Village Industries Committee : 8 pp.

Deodikar, G.B.; Thakar, C.V. (1966). Utilization and improvement of the local floras as bee pasturage. Indian Bee J.; *28* : 11-66.

Deodikar, G.B.; Thakar, C.V.; Phadke, R.P. (1957). High nectar concentration in floral nectaries of silver oak, *Grevillea robusta*. Indian Bee J.; *19*(7/8) : 84-85.

Deodikar, G.B.; Thakar, C.V.; Tonapi, K.V. (1958). Evolution in the genus *Apis* and its bearing on breeding better strains of Indian bees. Proc. Int. Beekeep. Congr.; *17* : 245-250.

Deodikar, G.B.; Ghatge, A.I.; Phadke, R.P.; Mohindre, D.B.; Kshirsagar, K.K.; Mural, K.S.; Thakar, C.V. (1977). I. Nesting behaviour of Indian honeybees; II Nesting behaviour of *Apis dorsata* F. Indian Bee J.; *39* : 1-12.

Deshmukh, A.K.; Mohana Rao, G.; Karve, A.D. (1985). Studies on the effect of honeybee pollination on the yield of safflower. Indian Bee J.; *47* : 1-2.

Devadason, A. (1971). Migratory beekeeping in Kerala. Indian Bee J.; *33*(3/4) : 35-38.

Devadason, A. (1972). Rubber plantation and beekeeping. Indian Bee J.; *34*(1/2) : 38-39.

Dhaliwal, G.S.; Atwal, A.S. (1969). Comparative aggressiveness of the four species of *Apis* under controlled conditions. Indian Bee J.; *31* : 17-21.

Dhaliwal, G.S.; Atwal, A.S. (1970). Interspecific relations between *Apis cerana indica* and *Apis mellifera*. *J.* Apic. Res.; *9*(2) : 53-59.

Dhaliwal, H.S.; Bhalla, O.P. (1981). Factors affecting optimal pollination of cauliflower, 28th Int. Cong. Apic. Acapulco, Mexico, 1981 ; 386-387. Bucharest : Apimondia Publishing House.

Dhaliwal, H.S.; Sharma, P.L. (1972). Homing ability of and the foraging range of the Indian honeybee, *Apis indica* F. Indian J. Ent.; *34*(3) : 225-227.

Dhaliwal, H.S.; Sharma, P.L. (1974). The foraging range of the Indian honey bee on two crops. J. Apic. Res.; *12* : 131-134.

Dhaliwal, H.S.; Sharma, P.L. (1974). Foraging range of the Indian honeybee. J. Apic Res.; *13* : 137-141.

Dhaliwal, H.S.; Srivastva, S.; Adlakha, R.L. (1977). Insect pollination of Lychee, *Litchi chinensis* Sonn., in the valley areas of Indian Himalayas. Proc. 26th Int. Apic. Congress, Adelaide; 396. Bucharest, Apimondia Publishing House.

Dhaliwal, J.S. (1975). Insect pollination in ber (*Zizyphus mauritiana* Lank.) Curr. Sci.; *44*(14) : 527.

Dhaliwal, J.S.; Atwal, A.S. (1974). Foraging range of *Apis mellifera* L. on barseem (*Trifolium alexandrinum* L.) at Ludiana. Indian Bee J.; *36*(1/4) : 17-18.

Dhaliwal, J.S.; Atwal, A.S. (1975). Effect of aphid population on the number of flowers, nectar sugar concentration and bee abundance on raya, *Brassica juncea* Coss. Indian Oilseed J.; *5* : 23-25.

Dhaliwal, J.S.; Atwal, A.S. (1976a). Competition among honeybees, wild honeybees, and wild bees for collecting nectar and pollen from toria (*Brassica campestris* var. toria). Indian Oilseed J.; *6*(1) : 7-10.

Dhaliwal, J.S.; Atwal A.S. (1976b). Note on the effect of air temperature, relative humidity and wind velocity on bees visiting barseem (*Triflium alexandrinum* L.) at Ludhiana. Indian J. Agric. Sci.; *46*(1) : 50-51.

Dhaliwal, J.S.; Atwal, A.S. (1980). Flight range, transportation of nests and the number of trips made by *Caratina* (*Pithitis*) *smaragdula* Fab. (Anthrophoridae, Hym.) in relation to lucern pollination at Ludhiana. Indian J. Ent.; *42*(3) : 434-437.

Dhaliwal, J.S.; Atwal, A.S. (1986). Factors influencing tripping efficiency of bees on alfalfa (*Medicago sativa*). In Pollination Biology : An Analysis, ed. R.P. Kapil. Inter-India Publ. New Delhi : 79-90 pp.

Dietz, A. (1992). Honeybees of the world. Chapter 2, pp. 23-71, from the Hive and the Honey Bee, Hamilton, Dadant & Sons, Ill., USA.

Diggs, J.G. (1945). The Practical Bee Guide, A Manual of Modern Beekeeping. Dublin, The Talbot Press.

Dinabandhoo, C.L.; Dogra, G.S. (1979). Dichlorovos for the control of acarine disease of Indian honeybee, *Apis cerana* Fab. Indian Bee J.; *41*(1/2) : 1-4.

Dinabandhoo, C.L.; Dogra, G.S. (1980). The acarine induced decline in Indian honeybee *Apis cerana indica* F. and its control in Kulu valley. Int. Conf. Apic. Trop. Climate *2* : 434-441 : New Delhi, ICAR Publ.

Dines, A.M.; Dalton, S. (1968). Honeybees from Close-up. New York, T.Y. Crowell Co.

Diwan, V.V.; Rao, S.K. (1972). Altitudinal variation in bee forage of West Coorg. Indian Bee J.; *33* : 39-50.

Diwan, V.V.; Satvi, S.R. (1965). Some interesting behaviour features of *Apis dorsata* Fab. Indian Bee J.; *27* : 52.

Diwan, V.V.; Vartak, V.D. (1980). Studies on the bee forage of Castle Rock area in Karnataka state. Proc. 2nd Int. Apic. Trop. Climate, New Delhi.; ICAR Publ. pp. 528-536.

Diwan, V.V.; Kshirsagar, K.K.; Raman Rao, A.V.; Roughunath, D.; Bhambure, C.S. (1971). Occurrence of a new bacterial disease of Indian honeybee, *Apis indica,* F. Curr. Sci.; *40*(8) : 196-197.

Douglas, J.C. (1884). Handbook of Beekeeping in India. Calcutta, Govt. Printing Press, India.

Douglas, J.C. (1886). The hive-bee indigenous to India and the introduction of the Italian bee. J. Asiat. Soc. Bengal II Nat. Sci. *55* : 83-96.

Douhet, M. (1965). Beekeeping in Madagascar. Proc. Int. Beekeep. Congr.; *20* : 690-710.

Dulta, P.C.; Verma, L.R. (1987). Role of insect pollinators on yield and quality of apple fruit. Indian J. Hort.; *44* : 274-279.

Dulta, P.C.; Rana, B.S.; Verma, L.R.; Mattu, V.K. (1988). Absconding behaviour of Indian honeybee. Indian Bee J.; *50* : 67.

Dutta, T.R.; Ahmad, R.; Abbas, R. (1983). The discovery of a plant in the Andaman Islands that tranquillizes *Apis dorsata*. Bee Wld.; *64* : 158-163.

Dyer, F.C. (1985a). Mechanisms of dance orientation in the Asian honeybee *Apis florea*. J. Comp. Physiol A : 183-198.

Dyer, F.C. (1985b). Nocturnal orientation by the Asian honeybee, *Apis dorsata*. Anim. Behav.; *33* : 769-774.

Dyer, F.C.; Gould, J.L. (1983). Honeybee navigation. Am. Scient.; *71* : 587-597.

Dyer, F.C.; Seeley, T.D. (1987). Interspecific comparisons of endothermy in honeybees (Apis) : deviations from the expected size related patterns. J. Exp. Biol.

Eckert, J.E. (1933). The flight range of the honeybee. J. Apic. Res.; *47* : 257-285.

Eckert, J.E.; Shaw, F.R. (1960). Beekeeping. New York, MacMillan : 536 pp.

Eisikowitch, D. (1999). Honeybee movement among trees in an almond orchard in Isreal. Proc. 36th Apimondia Congr. Vancouver : 184-187.

Elkon, B. (1955). The Honey Cookbook. New York, Alfred A Knopf : 174 pp.

Evans, J.; Barrett, S. (1989). A Complete Guide to Beekeeping. Unwin & Hyman; 192 pp.

Evans, H.E.; Eberhard, M.J.W.,(1970). The Wasps. Ann Arbor, University of Michigan Press : 266 pp.

Fahrenhorst, H. (1977). Chromosome number in the tropical honeybee species *Apis dorsata* and *Apis florea*. J. Apic. Res.; *16* : 56-58.

Fakhim-Zadeh, K. (1998). Innovation : improved device for venom extraction. Bee Wld.; *79*(1) : 52-56.

Fang Yue-Zhen (1984). The present status and development of keeping with European bees (*Apis mellifera*) in tropical and subtropical regions of China. Proc. Expert consult., Bangkok; 142-147. FAO Rome.

FAO (1984). Proceedings of the expert consultation on beekeeping with *Apis mellifera* in tropical and subtropical Asia. Rome, FAO : 238 pp.

FAO (1986). Tropical and Sub-Tropical Apiculture. Rome, FAO Agr. Serv. Bull. *68* : 283 pp.

Farrar. C.L. (1931). The evaluation of bees for pollination. J. Econ. Ent.; *24* : 622-627.

Fleming, R.L. Sr.; Fleming, R.L. Jr.; Bangdel, L.S. (1976). "Birds of Nepal" Published by Flemings. P. Box 229, Kathmandu, Nepal.

Fletcher, D.J.C. (1975). Significance of dorso-ventral abdominal vibration among honeybees (*Apis mellifera*). Nature; *256* : 721-723.

Fletcher, D.J.C. (1978). Management of *Apis mellifera adansonii* in Africa. Annu. Rev. Entomol.; *23* : 151-157.

Frankel, R.; Galun, E. (1977). Pollination Mechanisms, Reproduction and Plant Breeding. Berlin; Springer-Verlag.

Free, J.B. (1960). The behaviour of honeybees visiting flowers of fruit trees. J. Anim. Ecol.; *29* : 385-395.

Free, J.B. (1962). The effect of distance from pollinator varieties on the fruit set on trees in plum and apple orchards. J. Hort. Sci.; *37* : 262-271.

Free, J.B. (1963). The flower constancy of honeybees. J. Anim. Ecol.; *32* : 119-131.

Free, J.B. (1964). Comparison of the importance of insect and wind pollination of apple tree. Nature, Lond.; *201* : 726-727.

Free, J.B. (1966). The pollinating efficiency of honeybee visits to apple flowers. J. Hort. Sci.; *4* : 91-94.

Free, J.B. (1967). Factors determining the collection of pollen by honeybee foragers. Anim. Behav.; *15* : 134-144.

Free, J.B. (1977). The Social Organization of Honeybees. London, Edward Arnold : 68 pp.

Free, J.B. (1981). Biology and behaviour of the honeybee *Apis florea,* and possibilities for beekeeping. Bee Wld.; *62*(2) : 46-59.

Free, J.B. (1982). Bees and Mankind. London, George Allen & Unwin : 155 pp.

Free, J.B. (1987). Pheromones of Social Bees. London, Chapman & Hall : 218 pp.

Free, J.B. (1993). Insect Pollination of Crops. London, Academic Press : 684 pp.

Free, J.B.; Simpson, J. (1964). The pollination requirements of sunflowers (*Helianthus annus* L.). Emp. J. Exp. Agric.; *32* : 340-342.

Free, J.B.; Spencer-Booth, Y. (1961). Effect of temperature on *Apis indica* workers. Nature; *190* : 933.

Free, J.B.; Spencer-Booth, Y. (1963). The pollination of mustard by honeybees. J. Apic. Res.; *2* : 69-70.

Free, J.B.; Williams, I.H. (1979). Communication by pheromones and other means in *Apis florea* colonies. J. Apic. Res.; *18* : 16-25.

Free, J.B.; Raw, A.; Williams, I.H. (1975). Pollination of coconut (*Cocos nucifera* L.) in Jamaica by honeybees and wasps. Appl. Anim. Ethol.; *1* : 213-223.

Friedmann, H. (1955). The honey-guides (Washington : U.S. National Museum) Bull. U.S. nat. Mus. No. *208* : 292 pp.

Frisch, K. Von (1953). The Dancing Bee : An Account of the Life and Senses of the Honeybee. New York. Harcourt Brace and World Inc.

Frisch, K. Von (1967a). The Dance Language and Orientation of Bees. Cambridge, MA, USA : Belknap Press : 566 pp.

Frisch, K. Von (1967b). Honeybees : Do they use direction and distance information provided by their dancers. Science; *158* : 1072.

Frisch, K. Von (1971). Bees : Their Vision, Chemical Senses and Language. Ithaca, Cornell. University Press.

Fry, C.H. (1983). Honeybee predation by bee-eaters with economic considerations. Bee Wld.; *64*(2) : 65-87.

Fry, C.H. (1984). The Bee-eaters. Calton, UK, Poyser : 304 pp.

Fuchs, S.; Koeniger, N. (1974). Sound preduction as a means of defense of the colony of honeybee (*Apis cerana* F.). Apidologie ; *5* : 271-287.

Fukuda, H.; Ohtani, T. (1977). Survival and life span of drone honeybee (*Apis mellifera*). Res. Popul. Ecol.; *10* : 31-39.

Fukuda, H.; Sakagami, S.F. (1968). Worker brood survival in honeybees. Rese. Popul. Ecol. Kyoto Univ.; *10*(1) : 31-39.

Furgala, B. (1954a). The effect of the honeybee *Apis mellifera* L. on the seed set, yield and hybridisation of the cultivated sunflower, *Helianthus annus* L. Ph.D. Thesis, Manitoba University.

Furgala, B. (1954b). Honeybees increase seed yield of cultivated sunflower. Glean. Bee Cult.; *82* : 532-534.

Gandheker, S.S. (1959). Some common bee enemies in India. Glean. Bee Cult.; *87* : 360-363.

Ganguly, P.; Chanda, S. (1980). A calendar of bee plants of India with reference to melisso-palynological studies. Proc. 4th Int. Symp. Pollination, Bltsville, MA, USA : 477-481 pp.

Ganguly, P.; Chanda, S. (1980). A calender of bee plants of India with reference to melissopalynological study. Proc. 5th Int. Palynol. Conf., Cambridge : 146 pp.

Gary, N.E. (1963). Observation of mating behaviour in the honeybee. J. Apic. Res.; *2* : 3-13.

Gary, N.E. (1967). Diurnal variation in the intensity of flight activity from honey bee colonies. J. Apic. Res.; *6* : 65-68.

Gary, N.E. (1971). Magnetic retrieval of ferrous labels in a capture recapture system for honeybees and other insects. J. Econ. Ent.; *64* : 961-965.

Gary, N.E. (1992). Activities and behaviour of honeybees. Chapter 8, pp 269-372, from the Hive and the Honey Bee. Hamilton, Dadant & Sons, III., USA.

Gary, N.E.; Witherell, P.C.; Lorenzen, K. (1979). Distribution of honeybees during water collection. J. Apic. Res.; *18* : 26-29.

Gary, N.E.; Witherell, P.C.; Marston, J.M. (1973). Distribution of foraging bees used to pollinate alfalfa. Environ. Entomol.; *2* : 573-578.

Garg, R.; Sharma, O.P.; Dogra G.S. (1984). Formic acid, an effective acaricide against *Tropilaelaps clareae* Delfinado and Baker (Laelaptidae; Acarina) and its effect on the brood and longevity of honeybees. Am. Bee J.; *124*(10) : 736-738.

Gayre, G.R.; Papazian, C. (1986). Brewing Mead-Wasail in Makers of Mead. Boulder, Co. USA; Brewers Publication : 200 pp.

Gautam, S.K. (1984). Country Status Report on Beekeeping in Nepal. Proc. Expert. Consult. Bangkok : 198-202. FAO, Rome.

Gentry, C. and three others (1985). A Manual for Trainers of Small Scale Beekeeping Development Workers. Washington DC; Peace Crops : 407 pp.

Gerstung, F. (1891-1926). Der Bien und Seine Zucht. 7 edns. Berlin, Pienningstroff; Cited by Ruttner, F. (1988).

Getz, W.M.; Smith, K.B. (1983). Genetic kin recognition : honeybees discriminate between full and half sisters. Nature, Lond.; *302* : 147-148.

Ghatage, A.I. (1949). Some research work on *Apis florea*. Indian Bee J.; *11* : 5-9.

Ghatge, A.I. (1956). A scientific approach to (wild) rock bees. Ujjain, Madhya Pradesh; All-India Khadi and Village Industries Board : 17 pp.

Ghosh, C.C. (1920). Beekeeping in India. Proc. III Entomol. Meeting, Pusa; *2* : 770-782.

Ghosh, C.C. (1939). Beekeeping, Calcutta, Govt. Printing Press, India.

Girish, P.P. (1981). Role of bees in the pollination of summer squash (*Cucurbita pepo Linne.*) with special reference to *Apis cerana* F. (Hym.; Apidae). M.Sc. Thesis, Uni. of Agric. Sciences, Banglore.

Gill, N.Y.; Vear, K.C. (1958). Agricultural Botany. London, Gerald Duckworth.

Gladis T. (1996). Bees versus flies ? - rearing methods and effectiveness of pollinators in crop germplasm regeneration. Proc. 7th Int. Symp. on Pollination, Lethbridge; 235-238.

Gojmerac, W.L. (1980). Bees, Beekeeping, Honey and Pollination Westport, C.T., USA; AVI Publishing Co. : 192 pp.

Gong, Y.F. (1983). The natural Beekeeping conditions and honeybee races in China. J. Fujjan Agric. Coll.; *12* : 241-249.

Gong, Yi-Fei (1983). The natural beekeeping conditions and honeybee races in China. J. Fujian Agric. Coll.; *12* : 244-249.

Gopal Singh (1995). A Geography of India. Atma Ram & Sons, Delhi : 569 pp.

Gosavi, K.R. (1939). My bee-keeping tour in South India. Indian Bee J.; *1*(5) : 75-76.

Gould, J.L. (1976). The dance-language controversy. Quart. Rev. Biol.; *51* : 211-244.

Gould, J.L. (1980). The case for magnetic sensitivity in birds and bees (Such as it is). Am. Scient.; *68*(3) : 256-267.

Gould, J.L. (1982). Why do honeybees have dialects ? Behav. Ecol. Sociobiol.; *10* : 53-56.

Gould, J.L.; Gould, G.G. (1988). The Honeybee. New York, Scientific American Library : 239 pp.

Goyal, N.P. (1974). *Apis cerana indica* and *Apis mellifera* as complimentary to each other for the development of apiculture. Bee Wld.; *55*(3) : 98-101.

Goyal, N.P. (1978). Performance of *Apis mellifera* L. over *A. indica* F. as observed in Punjab plains. Indian Bee J.; *40*(1) : 3-5.

Goyal, N.P.; Atwal, A.S. (1971). Comparison of the way of communication with *Apis indica* F. and *Apis mellifera* L. Proc. Int. Beekeep. Congr.; *23* : 341-342.

Goyal N.P.; Atwal, A.S. (1973a). Ecology of sunflower pollination. Indian Oilseed J.; *3*(2) : 30-37.

Goyal, N.P.; Atwal, A.S. (1973b). Role of honeybees in sunflower pollination. Indian Bee J.; *35* : 41-47.

Goyal, N.P.; Atwal, A.S. (1975). Studies on the relation of population of insect pollinators with seed production in lucerne. Indian J. Ecol. *2*(1) : 58-61.

Goyal, N.P.; Atwal. A.S. (1976). Comparative communication behaviour among *Apis cerana indica* and *Apis mellifera;* dancing behaviour. Indian Bee J.; *38*(1/4) : 1-2.

Goyal, N.P.; Atwal, A.S. (1977). Wing beat frequencies of *Apis cerana indica* and *Apis mellifera* J. Apic. Res.; *16*(1) : 47-48.

Goyal, N.P.; Atwal, A.S. (1979). Introduction of exotic honeybees (*Apis mellifera*) in India. Indian Bee J.; *41*(3-4) : 39-45.

Goyal, N.P.; Kandoria, J.H. (1988). Pollination of crops by honeybees and isolation distances required for pure seed production. Indian Bee J.; *50* : 44-46.

Graham, Joe M. ed. (1992). The Hive and the Honey Bee Dadant & Sons. Hamilton, lll. : 1324 pp.

Grewal, G.S.; Atwal, A. S.; Kumar, S.; Kapil, R.P. (1969). Nesting behaviour of *Andrena leaena* Cam. (Andrenidae, Hym.) Proc. 22nd. Int. Beekeeping Congr., Munich, *1969* : 422-24.

Grewal, G.S.; Atwal, A.S.; Kumar, S.; Kapil, R.P. (1970a). Some observations on the nest architecture of *Collectes nursi* cam. (Collectidae; Hym.). Indian J. Ent.; *32*(2) : 184-186.

Grewal, G.S.; Kumar, S.; Kapil, R.P.; Atwal, A.S. (1970b). Nesting behaviour of *Megachile fasiculata* smith (Megachilidae, Hym.) (in Atwal, 1970).

Grewal, G.S.; Sidhu, A.S. (1978). Insect pollinators of some cucurbits in Punjab. Indian J. Agric. Sci.; *48*(2) : 79-83.

Grewal, G.S.; Sidhu, A.S. (1980). Studies on insect pollination in *Cucumis melo, Linn.,* Proc. 2nd. Int. Conf. Apic. Trop. Climates, New Delhi. ICAR Publ. *1983* : 537-544.

Grewal, K.S. (1953). Role of honeybees in pollination of *Brassica* crops. M.Sc. Thesis Punjab Agric. College, Ludhiana (India).

Griffiths, D.A.; Bowman, C.E. (1981). World distribution of the mite *Varroa jacobsoni,* a parasite of honeybees. Bee Wld.; *62*(4) : 154-163.

Grout, R.A. (1937). The influence of size of brood cell upon the size and variability of the honeybees (*Apis mellifera* L.) Res. Bull. Agr. Exp. Sta. Iowa state Coll.; *218* : 260-280.

Grzimek's Animal Life Encyclopedia (1975). Vol. 2 Insects. New York, Nostrand Reinhold Co.

Guanghung, Y. (1990). Royal jelly, pollens, venom and propolis : production technology and marketing potentials in China. Proc. ICIMOD Inten. Expert Meeting on Beekeeping, Kathmandu, Nepal. (Cited by Verma 1990).

Gupta, J.K.; Thakar, R.K. (1987). Nectar sugar production and flower visitors of the bramble, *Rubus ellipticus* Smith (Rosaceae) at Solan, India. Apidologie; *18* : 223-230.

Gupta, J.K.; Kumar, J.; Mishra, R.C. (1984). Nectar sugar production and honeybee foraging activity in different cultivars of cauliflower, *Brassica oleracea* var. botrytis. Indian Bee J.; *47* : 21-22.

Gupta, J.K.; Mishra, R.C.; Kumar, J. (1984). *Plectranthus* as forage for *Apis cerana indica* F. and *Apis mellifera* F. Apidologie; *15*(1) : 75-82.

Habermann, E. (1972). Bee and wasp venoms. Science; *177* : 314-322.

Hagmeister, A. (1973). Immunologischer Vergleich der Haemolymphproteine der vier rzenten *Apis-Arten mellifera, cerana, florea* and *dorsata*. Proc. Int. IUSSI Congr. *7* : 151-153.

Hamankawa, H.; Morimoto, H. (1967). Studies on honeybee foragers, *Apis mellifera* L. and Apis cerana F. Jap. J. Zootch. Sci.; *38* : 11.

Hambleton, J. I. (1944). The role of bees in the production of fruits and seeds. J. Econ. Ent.; *34* : 522-525.

Hameed, S.F.; Adlakha, R.L. (1973). Preliminary comparative studies on the brood rearing activities of *Apis mellifera* and *Apis cerana indica* in Kulu Valley. Indian Bee J.; *35*(1/4) : 27-35.

Hanel, H.; Ruttner, F. (1985). The origin of the pore in drone cell capping of *Apis cerana Fabr*. J. Apic. Res.; *16*(2) : 157-164.

Haragsim, O.; Vesley, V.; Sedivy, J.; Taimar, L.; Dockal, J.; Balcar, J. (1965). Activity of Honey Bees marked with Radioisotopes and moved to fields of lucerne (*Medicago sativa*) 20th Int. Beekeepers Congr.; 11/4.

Hartwig, A.; and Przelck, A. (1971). Nucleic acid in the intestine of *Apis mellifera* infected with *Nosema apis* and treated with Fumagillin DCH : Cytochemical and autoradiographic studies. J. Invert. Pathol.; 18 : 331-336.

Haseman, L. (1961). How long can spores of American foulbrood live ? Am. Bee J.; *101* : 298-299.

Hassanein, M.H. (1953). Studies on the effect of pollinating insects, especially the honeybee on the seed yield of clover in Egypt. Bull. Soc. Fouad Im. Ent.; *37* : 337-344.

Haydak, M.H. (1967). Bee nutrition and pollen substitutes. Apiacta *1967*(1) : 3-8.

Haydak, M.H. (1970). Honeybee nutrition Annu. Rev. Entomol.; *15* : 143-156.

Heinrich, B. (1979). Thermoregulation of African and European honeybees during foraging, attack, and hive exits and returns. J. Exp. Biol.; *80* : 217-229.

Heinrich, B. (1981a). The regulation of temperature in the honeybee swarm. Scient. Am.; *244*(6) : 120-129.

Heinrich, B. (1981b). Insect Thermoregulation. New York, John Wiley & Sons : 327 pp.

Heinrich, B. (1996). How honeybee regulates its body temperature. Bee Wld. *77*(3) : 130-137.

Hendrickson, A.N. (1916). The common honeybee as an agent in prune pollination. Bull. Calif. Agric. Exp. Stn. No : *274* : 127-132.

Hepburn, H.R. (1986). Honeybees and Wax. Berlin, Springer-Verlag.

Herbert, E.W. (1992). Honeybee nutrition. Chapter 6, pp. 197-233, from the Hive and the Honeybee. Hamilton, Dadant and Sons, lll., USA.

Hirschfelder, H.(1951). Quantitative undtersuchungen zum Polleneintragen der Bienenvolker 2.Bienenforsch 1; 67-77. In Hive & the Honeybee : Gary N.E 1992 Dadant & Sons Hamilton lll. Gary, N.E (1992) Activities and Behaviour of honeybees. pp. 269-372.

Hirschfelder, H. (1952). Das Verhalten der Tracheenmille *Acarapis woodi R.* beim verlassen und Aufsuchen der Wirtsbienen. Zeitschrift fur Bienenforschung; *1* : 141-170.

Hobbs, G.A. (1964). Importing and managing the alfalfa leaf-cutter bees. Publ. Dep. Agric. Can.; 1209.

Hobbs, G.A. and Lilly, C.E. (1955). Factors affecting efficiency of honeybees as pollinators of alfalfa in southern Alberta. Can. J. Agric. Sci.; *35* : 422-432.

Holst, E.C. (1946). Newer knowledge of American foul brood. Glean. in Bee Cult.; *74* : 138-139.

Hopper, T. (1977). A Guide to Bees and Honey. Emmaces, Rodale Press Inc.

Hoopingarner, R.A.; Waller, G.D. (1992). Crop pollination. Chapter 24, pp. 1044-1082; In the Hive and the Honey Bee. Hamilton, Dadant & Sons, Ill., USA.

Hoppe, H.; Wolfgang, R.; Stephen, E.W.C. (1989). The control of parasitic mites. *Varroa jacobsoni*, Acarapis woodi and Tropilaelaps clareae with formic acid. Am. Bee J. : *129* : 739-742.

Hoshiba, H.; Okada, J.; Kusanagi, A. (1981). The diploid drone of *Apis cerana japonica* and its chromosomes. J. Apic. Res.; *20* : 143-147.

Howard, A.; Howard, G.L.C; Abdur Rahman, K. (1916). Studies in Indian oilseeds. I-Safflower and Mustard. Mem. Dep. Agric. India. Bot. Ser.; *7* : 237-272.

Howard, A.; Howard, G.L.C.; Abdur Rahman, K. (1919). Studies on the pollination of Indian crops Mem. Dep. Agric. India, Bot. Ser.; *10* : 195-220.

Howes, F.N. (1979). Plants and Beekeeping. London, Faber & Faber..

Huber, F. (1814). New observations upon bees. Book (1926) published by Dadant & Sons (1926). 111.

Hurd, P.D. (1978). An Annotated Catalog of the Carpenter Bees (*Xylocopa*) of the Western Hemisphere. Washington D.C., Smithsonian Institution Press.

Hurd, P.D.; La Berge, W.E.; Linsley, E.G. (1980). Principal Sunflower Bees of North America with Emphasis on the South Western-USA. Washington DC; Smithsonian Institution Press.

Husain, S.W. (1938). How honey is extracted from the combs of the giant bee of India. Bee Wld.; *19* : 139-140.

Hussain, S.A.; Ali, S. (1983). Some notes on the ecology and status of the Orange-rumped Honeyguide, *Indicator xanthonotus,* in the Himalayas. J. Bombay Nat. Hist. Soc.; *80*(3) : 564-574.

Imms, A. D. (1925). A General Textbook of Entomology, including the Anatomy, Physiology, Development and Classification of Insects-see the subsequent revisions. London, Methuen; 890 pp.

Inoue, A. (1962). Preliminary report on rearing of Japanese honeybee queens in colonies of the European honeybee. Bee Sci. Nagoya; *3* : 10.

Inoue, A. (1985). Relationship between *Apis cerana japonica* and *Apis mellifera* with special reference to beekeeping technology. Abstr. Int. Beekeep. Congr.; *30* : 51-52.

Inoue, A.; Nakajima, T. (1985). A comparison of the venom composition in *Apis cerana japonica* and *Apis mellifera*. Proc. 30 Int. Apic Congr. : 441-443.

International Bee Res. Association. (1977). Honey standards, laws and regulations (world wide). (London, IBRA) : 7 pp.

International Bee Res. Association. (1979). World wide standards of hive products except honey, and for equipment used in beekeeping and in processing hive products. London, IBRA : 13 pp.

International conference on Apiculture in Tropical Climates (1980) (Proceedings of, 1983) New Delhi. Indian Council of Agricultural Research Publication; 728 pp.

Indian Standards Institution, New Delhi (1970). Specification for beehives 1515 (1969) : 20 pp. 1973, code for conservation and maintenance of honeybees 6695(1972) : 28 pp.1975, Layout for honey processing plant. 3891(1974) : 8 pp. 1975, Layout for a honey house. 7849 (1975) : 8 pp. 1976, specification comb foundation sheet 2072 (1976) 3 pp. 1977, specification for squeezed honey 8464 (1977) : 8 pp. 1979, specification for comb foundation mill. 3894 (1978) : 8 pp.

International Symposium on Melliferous Flora (1977). Honey plants-basis of Apiculture. International Symposium on Melliferous flora, Budapest. Bucharest, Apimondia Publishing House : 224 pp.

International Symposium on Pollination. Ist held in 1960, Proceedings (1962) ed. T.E. Mittler (Stockholm : Swedish Seed Growers Association) : 224 pp. 2nd, 1964, Proceedings (1966) ed. Ekerberg & E. Crane, Bee Wld. 47(1) Suppl : 216 pp. 3rd, l974 Compterendu (1975). Bull. tech. apic. 2 suppl. : 331 pp. 4th, 1978, Proceedings (1979) ed. D.M. Caron (Beltsville, USA, Agric. Expt. Sta. : 541 pp.).

Jadhav, L.D. (1981). Role of insects in the pollination of onion (*Allium cepa*) in the Maharastra State, India. Indian Bee J.; *43*(3) : 61-63.

Jadhav, L.D.; Ajri, D.S. (1981). Insect pollinators of onion (*Allium cepa*) in Ahmednagar district of Maharastra India. Indian Bee J.; *43*(4) : 109.

Jagannadham, B.; Goyal, N.P. (1980). Morphological and behavioural characteristic of honeybee workers reared in combs with larger cells. Khadigramodyog : 369-377. conf. Apic. Tropic. Climate; 2 : New Delhi, ICAR publ.; 233-253.

Jaiswal, P.L. ed. (1980). Handbook of Agriculture. New Delhi, Indian Council of Agricultural Research : 1303 pp.

Jarlan, A; Oliveira, D. de; Gingras J. (1996). Pollination of sweet pepper (*Capsicum annum L*) in green-house by the Syrphid fly, *Eristalis tenax* (L.). Proc. 7th Int. symposium on pollination, Lethbridge; 335-340.

Jay, S.C. (1963). The development of honeybees in their cells. J. Apic. Res.; *2* : 117-134.

Jay, S.C. (1974). Nectar and pollen collection by honeybees from coconut flowers. Bee Wld.; *55* : 105-111.

Jay, S.C. (1986). Spatial management of honeybees on crops. Annu. Rev. Entomol.; *31* : 49-65.

Jay, S.C.; Frankson, C.F. (1972). Effect of various hive covers, shading and ventilation on honeybees in the tropics. J. Apic. Res.; *11*(2) : 111-115.

Jayarathnam, K. (1970). *Hevea brasiliensis* as a source of honey. J. Palynol.; *6* : 101-103.

Jaycox, E.R. (1981). Beekeeping in Mid-West. Champaign, University of Illinois Press, Urbama.

Jhajj, H.S.; Goyal, N.P. (1979). Comparative studies on the flower constancey of *Apis cerana indica* F. and *Apis mellifera* L. Proc. 4th Int. Symp. Pollen, Beltsville, MA, USA, *1979* : 333-339.

Jhajj, H.S.; Goyal, N.P. (1979). Comparative behaviour of pollen foragers of *Apis cerana* F. and *Apis mellifera* L. J.Apic. Res.; *18*; 279-284.

Jitendra, Mohan, K.V. (1973). Some observations on the bee pollination in sunnhemp (*Crotalaria juncea*) and their importance in breeding work. Jute Bull.; *35* : 214-216.

Johanson, T.S.; Johanson, M.P. (1978). Some Important Operations in bee Management. London, IBRA.

Johanson, T.S.K.; Johanson, M.P. (1971). The honeybee colony in winter. Bee Wld.; *60* : 155-170.

Johansen, C.A.; Mayer, D.F.; Eves, J.D. (1978). Biology and management of the alkali bee, *Nomia melanderi* Cockerell (Hym. Halictidae). Melanderia; *28*(23) : 25-46.

Johnson, D.L. (1967). Honey bees; Do they use the direction information contained in their dance manoeuvres. Science; *155* : 844-847.

Jones, W.A. (1986). New bee blower developed. Australas. Beekeeper.; 88(1) : 7-9.

Joseph, D.; Mohandas, N. (1985). Role of honeybees in the pollination of cardamom. Indian Bee J.; *47* : 12-13.

Joshi, M.A. (1982). Honey and beeswax trade in India. Indian Bee J.; *44*(1) : 1-6.

Joshi, M.A.; Diwan, V.V.; Suryanarayana, M.C. (1980). Bees and honey in ancient India. Int. Conf. Apic. Trop. Climate, New Delhi *2* : 143-149.

Kadam, B.S.; Patankar, V.K. (1962). Natural cross-pollination in safflower. Indian J. Genet. Pl. Breed; *2* : 69-70.

Kakkar, K.L. (1980). Prospects of bees as pollinating agents of cauliflower. Proc. 2nd. Int. Conf. Apic. Trop. Climate. New Delhi. ICAR publ. : 445-56.

Kakkar, K.L. (1981). Foraging behaviour of insect pollinators of cauliflower bloom. Indian J. Ecol.; *8* : 126-130.

Kallapur, S.K. (1950). An experiment in the collection of wild honey. Indian Bee J.; *12* : 122-124.

Kalmer, F. (1996). Sunflower pollination in Czech Republic. Proc. 7th Int. Symp. on Pollination, Lethbridge, 407-412.

Kapil, R.P. (1956). Variation in biometrical characters of the Indian honeybee (*Apis indica* F.). Indian J. Entomol.; *18* : 440-457.

Kapil, R.P. (1957). The length of life and the brood rearing cycle of the Indian honeybee (*Apis indica* F.). Bee Wld.; *38*(10) : 258-263.

Kapil, R.P. (1962). Anatomy and histology of the female reproductive system of *Apis indica*. Ins. Sociaux; *9* : 145-153.

Kapil, R.P. (1971). A hive for the Indian honeybees. Apiacta; *6*(3) : 107-109.

Kapil, R.P. and Brar, H.S. (1971). Foraging behaviour of *Apis florea* Fab. in relation to *Brassica campestris* var toria. Proc. 23rd. Congr. Apic. Moscow; pp 335-339.

Kapil, R.P.; Dhaliwal, J.S. (1968). Biology of *Xylocopa* spp. I Seasonal activity, nesting behaviour and life cycle. J. Res. Punjab Agric. Univ.; *5* : 406-419.

Kapil, R.P.; Dhaliwal, J.S. (1969). Biology of *Xylocopa* spp. II Field activities, flight range and trials on transportation of nests. J. Res. Punjab Agric. Univ.; *6* : 262-271.

Kapil, R.P.; Jain, K.L. (1980). Biology and Utilization of Insect Pollinators for Crop Production. Hissar (India), Haryana Agric. University Press : 81 pp.

Kapil, R.P.; Kumar, S. (1969). Biology of *Ceratina binghami* Ckll. J. Res.; *6* : 360-72.

Kapil, R.P.; Kumar, S. (1975). Foraging activity of *A. dorsata* Fab. on *Brassica juncea* Hook. J. Bombay Nat. Hist. Soc.; *71*(2) : 327-331.

Kapil, R.P.; Grewal, G.S.; Kumar, S. (1968). Increase in production of lucerne seed by carpenter bees. Progr., Fmg.; *5* : 14.

Kapil, R.P.; Grewal, G.S.; Kumar, S.; Atwal, A.S. (1969a). Biology and nesting behaviour of *Megachile lanata* F. (Megachilidae, Hym.) Proc. XXII Int. Beekeeping Congr. Munich *1969* : 449-458.

Kapil, R.P.; Grewal, G.S.; Kumar, S.; Atwal, A.S.(1969b). Insect pollinators of rape-seed and mustard. Proc. 56th Indian Sci. Congr. Assoc., Bombay; p. 509.

Kapil, R.P.; Grewal, G.S.; Kumar, S.; Atwal, A.S. (1970). Role of *Ceratina binghami* Ckll in seed setting of *Medicago sativa* L. Seminar on forage production and its utilization, Jhansi (India) : 10 pp.

Kapil, R.P.; Grewal, G.S.; Kumar, S.; Atwal, A.S. (1971). Insect pollinators of rape-seed and mustard. Indian J. Ent.; *33* : 61-66.

Karmo, E.A. (1958). Honeybees as an aid in orchard and blueberry pollination in Nova Scotia *X* Int. Congr. Ent. 1956; *4* : 955-959.

Katznelson, H. (1950). The influence of antibiotics and sulpha drugs on *Bicillus larvae,* cause of American foulbrood of the honeybee in vitro and vivo. J. Bacteriol.; *59* :153-155.

Keeling, C.I.; Slessor, K.N.; Winston, M.L. (1999). Pheromones, bee biology, and management : the essence of royalty, the honeybee queens pheromone arsenal. Proc. 36th Apimondea Congr., Vancouver; 207-208.

Kellogg, C.R. (1941). Some characteristic behaviour of the oriental honeybee *Apis indica* in China. J. Econ. Entomol.; *34* : 717-719.

Kellogg, C.R. (1959). Chinese bees. In ABC-XYZ of Bee Culture. Medina, AI Root & Co. : 565-566.

Kellogg, C.R. (1967). Honey in Chinese history. Entomological excerpts from South eastern China (Fukien Province). Aborigenes : Silkworm, Honeybees and other insects. Clarvemont, Calif; 89 pp.

Kendall, D.A. (1970). Effectiveness of Pollinating Insects. Report of Long Ashton Res. Sta., University of Bristol : 99-100.

Kerr, W.E.; Bueno, D. (1970). Natural crossing between *Apis mellifera adansonii* and *Apis mellifera ligustica*. Evolution; *24* : 145-148.

Kerr, W.E.; Mouole, V. (1964). Geographic distribution of stingless bees and its implication (Hymenoptera; Apidae) J. Emtomol. Soc. N.Y.; *72* : 2-17.

Kevan, P.G. (ed) (1989). Beekeeping with *Apis cerana* in Tropical and Subtropical Asia. London, IBRA.

Kevan P.S. (1999). Crop pollination today and tomorrow : problems confronting apiculture and crop pollination. Proc. 36th Apimondia Congr., Vancouver; 181.

Khadi & Village Industries Commission, Bombay (1959). Bee flora of Karnataka and Kerala. Indian Bee J.; *21* : 90-92.

Khan, M.S.A. (1948). Some important nectariferous plants and pollen sources of Bhopal state (Central India). Indian Bee J.; *10* : 107-108.

Khan, A.H.; Afzal, M. (1950). Vicinism in Cotton. Indian Cotton Grow. Rev.; *4* : 227-239.

Killion, E.E. (1992). The production of comb and bulk comb honey. Chapter 16, pp. 706-722, from the Hive and Honey Bee. Hamilton, Dadant & Sons, Ill., USA.

Kitamura, T.; Maeta, Y. (1968). Flower-Visiting activity of *Osmia cornifrons* (Rodosz-Kowski) in relation to the varieties of peach, (In Japanese), Kontyu; 7 : 13-18.

Klungness, M.; Thorp, R.; Briggs, D. (1983). Field testing the germinability of almond pollen (*Prunus dulcis*) J. Hort. Sci.; *58* : 229-235.

Koeniger, N. (1975). Observations on alarm behaviour and colony defense of *Apis dorsata*. IUSSI Symp. Pheromones Def. Secretions Soc. Ins; 153-154.

Koeniger, N. (1976b). Interspecific competition between *Apis florea* and *Apis mellifera* in the tropics. Bee Wld.; *57*(3) : 110-112.

Koeniger, N. (1976c). The Asiatic honeybee *Apis cerana*. Proc. Ist Int. Apic. Conf. in Trop. Clim. London : 47-49.

Koeniger, N. (1982). Interactions among the four species of the genus *Apis*. In Breed, Mitchener, Evans (ed), The Biology of Social Insects. Westview Press, Boulder Co.

Koeniger, N.; Fuchs, S. (1973). Sound production as colony defense in *Apis cerana* Fabr. Proc. Int. IUSSI Congr.; *7* : 199-204.

Koeniger, N.; Koeniger, G. (1980). Observations and experiments on migration and dance communication of *Apis dorsata* in Sri Lanka. J. Apic. Res.; *19*(1) : 21-34.

Koeniger, N.; Koeniger, G.; Delfinado-Baker, M. (1983). Observations on mites of the Asian honeybee species (*Apis cerana, Apis dorsata, Apis florea*). Apidologie; *14* : 197-204.

Koeniger, N.; Koeniger, G.; Punchihewa, R.K.W.; Fabrititus Mo.; Fabritius Mi. (1982). Observations and experiments on dance communication in *Apis florea* F. J. Apic. Res.; *21* : 45-52.

Koeniger, N.; Koeniger, G.; Tingek, S. (1999). Living together in harmony : the Asian bee species. Proc. 36th Apimondia Congr., Vancouver; 115.

Koeniger, N.; Vorwohl, G. (1979). Competition for food among four sympatric species of Apini in Sri Lanka (*A.d., Apis cerana, A.f. and Trigona iridipennis*). J. Apic. Res.; *18*(2) : 95-109.

Koeniger, N.; Wijayaganasekra, H.N.P. (1976). Time of drone flights in the three Asiatic honeybee species (*A.c, A.f., Apis dorsata*). J. Apic. Res.; *15*(2) : 67-71.

Koeniger, N.; Weiss, J.; Maschwitz, U. (1979). Alarm pheromones of the sting in the genus Apis. J. Insect Physiol.; *25* : 467-426.

Koeniger, N.; Weiss, J.; Ritter, W. (1975). Catch, transport and flight cage maintenance of the Giant honeybee *Apis dorsata*. Proc. Int. Beekeeping Congr.; *25* : 313-317.

Kohli, N. (1958). Bee flora of northern India. Indian Bee J.; *20* : 113-118, 132-34, 150-151, 178-179, 192-193.

Kohli, N. (1959). Indian Bee J.; *21* : 7-8, 31-32, 61-62, 83-85, 106-107, 127-128.

Koltermann, R. (1973). Retroaktive Hemmung nach successiver Information seingabe bei *Apis mellifera* and *Apis cerana* (Apidae). J. Comp. Physiol.; *84* : 299-310.

Koltowspi; Z. (1999). Pollination requirements and beekeeping value of new rape cultivars. Proc. 36th Apimondia Congr., Vancouver; 236.

Kotze, W.A.G. (1949). Ton of honey from one colony in one year. Glean. Bee Cult.; *77*(8) : 487-490.

Kozin, R.B. (1976). Pollination of Entomophilous Agricultural Crops by Bees. New Delhi, Amerind Publishing Co.

Kramer, F.E. (1985). A technique for protecting colonies during insecticide applications and moving bees in high temperatures. Am. Bee J.; *125*(3) : 178-180.

Krishnaswamy, S.V. (1970). Soapnut trees; a nectar source. Indian Bee J.; *32*(3/4) : 83.

Krishnaswamy, S.V. (1980). Introduction and propagation of bee plants in Chickmeglur Distt. (Karnataka). Indian Bee J.; *42* : 72-73.

Kshirsagar, K.K. (1968). Introduction of *Apis mellifera* - its merits and demerits. Indian Bee J.; *30* : 64-67.

Kshirsagar, K.K. (1982). Current incidence of honeybee diseases and parasites in India. Bee Wld.; *63* : 162-164.

Kshirsagar, K.K. (1983). Spread of sacbrood disease in Uttar Pradesh, Indian Bee J.; *45* : 41-42.

Kshirsagar, K.K.; Diwan, V.V.; Chauhan, R.M. (1981). Occurence of sacbrood disease in *Apis cerana indica*. Indian Bee J.; *43* : 44.

Kshirsagar, K.K.; Mahindre, D.B. (1975). Some notes on bee predatory wasps in India. Indian Bee J.; *37*(1/4) : 4-9.

Kshirsagar, K.K.; Mittal M.C.; Mahindre, D.B.; Salvi, S.R.; Chauhan, R.M. (1980). Seasonal fluctuations in the incidence of acarine disease in bees and colony losses in India. Proc. 2nd. Int. conf. Apic. Trop. Clim. New Delhi, ICAR Publ. 1983 : pp. 442-453.

Kshirsagar, K.K.; Muvel, K.S.; Mittal, M.C.; Phadke, R.P. (1980a). Some observations on behaviour of *Apis florea F*. Proc. Int. conf. Apic. Trop. Climate ICAR publ.; 2 : 356-366.

Kshirsagar, K.K.; Salvi, S.R.; Mahindre, D.B.; Chauhan, R.M. (1980b). Menthol as an effective acaricidal fumigant. Proc. 2nd Int. Conf. Apic. Trop. Climates, New Delhi. ICAR Publ. *1983* : 254-256.

Kshirsagar, K.K.; Sexena, U.C.; Chauhan, R.M. (1982). Occurence of sac brood disease in *Apis cerana indica* F. in Bihar, India. Indian Bee J.; *44*(1) : 8-9.

Kshirsagar, K.K.; Suryanarayana, M.C.; Thakar, C.V. (1981). Import of exotic bees and quarantine facilities. Indian Bee, J.; *43* : 81-86.

Kullincevic, J.M.; Stairs, G.R.; Rothenbuhler, W.C. (1970). Virus causing paralysis of adult bees in Ohio. J. Invert. Pathol.; *16* : 423-426.

Kumar, J.; Gupta, J.K.; Dogra, G.S. (1988). Discovery of the ectoparasitic mite *Varroa jacobsoni on Apis mellifera* in Himachal Pradesh, India. Am. Bee J.; *128*(2) : 124.

Kumar, J.; Gupta, J.K.; Dogra, G.S. (1988). Pollination studies in some cultivars of cauliflower (*Brassica oleracea* var botrytis). Indian Bee J.; *50*(4) : 93-95.

Kumar, J.; Mishra, R.C.; Gupta, J.K.; Dogra, G.S. (1984). Pollination requirements of some peach cultivars. Indian Bee J.; *47* : 3-6.

Kumar, J.; Mishra, R.C.; Gupta, J.K. (1985). The effect of mode of pollination on *Allium* sp. with observation on insects as pollinators. J. Apic. Res.; *24*(1) : 62-66.

Kumar, J.; Mishra, R.C.; Gupta, J.K.; Dogra, G.S. (1985). Pollination requirements of some peach cultivars. Indian Bee J.; *47* : 3-6.

Kumar, S.; Kapil, R.P.; Atwal, A.S. (1968). The carpenter bee, *Ceratina binghami*. Proc. 13th Int. cong. Ent., Moscow; p. 108.

Laidlaw, H. H. (1987). Instrumental insemination of honeybee queens; its origin and development. Bee Wld.; *68* : 17-36, 71-88.

Laidlaw, H.H. (1992). Production of queens and package bees. Chapter 23, pp.989-1043, from the Hive and the Honeybee. Hamilton, Dadant & Sons, lII., USA.

Laidlaw, H.H.; Eckert, J.E. (1962). Queen Rearing. Berkeley, University of California Press.

Laidlaw, H.H.; Page, R.E. (1984). Polyandry in honeybees (*Apis mellifera* L.) sperm utilization and intracolony genetic relationship. Genetics; *108* : 985-997.

Laidlaw, H.H.; Page, R.E. (1989). Apiculture : Introductory Biology and Husbandry. Davis, CA, USA; University of California Press.

Laigo, F.M.; Morse, R.A. (1968). The mite *Tropilaelaps clareae* in *Apis dorsata* colonies in the Phillippines. Bee Wld.; *49* : 116-118.

Langridge, D.F.; Goodman, R.D. (1975). A study on pollination of oil seed rape (*Brassica campestris*). Aust. J. Exp. Agric. Anim. Husb.; *15* : 285-288.

Latif, A.; Qayyum, A.; Abbas, M. (1960). The role of *Apis indica* in the pollination of toria and sarson (*Brassica campestris* var. toria and dicotoma). Bee Wld.; *41* : 283-286.

Latif, A.; Qayyum A.; Manzoor-ul-Haq. (1956). Role of *Apis indica* F. in the pollination of Egyptian clover (*Trifolium alexandrinum* L.). Pakist. J. Scient. Res.; *8* : 48-50.

Latif, A.; Qayyum, A.; Manzoor-ul-Haq. (1960a). Multiple and two queen system in *Apis indica* F. Bee Wld.; *41* : 201-209.

Lensky, Y.; Cassier, P.; Notkin, M.; Delorme-Joulie, C.; Levinsohn, M. (1985). Pheromonal activity and fine structure of the mandibular glands of honeybee drones (*Apis mellifera* L.) (Hym. Apidae). J. Insect Physiol; *31* : 265-276.

Lesik, F. (1953). Microfertilization increases the nectar secretion of honey plants. Pchelovodstvo, Mosk.; *30* : 42-45.

Levin, M.D. (1959). Distribution patterns of young and experienced honeybees foraging on alfalfa. J. Econ. Entomol.; *54* : 482-484.

Levin, M.D. (1961a). Distribution of foragers from honeybee colonies placed in the middle of a large field of alfalfa. J. Econ. Entomol.; *54* : 431-434.

Levin, M.D. (1961b). Dispersion of field bees in alfalfa in relation to a neighbouring apiary. J. Econ. Entomol.; *54* : 482-484.

Lindauer, M. (1954). The water economy of temperature regulation of the honeybee colony. Z. Vergl. Physiol.; *36* : 391-432 (in German)

Lindauer, M. (1955a). The water economy and temperature regulation of the honeybee colony. Bee Wld.; *36* : 62-72, 81-92.

Lindauer, M. (1955b). Schwarmbienen auf Wohnungssuche. Z. Vergl. Physiol.; *37* : 263-324.

Lindauer, M. (1956). Uber die Verstandingung bei indischen Bienen. Z. Vgl. Physiol.; *38* : 521-557.

Lindauer, M. (1957). Communication among the honeybees and stingless bees of India. Bee Wld.; *38* : 3-14, 34-39.

Lindauer, M. (1971). Communication Among Social Bees. Harvard Books on Biology; Mass. USA.

Lindauer, M.; Kerr, W.E. (1960). Communication between the workers of stingless bees, Bee Wld.; *41* : 29-41, 65-71.

Lineburg, B. (1924). The feeding of honeybee larvae, Bull. U.S. Dep. Agric.; *1222* : 25-37.

Lindzon, R.D.; Silver, W.S. (1988). Anaphylaxis. Stings of bees and hornets and how to treat the victim. In; Emergency Medicine, concepts and clinical practice Vol I (R. Rosen et al. eds.); pp. 203-231. St. Louis, C.V. Mosby.

Lingappa, S.; Viraktamath, S.; Vastrad, A.S.; Rajasekhar, W. (1999). Utilization of honeybee *Apis cerana* Fab. for pollination of watermelon and safflower. Proc. 36th Apimondia Congr., Vancouver; 235.

Liu Xianshu (1985). Advancing Chinese apiculture. Proc. 3rd. Int. Conf. Apic. Trop. Clim. Nairobi; pp. 93-95.

Loper, G.M.; Wolf, W.W.; Taylor, O.R. (1987). Detection and monitoring of honeybee drone congregation areas by radar. Apidologie; *18*(2) : 163-172.

Lord, W.G.; Nagi, S.K. (1987). *Apis florea* discovered in Africa. Bee Wld.; *68*(1) : 39-40.

Louveaux, J. (1958). Recherches sur la recolte du pollen par les abeilles (*Apis mellifera* L.). Ann. Abeille; *1* : 113-188, 197-221.

Mackensen, O.; Roberts, W.C. (1948).A manual for the artificial Insemination of queen bees. USDA Agric. Res. Admin. Bur. of Entomol. & Plant Quarantine; E.T. 250.

Madhusoodanan, K.J.; Dandin, S.B. (1981). Flower bilogy of cardamom (*Elettaria cardamomum* (L.) Maton) in relation to the foraging behaviour of honeybees (*Apis* spp.) Indian Bee J.; *43* : 104-107.

Mahindre, D.B. (1968). *Apis dorsata*—the giant bee of India. Glean. Bee cult.; *96*(2) : 902-108.

Mahindre, D.B. (1983a). Handling rockbee colonies. Indian Bee J.; 45(2/3) : 72-73.

Mahindre, D.B. (1983b). A suitable hive for *Apis cerana*. Indian Bee J.; *45* : 80-81.

Mahindre, D.B.; Muvel, K.S.; Kshirsagar K.K.; Thakar C.V.; Phadke R.P.; Deodikar, G.B. (1977). Nesting behaviour of *Apis dorsata*. Proc. Int. Apic. Congr.; *26* : 299-300.

Makhdzir, M.; Osman, M.S. (1980). Beekeeping in coconut small holdings in Pontian, Johar, West Malaysia. 2nd Int. Conf. Apic. Tro. Clim. New Delhi; ICAR Publ. 1983.

Mani, M.S. (1962). Introduction to high Altitude Entomology. Methuen Co. London : 302 pp.

Manjo, G. (1975). The Healing Hand : Man and Wound in the Ancient World. Cambridge, MA, USA; Harvard University Press : 571 pp.

Mann, G.S.; Singh, G. (1980). Period of activity and comparative abundance of flower visiting insects on pear at Ludhiana (Punjab). Entomon.; *5* : 65-66.

Mann, G.S.; Singh, G. (1981). A note on activity and abundance of flower visiting insects of peach (*Prunus persica* L.) at Ludhiana (Punjab). Prog. Hort.; *13*(3/4) : 25-27.

Mann, G.S.; Singh, G. (1983). Activity and abundance of pollinators of plum at Ludhiana (Punjab). Am. Bee J.; *123*(8) : 595.

Mann, G.S.; Tanda, A.S. (1984). Activity and abundance of insects visiting grape flowers in Punjab (India). Proc. 3rd. Int. Conf. Apic. Trop. Climates, Nairobi : 139.

Mann, S.K.; Narang, A.S.; Atwal, A.S. (1970). Method of Sample collection of dorsata honey and its analysis. J. Food Sci. Technol.; 7 (3) : 142-143.

Manzoor-ul-Haq; Muhammed, F. (1980). Insect pollination of sunflower (*Helianthus annus*) in Pakistan. J. Apic. Res.; *19* : 83-87.

Manzoor-ul-Haq; Rafie-ul-Din; Ghaffar, A. (1978). Effect of insect pollination on fruit bearing in Kinnow mandrin (*Citrus reticulata*) and physical and chemical properties of fruit; J. Apic. Res.; *17* : 47-49.

Marcovic, O.; Molnar, L. (1955). Prispevok K izolacii a stanoveniu vcelieho jedu Chem. Zvesti; *8* : 80-90.

Materlinck. M. (1960). The Life of the Bee. New York, Dodd, Mead & Co.

Matheson, A. (1984). Practical Beekeeping in New Zealand. Wellington, Government Printer : 185 pp.

Mathew, T.J. (1975). The rubber tree as a source of honey in Kerala, India. Glean. Bee Cult.; *103*(6) : 190.

Mathpal, Y. (1984). Newly discovered paintings in Central India showing honey collection. Bee Wld.; *65*(3) : 121-126.

Matsuura, M.; Sakagami, S.F. (1973). A bionomic sketch of the Giant hornet, *Vespa mandarina,* a serious pest for Japanese apiculture. J. Fac. Sci. Hokkaido Univ. Ser. VI Zool.; *19* : 125-162.

Mattu, V.K.; Verma, L.R. (1985). Studies on the annual foraging cycle of *Apis cerana indica* F. in Simla Hills of North-West Himalayas. Apidalogie; *16*(1) : 1-18.

McCarthy, J. Pheromone based management for beekeeping and crop pollination. Proc. 36th Apimondia Congr., Vancouver; 166-167.

McDonald, J.H. (1930). Coffee Growing with special Reference to East Africa. London : East Africa Ltd.

McDowell, R. (1984). The Africanized Honeybee in the United States. USDA Report # 519.

McGregor, S.E. (1976). Insect Pollination of Cultivated Crop Plants. Washington DC, USDA Agric. Handb. No : 496.

McGregor, S.E.; Todd, F.E. (1956). Honeybees and cotton production. Glean. Bee Cult.; *84* : 649-652, 701.

Menzel, R.; Mercer, A. eds. (1987). Neurobiology and Behaviour of Honeybees. Berlin, Springer : 334 pp.

Meyer, W. (1956). Propolis, bees,and their activities. Bee Wld.; *37* : 25-26.

Michael, A.S. (1962). *Tropilaelaps clareae*,a mite infesting honeybee colonies. Bee Wld.; *43*(3) : 81-82.

Michener, C.D. (1944). Comparative external morphology, phylogeny, and classification of the bees (Hymenoptera) Bull. Am. Mus. Nat. Hist.; *82* : 157-326.

Michener, C.D. (1975). The Brazilian bee problem. Annu. Rev. Entomol.; *20* : 399-416.

Michener, C.D. (1979). Biogeography of the bees. Ann. Mo. Bot. Gdn.; *66*(3) : 277-347.

Michener, C.D.; Winston, M.L.; Jander, R.(1978). Pollen manipulation and related activities and structures in the Apidae. Univ. Kans. Sci. Bull.; *51* : 575-601.

Michener, G.E. (1974). The Social Behaviour of the Bees. Cambridge, MA, USA; Harvard University Press.

Millen, T.W. (1942). *Apis dorsata* queens. Indian Bee J.; *4* : 6-9.

Milum V.G. (1947). Grooming dance and associated activities of the honeybee colony. Trans Ill St. Acad. Sci.; *40* : 194-196.

Milum, V.G. (1955). Honeybee communication. Am. Bee J.; *95* : 97-104.

Milum, V.G. (1956). The significance of some honeybee dances. Proc. X Int. Congr. Entomol.; pp. 1085-1088.

Minderhoud, A. (1931). Undersuchungen uber das Betragen der Honigbiene als Blutenbestauber in Gartenhauwrssenschaft.; *4* : 342-362.

Mishra, R.C. (1995a). Honeybees and their Management in India. New Delhi, ICAR Pub1. : 168 pp.

Mishra, R.C. (1995b). Spread of Sacbrood virus disease among *Apis cerana indica* colonies in India. Personal communication.

Mishra, R.C.; Dogra, G.S. (1980). Post-embryonic development of *Apis cerana indica* F. worker bee. Int. Conf. Apic. Trop. Climate; *2* : 278-288.

Mishra, R.C.; Dubey, O.P. (1996). Aftermath of Thai Sac Brood Virus disease in *Apis cerana indica* F. In India. J. Anim. Morphol. Physiol.; *43* : 143-148.

Mishra, R.C.; Kaushik, H.D. (1992). Bee flora and beekeeping maps of India. Changing villages : Rural News Views; *11*(3) : 88-108.

Mishra, R.C.; Kumar, Y. (1995). Introduction and spread of *Apis mellifera* L. in India and the possibility of inbreeding harm. Indian Bee J.; *57* : 131-134.

Mishra, R.C.; Sharma, S.K. (1989). Growth regulators affect nectar pollen production and insect foraging in *Brassica* seed crop. Curr. Sci.; *57* : 1297-1299.

Mishra, R.C.; Sihag, R.C. (1987). Apicultural Research in India. New Delhi ; ICAR Report.

Mishra, R.C.; Singh, B.P. (1986). *Apis mellifera* L. for developing Indian Apiculture. Apiacta XXII : pp. 83-87.

Mishra, R.C.; Dogra, G.S.; Gupta, P.R. (1976). Some observations on insect pollinators of apple. Indian Bee J.; *38* : 22.

Mishra, R.C.; Gupta, J.K.; Kumar, J. (1985). Nectar sugar production in different cultivars of peach *Prunus persica* L. Indian Bee J.; *47* : 37-38.

Mishra, R.C.; Gupta J.K.; Kumar, J. (1987). Effect of modes of pollination on fruit characteristics of okra, *Abelmoschus esculentum (*L.*)* Moench. Proc. Indian National Sci. Acad.; *53* B (2) : 157-60.

Mishra, R.C.; Gupta, J.K.; Kumar, J. (1988). Effect of modes of pollination on yield and oil potential of *Brassica campestris* var. sarson with observations on insect pollinators. J. Apic. Res. *27* : 186-189.

Mishra, R.C.; Gupta, J.K.; Kumar, J. (1989). A new approach to the control of predatory wasps (*Vespa* spp.) of the honeybee (*Apis mellifera* L.). J. Apic. Res.; *28* : 126-130.

Mitchener, A.V. (1950). Honeybees and sunflowers. Ann. Con. Man. Agron.; p. 27.

Miyamoto, S. (1958). Biological studies on Japanese bee X. Differences in flower relationship between a Japanese and European honeybee. Sci. Rep. Hyogo Univ. Agric. Ser. agric. Biol. : 99-101.

Mohammed, Ali (1935). Pollination studies in tara (*Brassica napus* L. var. *dichotsma*) and sarson (*Brassica campestris* L. var. *sarson*). Indian J. Agric. Sci.; *5* : 125-154.

Mishra, R.C. (2000). [Beneficial Beekeeping in India] in Hindi, Kalyani, New Delhi, Ludhiana : 391 pp.

Mohanrao, G.; Lazar, M. (1980). Studies on bee behaviour and pollination in Onion. Khadi Gramodyog; *26* : 378-383,

Mohr, G.; Jay, S.C. (1988). Nectar and Pollen-collecting behaviour of honeybees on canola (*Brassica campestris* L. and *Brassica napus* L.) J. Apic Res.; *27* : 131 - 136.

Montgomery, B.E. (1958). Preliminary studies of the composition of some Indiana nectors. Proc. Indian Acad. Sci.; *68* : 159-163.

More, D. (1976). The Bee Book- the History and Natural History of Honey Bees. London, David and Charles Newton.

Morgenthaler, O. (1926). Bienenkrankheiten in Jahre 1926. Schweizerische Bienenzeitsung ; *49* : 176-224. (quoted by Bailey & Ball, 1991).

Morgenthaler, O. (1930). New investigations on Acarine disease. Bee Wld. ; *11* : 49-50.

Morris, L.E. (1929). Field Observations and experiments on the pollination of *Hevea brasiliensis*. J. Rubb. Res. Inst. Malaya; *1* : 41-49.

Morse, R.A. (1954). History of use of smoke for handling bees. Glean. Bee Cult.; *82* : 592-593, 668-670, 637.

Morse, R.A. (1969). The biology of *Apis dorsata* in the Philippines. Proc. Int. IUSSI Congr.; *6* : 165-167.

Morse, R.A. (1970a). Annotated Bibliography on *Apis dorsata*. London, IBRA.

Morse, R.A. (1970b). Annotated Bibliography on *Apis florea*. London, IBRA.

Morse, R.A. (1973). The Complete Guide to Beekeeping. London, Pelham Books : 204 pp.

Morse, R.A. (1975). Bees and Beekeeping. Ithaca, Cornell University Press.

Morse, R.A. (1978). Honeybee Pests, Prodators and Diseases. New York. Cornell University Press : 430 pp.

Morse, R.A. (1979). Rearing Queen Honeybees. Ithaca, NY, USA; Wicwas Press : 128 pp.

Morse, R.A. (1982). The beekeeping programme in Burma. Glean. Bee Cult. : *110* : 454-55, 467.

Morse, R.A.; Boch, R. (1971). Pheromone concert in swarmming honeybees. Ann. Ento. Soc. Am.; *64* : 1414-1417.

Morse, R.A.; Flottum, K. (1997). Honey Bee Pests, Predators, and Diseases. A.I. Root co., Medina, USA; 718 pp.

Morse, R.A.; Flottum, K. Eds. (1990). ABC and XYZ of Bee Culture. A.I. Root Co., USA.: 718 pp.

Morse, R.A.; Laigo, F.M. (1969). *Apis dorsata* in the Philippines. Monogr. Philipp. Assoc. Entomol.; *1* : 1-96.

Morse, R.A.; Shearer, D.A.; Boch, R.; Benton, A.W. (1969). Observations on alarm substances in the genus *Apis*. J. Apic. Res.; *6* : 113-118.

Mound, L.A. (1962). Extra-floral nectaries of cotton and their secretion. Emp. Cott. Grow. Rev.; *39* : 254-261.

Mukherjee, S.K. (1953). The Mango-its botany, cultivation, uses and future improvement, especially as observed in India. Econ. Bot. : 130-162.

Muller, H. (1983). The Fertilization of Flowers. London. Trans. D' Arcy W. Thompson.

Murray, J.A. (1964). A case of multiple bee stings in central Africa. J. Med. *10* : 249-251. 2243 stings of *Apis mellifera* in central Africa-the victim survived- the highest recorded number of stings-survival.

Murrell, D.C.; McDonald, D.N. (1986). The Prairie Beekeeping Manual. Edmonton, Alberta Agriculture (1984). The Alberta Beekeeping Manual. Alta. Agric. Dep. Edmonton.

Murrell, D.C.; Nash, W.T. (1981). Nectar secretion by toria (*Brassica campestris* L. vartoria) and foraging behaviour of three *Apis* species on toria in Banglash. J. Apic. Res.; *20* : 34-38.

Muttoo, R.N. (1939). Editorial on *Apis dorsata*. Indian Bee J.; *1*(1) : 1,3,43.

Muttoo, R.N. (1949). Causes of low honey yields in India. Indian Bee J.; *11*(11/12) : 124-132.

Muttoo, R.N. (1951). The correct scientific nomenclature for our Indian hive bees. Indian Bee J.; *13* : 150-153.

Muttoo, R.N. (1956). Facts about beekeeping in India. Bee Wld.; *37* : 125-133, 154-157.

Muttoo, R.N. (1957). Some so called 'peculiarities' of behaviour of Indian honeybees as compared to the European bees. Indian Bee J.; *19*(3/4) : 62-64.

Naik, K.C.; Rao, M.M. (1943). Studies on blossom biology and pollination in mangoes (*Mangifera indica* L.). Indian J. Hort.; *1* : 107-109.

Naile, F. (1942, 1976 ed.). American Master of Bee Culture-The Life of L.L. Langstroth. Cornell University Press. Ithaca : 215 pp.

Naim, M.; Phadke, K.G. (1976). Bee flora and seasonal activity of *Apis cerana indica* at Pusa (Bihar). Indian Bee J.; *38*(1/4) : 13-19.

Nair, K.S. (1983). Evaluation of importance of litchi as a nectar source. Indian Bee J.; *45*(4) : 108-109.

Nair, P.K.K. (1964). A pollen analytical study of Indian honeys. J. Indian Bot. Soc.; *43* : 179-191.

Nair, P.K.K. (1965). Pollen Grains of Western Himalayan Plants. London, U.K., Asia Publishing House; 102 pp.

Nair, P.K.K.; Singh, K.N. (1974). A study of two honey plants *Antigonon leptopus Hook* and *Moringa pterigosperma* Gaertn. Indian J. Hort.; *31*(4) : 375-379.

Nair, A.G.R.; Nagarajan, S.; Subramaniam, S.S. (1962). Chemical composition of nectar in *Thunbergia grandiflora*. Curr. Sci.; *33*(13) : 401.

Narayanan, E.S.; Sharma, P.L.; Phadke, K.G. (1960). Studies on requirement of various crops for insect pollinations. I, Insect pollinators of saunf (*Foeniculum vulgare)* with particular reference to the honeybees at Pusa (Bihar). Indian Bee J.; *22* : 7-11.

Narayanan, E.S.; Sharma, P.L; Phadke, K.G. (1961). Studies on requirements of various crops for insect pollinators. Insect pollinators of barseem- Egyptian clover (*Trifolium alexandrinum* L.) with particular reference to honeybees and their role in seed setting. Indian Bee J.; *23* : 23-30.

Nehru, C.R.S.; Thankamony, K.; Jayarathnam, K.; Joseph P.M.L. (1984). Studies on off seasonal bee forage in rubber plantations. Indian Bee J.; *46* : 23-25.

Niiranen, V. (1985). An efficient honey mixer to make creamed or spun honey. Am. Bee J.; *125*(12) : 809-811.

Nijjar, G.S.; Sandhu, B.S. (1971). A study on the fruit set problem in sweeet lime (*Citrus limethoides Tanaka*). J. Res. Punjab Agric. Univ.; *8*(4) : 411-15.

Nandi, N.C.; Mahabel, A. (1974). Nesting behaviour of *Apis dorsata* F. in relation to local wind directions around Poona, India. Indian Bee J.; *36* : 19-20.

Nandpuri, K.S.; Brar, J.S. (1966); Studies on floral biology in muskmelon (*Cucumis melo* L.). J. Res. Punjab Agric. Univ.; *3* : 395-399.

Nixon, M. (1983). World maps of *Varroa jacobsoni and Tropilaelaps clareae* with additional records for honeybee diseases and parasites previously mapped. Bee Wld.; *64*(3) : 124-131.

Norden, B.B. (1991). The Bee. Steward, Tabori and Chang.

Northwood, P.J. (1966). Some observations on flowering and fruit setting in cashew. *Anacardium occidentale* L. Trop. Agric. Trin. *43* : 35-42.

Nur, N. (1976). Studies on pollination in Musaceae (banana). Annals of Botany; *40* : 167-177.

Nye, W.P. (1962). Extra supering and shading as factors in honey production in Northern Utah. Glean. Bee Cult.; *90*(7) : 396-399, 436.

Nye, W.P.; Mackenson, O. (1970). Selective breeding of honeybees for alfalfa, pollen collection with tests in high and low alfalfa pollen collection region. J. Apic. Res.; *9* : 61-64.

Okada, I.; Sakai, T. (1960). A comparative study on natural comb of Japanese and European honeybee with special difference in cell number. Bull. Fac. Agr. Tamagawa Uni.; *1* : 1-11.

Olsson, J. (1988). *Apis cerana :* management methods for prevention of absconding. Proc. 4 Int. Conf. Apic. Trop. Climates. IBRA; 453-457.

O'Toole, C.; Raw, A. (1991). Bees of the World. New York, Facts on File.

Owen, M.D. (1978). Histamine in the venom of queen and worker bees (*Apis mellifera*). In Toxins : animal, plant, microbial. ed. P. Rosenberg. Pergamon Press, Oxford.

Page, R.E.; Laidlaw, H.H. (1982). Closed population honeybee breeding. J. Apic Res.; *21*(1) : 30-37, 38-44.

Page, R.E.; Laidlaw, H.H. (1992). Honeybee genetics and breeding. Chapter 7, pp. 235-267, from the hive and the Honey Bee. Hamilton, Dadant and Sons, III., USA.

Page, R.E.; Metcalf, R.A. (1984). A population investment sex-ratio for the honeybee (*Apis mellifera* L.). Am. Nat.; *124* : 680-702.

Panchabhavi, K.S.; Deviah, M.A.; Basavanna, G.P.C. (1976). The effect of keeping *Apis cerana indica* F. colonies on the seed set of sunflower, *Helianthus annus* L. Mysore J. Agric. Sci.; *10* : 631- 636.

Pande, Y.D.; Bndyopadhyay, S. (1985). The foraging behaviour of honeybees on flowers of pigeon pea (*Cajanus cajan)* in Agartala, Tripura. Indian Bee J.; *47* : 13-15.

Pande, P.K.; Sontakke, B.K.; Sarangi, P.K. (1988). Preliminary studies on the effect of bee (*Apis cerana indica* F.) pollination on yield of seasamum and niger. Indian Bee J.; *50* : 63-64.

Pandey, R.S. (1967). *Varroa jacobsoni;* a new mite infesting honeybee (*Apis indica*) colonies in India. Bee Wld.*; 48* : 16.

Pandey, R.S. (1977). Behaviour of the Indian Honey Bee in double brood chamber hives. Am. Bee J.; *117* : 627.

Pandey, R.S.; Yadava, R.P.S. (1970). Pollination of litchi (*Litchi chinensis)* by insects with special reference to honeybees J. Apic. Res.; *9*(2) : 103-105.

Pankiw, P.; Bolton, J.L.; McMohan, H.A; Foster, J.R. (1956). Alfalfa pollination by honeybees on the Regina plains of Saskatchewan. Can. J. Agric. Sci.; *36* : 114-119.

Pant, N.C. (1980). Proc. 2nd. Int. Conf. Apic. Trop. Climate; New Delhi, ICAR Publication.

Parks, H.B. (1929). Water storage by bees. Iowa St. Apiarist, Rep., pp. 53-56.

Park, O.W. (1923). Behaviour of Water carriers. Am. Bee J.; *63* : 553.

Park, O.W. (1925). Field bees and their work VII Int. Beekeep. Congr. Quebec; pp. 472-478.

Park, O.W. (1949). Activities of honeybees. In the hive and the Honey Bee, Grout. ed. Dadant & Sons, Hamilton, III.; pp. 79-152.

Parker, R.L. (1926). The collection and utilization of pollen by the honeybee. N.Y. Exp. Sta. Mem.; *98* : 3-55.

Parker, F.D. (1981a). Sunflower pollination; abundance, diversity and seasonality of bees and their effect on seed yields. J. Apic. Res.; *20* : 49-61.

Parker, F.D. (1981b). Sunflower pollination : abundance, diversity and seasonality of bees on male sterile and male fertile cultivars. Env. Ent.; *10* : 1011-1017.

Parker. F.D. (1981c). How efficient are bees in pollinating sunflowers ? J. Kansas Ent. Soc.; *54* : 61-67.

Pattanshetti, H.V.; Prasad, B.N. (1973). Bees help pollination of cardamom flowers. Curr. Res.; *2* : 56-57.

Percival, M.S. (1965). Floral Biology. Oxford, Pergamon Press : 243 pp.

Phadke, R.P. (1961). Some physico-chemical constants of Indian beeswax. Bee Wld.; *42* : 149-153.

Phadke, R.P. (1962a). Physico-chemical composition of major unifloral honeys from Mahabaleshwar (Western Ghats). Indian Bee J.; *24* (7/9) : 59-65.

Phadke, R.P. (1962b). Nectar concentration in *Carvia callosa* Bremk. Indian Bee J.; *26*(2) : 22-25.

Phadke, R.P. (1964). Nectar concentration in *Thelepaepale ixiocephala* Bremk. Indian Bee J. *27*(2) : 73-76.

Phadke, R.P. (1967a). Studies on Indian honeys. I, Proximate composition and physico-chemical characteristics of Indian multifloral apiary honeys from *Apis indica*. Indian Bee J.; *29* : 14-26.

Phadke, R.P. (1967b). 2 unifloral honeys of Mahabaleshwar. Indian Bee J. *29* : 33-46.

Phadke, R.P. (1968). Studies on Indian honey. 3, Proximate composition and physico-chemical characterizations of honeys from the wild honeybees *A.d., A.f., Trigona*. Indian Bee J.; *30*(1) : 3-8.

Phadke, K.G.; Naim, M.(1974). Observations on the honeybee visitation to the litchi (*Nephelium litchi)* blossoms of Pusa (*Bihar, India*). Indian Bee J.; *36* : 9-12.

Phadke, K.G.; Bisht, D.S.; Sinha, R.B.P. (1966). Occurence of mite, *Verroa jacobsoni Oudemans* in the brood cells of honeybees, *Apis indica* F. Indian J. Ent.; *28* : 411-412.

Phadke, R.P.; Nair, K.S.; Nandedkar, K.U. (1969). Indian beeswaxes I- Their physico-chemical constants. Indian Bee J.; *31*(2) : 52-55.

Phadke, R.P.; Nair, K.S.; Nandedkar, K.U. (1972). II- The nature of their chemical constituents. Indian Bee J.; *37*(1-4) : 15-18.

Phadke, K.G.; Pant, N.C.; Sinha, R.B.P. (1967). The foraging behaviour of *Apis indica* F. (Hym., Apidae) visiting flowers of sesame (*Sesamum orientale* L.). Indian J. Ent.; *29* : 157-163.

Phoon, A.C.G. (1983). Beekeeping in Malaysia. Pertanika; 6(Rev. Suppl.) 3-17.

Pickett, J.A.; Williams, I.H.; Martin, A.P.; Smith, M.C. (1980 and 1981). Nasanov pheromone of the honeybee *Apis mellifera* L. (Hymenoptera, Apidae). J.Chem. Ecol.; *6* : 425-434 and *7* : 543-554.

Phillips, E.F. (1930). Honeybees for the orchard. N.Y. Expt. Bull.; *190* : 3-24.

Pillai, P.K.T.; Pillai, G.B. (1975). Note on the shedding of immature fruits in cashew. Indian J. Agric. Sci.; *45* : 233-234.

Poncini, L. (1987). Beeswax-A review. Indian Bee J.; *49* : 14-24.

Popolizio, E.R.; Pailhe, L.A. (1973). Storing combs in Wax-moth safe storage rooms. Proc. 24 Int. Apic. Congr.; 381-382.

Popravko, S.A. (1977). Chemical organization of propolis and its standardization. Pchelovodstvo; 97(8); 21-23 (Russian -IBRA English E. 1537).

Prasanna Kumari, T.O. (1963). Sugar composition of nectar in coconut Curr. Sci.; *32* : 460-461.

Prell, H. (1927). Beitrage zur Kenntnis der Amobenseuche der erwachsenen Honigbiene. Archive fur Bienenkunde; 7 : 113-121 (quoted by Bailey & Ball, 1991).

Pringle, J.W.S. (1974). Locomotion flight. In the Physiology of Insects. ed. M.Rockstein Vol. III. New York, Acadamic Press : pp. 433-472.

Punchihewa, R.W.K.; Koeniger, N.; Kevan, P.G.; Gadawski, R.M. (1985). Observations in the dance communication and natural foraging ranges of *Apis cerana, Apis dorsata* and *Apis florea* in Sri Lanka. J. Apic. Res.; *24*(3) : 168-175.

Purewal, S.S.; Randhawa, G.S. (1947). Studies in *Hibiscus esculentus* (Ladyfinger) I Chromosome and pollen studies. Indian J. Agric. Sci.; *17* : 129-136.

Pyramarn, K.; Wongsiri, S. (1986). Bee flora of four species of *Apis* in Thailand. J. Sci. Res. Chula Univ.; *11*(2) : 95-103.

Qayyum, H.A.; Ahmad, N. (1967). Biology of *Apis dorsata* F. Pak. J. Sci.; *19* : 109-113.

Radchenko, T.G. (1964). The influence of pollination on the crop and the quality of seed of winter rape. Bdzhil Mistro (1) : 68-74.

Rafiq, A. (1984). Country status report on beekeeping in Pakistan. Proc. Expert. Consult.; 203- 210, FAO Rome.

Rahman, K.A. (1940). Insect pollinators of toria (*Brassica napus* L. var. *dichotoma*) and Sarson (*Brassica campestris* L. var. *Sarson*) at Lyallpur. Indian J. Agric. Sci.; *10* : 422-447.

Rahman, K.A. (1945). Progress of beekeeping in the Punjab. Bee Wld.; *26* : 42-44, 50-52.

Rahman, K.A. (1946). Size of comb cells of the Indian honeybee. Bee Wld.; *28* : 68-69.

Rahman, K.A.; Singh, S. (1940). Italian versus Indian Bees in the Punjab. Indian Bee J.; 2 : 20-22.

Rahman, K.A.; Singh, S. (1941). Nectar and pollen plants of the Punjab. Indian Bee J.; *4* : 32-35.

Rahman, K.A.; Singh, S. (1945). American bees by air. Indian Bee J.; 7 : 106.

Rahman, K.A.; Singh, S. (1946). Size of the cell of the brood comb of the Indian honeybees. Indian Bee J.; *8*(11) : 154-156.

Rahman, K.A. (1948). Variation in the tongue length of the honeybee. Indian J. Entomol.; *10* : 63-73.

Rai, K.M.; Gupta, B.P. (1983). Role of honeybees and other insects as pollinators in apple and pear. Indian Bee J.; *45* : 56-57.

Rajan, B.K. (1980). Apiculture and farm forestry in semi-arid tracts of Karnatka. Proc. 2nd Int. Conf. Trop. Clim. New Delhi, ICAR Public; pp. 187-196.

Ramsay, J. (1987). Plants for Beekeeping in Canada and the Northern USA. London, IBRA.

Rana, R.S. (1989). Aggressive and hoarding behaviour of *Apis mellifera* and *Apis cerana* F. and their role in the pollination of plum and apple bloom. Ph.D. Thesis, H.P. Uni. Simla.

Rana, B.S.; Garg, L.D.; Khurana, S.M.P.; Verma, L.R.; Agarwal, H.O. (1986). Thai sacbrood virus of honeybee (*Apis cerana indica* F.) in northwest Himalayas. Indian J. Virol. *2* : 127-131.

Rana, et al., (1987). Incidence of Thai Sacbrood Virus disease in *Apis cerana indica* F. in Southeast Asia. In chemistry and Biology of Social insects, ed. J. Eden & H. Rembold. Munchen, Peperney Verlag : 640-641.

Rao, B.S. (1961). Pollination of hevea in Malaya. J. Rubb. Res. Inst. Malaya; *17* : 14-18.

Rao, V.N.M. and Hassan, M.V. (1957). Variations in the seed characters of cashew (*Anacardium occidentale* L.) Indian J. Agric. Sci.; *26* : 211-216.

Rao, G.M.; Suryanarayana, M.C. (1988). Studies on pollination of watermelon *Citrullus lunatus* (Thunb). Manst. Indian Bee J.; *50* : 5-8.

Rao, G.M.; Suryanarayana, M.C. (1989). Effect of honeybee pollination on seed yield in onion (*Allium cepa* L.). Indian Bee J.; *51* : 9-11.

Rao, G.M.; Lazar, M.; Suryanarayana, M.C. (1981). Foraging behaviour of honeybees in sesame (*Sesamum indicum* L.) Indian Bee J.; *43* : 97-100.

Rao, M.M.; Lazar, M. (1980). Studies on the behaviour and pollination on onion (*Allium cepa* L.) Proc. 2nd Int. Conf. Apic. Trop. Climates. New Delhi. ICAR Publ. 1983 : 580-89.

Rao, M.M.; Lazar, G.M.; Suryanarayana, M.C. (1984). Studies on floral biology and pollination of litchi (*Litchi chinensis Sonn*). Indian Bee J.; *46* : 7-11.

Rao, M.M.; Nair, P.K.K. (1985). Mango tree a source of bee forage. Indian Bee J.; 49-50.

Rauala, T.S. (1972). Pollination studies in cauliflower (*Brassica oleracea* V. *botrytis* L.). J. Res. Punjab Agric. Univ.; *9* : 580-585.

Raw, A.; Free, J.B. (1977). The pollination of coffee (*Coffea arabica*) by honeybees. Trop. Agric. (Trinidad); *54* : 365-369.

Rawat, B.S. (1982). Bee farming in India. Ranikhet, Rawat Apiaries : 258 pp.

Reddi, C.S.; Reddi, E.U.B.; Reddi, N.S. (1980). Floral rewards and honeybee visitation rates in soapnut tree. Proc. 2nd Int. Cof. Apic. Trop. Climates. New Delhi, ICAR Publ. *1983* : 600-602.

Reddi, E.U.B. (1987). Under-pollination a major constraint of cashewnut production. Proc. Indian Nat. Sci. Acad. B; *53* : 249-251.

Reddy, C.C. (1980a). Observations on the annual cycle of foraging and brood rearing by *Apis cerana indica* colonies. J.Apic. Res.; *19* : 17-20.

Reddy, C.C. (1980b). Studies on the nesting behaviour of *Apis dorsata* F. Proc. 2nd Int. Conf. apic. trop. climates. New Delhi, ICAR Publ. *1983* : 595-599.

Reddy, M.C.M.; Gupta, J.K. (1987). Wild cherry, *Prunus puddum* Roxb., a honey plant. In Social Forestry for Rural Development (ed. P.K. Khosla and R.K. Kohli) I.S.T.S.; pp. 207-14.

Rennie, J.; White, P.B.; Harvey, F.J. (1921). Isle of Wight disease in hive bees. Trans. R. Sec. Edinb.; *52* : IV (29) 737-779.

Ribbands, C.R. (1949). The foraging method of individual honeybees. J. Anim. Ecol.; *18* : 47-66.

Ribbands, C.R. (1953). The Behaviour and Social Life of Honeybees. London. IBRA : 352 pp. and Dover Publication Inc. N.Y.

Richards, K.W. (1984). Alfalfa leaf cutter bee management in Western Canada. Ottawa, Agriculture Canada Publication No : 1495E; 54 pp.

Rinderer, T.E. ed. (1986). Bee Genetics and Breeding. New York Acadamic Press : 425 pp.

Ritter, W. (1981). Varroa disease of the honeybees, *Apis mellifera* L. Bee Wld.; *62* : 141-153.

Robinson, G.E. (1986). The dance language of the honeybee; the controversy and its resolution. Am. Bee J.; *126* (3) : 184-189.

Robinson, G.E.; Page, R.E. (1988). Genetic determination of guarding and undertaking in honey-bee colonies. Nature, Lond.; *333*(6171) : 356-358.

Rodionov, V.V.; Shabarshov, I.A. (1986). The Fascinating World of Bees. Moscow, Mir Publishers.

Rosa, J.T. (1924). Fruiting habit and pollination of cantaloupe. Proc. Am. Soc. Hort. Sci.; *21* : 51-57.

Rose, A. (1999). Progress in bee venom therapy for HIV/AIDS. Proc. 36th Apimondia Congr., Vancouver; 73-74.

Rothenbuhler, W.C. (1964). Behaviour genetics of nest cleaning in honeybee IV. Am. Zool., *4* : 111-123.

Roubik, D.W. (1989). Ecology and Natural History of Tropical Bees. Cambridge, Cambridge University Press.

Roubik, D.W.; Sakagami, S.F.; Kudo, I. (1985). A note on distribution and nesting of the Himalayan honeybee *Apis laboriosa* Smith (Hym. Apidae). J. Kansas Ento. Soc.; *58*(4) : 746-749.

Rudnev, V.Z. (1941a). The effect of pollination by bees on yield of sunflower seeds. Soc. Zern. Hox.; *2* : 134-140.

Rudnev, V.Z. (1941b). The technique of utilization of bees for the pollination of sunflower. Soc. Zern. Hox.; *3* : 77-84.

Ruttner, F. (1956). The mating of the honeybee. Bee Wld.; *57* : 3-15, 23-24.

Ruttner, F. (1966). The life and flight activity of drones. Bee Wld.; *47*(3) : 93-100.

Ruttner, F. (1969). The cause of hybridization barrier between *Apis mellifera* L. and *Apis cerana* F.; Experiment with natural and artificial insemination. Proc. XXII Int. Beekeeping Conf. Prelim. Sci. Meet.

Ruttner, F. (1976). Isolated populations of honeybees in Australia. J. Apic. Res.; *15* : 97-104.

Ruttner, F. ed. (1983). Queen Rearing : Biological Basis and Technical Instruction. Bucharest, Apimondia Publishing House : 358 pp.

Ruttner, F. (1985). Reproductive behaviour in honeybees. Fortschr Zool 31. Holldobler, B.; Lindauer, M. (eds) Exper. Behav. Ecol. Sociobiol; 225-236.

Ruttner, F. (1986a). Geographic variation and taxonomy of *Apis cerana*. Proc. Int. Beekeep. Congr.; *30* : 130-133.

Ruttner, F. (1986b). Geographical variability and classification. In Th. Rinderer (ed.) Bee Genetics and Breeding : 23-56. Orlando FA, Acadamic Press.

Ruttner, F. (1987). Breeding Techniques and Selection for Breeding of the Honeybee. Northern Bee Books, Mytholmroyd, U.K.

Ruttner, F. (1988). Biogeography and Taxonomy of Honeybees. Berlin, Springer-Verlag : 284 pp.

Ruttner, F.; Maul, V. (1983). Expertimental analysis of the reproductive interspecific isolation of *Apis mellifera* L. and *Apis cerana* F. Apidologie : *14* : 309-327.

Ruttner, F.; Woyke, J.; Koeniger, N. (1972). Reproduction in *Apis cerana* I. Mating behaviour. J. Apic. Res.; *11* : 141-146.

Ruttner, F; Woyke, J.; Koeniger, N. (1973). Reproduction in *Apis cerana 2*. Reproductive organs and natural insemination. J. Apic. Res.; *12* : 21-34.

Sakai, T.; Matsuka, M.C. (1982). Beekeeping and honey resources in Japan. Bee Wld.; *63* : 63-71.

Sakagami, S.F. (1959). Some interspecific relations between Japanese and European honeybees. J. Anim. Ecol.; *28* : 51-68.

Sakagami, S.F. (1960). Preliminary report on the specific difference of behaviour and other ecological characters between European and Japanese honeybees. Acta Hymenopteralogica; *1* : 171-198.

Sakagami, S.F. (1973). Additional observations on the nest of the dwarf honeybee *Apis florea*. Kont.; *41* : 217-219.

Sakagami, S.F. (1982). Stingless bees. pp. 362-423 from Social Insects Vol. III ed. H.R. Hermann, London, Acadamic Press.

Sakagami, S.F.; Akahira Y. (1958). Comparison of ovarian size and number of ovarioles between workers of Japanese and European honeybees (Studies on the Japanese honeybee *Apis indica cerana* F.). Kontyu; *26* : 103-107.

Sakagami, S.F.; Fukuda, H. (1968). Life tables for worker honeybees. Researches Popul. Ecol. Kyoto Univ.; *10*(2) : 127-139.

Sakagami, S.F.; Yoshikawa (1973). Additional observation on the nest of the dwarf honeybee, *Apis florea*. Kont.; *41* : 217-219.

Sakagami, S.F.; Matsumura, T.; Ito, K. (1980). *Apis laboriosa* in Himalaya, the little known world's largest honeybee (Hym. apidae). Insecta Matsumurana; *19* : 47-77.

Sagar, P. (1981). Role of insects in cross pollination of fennel crop at Ludhiana. J. Res. Punjab Agric. Univ.; *18*(4) : 388-92.

Sagar, P. and Brar, H.S. (1983). Activity and population abundance of flower visiting insects of coriander, *Coriandrum sativum* crop in Punjab. Indian Bee J. 45.

Salim Ali (1977). Field Guide to the Birds of the Eastern Himalayas. Bombay Natural History Society; Oxford University Press, Bombay : 265 pp.

Salim Ali (1979). The Book of Indian Birds. /Bombay Natural History Society; Oxford University Press, Bombay : 187 pp.

Salvi, S.R. (1975). Seasonal management of the Indian hive bee at Pathankot, Punjab : A preliminary report. Indian Bee J.; *37*(1/4) : 10-14.

Sammataro, D. (1970). The Beekeepers Handbook. Dexter, Michigan; Peach Mountain Press.

Sammataro, D.; Erickson, E.H.; Garment, M.B. (1985). Ultra structure of the sunflower nectary. J. Apic. Res.; *24*(3) : 150-160.

Sandhu, A.S.; Singh, S. (1960). The biology and brood rearing activity of the little honeybee, *Apis florea* F. Indian Bee J.; *22* : 27-35.

Saraf, S.K. (1972). Bee flora in Kashmir. Indian Bee J.; *34*(1/2) : 1-10.

Saraf, S.K. (1985a). Beekeeping in Gurej. Jammu & Kashmir, India. Indian Bee J., *47* : 36-37.

Saraf, S.K. (1985b). Buckwheat, (*Fagopyrum sagittatum*), a major nectar source in Gurej (India). Indian Bee J.; 47.

Saraf, F.K.; Wali, J.L. (1972). Preliminary studies on the egg laying capacity of the queen bee of *Apis indica* F. (hill strain) in Kashmir. Indian Bee J.; *34* : 27-31.

Sasagawa, H.; Matsyama, S. (1999). Semiochemicals inducing social behaviour in honey Bees. Proc. 36th Apimondia Congr., Vancouver; 209-210.

Satyanarayana, I. (1975). Sugar concentration in nectar from *Syzgium cumini* Skeels. Indian Bee J.; *37*(1/4) : 21-24.

Sawyer, R. (1988). Honey Identification. Cardiff, Cardif Academic Press : 115 pp.

Schmidt, J.O. (1992). Allergy to vanomous insects. Chapter 27, pp. 1209-1255, from the Hive and the Honey Bee. Hamilton, Dadant & Sons, III., USA.

Schmidt, J.O. (1982). Biochemistry of insect venoms, Annu. Rev. Entomol.; *27* : 339-368.

Schmidt, J.O.; Schmidt, P.J.; Starr, C.K. (1985). Investigating the Giant honeybee *Apis dorsata* in Sabah. Am. Bee J.; *125* : 749-751.

Schneider, S.S.; Stamps, J.A.; Gary, N.E. (1986a). The vibration dance of the honeybee, I. Communication regulating foraging on two time scales. Anim. Behav.; *34* : 366-385.

Schneider, S.S.; Stamps, J.A.; Gary, N.E. (1986b). The vibration dance of the honeybee , II. The effects of foraging success on daily patterns of vibration activity. Anim. Behav.; *34* : 386-391.

Schumacher, H.H. (1969). Communist China honey production. Am. Bee J.; *109* : 302-305.

Schumacher, H.H. (1979). The Chinese honey trade and what goes on in the Peoples Republic of China. Am. Bee J. *119* : 634, 636.

Schumacher, H.H. (1980). The giant awakes- Chinese honey trade. Am. Bee J.; *120* : 605-607, 611.

Seeley, T.D. (1978). Life history strategy of the honeybee, *Apis mellifera*. Oecologia (Berlin); *32* : 109-118.

Seeley, T.D. (1982). How honeybees find a home ? Scient. Ame.; *247* : 158-168.

Seeley, T.D. (1983a). Division of labour between scouts and recruits in honeybee foraging. Behav. Ecol. Sociobiol.; *12* : 253-259.

Seeley, T.D. (1983b). The ecology of temperate and tropical honeybee societies. Am. Sci.; *71* : 264-272.

Seeley, T.D. (1985a). Honeybee Ecology : A Study of Adaptation to Social Life. Princeton, N.J., USA; Princeton University Press : 202 pp.

Seeley, T.D. (1985b). The information center strategy of honeybee foraging. Fortsch Zool.; *31* : 75-90.

Seeley, T.D.; Heinrich, R. (1981). Regulation of temperature in the nests of social insects. In : Heinrich B(ed). Insect Thermoregulation; 159-234. Wiley, N.Y.

Seeley, T.D.; Morse, R.A. (1976). The nest of the honeybee. Ins. Sociaux; *23* : 495-512.

Seeley, T.D.; Morse, R.A. (1977). Dispersal behaviour of honeybee swarms. Psyche.; *84* : 199-209.

Seeley, T.D.; Visscher, P.K. (1985). Survival of honeybees in cold climates : the critical timing of colony, growth and reproduction. Ecol. Entomol.; *10* : 81-88.

Seeley, T.D.; Seeley, R.H.; Akratanakul, P. (1982). Colony defense strategies of the honeybee in Thailand, Ecol. Manogr.; *52* : 43-63.

Schneider, S.S.; Stamps, J.A.; Gary, N.E. (1986). The vibration dance of the honeybee communication regulating foraging on two time scales. Anim. Behav.; *34* : 366-385.

Shah, A.M. (1980). Beekeeping in Kashmir. Proc. Int. Conf. Apic. Trop. Climate; *2* : 197-204. New Delhi, ICAR, Publ.

Shah, F.A. (1976). "Mags bani-in urdu". Rajbagh, Srinagar; Shah Beekeepers.

Shah, F.A. (1978). Bread Mushik (*Sali aegyptica* L.) the saviour bee plant of Kashmir. Indian Bee J.; 40-69.

Shah, F.A. (1983). Fundamentals of beekeeping. Srinagar, Shah Beekeepers : 64 pp.

Shah, F.A.; Shah, T.A. (1976). A note on the bee activity and bee flora of Kashmir. Indian Bee J.; *38*(1/4) : 29-33.

Shah, F.A.; Shah, T.A. (1980a). Flight range of *Apis cerana* from Kashmir. Indian Bee J.; *42* : 48.

Shah, F.A.; Shah, T.A. (1980b). Early life, mating and egg laying of *Apis cerana* queens in Kashmir. Bee Wld.; *61* : 137.

Shah, F.A.; Shah, T.A. (1981). Egg laying capacity of Kashmir bee, *Apis cerana* F. Bee Wld.; *62*(3) : 114-115.

Shah, F.A.; Shah, T.A. (1982). The role of Kashmir bees in exploiting beekeeping potential in India. Indian Bee J.; *44*(2) : 37-42.

Shahid, M.; Qayyum, A. (1977). Bee flora of the N.W.F.P. Pak. J. For.; *27* : 1-10.

Shakuntala Nair, K.; Wakle, D.M. (1983). A qualitative and quantitative study of foliar nectar of rubber tree. Indian Bee J.; *45* : 113-114.

Sharma, H.C. (1959). Taxonomic studies on Indian Apoidea. Unpublished thesis; Associate Indian Agric. Res. Inst.; New Delhi; 144 pp.

Sharma, M. (1970). An analysis of pollen loads of honeybees from Kangra, India. J. Polymol.; *6* : 104-110.

Sharma, M. (1972). Studies in the flowers of *Datura stramonium* Linn. in relation to bee botany. J. Palymol.; *8* : 17-21.

Sharma, O.P.; Raj, D. (1985). Diversity of bee flora in Kangra Shivaliks and its impact on beekeeping. Indian Bee J.; *47* : 21-24.

Sharma, O.P.; Thakur, A.K. (1982). Beekeeping in Kangra Valley. Indian Fmg; *31*(12) : 33-35.

Sharma., O.P.; Garg, R.; Dogra, G.S. (1982) Efficacy of formic acid against *Acarapis woodi* (Rennie). Indian Bee J.; *45*(1) : 1-2.

Sharma, O.P.; Mishra, R.C.; Dogra, G.S. (1980). Management of *Apis mellifera* in Himachal Pradesh. Proc. 2nd Int. Conf. Apic. Trop. Climate; New Delhi, ICAR Publ. : 205-210.

Sharma, P.L. (1945). Distribution of *Apis indica* F. in the plains of northern India. Indian Bee J.; *1* : 53-54.

Sharma, P.L. (1944). Queen introduction. Indian Bee J.; *6*(9/10) : 152-154.

Sharma, P.L. (1958a). Sugar concentration of nectar of some Punjab honey plants. Indian Bee J.; *20*(7) : 81-91.

Sharma, P.L. (1958b). Brood rearing activity of *Apis indica* F. and egg laying capacity of its queen. Indian Bee J.; *20*(11) : 166-173.

Sharma, P.L. (1960a). Observations on the swarming and mating habits of Indian honeybee. Bee Wld.; *41* : 121-123.

Sharma, P.L. (1960b). Experiments with *Apis mellifera* in India. Bee Wld.; *41* : 230-232.

Sharma, P.L. (1961). The honeybee population among insects visiting pome fruit flowers and their role in setting fruit. Bee Wld.; *42*(1) : 6-8.

Sharma, P.L.; Dhaliwal, H.S.; Kakar, K.L. (1974). Insect visitors and pollinators of cauliflower (*Brassica oleracea* var. botrytis) seed crop bloom. Himachal J. Agric. Res.; *2*(2) : 64-78.

Sharma, P.L.; Dogra, G.S.; Mishra, R.C. (1979). Evaluation of methods of control of predatory wasps, *Vespa* spp. in apiaries. Indian Bee J.; *41*(3/9) : 10-16.

Shearer, D.A.; Boch, R.; Morse, R.A; Laigo, F.M. (1976). Occurrence of 9-oxodec-trans-2-enoic acid in queens of *Apis dorsata, Apis cerana,* and *Apis mellifera*. J. Insect Physiol. *16* : 1437-1441.

Shelar, D.G.; Suryanarayana, M.C. (1981). Preliminary studies on pollination of coriander (*Coriandrum sativum* L.). Indian Bee J.; *43* : 110-111.

Shimanuki, H. (1981). Controlling the greater wax moth—a pest of honeycombs. USDA, Farmers Bull. # 2217. (1999). American Foulbrood Disease. Proc. Apimondia; p. 45.

Shimanuki, H.; Knox, D.A.; Furgala, B.; Caron, D.M.; Williams, J.L. (1992). Diseases and pests of honeybees. Chapter 25, pp. 1083-1151; from the Hive and the Honeybee. Hamilton, Dadant & Sons, III., USA.

Shipolini, R.A. (1984). Biochemistry of bee venom. Chap. 2, pp. 49-85 : In Handbook of Natural Toxins Vol. 2, ed. A.T.Tu. New York; Marcel Dekker Inc.; 732 pp.

Showler, K. (1975). The Observation Hive. Bee Books New and Old, Steventon Hants.

Shuel, R.W. (1992). The production of nectar and pollen. Chapter 10, pp. 401-436, from the Hive and the Honey Bee. Hamilton, Dadant & Sons, III., USA.

Siddappaji, C.; Channa-Basavanna, G.P (1980). Role of Honeybees in the pollination of cardamom (*Elettaria cardemomum* L.) Proc. 2nd Int. Conf. Apic. Trop. Clim.; New Delhi, ICAR Publ. *1983* : 640-648.

Sidhu, A.S. (1959). Role of honeybee in cotton Pollination. M.Sc. Thesis; Punjab Agric. College, Ludhiana (India).

Sidhu, A.S.; Atwal, A.S. (1970). The biology and nesting behaviour of *Andrena ilerda* (in Atwal, 1970).

Sidhu, A.S.; Singh, S. (1961). Studies on the agents of cross pollination in cotton. Indian Cott. Gr. Rev.; *15*(6) : 341-53.

Sidhu, A.S.; Singh, S. (1962). Role of honeybees in cotton production. Indian Cott. Gr. Rev.; *16*(1) : 18-23.

Sihag, R.C. (1982a). Effect of competition with *Parkinsonia aculeata* L. on pollination and seed production in *Medicago sativa* L. Indian Bee J.; *44*(4) : 89-90.

Sihag, R.C. (1982b). Problem of wax moth (*Galleria mellonella*)infestation in Giant honeybee (*Apis dorsata)* in Haryana. Indian Bee J.; *44* : 107-109.

Sihag, R.C. (1986). Insect pollination increases seed production in cruciferous and umbelliferous crops. J.Apic. Res.; *25* : 121-126.

Sihag, R.C.; Kapil. R.P. (1983). Foraging strategies of honeybees as determined by quality and quantitiy of nectar. Proc. 5th Int. Symp., Versailles; pp. 51-59.

Sikka, S.M.; Gupta, N.P. (1949). Pollination studies in *Sesamum orientale* L. Indian J. Genet.; *9* : 33-191.

Simpson, J. (1960). The age of queen honeybees and the tendency of their colonies to swarm. J. Agric. Sci.; *54*(2) : 195.

Simpson, J.; Greenwood, S.P. (1974). Influence of artificial queen-piping sound on the tendency of honeybee *Apis mellifera* colonies to swarm, Insects Soc.; *21*(3) : 283-287.

Singh, G. (1972). Defensive behaviour of *Apis indica* F. (hill strain) against predatory hornets in Kashmir. Indian Bee J.; *34*(3/4) : 65-69.

Singh, G.; Singh, G. (1971). *Plectranthus rugosus* wall; the major honey plant of Kashmir Valley. Indian Bee J.; *33* : 58-59.

Singh, L.B. (1960). The Mango. London : Leonard Hill.

Singh, M.P. (1984). Studies on the activity of some insect pollinators on jujube (*Zizyphus mauritiana Lamk*) Entomon.; *9* : 177-180.

Singh, R.N. (1978). Mango. New Delhi, ICAR Publ.

Singh, S. (1940). Tearing down of combs. Bee Wld.; *21*(6) : 66.

Singh, S. (1948). Some important honey plants of the Punjab (India). Rep. Iowa State Apiar.; 34-42.

Singh, S. (1950). Behaviour studies of honeybees in gathering nectar and pollen. Bull. Cornell Agric. Exp. Sta.; *288* : 1-59.

Singh, S. (1953). Activities of Insect pollinators and orchard planning. Indian J. Hort.; 10.

Singh, S. (1954). Insect pollinators and the breeding of fruit varieties. Indian J. Hort.; 11.

Singh, S. (1957). Acarine disease in the Indian bee (*Apis indica* F.); Indian Bee J.; *19*(3/4) : 27-28.

Singh, S. (1962). Beekeeping in India, New Delhi. ICAR; 214 pp. Reprinted 1982.

Singh, S. (1964). Beekeeping and entomology in India. Indian J. Entomol.; *25* : 497-511.

Singh, S.; Adlakha, R.L. (1958). Acarine disease of adult honeybee. Indian Bee J.; *20*(6) : 64-78.

Singh, S.; Adlakha, R.L. (1960). Survey of acarine disease of adult honeybee in Simla hills. Indian Bee J.; *22*(1/3) : 3-6.

Singh, J.P.; Dharamwal, S.S. (1970). The role of honeybees in seed setting on onion at Pantnagar, District Nainital, U.P., India. Indian Bee J.; *32*(1) : 23-27.

Singh, B.P.; Phadke, R.P.; Mittal, M.C. (1983). Beekeeping potential and pattern of management in Uttar Pradesh. Indian Bee J.; *43* : 35-37.

Singh, R.P.; Singh, P.N. (1987). Bee foraging plants of Gorakhpur. Indian Bee J.; *49* : 41-43.

Singh, R.P.; Singh, P.N. (1988*a*). Bee forage and bee pollinated plants of Allahabad. Proc. Nat. Acad. Sci. India; 58 (Sec. B) : *437* : 442.

Singh, R.P.; Singh, P.N. (1988*b*). Proline an early index of Thai Sac Brood Virus disease of honey bee. Nat. Acad. Sci. Letters; 11 (5).

Singh, R.P.; Singh, P.N. (1992). Impact of bee pollination on seed yield, carbohydrate composition and lipid composition of mustard seed. J. Apic. Res., *31* : 128-133.

Singh, Y. (1983). Beekeeping in Uttar Pradesh : A review. Indian Bee J.; *45* : 84-92.

Smith, F.G. (1960). Beekeeping in the Tropics. London, Longman : 265 pp.

Smith, F.G. (1964). The hive environment in hot climates. J. Apic. Res.; *3*(2) : 117-122.

Smith, M.V.; Townsend, G.F. (1952). A method of measuring pollination population on field crops. Can. Ent.; *134* : 314.

Snelgrove, L.E. (1972). Swarming : Its Control and Prevention. Author Publisher : 95 pp.

Snodgrass, R.F. (1925). (2nd ed. 1956). Anatomy of the Honeybee. Ithaca, N.Y., USA. Cornell University Press : 327 pp.

Snodgrass, R.E.; Erickson, E.H. (1992). Anatomy of the honeybee Chapter 4, pp. 103-169, from the Hive and the Honey Bee. Hamilton, Dadant and Sons, III., USA.

Southwick, E.E. (1985). Allometric relations, metabolism and heat conductance in clusters of honeybees at cool temperatures. J. Comp. Physiol. B; *156* : 143-149.

Southwick, E.E. (1987). Cooperative metabolism in honeybees : an alternative to antifreeze and hiberantion. Int. Conf. Afr. Bees and Mites; Columbus, OH. U. Press.

Southwick, E.E. (1992). Physiology and social physiology of the honeybee. Chapter 5, pp. 171-196. The Hive and the Honey Bee. Hamilton, Dadant & Sons, III. USA.

Sovoleva, E.M. (1952). Bees and lucern production. Pchelovodstvo, Mosk.; *29* : 39-41. (quoted by Free, 1970).

Spacznska, J.O.I. (1975). The World of the Wasp. New York, Crane, Russak & Co.

Srivastava, V.K. (1969). Studies on floral biology of *Abelmoschus esculentum* (L.) Moench. Indian J. Hort.; *26* : 165-170.

Srivastave, D.C.; Tripathi, R.L. (1980). Prospects of apiculture in tribal areas of chota Nagpur. Proc. 2nd Int. Conf. Apic. Trop. Clim.; New Delhi, ICAR Publ. *1983* : 227-232.

Stanley, R.G.; Linskens, H.F. (1974). Pollen : Biology, Biochemistry, and Management. New York, Springer-Verlag.

Stein, I. (1989). Royal Jelly : The New Guide to Nature's Richest Food, Wellingborough, U.K.; Thorsons Publishing Group.

Steinhaus, E.A. (1949). Principles of Insect Pathology. McGraw-Hill, New York.

Stejskal, M. (1955). Gregarines found in the honeybee (*Apis mellifera* L.) in Venezuela. J. of Protozoology; *2* : 185-188.

Stephen, W.A. (1968). Mites : a beekeeping problem in Vietnam and India. Bee Wld.; *49*(3) : 119-120.

Stephen, W.P. (1962). Propagation of the leaf-cutter bee for alfalfa seed production. Stn. Bull. Ore. Agric. Exp. Stn.; 586.

Stephen, W.P. (1965). Temperature effects on the development and multiple generation in the alkali bee (*Nomia melanderi*). Ent. Exp. Appl.; *8* : 228-240.

Strickland, S.S. (1982). Honey hunting by the Gurungs of Nepal. Bee Wld. ; *63* : 153-161.

Stuart, F.S. (1947). Beekeeping Practice. London, C. Arthur Pearson Ltd.

Style, S. (1992). Honey from Hive to Honey Pot. San Francisco, Chronicle Books.

Subbareddi, C.E.; Reddi, U.B.; Reddi; N.S. (1980). Floral reward and honeybee visitation rates in the soapnut tree. Proc. 2nd Int. conf. Apic. Trop. climates. New Delhi, ICAR Publ.; *1983* : 600-608.

Subbiah, M.S.; Mahadevan, V. (1957). *Vespa cincta* Fab. a predator of honeybee, and its control. Indian J. Vet. Sci.; *27* : 153-154.

Sugden, E.A.; Thorp, R.W.; Buckmann, S.L. (1996). Honeybee-native bee competition : focal point for environmental response in Australia. Bee Wld.; *77* : 26-44.

Sukhija, B.S.; Saimbhi, M.S.; Nandpuri, K.S. (1981). Studies on floral ecology of Okra. J. Res. Punjab Agric. Univ.; *18* : 365-371.

Suryanarayana, M.C. (1966). Studies on flora of Coorg with emphasis on the bee forage plants I. General ecological and botanical features. Indian Bee J.; *28* : 59-75.

Suryanarayana, M.C. (1980). The rubber tree, *Hevea brasiliensis* an important nectar source in the tropics. Proc. 2nd Int. Conf. Apic. Trop. Climate, New Delhi; ICAR Publ. *1983* : 656-658.

Svensson, B. (1977). Apiculture in China. Bitidningen; *76* : 232-235, 337-341.

Syed, R.A. (1979). Studies on oil palm pollination by insects. Bull. Ent. Res.; *69* : 213-224.

Syed, R.A. (1982). Insect pollination of oil palm : feasibility of introducing *Elaeidobius* spp. into Malaysia. In Pushparajah, E. and Chew Poh Soon (eds.). The oil palm in the eighties. A report of the Proceedings of the International Conference on Oil Palm in Agriculture in the Eighties. Kuala Lumpur, June 1981, Vol I, 263-289. Kuala Lumpur Incorporated Society of Planters.

Synge, A.D. (1947). Pollen collection by honeybees, (*Apis mellifera*). J. Anim. Ecol.; *16* : 122-138.

Tanda, A.S. (1983). Assessing the role of honeybees in a field of Asiatic cotton (*Gossypium arboreum* L.). Am. Bee J.; *123* : 593-594.

Tanda, A.S. (1984a). Bee pollination increases yield of 2 interplanted varieties of Asiatic cotton (*Gossypium arboreum* L.). Am. Bee J.; *124* : 539-540.

Tanda, A.S. (1984b). Insect pollination in Okra (*Abelmoschus esculentum*). Proc. Ill Orient Ent. Symp.; Kariavattam, India; 25.

Tanda, A.S. (1984c). Foraging behaviour of three species of *Apis* on raya in relation to the sugar concentration in its nectar. Indian Bee J.; *46* : 5-6.

Tanda, A.S. (1985). Floral biology, pollen disposal, and foraging behaviour of honeybees in Okra (*Abelmoschus esculentum*). J. Apic. Res.; *24* : 225-227.

Tanda, A.S.; Goyal, N.P. (1979a). Insect pollination in Asiatic cotton (*Gossypium arboreum*). J. Apic, Res.; *18*(1) : 64-72.

Tanda, A.S.; Goyal, N.P. (1979b). Some observations on the behaviour of *Apis mellifera* L. and *Apis cerana* F. workers in a field of desi cotton. (*Gossypium arboreum* Linn). Am. Bee J.; *119*(2) : 106.

Tarr, H.L.A. (1937a). Studies on American foulbrood of bees I. the relative pathogenicity of vegetative cells and endospores of *Bacillus larvae* for the brood of the bee. Annu. of Appl. Biol.; *24* : 377-384.

Tarr, H.L.A. (1937b). Certain factors influencing the germination of the endospores of *Bacillus larvae*. Bee Wld.; *18* : 57-58.

Taylor, R. (1977). How to Raise Beautiful Comb Honey. New York, Lindon Books : 251 pp.

Taylor, R. (1980). The Book of Beekeeping—How to do it ? New York, Lindon Books : 251 pp.

Teaotia, S.S.; Chauhan, R.S. (1964). Flowering, pollination, fruitset, and fruit drop studies in Ber (*Zizyphus mauritiana* Lam.). Indian J. Hort.; *21* : 40-41.

Tew, J.E. (1992). Honey and wax—a consideration of production, processing and packaging techniques. Chapter 15, pp. 657-705, from the Hive and the Honey Bee. Hamilton, Dadant & Sons, Ill., USA.

Tewari, G.N.; Kirti Singh (1983a). Role of insect pollinators in vegetable seed production. Indian Bee J.; *45* : 51.

Tewari, G.N.; Kirti Singh (1983b). Studies on insect pollinators in relation to seed production in cauliflower (*Brassica oleracea* var botrytis). Indian Bee J.; *45* : 54-55.

Tewari, G.N.; Singh, K. (1983). Studies on insect pollinators in relation to seed production in cauliflower (*Brassica oleracea* var botrytis). Indian Bee J.; *45*(2/3) : 54-55.

Thakur, A.K.; Sharma, O.P.; Garg, R.; Dogra, G.S. (1982). Comparative studies on foraging behaviour of *Apis mellifera* and *Apis cerana indica* on mustard. Indian Bee J.; *44* : 91-92.

Thakar, C.V. (1973). A preliminary note on hiving *Apis dorsata* colonies. Bee Wld., *54* : 24-27.

Thakar, C.V. (1976a). Practical aspects of bee management in India with *Apis cerana indica*. Proc. Int. Conf. Apic. Trop. Climate 1 : 51-59. London, IBRA.

Thakar, C.V. (1976b). The beekeeping development and research programme of the Khadi and Village Industries Commission, India. Proc. 1st Int. Conf. Apic. Trop. Climates : 125-134. London, IBRA.

Thakar, C.V.; Tonapi, K.V. (1961). Nesting behaviour of Indian honeybees 1. Differentiation of worker queen and drone cells of the comb of *Apis dorsata* Fab. Bee Wld.; *42* : 61-62.

Thakar, C.V.; Tonapi, K.V. (1962). Nesting behaviour of Indian honeybee II. Nesting habits and comb cell differentiation in *Apis florea* Fab. Indian Bee J.; *24* (1/3) : 27-31.

Thakar, C.V.; Diwan, V.V.; Salvi, S.R. (1962). Floral calendar of major and minor bee forage plants in Mahabaleshwar hills (Western Ghat). Indian Bee J.; *24* : 35-48.

Tompson, C.R. (1940). Presidental address to First Annual General Meeting of the All India Beekeepers Association. Indian Bee J.; *2*(1/2) : 10-13.

Tompson, C.R. (1944). Research work on *Apis mellifera*. Indian Bee J.; *6*(11/12) : 163-164, 168.

Tirgari, S. (1971). Biology and behavioural characteristics of the Iranian Dwarf honeybee (*Apis florea*). Proc. Int. Beekeep. Congr.; *23* : 344-345.

Tirgari, S.; Miralizadeh, E.A. (1987). Exploitation and protection measures of *Apis florea* in Iran Proc. IUSSI Congr.

Todd, F.E.; Bishop, R.K. (1941). The role of pollen in the economy of the hive. Circ. U.S. Burr. Ent., No. E 536 : pp. 1-9.

Towne, W.F. (1985). Acoustic and visual cues in the dance of four honeybee species, Behav. Ecol. Sociobiol.; *16* : 185-187.

Tripathi, K.L. (1971). Bee-keeping in West Bengal. Indian Bee J.; 33 : 23-28.

Treherne, R.C. (1923). The relation of insects to vegetable seed production. Rep. Queb. Soc. Prot. Pl.; *15* : 47-59.

Triasco, V.V. (1956). The mating sign of the queen and its characteristics. Pchelovodstvo; *33* : 43-50 (in Russian).

Trump, R. (1987). Bees and their Keepers. Ames, Iowa State University Press.

Tufts, W.P. (1919). Almond Pollination. Univ. Calif. Bull. No : *306* : 337-366.

Tulloch, A.P. (1973). Factors affecting analytical values of bees wax and detection of adulteration. J. Am. Oil. Chem. Soc.; *50* : 269-272.

Tulloch, A.P. (1980). Beeswax-composition and analysis Bee Wld.; *61* : 4762.

USDA (1980). Beekeeping in the United States. Washington DC., USDA Handbook No.; *335* : 193 pp.

USDA Foreign Agricultural Service (1988). Chinese honey production on the rise, but exports to US decline. Am. Bee J.; *128* : 668-671.

Valli, E. (1998). Golden harvest of the Raji. National Geographic; *1993*(6) : 84-101.

Valli, E.; Summers, D. (1988). Honey Hunters of Nepal. London, Thames & Hudson : 104 pp. New York, Harry N. Abrams Inc.

Vansell, G.H. (1942). Factors affecting the usefulness of honeybees in pollination. USDA, Cir. 650.

Vansell, G.H. (1944). Cotton nectar in relation to bee activity and honey production. J. Econ. Ent.; *37* : 528-530.

Vansell, G.H.; Watkins, W.G.; Bishop, R.K. (1942). Orange nectar and pollen in relation to bee activity. J. Econ. Ent.; *35* : 321-323.

Vansell, G.H.; Todd, F.E. (1946). Alfalfa tripping by insects. J. Am. Soc. Agron.; *38* : 470-488.

Varma, S.K.; Joshi, N.K. (1983). Studies on the role of honeybees in the pollination of cauliflower (*Brassica oleracea* var botrytis). Indian Bee J.; *45*(2/3) : 52-53.

Vashishtha, B.B.; Pareek, O.P. (1979). Flower morphology, fruit set and fruit drop in some ber (*Zizyphus mauritiana*) cultivars. Ann. Arid. Zone; *18* : 165-169.

Vasu, H.D. (1967). Studies on beekeeping at Delhi : possibilities, and the plants visited by bees for pollen and nectar. Indian Bee J.; *29* : 63-64.

Vats, B.R. (1953). Breeding Italian bees in Kashmir. Kashmir *3* : 23-24; 33.

Velthuis, H.H.W.; Clement, J.; Morse, R.A.; Laigo, F.M. (1971). The ovaries of *Apis dorsata* workers and queens from the Philippines. J. Apic. Res.; *10* : 63-66.

Verma, L.R. (1970). A comparative study of temperature regulation in *Apis mellifera* L. and *Apis indica* F. Am. Bee. J.; *110*(10) : 390-391.

Verma, L.R. (1973). An ionic basis for a possible mechanism of sperm survival in the spermatheca of the queen honeybee, *Apis mellifera* L. Comp. Biochem. Physiol.; *44* : 1325-1331.

Verma, L.R. (1974). Honeybees spermmatozoa and their survival in the queens spermatheca. Bee Wld.; *55* : 53-61.

Verma, L.R. (1978a). Biology of honeybee (*Apis mellifera* L.) spermatozoa 1- Effect of different diluents on mortility and survival. Apidologie; *9* : 167-173.

Verma, L.R. (1978b). Effect of Washing and sugar on respiration as studied by Cartesian diver technique. J. Apic. Res.; *17* : 176-181.

Verma, L.R. (1983). Effect of deep freezing on survival of the honeybee (*Apis mellifera*) spermatozoa. Am. Bee J.; *123* : 851-952.

Verma, L.R. (1984). Beekeeping in northern India : major constraints and potentials; pp. 148-155 from Proc. Expert. Consultation ed., FAO.

Verma, L.R. (1986a). Biological and economic characters of *Apis cerana indica* F., honeybee. Science; *7* : 151-157.

Verma, L.R. (1986b). Studies on genetic improvement of Indian honeybee *Apis cerana indica* F. by selection and breeding. Final Project Report DST, Govt. India : pp 153.

Verma, L.R. (1987a). Current status of parasitic mites in relation to beekeeping with *Apis cerana* F. and *Apis mellifera* in India. Proc. FAO Workshop on Parasitic Bee Mites and their Control; Pulawy, Poland : 195-198.

Verma, L.R. (1987b). Pollination ecology of apple orchards by hymenopterous insects in Matiana Narkanda temp. zone. Final Report Min. Environ. & Forests.; *G01* : 118 pp.

Verma, L.R. (1991). Beekeeping in Integrated Mountain Development : Economic and Scientific Perspectives. Edinburgh, Aspect Publications : 387 pp.

Verma, L.R.; Chauhan, P. (1985). Distribution, abundance and diversity of insect pollinators in apple orchards of Simla hills. Indian J. Ecol.; *12* : 286-298.

Verma, L.R.; Dulta, D.K. (1986). Foraging behaviour of *Apis cerana indica* and *Apis mellifera* in pollinating apple flowers. J. Apic. Res.; *25* : 197-201.

Verma, L.R.; Edwards, D.K. (1971). Metabolic acclimatization to temperature, and temperature tolerance in *Apis mellifera* L. and *Apis cerana indica* F. J. Apic. Res.; *10* : 105-108.

Verma, L.R.; Mattu, V.K.; Singh, M.P. (1984). Races of the Indian honeybee in the Himalayas. Abstr. Int. Congr. Entomol.; *17* : 508.

Verma, Savitri (1988). Chromosomal analysis of mitotic and meiotic divisions of four species of genus *Apis* (honeybees) Int. Symp. Recent Advances in Cytogenetic Res.; Kurukshetra, India : 100.

Verma, S.; Ruttner, F. (1983). Cytological analysis of the thelytokous parthenogenesis in the cape honeybee (*Apis mellifera capensis* Eschltz). Apidologie; *14* : 41-58.

Vishwanathan, H. (1950a). Note on *Apis dorsata* queen cells. Indian Bee J.; *12* : 55.

Viswanathan, H. (1950b). Temperature reading on *Apis dorsata* combs. Indian Bee J.; *12* : 72.

Vivian, J. (1986). Keeping Bees. Charlotte, Vermont, Williamson Publ. : 230 pp.

Wafa, A.K.; Ibrahim, S.H. (1959). Pollinators of the chief sources of nectar and pollen grain plants in Egypt. Bull. Soc. Ent. Egypte ; *43* : 133-154.

Wakhle, D.M.; Desai, D.B. (1980). Enzymes in Indian honeys. Proc. 2nd Int. Conf. Apic. Trop. Climates; 703-710. New Delhi, ICAR Publ.

Wakhle, D.M.; Nair, K.S.; Phadke, R.P. (1978). Effect of bee pollination on the oil and protein content in the seeds of sunflower (*Helianthus annus* L.). Indian Bee J.; *40*(1) : 1-2.

Wakhle, D.M.; Nair, K.S.; Ramesh, B. (1981). Sugar Composition in nectars of certain plants. Indian Bee J.; *43*(1) : 6-8.

Wali-ur-Rahman, M.I.C. (1985). Bee foraging plants at Peshawar, Pakistan. Pakistan J. For.; *35*(2) : 71-76.

Walker, G.D. (1975). Evaluating the foraging behaviour of honeybees as pollinators of hybrid onions. Proc. III Int. Symposium on Pollination, Prague; *1974* : 175-180.

Walker, P.; Crane, E. (1987). Constituents of propolis. Apidologie ; *18* : 327-334.

Warren, C.; Warren, B. (1982). Plastic frames and foundation. Bee Wld.; *63*(3) : 104-105.

Watmough, R.H. (1987). A leaf-cutter bee (*Megachilidae*) and a carpenter bee (*Anthrophoridae*) as possible pollinators of lucern (*Medicago sativa* L.) in the Oudtshoorn District. S. Afr. Bee J.; *59*(5) : 114.

Weaver, N.; Weaver, E.C. (1981). Beekeeping with the stingless bee *Melipona beecheii* by the yacatecan Maya. Bee Wld.; *62* : 7-19.

Wedmore, E.B. (1979). A Manual of Beekeeping. Bee Books, New and Old. Steventon, Hempshire : 389 pp.

Weiss, K. (1983). Experiences with plastic combs and foundation. Bee Wld. *64*(2) : 56-62.

Wenner, A.M. (1962). Sound production during the waggle dance of honeybee? Anim. Behav.; *10* : 79-95.

Wenner, A.M. (1967). Honeybees : Do they use the distance information contained in their dance maneouver ? Science, *158* : 1076-1077.

Wenner, A.M. & Wells, P.H. (1990). Anatomy of Controversy- the Question of a Language among the Bees. New York, Columbia University Press.

Wheeler, W.W. (1923). Social Life among the Insects. Harcourt, Brace & Co.

Whitcombe, H.J.; Douglas, J.S. (1955). Bees are my Business. New York, G.P. Putnams & Sons : 245 pp.

Whitcombe, R.P. (1981). Temperature regulation and foraging activity of two species of honeybees (*Apis mellifera* and *Apis florea*) in Oman. Preliminary report, Honey production project. Apic. Res. Programme, Durham Univ. Khabura Dev. Project; 45 pp.

Whitcombe, R.P. (1982). Experiments with a hive for little bee : some observations on manipulating colonies of *Apis florea* in Oman (Part I). Indian Bee J.; *44*(3) : 57-63.

Whitcombe, R.P. (1984). The biology of *Apis* spp. in Oman with special reference to *Apis florea* Fab. Univ. Durham; Ph. D. Thesis : 621 pp.

White, G.F. (1907). The cause of American foulbrood. Circular, U.S. Bureau of Entomology ; No : 94.

White, G.F. (1917). Sacbrood. Bulletin, U.S. Department of Agriculture ; No. 431.

White, J.W. (1992). Honey. Chapter 21, pp. 869-925, from the Hive and the Honey Bee. Hamilton, Dadant & Sons, III; USA.

White, J.W. & others (1962). Composition of American honeys. Washington DC, USDA Tech. Bull. No. *1261* : 124 pp.

Whynott, D. (1991). Following the Bloom-Across America with the Migratory Beekeepers. Hassisburge, P.A., Stackpole Books.

Wickramasinghe, R.H. (1983). Beekeeping and its management on coconut, rubber, cashew, citrus, gingelly fruit, timber and other plantations. J. Nat. Inst. Plantn. Management ; *2*(1) : 94-104.

Wigglesworth, V.B. (1984). The Principles of Insect Physiology. London, Methuen & Co.

Wille, A. (1983). Biology of the Stingless bees. Annu. Rev. Entomol.; *28* : 41-64.

Wille, H. (1967). Mischinfektionen in der Honigbiene (*Apis mellifera* L.) nach Ermittlungen in Schweizerischem Material der Jahre 1965/66. Zeitschrift fur Bienen forschung; *9* : 150-171 (quoted by Bailey & Ball, 1991).

Williams, I.H. (1977). Behaviour of insects foraging on pigeon pea (*Cajanus cajan (L.)* Millsp. in India. Trop. Agric.; *54* : 353-363.

Williams, I.H. (1980). The pollination of pigeon pea (*Cajanus cajan* (L) Millsp) in India. 2nd Int. conf. Apic. Tropic. Clim. New Delhi, ICAR Publ.; pp. 661-666.

Williams, J.L. (1978). Insects : Lepidoptera (Moths). In Honey Bee Pests, predators, and Diseases, R.A. Roger ed. Cornell Univ. Press; Ithaca, New York : pp. 105-127.

Williams, P.H. (1985). A preliminary cladistic investigation of relationships among the bumblebees (Hymenoptra, Apidae). System. Entomol.; *10* : 239-255.

Wilson, E.O. (1971). The Insect Societies. Cambridge, M.A., USA : Harvard University Press : 548 pp.

Winston, M.L. (1979). The potential impact of the Africanized honeybee on apiculture in Mexico and central America. Am., Bee J.; *119* : 584-586, 642-645.

Winston, M.L. (1987). The Biology of the Honeybee. Cambridge, MA, USA; Harvard University Press : 281 pp.

Winston, M.L. (1992*a*). Killer Bees—the Africanized Honey Bee in America. Boston, Harvard University press.

Winston, M.L. (1992*b*). The honeybee colony : life history. Chapter 3, pp. 73-101, from the Hive and the Honey Bee. Hamilton, Dadant and Sons, III, USA.

Winston, M.L.; Michener, C. (1977). Daul origin of highly social behaviour among bees. Proc. Nat, Acad, Sci. USA; *74* : 1135-1137.

Wongsiri, S.; Polnurak, P.; Sylvester, H.A. (1985). Honey yield of *Apis mellifera* in a rubber plantation in Thailand Proc. 30th Int. Beekeep. Congr. Nagoya, Japan; pp. 417-420.

Wongsiri, S.; Lai, Y-S; Liu Z-S (1986). Beekeeping in the Guangdong Province of China and some observations on the Chinese honeybee *Apis cerana cerana* and the European honeybee *Apis mellifera ligustica*. Am. Bee J.; *126*(11) : 748-752.

Wongsiri, S.; Tangkanasing, P. (1986). *Apis cerana* F. beekeeping in Thailand : problems and research needs. J. Sci. Res.; Chula Univ. : 1-6.

Woodrow, A.W. (1933). The comparative value of different colonies of bees for fruit pollination. N.Y. Exp. Sta. Mem.; *147* : 3-28.

Woodrow, A.W. (1941). Behaviour of honeybees towards brood infected with American foulbrood. Am. Bee J. ; *81* : 363.

Woodrow, A.W. (1942). Susceptibility of honeybee larvae to individual inoculation with spores of *Bacillus larvae* J. Econ. Entomol.; *35* : 892-895.

Woyke, J. (1964). Causes of repeated mating flights by queen honeybees. J. apic. Res.; *3*(1) : 17-23.

Woyke, J. (1973). Instrumental insemination of *Apis cerana indica* queens. J. Apic. Res.; *12* : 151-158.

Woyke, J. (1975a). The eggs in the comb cells of three honeybee species in India. Proc. Int. Beekeep. Congr.; *25* : 340.

Woyke, J. (1975b). Natural and instrumental insemination of *Apis cerana indica* in India. J. Apic Res.; *14*(3/4) : 153-159.

Woyke, J. (1976). Brood-rearing efficiency and absconding in Indian honeybees. J. Apic. Res.; *15* : 133-143.

Woyke J. (1979). Sex determination in *Apis cerana indica*. J. Apic. Res; *18* : 122-127.

Woyke, J. (1980). Evidence and action of cannibalism substance in *Apis cerana indica*. J. Apic. Res.; *19* : 6 -16.

Woyke, J. (1981). Comparative study of bees throughout the year in temperate and tropical zones. Apiacta ; *16*(3) : 107-110.

Woyke, J. (1984). Survival and prophylactic control of *Tropilaelaps clareae* infesting *Apis mellifera* colonies in Afghanistan. Apidologie; *15*(4) : 421-434.

Yek.; Zhong, S.Y. (1981). Flowering and nectar secretation of citrus trees. Zhon Yang; *6* : 22-23.

Zeuner, F.E.; Manning, F.J.; Morris, S.F. (1976). A monograph on fossil bees (Hymenoptera : Apoidea). Bull. Brit. Mus. Nat. Hist. *27* : 149-268.

Zmarlicki, C. (1984). Evaluation of honey plants in Burma—case study; pp. 57-76 : Proc. Expert Consultation, FAO.

Index

C

N

O

P

Shrimati Indira Gandhi, the then Prime Minister of India, discussing the Honeybee Project with Atwal family.

The author, Dr A.S. Atwal and wife Dr Sarjit K. Siddoo, M.D.

(1)

PLATE 1. Four species of *Apis* foraging ; from top left clockwise, *Apis dorsata* F., *Apis mellifera* L., *Apis cerana* F., *Apis florea* F.

PLATE 2. *Apis dorsata* F.; nest close to the ground (above), and a part of comb showing honey cells and brood cells (below).

(3)

PLATE 3. *Apis cerana indica* F.; honeybees trying to set up their colony in a Langstroth hive; note, comb building of their own cell size.

PLATE 4. A newly fixed comb foundation sheet being raised into a comb by bees, a complete comb, with sealed honey (white), brood (straw colour).

PLATE 5(*a*). Hive activity, bees fanning and guarding outside the enterance; below, bees active on a frame containing sealed honey, pollen and brood.

PLATE 5(*b*). Propolis deposits on top of frames and on the sides cementing them together.

PLATE 6(*a*). Two *A.m.* workers transferring food from one to another.

PLATE 6(*b*). After opening her Nasonov gland, *A.m.* spreading her pheromone that keeps the colony in order.

PLATE 7. Opening a hive after removing the top covers, and giving puffs of smoke.

PLATE 8. Migratory Beekeeping; Top—boxes ready to be transported in a truck; Below—colonies lying in farmers' fields (*Brassica* in distant front and clover on left).

PLATE 9. Uncapping the combs and honey extraction.

(10)

PLATE 10. Honey processing plant.

PLATE 11. Above—Hibernating larvae (in cocoons) of the Greater Wax Moth; Below—Adult of the Lesser Wax Moth.

PLATE 12. From top left, clockwise : *Apis mellifera* L. ; a solitary bee ; *Apis dorsata* F. ; *Apis cerana* F.—all foraging on *Brassica* flowers (body sizes are not comparable).

PLATE 13. Top— *Apis dorsata* F. on *Brassica campestris* var. sarson; Below— *Apis mellifera* L. on *Chrysanthemum.*

PLATE 14. *Apis dorsata* F. and *Apis florea* F. on rose varieties, Charlston and Kiss of Fire, respectively.

PLATE 15. Top— *Apis cerana* F. on Callendula ; Below— *Apis florea* F. on rose.

PLATE 16. Preponderance of syrphids on dahlia (top) and on saunf (bottom).

PLATE 17. *Apis mellifera* L. foraging on sunflower.

PLATE 18. A grove of *Eucalyptus* and a field of *Brassica.*

PLATE 19. *Apis mellifera* L. foraging on saunf and black ants on carrot.

PLATE 20. Plum flowers.

PLATE 21. Peach flowers.